任务型语码转换式双语教学系列教材

总主编 刘玉彬 副总主编 杜元虎 总主审 段晓东

生物与制药工程

BIOENGINEERING AND PHARMACEUTICAL ENGINEERING

主 编 刘 秋
主 审 刘英蘋

大连理工大学出版社

图书在版编目(CIP)数据

生物与制药工程 / 刘秋主编. 一 大连 ：大连理工大学出版社，2014.6(2015.5 重印)

任务型语码转换式双语教学系列教材

ISBN 978-7-5611-9137-8

Ⅰ.①生… Ⅱ.①刘… Ⅲ.①生物工程一双语教学一高等学校一教材一英、汉②制药工业一化学工程一双语教学一高等学校一教材一英、汉 Ⅳ.①Q81②TQ46

中国版本图书馆 CIP 数据核字(2014)第 095300 号

大连理工大学出版社出版

地址：大连市软件园路 80 号 邮政编码：116023

发行：0411-84708842 邮购：0411-84703636 传真：0411-84701466

E-mail：dutp@dutp.cn URL：http://www.dutp.cn

大连理工印刷有限公司印刷 大连理工大学出版社发行

幅面尺寸：183mm×233mm 印张：22 字数：697 千字

2014 年 6 月第 1 版 2015 年 5 月第 2 次印刷

责任编辑：邵 婉 责任校对：诗 宇

封面设计：波 朗

ISBN 978-7-5611-9137-8 定 价：38.00 元

总序

PREFACE

2014 年的初夏,我们为广大师生奉上这套“任务型语码转换式双语教学系列教材”。

“任务型语码转换式双语教学”是双语教学内涵建设的成果,主要由两大模块构成:课上,以不影响学科授课进度为前提,根据学生实际、专业特点、学年变化及社会需求等,适时适量地渗透英语专业语汇、语句、语段或语篇,“润物细无声”般地扩大学生专业语汇量,提高学生专业英语能力;课外,可向学生提供多种选择的“用中学”平台,如英语科技文献翻译、英语实验报告、英语学术论文、英语小论文、英语课程设计报告、模拟国际研讨会、英语辩论、工作室英语讨论会等,使学生的专业英语实践及应用达到一定频度和数量,激活英语与学科知识的相互渗透,培养学生用英语学习、科研、工作的能力及适应教育国际化和经济一体化的能力。

为保证“任务型语码转换式双语教学”有计划、系统、高效、科学地持续运行,减少教学的随意性和盲目性,方便师生的教与学,我们编写了这套“任务型语码转换式双语教学系列教材”。

本套教材的全部内容均采用汉英双语编写。

教材按专业组册,涵盖所有主干专业课和专业基础课,力求较为全面地反映各学科领域的知识体系。

分册教材编写以中文版课程教材为单位,即一门课为分册教材的一章,每章内容以中文版教材章节为序,每门课以一本中文教材为蓝本,兼顾其他同类教材内容,蓝本教材绝大部分是面向 21 世纪的国家规划教材。

教材的词汇短语部分,注意体现学科发展的新词、新语,同时考虑课程需求及专业特点,在不同程度灵活渗透了各章节的重要概念、定义,概述了体现章节内容主旨的语句及语段。分册教材还编写了体现各自专业特点的渗透内容,如例题及解题方法,课程的发生、发展及前沿简介,图示,实验原理,合同文本,案例分析,法条,计算机操作错误提示等。

部分教材补充了中文教材未能体现的先进理论、先进工艺、先进材料或先进方法的核心内容,弥补了某些中文教材内容相对滞后的不足;部分教材概述了各自专业常用研究方法、最新研究成果及学术发展的趋势动态;部分教材还选择性地把编者的部分科研成果转化为教材内容，以期启发学生的创新思维，开阔学生的视野，丰富学生的知识结构，从教材角度支持学生参与科研活动。

本套教材大多数分册都编写了对“用中学”任务实施具有指导性的内容，应用性内容的设计及编写比例因专业而异。与专业紧密结合的应用性内容包括英语写作介绍，如英语实验报告写作，英语论文写作，英语论文摘要写作，英语产品、作品或项目的概要介绍写作等。应用性内容的编写旨在降低学生参与各种实践应用活动的难度，提高学生参与“用中学”活动的可实现性，帮助学生提高完成“用中学”任务的质量水平。

考虑学生英语写作和汉译英的方便，多数分册教材都编写了词汇与短语索引。

“任务型语码转换式双语教学系列教材”尚属尝试性首创，是多人辛勤耐心劳作的结果。尽管在编写过程中，我们一边使用一边修改，力求教材的实用性、知识性、先进性融为一体，希望教材能对学生专业语汇积累及专业资料阅读、英语写作、英汉互译能力的提高发挥作用；尽管编者在教材编写的同时也都在实践“任务型语码转换式双语教学”，但由于我们缺乏经验，学识水平和占有资料有限，加上为使学生尽早使用教材，编写时间仓促，在教材内容编写、译文处理、分类体系等方面存在缺点、疏忽和失误，恳请各方专家和广大师生对本套教材提出批评和建议，以期再版时更加完善。

在教材的编写过程中，大量中外出版物中的内容给了我们重要启示和权威性的参考帮助，在此，我们谨向有关资料的编著者致以诚挚的谢意！

编　者
2014 年 5 月

前言 FOREWORD

本教材本着以学生为本，因材施教，尊重学生个性需求，注重学生知识、能力、素质协调发展的教学理念，结合教育部提出的在高校推进双语教学，改革传统的采用原版英语教材的全英语式教学模式，探索出了可普遍实施、让学生大面积受益的任务型语码转换式双语教学模式。通过课内适时适量渗透，课外开展英语科技文献翻译、英语实验报告编写、英语学术论文和英语小论文撰写等英语应用能力的实践训练，经过全校师生多年的共同努力，解决了目前国内各高校英语不能作为一种有效的语言工具服务于专业教育和创新教育的问题，实现了英语学习的群体性参与，无障碍训练，教师学生英语水平共同提高的目的。“任务型语码转换式双语教学系列教材”正是我们学校双语教学改革的重要成果之一，也是提高学生英语水平的重要手段之一。本教材涵盖了生物工程和制药工程专业的主干课程，可满足生物与制药相关专业双语教学的需求。同时，本教材根据学生个性化需求，增加了很多实用内容。本教材具备以下特色：

1. 实用

本教材体现了因材施教的理念，以学生任务型需求为编写目的。共分为两篇，第一篇以生物工程和制药工程专业涉及的主要理论课程为主，在适量提供英语专业词汇的基础上，增加了更多的短语表达方式，以追求实用性效果为主。第二篇以实践应用为主，包括主干课程中的基本实验(实践)操作的全英文讲解、各种面试环节的自我介绍、专业介绍、学院介绍、英语专业论文的投稿说明、生物工程学科领域中的最新研究成果、国际会议日程安排等。

2. 新颖

本教材正是根据任务型语码转换式双语教学模式的特点，按照专业的课程体系，设计了涵盖生物工程和制药工程专业四年大学生涯各个教学环节的双语教学内容，使学生通过每个专业、每门课程、每个教师在每一堂课、课内外的每一个实践环节中的逐渐渗透，逐步适应双语教学，提高英语水平。

3. 全面

本教材也是生物工程专业和制药工程专业所有主干课程、主要实验实践环节的完整教学体系的具体体现。不仅涉及所有理论课程的重要词汇、短语以及概念、公式的原版英文注释，还包括主要实验课程和实践环节中实际操作内容的原版英文讲解。同时，本教材还涵盖了学生日常学习生活过程中所面临的英语实际操作环节的训练等。

4. 精炼

本教材虽然涵盖生物工程和制药工程两个专业的理论与实践课程，但本着精炼实用的原则，以国家教学指导委员会制定的专业课程体系为蓝本，选取更能体现专业特色的主干课程，精炼专业课程体系，具有面向全国同类专业的普适性特色。

本教材共分两篇：第一篇分为三十四章，参加编写的人员有崔韶晖（第八、十三、十九章）、高明波（第十八、二十章）、华瑞年（第一章）蒋本国（第五、三十一章）、姜波（第十二章）、金黎明（第二十九、三十三章）、李兴泰（第十、二十三章）、刘宝全（第二十四章）、刘巨涛（第二章）、刘长建（第二十六、三十四章）、刘俏（第四、十一、十七章）、刘秋（第七、十五、二十二章）、朴永哲（第十四章）、齐小辉（第二十五章）、权春善和金黎明（第十六章）、闫建芳（第六章）、许永斌（第二十七、二十八章）、张树彪和那立艳（第三章）、张艳梅（第三十、三十二章）、赵晶（第二十一章）、郑维（第九章）。第二篇分为十八章。参加编写的人员有华瑞年（第一章）、刘巨涛（第二章）、闫建芳（第三、九章）、刘俏（第四章）、刘秋（第五、七、十一、十二、十三、十四、十五、十六、十七章）、刘长建（第六、八、十章）、齐小辉（第十八章）。

本教材在编写过程中，虽然参考了国内外有关文献，但限于水平和时间仓促，篇幅有限，错误和不足之处在所难免，希望读者批评指正。

编　者
2014 年 5 月

使用说明

1. 正文中★的含义：表示该词条在本章节中为重要词条，要求学生必须掌握。
2. 中文词条有几种译法时，采取逐一并列。

例 生物转化 bioconversion [ˌbaiəukənˈvəːʃən]
biotransformation [ˌbaiəuˌtrænsfəˈmeiʃən]

3. 正文中各章节涉及的主要内容、概念、公式等，在每章节的最前面逐一列出。问题、计算等在每章节的最后列出。
4. 查阅方法：按照课程的章节顺序进行查阅。

目录 CONTENTS

目录 CONTENTS

目录 CONTENTS

第一篇　理论课程部分

第一章　普通化学

Chapter 1　General Chemistry

第一节　绪论

Section 1　Introduction

单质　elementary substance
国际纯粹与应用化学联合会　International Union of Pure and Applied Chemistry Association (IUPAC)
不确定度　uncertainty [ʌn'sə:tnti]
有效数字的计算规则　rules of significant figure in calculations
观测值　observed value
准确度　accuracy ['ækjurəsi]
修约　rounding ['raundiŋ]
阿伏伽德罗常数　Avogadro constant ['kɔnstənt]
等离子体　plasma ['plæzmə]
低压气体　low-pressure gases
经验定律　empirical laws
玻意耳定律　Boyle's law
等温线　isotherm ['aisəuθə:m]
摄氏温标　Celsius scale [skeil]
热力学温度　thermodynamic temperature
查理-盖·吕萨克定律　Charles-Gay-Lussac law
等压线　isopiestic line
摩尔气体常数　molar gas constant
玻耳兹曼常数　Boltzmann constant
状态方程式　equation of state
理想气体状态方程式　the ideal-gas equation of state
混合理想气体　mixture of ideal gases
气体分压定律　law of partial pressure
分体积定律　law of partial volume
气体的扩散　diffusion of gases
扩散定律　diffusion law
格雷姆定律　Graham's law
隙流定律　effusion law
气体动力学论　gas kinetic theory
气体分子运动论　gas molecular movement theory
根均方速率　root mean square velocity
速率分布函数　speed distribution function
最概然速率　most probable speed
实际气体　real gas, imperfect gas
范德华方程　van Der Waals Equation
临界温度　critical temperature
临界压力　critical pressure
临界体积　critical volume
临界摩尔体积　critical molar volume
临界点　critical point

1 A substance that can't be broken down by chemical means is called element. Elements are defined by the number of protons they possess.
一种物质不能再用化学方法分解，称为元素。元素是由质子数所决定的。

2 One mole of carbon is 6.022×10^{23} atoms of carbon (Avogadro's number).
1摩尔碳原子是 6.022×10^{23} 个原子(阿伏伽德罗常数)。

3 The volume of a gas sample at constant temperature is inversely proportional to its pressure.
恒温下气体的体积与压力成反比。

4 Thermodynamic temperature scale is named for Lord Kelvin (1848).
热力学温度的刻度是以开尔文来命名的。

5 The equation for the Ideal Gas Law is $PV = nRT$, where P is pressure, V is volume, n is number of moles, and T is temperature. The gas constant "R" equals 8.314 J/mol·K
理想气体方程为 $PV = nRT$，其中常数 R 为 8.314 J/mol·K，P 是压力，V 是体积，n 是摩尔数。

6 The total pressure of a mixture of gases is equal to the sum of the partial pressures of its component gases.
混合气体的总压等于组分的分压的和。

7 Movement of a fluid from an area of higher concentration to an area of lower concentration is called diffusion.
流体从高浓度向低浓度的流动称为扩散。

8 Diffusion is a result of the kinetic properties of particles of matter. The particles will mix until they are evenly distributed.
扩散是粒子的动力学性质。物质的粒子会扩散到均匀分布为止。

9 The effusion rate of a gas is inversely proportional to the square root of its density or molecular mass.
气体扩散的速率与它的密度或摩尔质量成反比。

第二节 溶液和溶液的依数性
Section 2 Solution and Colligative Properties

液体的蒸气压 vapor pressure of liquid
流体 fluid [ˈflu(:)id]
逃逸倾向 escaping tendency
蒸发 evaporation [iˌvæpəˈreiʃən]
凝结 condensation [kɔndenˈseiʃən]
饱和蒸气压 saturated vapor pressure
各向异性 anisotropy [ˌænaiˈsɔtrəpi]
蒸馏 distillation [ˌdistiˈleiʃən]
液体的沸点 boiling point of liquid
减压蒸馏 vacuum distillation
表面功 surface work
液体的表面张力 surface tension of liquid
蒸馏水 distilled water
电导水 conductivity water
痕量分析 trace analysis
超微量分析 supermicro-analysis [sju:pəˈmaikrə-əˈnælisis]
偶极矩 dipole moment
氢键 hydrogen bond
缔合 association [əˌsəusiˈeiʃən]
相图 phase diagram
水的相图 phase diagram of water
熔化曲线 melting curve
升华曲线 subliming curve
气化曲线 vaporization curve
三相点 triple point
重水 heavy water
超临界流体萃取 supercritical fluid extraction
超临界流体色谱 supercritical fluid chromatography
等离子体-物质的第四态 plasma- the forth state of matter
超高密度态-物质的第五态 super-high density state-the fifth state of matter
混合物 mixture [ˈmikstʃə]
溶液 solution [səˈlju:ʃən]
固溶体 solidsolution
溶剂 solvent [ˈsɔlvənt]
溶质 solute [ˈsɔlju:t]
非极性溶剂 nonpolarsolvent
极性溶剂 polarsolvent
涨落 fluctuation motion
溶解度 solubility [ˌsɔljuˈbiliti]
吸热的 endothermic [ˌendəuˈθə:mik]
放热的 exothermic [ˌeksəuˈθə:mik]
电解质 electrolyte [iˈlektrəulait]
非电解质 nonelectrolyte [ˈnɔniˈlektrəulait]
溶液浓度表示法 ways of expressing concentration
B的物质的量浓度 amount-of-substance concentration of B
B的质量摩尔浓度 molality of solute B
B的摩尔分数 mole fraction of B
B的摩尔比 mole ratio of solute B
B的质量浓度 mass concentration of B
蒸气压降低 vapor pressure lowing
沸点升高 boiling point elevation
凝固点降低 freezing point lowing
渗透压 osmotic pressure
依数性 colligative properties
理想溶液 ideal solution
正常沸点 normal boiling point
凝固点 freezing point
半透膜 semi-permeable membrane
反渗透现象 reverse osmosis phenomena

分散系统　dispersed system
分散相　dispersed phase
分散介质　dispersed medium
真溶液　true solution
溶胶　sol [səul]
乳状液　emulsion [i'mʌlʃən]
比表面　specific surface
不稳定系统　unstable system
正溶胶　positive sol
负溶胶　negative sol
丁铎尔效应　Tyndall effect
电泳　electrophoresis [i,lektrəfə'ri:sis]
电渗　electroosmosis [i,lektrəuɔz'məusis]
电动现象　electrokinetic phenomenon
双电层　double electric layer
溶胶粒子的结构　structure of colloid particle
布朗运动　Brownian motion
两性溶胶　amphoteric sol
聚结不稳定性　coagulation instability
动力学稳定性　kinetic stability
聚沉　coagulation [kəu,ægju 'leiʃən]
亲水凝胶和疏水凝胶　hydrophilic gel and hydrophobic gel
保护作用　protective effect

1 The pressure exerted by a vapor in equilibrium with the solid or liquid phase of the same substance is called vapor pressure, which also refers to the partial pressure of the substance in the atmosphere above the solid or the liquid.
蒸气压是指固相或液相与它的蒸气达到平衡时的压力，也指大气中该物质的分压。

2 The temperature and pressure at which solid, liquid, and vapor phases of a particular substance coexist in equilibrium is called triple point.
物质的固相、液相和气相共存时的温度和压力称为三相点。

3 Example: The triple point for water is at 0.01 degrees Celsius at 4.56 mmHg.
例如：水的三相点在 4.55 毫米汞柱下为 0.01 摄氏度。

4 Ethyl alcohol (ethanol) is a nonelectrolyte because it can not ionize when dissolved in water.
乙醇是一种非电解质，因为溶入水后，乙醇不能电离出任何离子。

第三节　原子结构
Section 3　Atomic Structure

原子结构　atomic structure
连续光谱　continuous spectrum
氢原子光谱　hydrogen atom spectrum
玻尔模型　Bohr's Model
线状光谱　line spectrum
不连续光谱　uncontinuous spectrum
里德伯常量　Rydberg constant
量子　quantum ['kwɔntəm]
光子　photon ['fəutɔn]
普朗克常量　Planck constant
玻尔半径　Bohr radius ['reidjəs]
波粒二象性　wave-particle dualism ['dju:əlizəm]
质能关系　mass-energy relation [ri'leiʃən]
测不准原理　uncertainty principle
波函数　wave function
原子轨道　atomic orbitals
驻波　standing waves
概率波　probability waves
概率密度　probability density ['densiti]
薛定谔方程　Schrödinger's equation
哈密顿算子　Hamiltonian [,hæmil'təuniən]
本征函数　eigenfunction ['aigən,fʌŋkʃən]
本征值　eigenvalue [aigən,vælju:]
单电子原子的波函数　wave function of single electronic atom
主量子数　principal quantum number
轨道角动量量子数　orbital angular momentum quantum number
磁量子数　magnetic quantum number
自旋角动量量子数　spin angular momentum quantum number
液晶　liquid crystal
电子云　electron cloud
关联勒让德函数　associated Legendre function
界面图　boundary surface
径向分布函数　radial distribution function
角度分布图　angle distribution figure
多电子原子结构　structure of many-electron atoms
屏蔽效应　shielding effect
钻穿效应　penetration effect

泡利不相容原理 Pauli exclusion principle
洪特规则 Hund's rule
元素的周期性 periodic properties of the elements
共价半径 covalent radius
金属半径 metallic radius
范德华半径 van der Waals radius
电离能 ionization energy
电子亲和能 electron affinity
电负性 electronegativity

1 Electronegativity is a property of an atom which attracts the electrons of a bond.
电负性是原子在化学键中吸引电子的能力。

2 The Cl atom has a higher electronegativity than the H atom, so the bonding electrons will be closer to the Cl than to the H in the HCl molecule.
氯原子比氢原子的电负性大，因此在 HCl 分子中，电子对靠近氯原子。

3 Electron cloud has an energy state described by given values of the n, l, and ml quantum numbers.
电子云具有的能级状态是由 n，l，ml 等量子数决定的。

4 An orbital can contain two electrons with paired spins and is often associated with a specific region of an atom.
一个轨道能够容纳两个电子，并自旋相反，在原子中占有一固定的区域。

5 The electrons in the $2p_x$ orbital of an atom are generally found within a dumbbell-shaped cloud about the x-axis.
原子中 $2p_x$ 轨道上的电子通常以哑铃形状出现在 x 轴。

6 Region of negative charge surrounding an atomic nucleus that is associated with an atomic orbital is called electron cloud.
原子轨道中，围绕原子核旋转的带有负电荷的区域称为电子云。

第四节 分子结构
Section 4 Molecular Structure

分子结构 molecular structure
范德华力 van der Waals force
电价的 electrovalent [i,lektrəu'veilənt]
共价的 covalent [kəu'veilənt]
离子键 ionic bond
正离子(阳离子) positive ion (cation) ['kætaiən]
负离子(阴离子) negative ion (anion) ['ænaiən]
离子型化合物 ionic compound
晶格能 lattice energy
玻恩-哈伯循环 Born-Haber cycle
升华能 sublimation energy
解离能 dissociation energy
电离能 ionization energy
电子亲和势 electron affinity
决定离子化合物性质的主要因素 primary factors that determined the properties of ionic compounds
离子电荷 ionic charges
离子半径 ionic radius
价键理论 valence-bond theory
分子轨道理论 molecular-orbital theory
轨道杂化理论 hybrid-orbital theory
价层电子对互斥理论 valence shell electron pair repulsion theory
电子对 electron pair
共价键 covalent bond
杂化 hybridization [.haibridai'zeiʃən]
杂化轨道 hybrid orbital
反键轨道 antibonding orbital
成键轨道 bonding orbital
自由电子 free electrons
金属键 metallic bond
导带 conduction band
禁带 forbidden band
非极性分子 nonpolar molecule
极性分子 polar molecule
诱导偶极 induced dipole
分子间力 intermolecular force
取向力 orientation force
诱导力 induced force
色散力 dispersion force
伦敦力 London force

1 ΔH (enthalpy change) for the process in which oppositely charged ions in the gas phase combine to

form an ionic lattice in the solid phase is called lattice energy.
极性相反的离子在气相中形成固态的晶格，这一过程的焓变称为晶格能。

2 Electron pairs around a central atom tend to orient themselves as far apart as possible.
在原子中，电子对之间总是保持尽可能远的距离。

3 Electron pair repulsion theory is used to predict the geometry of a molecule or a polyatomic ion.
电子对排斥理论可以用 theory 来预测分子或多原子离子的几何构型。

第五节 晶体结构
Section 5 Crystal Structure

无定形物质 amorphous substance
多晶体 polycrystal [ˌpɔliˈkristəl]
晶面角 interfacial angle
晶面角守恒定律 law of constancy of interfacial angle
各向异性 anisotropy [ˌænaiˈsɔtrəpi]
对称 symmetry [ˈsimitri]
各向同性 isotropy [aiˈsɔtrəpi]
空间点阵 space lattice
点阵 lattice [ˈlætis]
晶胞参数 unit cell parameter
立方晶系 cubic system
四方晶系 tetragonal system
正交晶系 rhombic system
三方晶系 rhombohedral system
六方晶系 hexagonal system
单斜晶系 monoclinic system
三斜晶系 triclinic system
离子晶体 ionic crystal
原子晶体 atomic crystal
分子晶体 molecular crystal
理想晶体 ideal crystal
晶体缺陷 crystal defect
非整比化合物 non-stoichiometric compound
固体化学 solid chemistry

第六节 化学热力学
Section 6 Chemical Thermodynamics

热化学 thermochemistry
热力学系统 thermodynamic system
敞开系统 open system
封闭系统 closed system
隔离系统 isolated system
状态函数 state function
等温过程 isothermal process
等压过程 isobar process
绝热过程 adiabatic process
循环过程 cyclic process
途径 path [pɑːθ]
体积功 volume work
膨胀功 expansion work
能量守恒和转化定律 law of energy conservation and transformation
热力学第一定律 first law of thermodynamics
热力学能 thermodynamic energy
内能 internal energy
容量性质 capacity property
恒容系统 isovolumetric system
绝热系统 adiabatic system
强度性质 intensive property
焓-定压下的热效应 enthalpy-the heat transferred under the constant pressure
比热容 specific heat capacity
摩尔热容 molar heat capacity
摩尔定容热容 molar heat capacity at constant volume
摩尔定压热容 molar heat capacity at constant pressure
反应进度 extent of reaction
反应物 reactant [riːˈæktənt]
生成物 product [ˈprɔdəkt]
化学计量数 stoichiometric number
国际标准化组织 International Organization of Standardization
化学反应 chemical reaction [ri(ː)ˈækʃən]
标准状态 standard state
盖斯定律 Hess's law
用生成焓求算反应焓 using enthalpies of formation to calculate enthalpies of reaction
标准摩尔生成焓 standard molar enthalpy of

formation
燃烧 combustion [kəm'bʌstʃən]
化学反应的方向和限度 the direction and limitation of chemical reaction
燃料电池 fuel cell
自发性 spontaneity [ˌspɔntə'niːiti]
自由膨胀 free expansion
可逆过程 reversible process
不可逆过程 irreversible process
热力学第二定律 second law of thermodynamics
亥姆霍兹函数 Helmholtze function
亥姆霍兹自由能 Helmholtze free energy
自发过程 spontaneous process
反应方向的预测 predicting the direction of reaction
吉布斯自由能 Gibbs free energy
标准摩尔吉布斯自由能 standard molar Gibbs free energy
凝聚态 condensed state
标准摩尔生成吉布斯自由能 standard molar Gibbs free energy of formation
熵 entropy ['entrəpi]
熵变的计算 calculation of entropy changes
热力学第三定律 third law of thermodynamics

1 The total energy of a system and its surroundings remains constant.
系统和环境的总能量保持恒定。

2 The change in the energy of a system equals to the heat flow in the system from the surroundings minus the work done by the system on the surroundings.
系统能量的变化等于从环境到系统的热量流动减去系统对环境所做的功。

3 A process that will occur without any energy input from the surroundings is called spontaneous process.
没有外界能量输入而能够发生的过程称为自发过程。

第七节 化学平衡
Section 7 Chemical Equilibrium

可逆反应 reversible reaction
化学平衡 chemical equilibrium
动态平衡 dynamic equilibrium
平衡常数 equilibrium constant
均相化学平衡 homogeneous phase chemical equilibrium
多相化学平衡 multiple phase chemical equilibrium
多重平衡规则 rules of multiple equilibrium
反应等温式 reaction isotherm equation
逸度 fugacity [fju(ː)'gæsiti]
多重平衡 multiple equilibrium
标准平衡常数 standard equilibrium constant
逸度系数 fugacity coefficient
活度系数 activity coefficient
勒夏特列原理 Le Chatelier's principle
分压(或浓度)的影响 effects of partial pressure (concentrations) changes
压力的影响 effects of pressure changes
温度的影响 effects of temperature changes
同时平衡 simultaneous equilibrium
耦合反应 coupling reaction

第八节 酸碱平衡和酸碱滴定
Section 8 Acid-Base Equilibria and Acid-Base Titration

强电解质 strong electrolyte
弱电解质 weak electrolyte
电离度 degree of ionization
离子氛 ion atmosphere
离子强度 ionic strength
德拜-休格尔极限定律 Debye-Huckel extreme law
平均活度系数 mean activity coefficient
阿仑尼乌斯电离理论 Arrhenius theory of dissociation
酸碱质子理论 proton theory of acid-base
质子给予体 proton donor
质子接受体 proton acceptor
共轭酸碱对 conjugated pair of acid-base
两性物质 amphoteric compound
酸碱半反应 half-reaction of acid-base
质子传递反应 protolysis reaction
解离常数 ionization constant

中和反应 neutralization reaction [ˌnju:trəlaiˈzeiʃən]
质子传递常数 constant of proton transfer
路易斯酸碱理论 Lewis's acid-base theory
滴定剂 titrant [ˈtaitrənt]
物料平衡式 mass or material balance equation
电荷平衡式 charge balance equation
质子平衡式 proton balance equation
参考水准 reference level
零水准 zero level
物种 species
一元弱酸 unary weak acid
多元酸 polyprotic acid
分布系数 distribution coefficient
H^+浓度的计算 calculation of H^+ concentration
强酸(碱)溶液 strong acid (base) solution
一元弱酸(碱)溶液 unary weak acid(base) solution
两种弱酸混合溶液 mixed solution of two types of weak acids
多元酸(碱)溶液 polyprotic acid (base) solution
两性物质溶液 amphoteric compound solution
缓冲容量 buffer capacity
蒸馏法 distillation method
标准缓冲溶液 standard buffer solution
滴定分析法 titrimetric analysis
标准滴定溶液 standard titration solution
化学计量点 stoichiometric point
质量分数 mass fraction
容量分析法 volumetric analysis
指示剂误差 indicator error
终点误差 end point error
直接滴定 direct titration
反滴定法 back titration
置换滴定法 replacement titration
间接滴定法 indirect titration
基准物质 primary standard substance
滴定分析的计算 calculation of titration analysis
酸碱指示剂 acid-base indicator
酸碱滴定曲线 acid-base titration curve
强酸-弱碱滴定 strong acid-weak base titration
多元酸的滴定 titration of polyprotic acid
酚酞 phenolphthalein [ˌfinɔlˈfθæli:n]
甲基橙 methyl orange
甲基红 methyl red
铵盐中氮的测定 measuring nitrogen in ammonium salt
甲醛法 formaldehyde method

A solution containing either a weak acid and its salt or a weak base and its salt is called buffer solution, which is resistant to changes in pH.
一种溶液中包含弱酸及其弱酸盐或弱碱及其弱碱盐称为缓冲溶液,它能够起到维持体系 pH 值恒定的作用。

第九节 配位平衡和配位滴定
Section 9 Coordination Equilibrium and Coordination Titration

配位化合物 coordination compound
配位平衡 coordination equilibrium
配位滴定法 complexometry [ˌkɔmpleksˈɔmitri]
中心原子 central atom
配位体 ligand [ˈligənd,ˈlaigənd]
配合物的组成 composition of coordination compound
内界 inner [ˈinə(r)]
外界 outer [ˈautə]
配位数 coordination number
单核配合物 mononuclear complex
多核配合物 polynuclear complex
单齿配位体 unidentate ligand
多齿配位体 multidentate ligand
螯合物 chelate [ˈki:leit]
异构体 isomer [ˈaisəumə]
结构异构 structural isomerism
空间异构 spatial isomerism
旋光异构体 rotatory isomer
手性 chirality [ˈtʃirəliti]
手征性的 chiral [ˈtʃirəl]
外消旋混合物 racemic mixture
逐级稳定常数 stepwise stability constant
累积稳定常数 cumulative stability constant
稳定常数 stability constant
乙二胺四乙酸 ethylenediaminetetraacetic acid
质子化常数 protonation constant
形成常数 formation constant
累积质子化常数 cumulative protonation constant
条件稳定常数 conditional stability constant

条件形成常数 conditional formation constant
副反应系数 side reaction coefficient
配位剂 Y 的副反应系数 side reaction coefficient of coordination agent Y
金属离子 M 的副反应系数 side reaction coefficient of metal ion M
金属指示剂 metallochromic indicator
金属指示剂的转变点 transition point of metallochromic indicator
封闭作用 blocking [ˈblɔkiŋ]
掩蔽剂 masking agent
解蔽法 demasking method
硬水 hard water
总硬度 total hardness
配位滴定曲线 coordination titration curve
配位化合物的应用 applications of coordination compound
配位化学 coordination chemistry
贵金属的湿法冶金 hydrometallurgy of precious metal
分离 separation [sepəˈreiʃ(ə)n]
提纯 purification [ˌpjuərifiˈkeiʃən]
配位催化 coordination catalysis
电镀 electroplate [iˈlektrəupleit]

第十节 沉淀平衡和沉淀滴定
Section 10 Precipitate Equilibrium and Precipitate Titration

沉淀平衡 precipitate equilibrium
难溶物，微溶物 slightly soluble substance
饱和溶液 saturated solution
条件溶度积 conditional solubility product
盐效应 salt effect
活度积 activity product
溶度积 solubility product
溶解度 dissolvability [diˈzɔlvəbiliti]
同离子效应 common-ion effect
酸效应 acid effect
配位效应 coordination effect
陈化 ageing [ˈeidʒiŋ]
凝聚 coagulation [kəuˌægju ˈleiʃən]
胶溶 peptization [ˌpeptaiˈzeiʃən]
影响沉淀溶解度的其他因素 additional factors that affect precipitate dissolvability
分级沉淀法 fractional precipitate
沉淀的转化 precipitate conversion
沉淀滴定法 precipitate titration
重量分析法 gravimetry [grəˈvimitri]
伏哈德法 Volhard's method
法扬司法 Fajangs' method
吸附指示剂 adsorption indicator
沉淀的形成 formation of precipitate
晶形沉淀 crystalline precipitate
凝乳状沉淀 gelatinous precipitate
无定形沉淀 amorphous precipitate
氢氧化物沉淀 precipitation of hydroxide
硫化物沉淀 precipitation of sulfide
晶核 nuclei [ˈnju:kliai]
临界核 critical nuclei
均相成核作用 homogeneous nucleation
晶种 seed [si:d]
异相成核作用 heterogeneous nucleation
相对过饱和度 relative supersaturation
定向速度 directional velocity
共沉淀 coprecipitation [ˈkəupriˌsipiˈteiʃən]
抗衡离子 counter ion
双电层 double electrical layer
吸附共沉淀 adsorption coprecipitation
包藏 occlusion [əˈklu:ʒən, ɔ-]
混晶 mixed crystal
固溶体 solid solution
后沉淀 postprecipitation
母液 mother liquor
均相沉淀法 homogeneous precipitation method
离子缔合物 ionic association
有机沉淀剂 organic precipitant
载体 carrier [ˈkæriə]
无机共沉淀剂 inorganic coprecipitator
惰性共沉淀剂 inert coprecipitator

第十一节 原电池和氧化还原反应
Section 11 Primary Cell and Oxidation-Reduction Reactions

电子传递反应 electron transfer reaction
氧化还原反应 oxidation-reduction reaction

氧化还原滴定　oxidation-reduction titration
氧化数　oxidation number
表观电荷数　apparent charge number
氧化还原方程式的配平　balancing oxidation-reduction equations
盐桥　salt bridge
原电池　primary cell
半电池　half cell
负极　negative pole
阳极　anode [ˈænəud]
正极　positive pole
阴极　cathode [ˈkæθəud]
电极电势　electrode potential
标准氢电极　standard hydrogen electrode
电动势　electromotive force
标准电极电势　standard electrode potential
甘汞电极　calomel electrode
氧化剂　oxidant [ˈɔksidənt]
还原剂　reductant [riˈdjuːsdənt]
法拉第常数　Faraday constant
能斯特方程　Nernst equation
直接碘量法　direct iodimetry
间接碘量法　indirect iodimetry
电极电势的应用　applications of electrode potential
拉蒂默图　Latimer diagram
歧化反应　dismutation reaction
超电势　overpotential [ˈəuvəpəˈtenʃəl]
电化学　electrochemistry [iˌlektrəuˈkemistri]
极化作用　polarization [ˌpəuləraiˈzeiʃən; -riˈz]
去极剂　depolarization agent
碱性电池　alkaline cell
蓄电池　storage cell
燃料电池　fuel cell
电解　electrolysis [ilekˈtrɔlisis]
电解池　electrolytic cell
分解电压　decomposition voltage
反电动势　back electromotive force
金属腐蚀　metal corrosion
电化学腐蚀　electrochemical corrosion
保护层　protective layer
氧化还原滴定法　oxidation-reduction titration method
钝化　passivation
条件电极电势　condition electrode potential
氧化还原指示剂　oxidation-reduction indicator
速率常数　rate constant
速率方程　rate equation
反应级数　reaction order
零级反应　zeroth order reaction
均相反应　homogeneous reaction
多相反应　heterogeneous reaction
活化分子　activating molecular
活化能　activation energy
负催化剂　negative catalyst
抑制剂　retarder [riˈtɑːdə]
均相催化反应　homogeneous catalysis
多相催化反应　heterogeneous catalysis
酶催化　enzyme catalysis
链反应　chain reaction
高锰酸钾法　potassium permanganate method
双氧水　perhydrol [pəˈhaidrɔl]
重铬酸钾法　potassium dichromate method
基元反应　elementary reaction
绿色化学　green chemistry
自由基　free radical

The passage of a direct electric current through an ion-containing solution which produces chemical reactions at the electrodes is called electrolysis.

在含有离子的溶液中通过直流电而在电极上产生化学反应的过程称为电解。

第十二节　元素化学
Section 12　Elementary Chemistry

碱土金属　alkaline earth metals
氧化物　oxide [ˈɔksaid]
氢氧化物　hydroxide [haiˈdrɔksaid,-sid]
含氧酸　oxacid [ɔkˈsæsid]
水的硬度　hardness of water
硼族元素　boron family element
碳族元素　carbon family element
温室效应　greenhouse effect
锗　germanium [dʒəːˈmeiniəm]
锡　tin [tin]
铅　lead [liːd]
氮族元素　nitrogen family element

氮 nitrogen ['naitrədʒən]
铵盐 ammonium salt
联氨 hydrazine ['haidrəzi:n, -zin]
磷 phosphorus ['fɔsfərəs]
砷 arsenic ['ɑ:sənik]
锑 antimony ['æntiməni]
铋 bismuth ['bizməθ]
氧族元素 oxygen family element
氧 oxygen ['ɔksidʒən]
臭氧 ozone ['əuzəun, əu'z-]
过氧化氢 peroxide [pə'rɔksaid]
硫 sulphur ['sʌlfə]
硫化物 sulfide ['sʌlfaid]
硫化氢 hydrogen sulfide
多硫化物 polysulfide
一氧化二硫 disulfide oxide
二氧化硫 sulphur dioxide
三氧化硫 sulfur trioxide
亚硫酸 sulphurous acid
硫酸 sulphuric acid/sulfuric acid
焦硫酸 pyrosulfuric acid
硫代硫酸钠 sodium thiosulfate
硒 selenium [si'li:niəm, -njəm]
碲 tellurium [te'ljuəriəm]
卤素 halogen ['hælədʒən]
氟 fluorine ['flu(:)əri:n]
氯 chlorine ['klɔ:ri:n]
溴 bromine ['brəumi:n]
碘 iodine ['aiədi:n;(US) 'aiədain]
卤化氢 hydrogen halide
卤化物 halide ['hælaid]
氰 cyanogen [sai'ænədʒin]
氰化物 cyanide ['saiənaid]
硫氰酸 sulfocyanic acid
稀有气体 noble gases
氦 helium ['hi:ljəm, -liəm]
氖 neon ['ni:ən]
氩 argon ['ɑ:gɔn]
氪 krypton ['kriptɔn]
氙 xenon ['zenɔn]
配位化合物的化学键理论 chemical bond theory of coordination compound
分裂能 break-up energy
晶体场稳定化能 crystal field stabilization energy
钛 titanium [tai'teinjəm, ti-]
二氧化钛 titanium dioxide
四氯化钛 titanium tetrachloride
钒 vanadium [və'neidiəm, -djəm]
铬 chromium ['krəumjəm]
钼 molybdenum [mə'libdinəm]
钨 wolframium [wul'freimiəm, vɔ:l-]
铬酸 chromic acid
氧化铬 chromic oxide
氢氧化铬 chromic hydroxide
硫酸铬 chromic sulfate
氯化铬 chromium chloride
锰 manganese [ˌmæŋgə'ni:z,'mæŋgəni:z]
绿矾 green vitriol
黄血盐 potassium ferrocyanide
二茂铁 ferrocene ['ferəˌsi:n]
钴 cobalt [kə'bɔ:lt, 'kəubɔ:lt]
镍 nickel ['nikl]
鉴定 identification [aiˌdentifi'keiʃən]
钌 ruthenium [ru:'θi:niəm]
铑 rhodium ['rəudiəm, -djəm]
铱 iridium [i'ridiəm]
钯 palladium [pə'leidiəm]
铂 platinum
铜 cuprum ['kju:prəm]
银 silver ['silvə]
金 gold [gəuld]
锌 zinc [ziŋk]
镉 cadmium ['kædmiəm]
汞 mercury ['mə:kjuri]
镧系元素 lanthanide [lænθə'nid,-'naid]
镧系收缩 lanthanide contraction
锕系元素 actinide ['æktinaid]
核素 nuclide ['nju:klaid]
同位素 isotope ['aisəutəup]
硼烷 borane ['bəurein]
缺电子化合物 electron deficiency compound

例 Atoms with the same number of protons, but different numbers of neutrons are called isotopes. Isotopes are different forms of a single element.

Examples: Carbon 12 and Carbon 14 are both isotopes of carbon, one with 6 neutrons and one with 8 neutrons (both with 6 protons).

原子具有相同的质子数和不同的中子数称为同位素。同位素是一种元素的不同形式。例如:碳 12 和碳 14 是碳的两种同位素,一种含有 6 个中子,另一种含有 8 个中子(二者都含有 6 个质子)。

2 Greenhouse effect is the effect of water and carbon dioxide absorbing outgoing infrared radiation, raising a system's temperature. The term is generally used with reference to the Earth's increasing temperature, although it can also be applied to other systems, such as greenhouses and automobiles. 水和二氧化碳吸收外来的红外辐射而导致体系温度升高的效应称为温室效应。在描述地球变暖以及温室、汽车等情况时经常会使用该术语。

第二章 有机化学

Chapter 2 Organic Chemistry

第一节 绪论

Section 1 Introduction

★ 有机化学 organic chemistry
★ 有机化合物 organic compound [ɔː'gænik]['kɔmpaund]
碳化合物 carbon compound
化合价 valence ['veiləns]
价键理论 valence bond theory
分子轨道理论 molecular orbital theory [məu'lekjulə]['ɔːbitl]['θiəri]
原子轨道 atomic orbital [ə'tɔmik]['ɔːbitl]
★ 有机化学反应 organic reaction
★ 分子式 molecular formula
实验式 empirical formula [em'pirikəl]['fɔːmjulə]
元素分析 elemental analysis [ˌeli'mentl][ə'nælisis]

1 Carbon has five atomic orbitals available for its six electrons; they are the *s* orbital in the first shell, the *s* orbital in the second shell and the three *p* orbitals in the second shell.
碳原子的 6 个电子有 5 个可占用的原子轨道,第一层的 *s* 轨道,第二层的 *s* 轨道和 3 个 *p* 轨道。

2 The 1*s* and 2*s* orbitals are spherical in shape.
1*s* 和 2*s* 轨道是球形的。

3 The 2*p* orbitals are dumbbell in shape and can be assigned $2p_x$, $2p_y$, or $2p_z$ depending on the axis along which they are aligned.
2*p* 轨道为哑铃形,根据轴的取向可以将它们分为 $2p_x$,$2p_y$,或 $2p_z$ 轨道。

4 The 1*s* orbital has a lower energy than the 2*s* orbital that has a lower energy than the 2*p* orbitals.
1*s* 轨道的能量低于 2*s* 轨道,而 2*s* 轨道的能量低于 2*p* 轨道。

5 Carbon is in the second line of the periodic table, and has six electrons which will fill up lower energy atomic orbitals before filling higher energy orbitals, this is known as the aufbau principle.
碳在周期表中位于第二周期,它的 6 个电子在填充较高能级轨道之前首先填充较低能级的原子轨道,这就是已知的能量最低原理。

6 Each orbital is allowed a maximum of two electrons of opposite spin, this is known as Pauli exclusion principle.
每个轨道最多可容纳 2 个自旋相反的电子,这就是已知的泡利不相容原理。

7 The 2*p* orbitals have equal energy.
2*p* 轨道的能级是相同的。

第二节 烷烃

Section 2 Alkanes

★ 烃 hydrocarbons ['haidrəu'kɑːbəns]
★ 烷烃 alkanes ['ælkeiniz]
通式 general formula
★ 甲烷 methane ['meθein]
★ 乙烷 ethane ['eθein]
碳-碳共价键 carbon-carbon covalent bond
国际命名法 IUPAC nomenclature [nəu'menklətʃə]
环烷烃 cycloalkanes [ˌsaikləu'ælkeiniz]
★ 异构现象 isomerism [ai'sɔmərizm]
★ 异构体 isomers
★ 构造异构体 constitutional isomers
脂肪族的 aliphatic [ˌæli'fætik]
芳香族的 aromatic [ˌærəu'mætik]
★ 电子云 electron cloud
★ 杂化轨道 hybrid orbital
★ sp^3 杂化 sp^3 hybridization [ˌhaibridai'zeiʃən]
★ σ 键 sigma bond

电子结构 electronic structure
键长 bond length
键角 bond angle
★ 构型 configuration [kən,figju'reiʃən]
★ 构象 conformation [,kɔnfɔ:'meiʃən]
构象分析 conformational analysis
物理性质 physical properties ['fizikəl]['prɔpətis]
烷烃的卤代 halogenation [hælədʒə'neiʃən] of alkanes
★ 自由基 free radical
★ 反应中间体 reaction intermediate [ri(:)'ækʃən][,intə'mi:djət]
★ 均裂 homolytic cleavage [hɔ'mɔlaitik]['kli:vidʒ]
★ 异裂 heterolytic cleavage ['hetərəlaitik]['kli:vidʒ]
★ 链反应 chain reaction
烷烃的燃烧 combustion of alkanes

1 Alkanes are organic molecules consisting solely of carbon and hydrogen atoms linked by single σ bond.
烷烃是由 σ 单键连接的只含有碳原子和氢原子的有机分子。

2 All the carbon atoms of alkanes are tetrahedral and sp^3 hybridized.
烷烃的碳原子呈四面体型,均为 sp^3 杂化。

3 Alkanes are stable molecules and unreactive to most chemical reagents.
烷烃是稳定的分子,与大多数化学试剂都不反应。

4 Alkanes have the general formula C_nH_{2n+2}.
烷烃的通式为 C_nH_{2n+2}。

5 Alkyl groups (C_nH_{2n+1}) are alkane portions of a more complicated structure.
烷基(C_nH_{2n+1})是较复杂结构的烷烃部分。

6 Alkyl groups can be drawn as a skeletal drawing, which only shows the C-C bonds or as CH_3, CH_2CH_3 etc.
烷基可以用骨架图表示,骨架图只显示碳 - 碳键,也可以用 CH_3, CH_2CH_3 等来表示。

第三节 烯烃
Section 3 Alkenes

★ 烯烃 alkenes ['ælkeiniz]
★ 双键 double bond
★ sp^2 杂化 sp^2 hybridization
★ π 键 π-bond
★ 顺反异构体 cis-trans isomerism
★ 乙烯 ethylene ['eθili:n]
★ 丙烯 propylene ['prəupili:n]
★ 次序规则 sequence rules
★ 偶极矩 dipole moments ['daipəul]['məumənts]
★ 反应机理 reaction mechanism
★ 催化氢化 catalytic hydrogenation [,kætə'litik][,haidrədʒə'neiʃən]
★ 亲电加成 electrophilic additions [i,lektrəu'filik][ə'diʃəns]
与卤化氢加成 addition of HX
★ 马氏规则 Markovnikov's rule
★ 反马(加成) anti- Markovnikov's rule
聚合 polymerization [,pɔlimərai'zeiʃən]

1 Alkenes can be prepared by the reduction of alkynes or the elimination of alkyl halides and alcohols.
烯烃可以通过炔烃的还原反应或卤代烷和醇的消除反应制备。

2 Vicinal dibromides can be dibrominated by treatment with zinc in acetic acid or with sodium iodide in acetone.
用乙酸溶液中的锌粒或丙酮中的碘化钠处理,邻二溴代物会被脱掉溴。

3 Alkenes are planar with bond angles of 120°.
烯烃为键角是 120 度的平面型结构。

4 The carbon atoms of the C-C double bond are sp^2 hybridized and the double bond is made up of one σ bond and one π bond.
烯烃中组成碳碳双键的碳原子呈现 sp^2 杂化,双键由一根 σ 键和一根 π 键构成。

5 The carbon-carbon double bond is stronger and shorter than a carbon-carbon single bond.
与碳碳单键相比,碳碳双键的键强高,键长短。

6 The π bond is weaker than the σ bond.
π键比σ键弱。

7 Bond rotation round a carbon-carbon double bond is impossible.
碳碳双键不能旋转。

第四节 炔烃
Section 4 Alkynes

★ 炔烃 alkynes ['ælkainiz]
★ 叁键 triple bond
★ sp 杂化 sp hybridization
★ 乙炔 ethyne ['eθain]
★ 丙炔 propyne ['prəupain]
★ 不对称电子分布 unsymmetrical ['ʌnsi'metrikəl] electron distribution
★ 还原 reduction
★ 林德拉催化剂 Lindlar's catalyst
选择性氢化 selective hydrogenation [si'lektiv][ˌhaidrədʒə'neiʃən]
亲核加成 nucleophilic addition [ˌnju:kliəu'filik][ə'diʃən]
硼氢化 hydroboration ['haidrəu,bɔ:'reiʃən]

1 Alkenes can be reacted with bromine to give a vicinal dibromide.
烯烃与溴发生反应生成邻二溴代物。

2 Treatment of the dibromide with a strong base results in the loss of two molecules of hydrogen bromide and the formation of an alkyne.
用强碱处理二溴化物,会导致其脱去两分子的溴化氢,形成炔烃。

3 Alkynes are linear with the triple bond and the carbons of which the triple bond is composed are *sp* hybridized.
炔烃是含叁键的线型结构,其中组成叁键的碳原子为 *sp* 杂化。

4 The triple bond is made up of one σ bond and two π bonds.
叁键由一个σ键和两个π键构成。

5 The triple bond of an alkyne is stronger than a carbon-carbon single bond or a carbon-carbon double bond.
炔烃中叁键的强度比碳-碳单键和碳-碳双键高。

6 The two π bonds present in the triple bond are weaker and more reactive than the σ bond.
叁键的两个π键的键能比σ键低,反应性能比σ键高。

第五节 二烯烃和共轭效应
Section 5 Dienes and Conjugative Effect

★ 二烯 dienes ['daii:niz]
★ 共轭二烯 conjugated diene ['kɔndʒugeitid]['daii:n]
★ 累积二烯 cumulated ['kju:mjulitid]diene
★ 隔离二烯 isolated diene
★ 共轭效应 conjugative effect
★ 超共轭效应 hyperconjugation [ˌhaipəkəndʒu'geiʃən]
★ 动力学控制 kinetic control [kai'netik][kən'trɔl]
★ 电子离域 electron delocalization [i'lektrɔn][di:'ləukəlaizeiʃən]
★ 1,4-加成 1,4-addition
★ 热力学控制 thermodynamic control ['θə:məudai'næmik][kən'trol]
★ D-A 反应 Diels-Alder reaction
★ 诱导效应 inductive effect
★ 共振效应 resonance effect ['rezənəns][i'fekt]

1 Alkyl groups can donate electrons towards a neighboring positive center and this makes to stabilize the ion since some of the positive charge is partially dispersed over the alkyl groups.
烷基可以对邻近的正电荷中心提供电子,这会增加离子的稳定性,使部分正电荷得到分散。

2 The more alkyl groups which are attached, the greater the electron donating power and the more stable the carbocation.
所连的烷基越多则提供电子的能力就越强，碳正离子就越稳定。

3 Hyperconjugation means the overlap of the vacant $2p$ orbital with a neighboring C-H σ-bond orbital.
超共轭效应就是 $2p$ 空轨道与相邻的 C-H σ 键的重叠。

4 Hyperconjugation means that the $2p$ orbital is not completely vacant since the σ electrons of the C-H bond can spend a small amount of time entering the space occupied by the $2p$ orbital. In other words, C-H bond becomes slightly electron deficient.
超共轭效应实际上就是 C-H 键的 σ 电子有时进入 $2p$ 轨道的区域，使 $2p$ 轨道不完全空，也就是 C-H 键呈轻微的缺电子状态。

5 Hyperconjugation makes the positive charge be delocalized and hence stabilized.
超共轭效应使正电荷发生离域，增加正离子的稳定性。

6 The reactions of a conjugated diene indicate that a conjugated diene should be considered as a functional group in its own right, rather than as two separate alkenes.
由共轭二烯发生的化学反应可以看出共轭二烯烃是一个特殊的官能团，而不是两个孤立烯烃的加合。

第六节 波谱法
Section 6 Spectroscopic Methods

★ 电磁波谱 electromagnetic spectrum [ilektrəu'mægnitik]['spektrəm]
波长 wavelength
频率 frequency ['fri:kwənsi]
★ 紫外可见光谱法 ultraviolet and visible spectroscopy
★ 朗伯-比尔定律 Lambert-Beer law
★ 电子迁移 electronic transition
★ 基态 ground state [graund][steit]
★ 激发态 excited state
★ 吸收光谱 absorption spectrum [əb'sɔ:pʃən]['spektrəm]
★ 摩尔吸光系数 molar absorptivity ['məulə][əbsɔ:p'tiviti]
电子光谱 electronic spectrum
★ 红外光谱法 infrared spectroscopy ['infrə'red][spek'trɔskəpi]
★ 波数 wave number
★ 对称伸缩 symmetric stretching [si'metrik]['stretʃiŋ]
★ 面内弯曲 an in-place bending vibration
★ 面外弯曲 an out-of-plane bending vibration
★ 核磁共振谱 nuclear magnetic resonance spectroscopy
^{1}H 核磁共振谱 proton NMR spectroscopy
^{13}C 核磁共振谱 carbon-13 NMR spectroscopy
★ 核自旋 nuclear spin
★ 低场 downfield ['daun'fi:ld]
★ 高场 upfield [ʌp'fi:ld]
★ 屏蔽 shielding ['ʃi:ldiŋ]
★ 去屏蔽 deshielding
★ 化学位移 chemical shift
★ 自旋耦合 spin-spin coupling
★ 偶合常数 coupling constant

1 Instruments called visible-UV spectrometers are used to measure the amount of light absorbed at each wavelength of the visible and UV regions.
紫外-可见分光光度计是用来测量物质吸收紫外及可见光的量的仪器。

2 Organic compounds also absorb electromagnetic energy in the infrared region of the spectrum.
有机化合物通常可以吸收红外波谱区的电磁能。

第七节 环烷烃和质谱
Section 7 Cycloalkanes and Mass Spectrometry

★ 环烷烃 cycloalkanes [ˌsaikləu'ælkeiniz]
★ 环丙烷 cyclopropane [ˌsaikləu'prəupein]

★ 环丁烷 cyclobutane [ˌsaikləuˈbjuːtein]
★ 拜尔角张力学说 Baeyer angle strain theory
角张力 angle strain
扭转张力 torsional [ˈtɔːʃənəl] strain
范德华张力 Van der Waals strain
偶极-偶极相互作用 dipole-dipole interaction
氢键 hydrogen bond
★ 椅式构象 chair form
★ 扭船式构象 twist-boat form
★ 平伏键 equatorial [ˌekwəˈtɔːriəl] bond
★ 直立键 axial bond
★ 质谱法 mass spectrometry
质谱仪 mass spectrometer
质谱图 mass spectrum
★ 分子离子 molecular ion
高能电子 high-energy electron
★ 质/荷比 mass charge ratio
★ 碎裂 fragmentation [ˌfrægmenˈteiʃən]

1 Cycloalkanes are cyclic alkane structures.
环烷烃是具有环状结构的烷烃。

2 Cycloalkanes have the general formula C_nH_{2n}.
环烷烃的通式是 C_nH_{2n}。

3 Most cycloalkanes are unreactive to chemical reagents. However, three- and four-membered rings are reactive due to ring strain and behave like alkenes.
大多数环烷烃与化学试剂不反应,但是三元环和四元环由于环张力的影响而很活泼,反应行为与烯烃类似。

4 Cycloalkanes are named according to the number of carbon atoms making up the ring, for example, cyclopropane, cyclobutane, cyclopentane, cyclohexane, etc.
环烷烃是根据成环的碳原子数来命名的,如环丙烷、环丁烷、环戊烷、环己烷等。

第八节 立体化学
Section 8 Stereochemistry

★ 立体化学 stereochemistry [ˌstiəriəˈkemistri]
手性试剂 chiral reagent
★ 手性中心 chiral center
★ 对映体 enantiomer [iˈnæntiəumə]
★ 立体异构现象 stereoisomerism [ˌstiːriəuaiˈsɔmərizəm]
★ 平面偏光 plane polarized light
★ 光学活性 optical activity
尼克尔棱镜 Nicol prism [niˈkəul][ˈprizəm]
★ 比旋光度 specific rotation
★ 手性碳原子 chiral carbon atom
★ 绝对构型 absolute configuration [ˈæbsəluːt][kənˌfigjuˈreiʃən]
★ 外消旋混合物 racemic mixture [rəˈsiːmik][ˈmikstʃə]
★ 内消旋化合物 meso compound [ˈmesəu][ˈkɔmpaund]
★ 费歇尔投影式 Fischer projection

1 Isomers are compounds which have the same molecular formula but have the atoms arranged in a different way.
异构体是分子式相同但原子排列方式不同的化合物。

2 There are three types of isomers: constitutional, configurational, and conformational.
有三种类型的异构体:构造异构体、构型异构体和构象异构体。

3 Constitutional isomers are compounds which have the same molecular formula but have the atoms joined together in a different way.
构造异构体是分子式相同但原子间连接方式不同的化合物。

4 Constitutional isomers have different physical and chemical properties.
构造异构体具有不同的物理和化学性质。

5 Configurational isomers have the same molecular formula and the same bonds.
构型异构体的分子式和化学键相同。

6 Substituted alkenes and cycloalkanes can exist as configurational isomers.
取代烯烃和取代环烷烃存在构型异构体。

7 Alkenes having two different substituents at each end of the double bond can exist as two configurational isomers.

双键两端分别连接两个不同取代基的烯烃存在两种构型异构体。

8 Optical isomers are configurational isomer, which have the ability to rotate plane-polarized light clockwise or counterclockwise.

光学异构体是指能够使平面偏振光向顺时针或逆时针方向偏转的构型异构体。

9 Optical isomers have identical chemical and physical properties but can have difference in biological properties.

光学异构体具有相同的化学和物理性质，但是具有不同的生物活性。

第九节 卤代烷
Section 9 Alkyl Halides

★ 卤代烷 alkyl halides [ˈælkil][ˈhælaidiz]
★ 范德华半径 Van der Waals radius
范德华引力 Van der Waals attractive force
★ 氯仿 chloroform [ˈklɔ(:)rəfɔ:m]
★ 二氯甲烷 dichloromethane [dai,klɔ:rəˈmeθein]
可极化度 polarizability [ˈpəulə,raizəˈbiləti]
★ 碳正离子 carbocation [ˌkɑ:bəˈkeiʃən]
★ 碳负离子 carbanion [ˈkɑ:bənaiən]
★ 亲电试剂 electrophile
★ 亲核试剂 nucleophile [ˈnju:kliəfail]
★ 亲核取代反应 nucleophilic substitution [ˌsʌbstiˈtju:ʃən] reaction
★ 离去基团 leaving group
★ S_N2 反应 S_N2 reaction
★ 双分子反应 biomolecular reaction
★ 过渡态 transition state
★ 构型转化 configuration conversion [kən,figjuˈreiʃən][kənˈvə:ʃən]
★ 瓦尔登转化 Walden conversion
★ S_N1 反应 S_N1 reaction
★ 单分子反应 unimolecular [ˌju:niməuˈlekjulə] reaction
★ 定速步骤 rate-determining step
★ 格氏试剂 Grignard reagent
溶剂化 solvolysis [sɔlˈvɔlisis]
立体效应 steric effect
可极化的 polarizable [ˈpəuləraizəbl]
★ 极性非质子溶剂 polar aprotic solvent [ˈpəulə][əˈprɔtik][ˈsɔlvənt]
★ 极性质子溶剂 polar protic solvent
★ 介电常数 dielectric constant [ˌdaiiˈlektrik][ˈkɔnstənt]
★ 扎依采夫规则 Saytzeff rule

1 Alkanes are converted to alkyl halides by reaction with hydrogen halides. Treatment with halogens results in dihaloalkanes.

烯烃与卤化氢反应生成卤代烷，与卤素反应生成二卤代烷。

2 Alkyl halides consist of an alkyl group attached to a halogen. The carbon linked to the halogen is sp^3 hybridized and tetrahedral.

卤代烷由烷基和卤素连接组成，和卤素连接的碳是 sp^3 杂化，呈四面体构型。

3 The carbon-halogen bond length increases and the bond strength decreases as the halogen increases in size.

随着卤原子原子序数的增加，碳卤键键长增加，键强减小。

4 The C-halogen bond is a polar σ bond where the halogen is slightly negative and the carbon is slightly positive.

C-X 键为极性 σ 键，其中卤原子带部分负电荷，碳原子带部分正电荷。

5 Alkyl halides have a dipole moment.

卤代烷具有偶极距。

6 Alkyl halides are poorly soluble in water, but dissolve in organic solvents.

卤代烷微溶于水，但能溶于有机溶剂。

7 Alkyl halides react as electrophiles at the carbon center.
卤代烷反应时卤代烷的中心碳表示出亲电性。

8 Alkyl halides perform nucleophilic substitution reactions and elimination reactions.
卤代烷能够进行亲核取代反应和消除反应。

第十节 醇和醚
Section 10 Alcohols and Ethers

★ 醇 alcohols [ˈælkəhɔlz]
★ 醚 ethers [ˈiːθəz]
★ 甲醇 methanol [ˈmeθənɔl]
★ 乙醇 ethanol [ˈeθənɔːl]
★ 丙醇 propyl [ˈprəupil] alcohol
★ 乙二醇 ethylene glycol [ˈeθiliːn][ˈglaikɔl]
★ 共振结构 resonance structure
质子化醇 protonated [ˈprəutəˌneitid] alcohol
质子化醚 protonated ether
烷氧离子 alkoxide ion
★ 威廉森法制醚 Williamson synthesis of ether
★ 锌盐 oxonium [ɔkˈsəunjəm] salt
★ 环氧化物 epoxide [eˈpɔksaid]
酸催化开环 acid-catalyzed ring opening
碱催化开环 base-catalyzed ring opening
★ 冠醚 crown ether [kraun][ˈiːθə]
★ 相转移催化剂 phase-transfer catalysis

1 Functional groups such as alkyl halides, carboxylic acids, esters, alkenes, aldehydes, ketones, and ethers can be turned into alcohols.
诸如卤代烷、羧酸、酯、烯烃、醛、酮和醚的官能团能够转变为醇。

2 The carbon and oxygen atoms of the alcohol groups are sp^3 hybridized so that the C-O-H bond angle is approximately 109°.
由于醇中的碳原子和氧原子都是 sp^3 杂化，所以 C-O-H 的键角近似 109°。

3 Hydrogen bonding means that alcohols have higher boiling points than comparable alkanes.
氢键的形成使醇比相近分子量的烷烃具有更高的沸点。

4 Alcohols of low molecular weight are soluble in water and act as weak acids and weak bases.
低分子量的醇能够溶解在水中，表现为弱酸或弱碱。

5 Alcohols are polar.
醇是极性分子。

6 The oxygen atom is a nucleophilic center.
氧原子是亲核中心。

7 Alcohols will not react with nucleophiles, but will react with strong bases in an acid-base reaction to give an alkoxide ion.
醇不与亲核试剂反应，但是能够与强碱发生酸 - 碱反应形成烷氧基负离子。

8 Ethers can be synthesized by the S_N2 reaction of an alkyl halide with an alkoxide ion.
醚可用卤代烷和烷氧基负离子通过 S_N2 反应来合成。

9 Alcohols and alkyl halides can be reacted with silver oxide to give an ether.
醇和卤代烷可与氧化银反应得到醚。

10 Ethers consist of an sp^3 hybridized oxygen lined to two carbon atoms by a single σ bond.
醚由一个 sp^3 杂化的氧原子与两个碳原子通过 σ 单键连接而形成。

11 Alkyl ethers are ethers where two alkyl groups are linked to the oxygen.
烷基醚是指两个烷基连接在氧上的醚。

12 Aryl ethers are ethers where one or two aromatic rings are attached to the oxygen.
芳基醚是指一个或两个芳香环连接在氧上的醚。

13 Ethers are relatively unreactive since they have weak nucleophilic and electrophilic centers.
醚有弱的亲核和亲电中心，因此相对不活泼。

第十一节 醛和酮

Section 11 Aldehydes and Ketones

★ 醛 aldehyde [ˈældihaid]
★ 酮 ketone [ˈkiːtəun]
★ 甲醛 formaldehyde [fɔːˈmældiˌhaid]
★ 丙酮 acetone [ˈæsitəun]
★ 醛基 formyl [ˈfɔːmil] group
★ 酰基 acyl [ˈæsil] group
★ 羰基 carbonyl [ˈkɑːbənil] group
★ 互变异构 tautomerization [tɔːˈtɔməraizeiʃən]
醇酮互变异构 keto-enol tautomerization
★ 亲核加成 nucleophilic addition
★ 半缩醛 hemiacetal [ˌhemiˈæsitæl]
★ 半缩酮 hemiketal [ˌhemiˈkiːtəl]
★ 缩醛 acetal [ˈæsitæl]
★ 缩酮 ketal [ˈkiːtæl]
水解 hydrolysis [haiˈdrɔlisis]
2,4-二硝基苯腙 2,4-dinitrophenylhydrazine [ˌdaiˌnaitrəuˌfenilhaidrəziːn]
氨基脲 semicarbazide [ˈsemiˈkɑːbəzaid]
羟胺 hydroxylamine [haiˌdrɔksiləˈmiːn]
★ 克莱门森还原 Clemmensen reduction
★ 乌尔夫-凯西纳还原 Wolff-Kishner reduction
★ 魏悌锡反应 Wittig reaction
★ 缩合 condensation [kɔndenˈseiʃən]
★ 醇醛缩合 aldol condensation
★ α-氢 α-hydrogen
★ 康尼扎罗反应 Cannizzaro reaction
★ 普尔金反应 Perkin reaction
★ 卤仿反应 haloform reaction
★ α,β-不饱和醛酮 α,β-unsaturated aldehyde and ketone
★ 共轭加成 conjugate addition

1 Aldehydes and ketones have higher boiling points than alkanes comparable molecular weight due to the polarity of the carbonyl group.
由于羰基具有极性,所以醛和酮的沸点比具有相近分子量的烷烃高。

2 The oxygen of the carbonyl group is a nucleophilic center.
羰基中的氧原子是亲核中心。

3 The carbonyl carbon is an electrophilic center.
羰基中的碳原子是亲电中心。

4 Ketones are in rapid equilibrium with an enol.
醛酮和烯醇异构体处于快速平衡之中。

5 The keto and enol forms are called tautomers.
醛酮及其烯醇式互为互变异构体。

6 Nucleophilic addition involves the addition of a nucleophile to an aldehyde or a ketone. The nucleophile adds to the electrophilic carbonyl carbon.
亲核加成是一个亲核试剂对醛或酮的加成,亲核试剂加到亲电的羰基碳原子上。

第十二节 羧酸及其衍生物

Section 12 Carboxylic Acid and Their Derivatives

★ 羧酸 carboxylic [ˌkɑːbɔkˈsilik] acid
★ 羧酸衍生物 carboxylic acid derivative
★ 甲酸 formic acid
★ 乙酸 acetic acid
★ 二元羧酸 dicarboxylic [daiˌkɑːbɔkˈsilik] acid
★ 酯 ester
★ 酸酐 carboxylic anhydride [ænˈhaidraid]
★ 酰氯 acyl chloride [ˈæsil][ˈklɔːraid]
★ 酰胺 amide [ˈæmaid]
★ 亲核加成-消除反应 nucleophilic addition-elimination reaction
★ 氢负离子 hydride ion
★ 烷基负离子 alkanide ion
★ 酯化反应 esterification [esˌterifiˈkeiʃən]
★ 皂化反应 saponification [səpɔnifiˈkeiʃən]
内酯 lactone [ˈlæktəun]
★ 乙酰乙酸乙酯 ethyl acetoacetate [ˈeθil][ˌæsitəˈæsiteit]

★ 克莱森酯缩合　Claisen condensation
★ 交叉酯缩合反应　cross Claisen condensation
★ 狄克曼反应　Dickmann reaction
★ 克瑙文盖尔反应　Knoevenagel reaction
★ 麦克尔加成　Michael addition

1 There are four common carboxylic acid derivatives derived from a parent carboxylic acid — acid chlorides, acid anhydrides, esters, and amides.
来自羧酸母体的常见的衍生物有四类，即酰氯，酸酐，酯和酰胺。

2 The functional groups of carboxylic acid derivatives are planar with bond angels of 120°.
羧酸及其衍生物的官能团的键角为 120°，呈平面型。

3 The carbonyl group is made up of a strong σ bond and a weaker π bond.
羰基由一个强 σ 键和一个弱 π 键组成。

4 The carbonyl group is polarized such that the oxygen acts as a nucleophilic center and the carbon acts as an electrophilic center.
羰基是个极性基团，其中氧原子是亲核中心而碳原子是亲电中心。

5 Carboxylic acids are polar and can take part in hydrogen bonding.
羧酸具有极性，可参与氢键的形成。

6 Carboxylic are soluble in water and have high boiling points.
羧酸能溶于水并具有较高的沸点。

7 Carboxylic acids are weak acids in aqueous solution and forms water soluble salts when treated with alkali.
在水溶液中羧酸是个弱酸，与碱反应生成水溶性的盐。

8 Carboxylic acids and acid derivatives undergo nucleophilic substitutions.
羧酸及其衍生物均可发生亲核取代反应。

第十三节　胺
Section 13　Amines

★ 胺　amine [ˈæmiːn]
★ 伯胺　primary amine
★ 仲胺　secondary amine
★ 叔胺　tertiary [ˈtəːʃəri] amine
★ 芳胺　aryl amine
★ 杂环胺类　heterocyclic [ˌhetərəuˈsaiklik] amine
★ 季铵盐　quaternary ammonium salt [kwəˈtəːnəri][əˈməunjəm][sɔːlt]
★ 重氮盐　diazonium [ˌdaiəˈzəuniəm] salt
★ 霍夫曼降级反应　Hofmann degradation [ˈhɔfmən][ˌdegrəˈdeiʃən]
★ 霍夫曼消除　Hofmann elimination
★ 霍夫曼规则　Hofmann's rule

1 Amines consist of an sp^3 hybridized nitrogen linked to three substituents by σ bonds.
胺分子是由 sp^3 杂化的氮原子以 σ 键与三个取代基相连。

2 The functional group of amines is pyramidal in shape with bond angels of approximately 109°.
氨基呈锥形结构，键角约为 109°。

3 Amines are polar compounds with higher boiling points than comparable alkanes.
胺是极性化合物，沸点比相应烷烃高。

4 Amines have similar water solubilities to alcohols due to hydrogen bonding.
由于胺能够形成氢键，所以其水溶性与醇相似。

5 Low molecular weight amines are completely miscible with water.
低分子量的胺能够和水完全混溶。

6 Low molecular weight amines have an offensive fishy smell.
低分子量的胺具有刺激性的腥臭味。

第十四节 芳香化合物

Section 14 Aromatic Compound

★ 芳香化合物 aromatic [ˌærəuˈmætik] compound
★ 苯 benzene [ˈbenziːn]
氯苯 chlorobenzene [ˌklɔːrəˈbenziːn]
甲苯 toluene [ˈtɔljuiːn]
苯酚 phenol [ˈfiːnəl]
苯胺 aniline [ˈæniliːn]
★ 邻位 ortho o-
★ 间位 meta m-
★ 对位 para p-
苯的凯库勒结构式 Kekulé structure for benzene
★ 离域 π 电子 delocalized [diːˈləukəlaizid] π electron
★ 苯基 phenyl [ˈfenəl]
★ 反芳香性化合物 antiaromatic [ˌæntiærəuˈmætik] compound
★ 非芳香性化合物 nonaromatic compound
★ 休克尔芳香性规则 Hückel $4n+2$ rule
苄基游离基 benzyl [ˈbenzil] radical
★ 芳香亲电取代 electrophilic aromatic substitution
卤代 halogenation [hælədʒəˈneiʃən]
硝化 nitration [naiˈtreiʃən]
磺化 sulfonation [ˌsʌlfəˈneiʃən]
付氏烷基化 Friedel-Crafts alkylation
付氏酰基化 Friedel-Crafts acylation
★ 活化基团 activating group
★ 钝化基团 deactivating group
★ 邻对位定位基 ortho-para director
★ 间位定位基 meta director

1 Aromatic compounds such as benzene are more stable than suggested from their structure.
苯等芳香族化合物比根据其结构推出的稳定性更强。

2 Aromatic compounds undergo reactions which retain the aromatic ring system and behave differently from alkenes or polyenes.
芳香族化合物的反应保持了芳环体系，与烯烃或多烯烃的性质不同。

3 Aromatic compounds are cyclic and planar with sp^2 hybridized atoms.
芳香族化合物是由 sp^2 杂化原子组成的环状平面型化合物。

4 Aromatic compounds also obey the Hückel rule and have $4n+2$ π electrons where $n=1,2,3,\ldots$
芳香族化合物遵守休克尔规则，都含有 $4n+2$ 个 π 电子，$n=1, 2, 3, \cdots$

第三章 物理化学
Chapter 3 Physical Chemistry

第一节 化学热力学基础
Section 1 Fundamentals of Chemical Thermodynamics

隔离系统 isolated system [ˈaisəleitid]
宏观性质 macroscopic property [ˈprɔpəti]
强度性质 intensive property [inˈtensiv]
广度性质 extensive property [iksˈtensiv]
状态函数 state function [ˈfʌŋkʃən]
热力学平衡 thermodynamic equilibrium [ˌiːkwiˈlibriəm]
相平衡 phase equilibrium
化学平衡 chemical equilibrium
定容过程 isochoric process
循环过程 cyclic process [ˈsaiklik]
自由膨胀过程 free expansion process [iksˈpænʃən]
凝聚相 condensed phase
流体相 fluid phase [ˈflu(ː)id]
饱和蒸气压 saturated vapor pressure [ˈsætʃəreitid]
标准沸点 standard boiling point [ˈstændəd]
临界参量 critical parameters [pəˈræmitə]
超临界流体 excess critical fluid [ikˈses, ˈekses]
化学计量数 chemical stoichiometric number
体积功 volume work [ˈvɔljuːm]
准静态过程 quasi-static process [ˈkwɑːzi(ː), ˈkweisai]
可逆过程 reversible process [riˈvəːsəbl]
热力学能 thermodynamic energy [ˈθəːməudaiˈnæmik]
第一类永动机 first kind of perpetual motion machine
理想气体 perfect gas
热容 heat capacity [kəˈpæsiti]
恒容摩尔热容 molar heat capacity at constant volume
恒压摩尔热容 molar heat capacity at constant pressure
吉布斯自由能 Gibbs free energy
相变焓 enthalpy of phase transition [trænˈziʒən, -ˈsiʃən]
热化学方程式 thermochemical equation [ˌθəːməˈkemikl]
节流过程 throttling process [ˈθrɔtl]
焦-汤系数 Joule-Thomson coefficient
卡诺定理 Carnot theorem [ˈθiərəm]
熵增原理 the principle of the increase of entropy
亥姆霍茨函数判据 Helmholtz function criterion [kraiˈtiəriən]
吉布斯函数判据 Gibbs function criterion
麦克斯韦关系式 Maxwell's relations

1 The standard state for a material is defined as being the substance at standard pressure.
物质的标准状态为在标准压力下物质的状态。

2 Work is a form of energy transfer, the most common forms of work are pressure-volume work and electrical work.
功是能量传递的一种形式，功最常见的形式是压力-体积功和电功。

3 The physical properties of a perfect gas are completely described by the amount of substance, temperature, pressure and volume. These four parameters are not independent, and the relations between them are expressed by the gas state law.
理想气体的物理性质由其物质的量、温度、压力和体积决定。这四个参数不是相互独立的，它们的关系可由气体状态方程描述。

4 Enthalpy is a state function whose absolute value cannot be known, whereas ΔH can be ascertained either by direct methods or indirectly.
焓是一个状态函数，它的准确值不能确定，但焓变可以通过一些直接或间接的方法来确定。

5 Like enthalpy the Gibbs and Helmoholtz free energies are state functions which do not have

measurable absolute values. The Gibbs free energy change represents the maximum amount of non-volume work.
和焓一样，吉布斯和亥姆霍茨自由能也是状态函数，它们的绝对值不能测量。自由能的变化代表着体积功的最大值。

第二节 相平衡热力学
Section 2 Thermodynamics of Phase Equilibrium

相平衡 phase equilibrium
克拉佩龙方程 Clapeyron equation
特鲁顿规则 Trouton rule
多组分系统 multicomponent system [ˌmʌltikəm'pəunənt]
质量浓度 mass concentration [ˌkɔnsen'treiʃən]
摩尔分数 mole fraction
偏摩尔量 partial molar quantity ['pɑːʃəl]
化学势 chemical potential [pə'tenʃ(ə)l]
物质平衡判据 equilibrium criterion of substance [krai'tiəriən]
拉乌尔定律 Raoult's law
亨利定律 Henrry's law
理想液态混合物 mixture of ideal liquid
分配定律 distribution law [ˌdistri'bjuːʃən]
依数性质 colligative properties
渗透压 osmotic pressure [ɔz'mɔtik]
正偏差 positive deviation [ˌdiːvi'eiʃən]
负偏差 negative deviation

1 Osmosis is the movement of a solvent from a solution of lower solute concentration (higher solvent concentration) to one of higher solute concentration (lower solvent concentration). In osmotic pressure measurements, a semi-permeable membrane, which is permeable only to solvent separates two liquids. This means that this technique is only used for relatively large solute molecules, often polymers or biological macromolecules.
渗透是溶剂从溶质浓度低(溶剂浓度高)的溶液向溶质浓度高(溶剂浓度低)的溶液的流动。在渗透压的测量中，采用一种只能透过溶剂分子的半透膜来隔离两种溶液。这意味着该技术只能用来测量较大分子，例如聚合物或生物大分子。

2 An ideal solution is a mixture of two species, A and B, which show similar molecular interactions between molecules of A, molecules of B and molecules of B and A.
理想液态混合物是指 A-A, A-B, B-B 相互作用力相似的液态混合物。

第三节 相平衡状态图
Section 3 Phase Diagram

相图 phase diagram ['daiəgræm]
相律 phase rule
相数 number of phase
物种数 number of substance ['sʌbstəns]
组分数 number of component [kəm'pəunənt]
自由度数 number of degrees of freedom [di'griː]
过冷现象 supercooled phenomenon [ˌsjuːpə'kuːl]
亚稳状态 metastable state
三相点 triple point ['tripl]
系统点 system point
冰点 ice point
液态完全互溶系统 liquid full miscible system ['misibl]
定温连接线 isothermal line [ˌaisəu'θəːməl]
泡点 bubble point ['bʌbl]
露点 dew point
最低恒沸点 minimum azeotropic point [əˌziːə'trɔpik]
最高恒沸点 maximum azeotropic point
杠杆规则 lever rule
液态部分互溶系统 liquid partially miscible system
临界会溶点 critical consolute point ['kɔnsəljuːt]
热分析法 thermal analysis
步冷曲线 cooling curve [kəːv]
低共熔点 eutectic point [juː'tektik]
转晶反应 transition crystal reaction ['kristl]

For an ideal miscible liquid mixture of A and B, temperature-composition phase diagram can be used to determine the composition of both vapor and liquid at any temperature in the two-phase region. The relative amounts of each phase can also be determined if the overall composition is known. When B is the more volatile component, the vapor is richer in B and liquid richer in A. Separation of pure A and B can be achieved by distillation. For some non-ideal systems, there is a low-boiling or high-boiling azeotropic point on the diagram. In this case separation of pure A and B by distillation is not possible.
对理想的液体混合物A和B来说,温度-组成相图可以用来在两相共存区的任一温度确定气相和液相的组成。如果总的组成是已知的,那么每一相的组成也就确定了。如果B是较易挥发液体,那么液相中A富集,气相中B富集。A和B可以用蒸馏来分离。在非理想体系中,相图中有低或高的共沸点存在。在这种情况下,A和B的分离就不能用蒸馏进行了。

第四节 化学平衡热力学
Section 4 Thermodynamics of Chemical Equilibrium

标准亲和势 standard affinity [ə'finiti]
生成吉布斯函数 Jibbs function of formation
分压比 ratio of partial pressure ['reiʃiəu]
逸度因子 fugacity factor
分逸度比 ratio of partial fugacity
主反应物 principal reactant ['prinsəp(ə)l, -sip-]
分解压 dissociation pressure [di,səuʃi'eiʃən]
同时反应 simultaneous reaction
独立反应 independent reaction [indi'pendənt]
耦合反应 coupling reaction ['kʌpliŋ]

An equilibrium is established when the rates of the forward and backword reactions are equal. At equilibrium, the Gibbs free energy change is zero.
当一个反应正逆向进行的速率相同时,该反应就达到了化学平衡。在平衡点,吉布斯自由能的变化为零。

第五节 统计热力学初步
Section 5 Introduction of Statistical Thermodynamics

统计热力学 statistical thermodynamics
粒子 particle ['pɑ:tikl]
能级分布 distribution of energy levels
量子态分布 distribution of quantum states
概率 probability [,prɔbə'biliti]
最概然分布 most probable distribution
粒子态 particle state
量子态 quantum state ['kwɔntəm]
独立子系统 system of independent particles [indi'pendənt]
相依子系统 system of interacting particles
定域子系统 system of localised particles
离域子系统 system of non-localised particles
平衡分布 equilibrium distribution
系综 ensemble [ə:n'sɑ:mbl]
正则系综 canonical ensemble [kə'nɔnikəl]
微正则系综 microcanonical ensemble
巨正则系综 grand canonical ensemble
玻耳兹曼分布 Boltzmann distribution
残余熵 residual entropy [ri'zidjuəl]
粒子配分函数 partition function of particles
转动 rotation [rəu'teiʃən]
振动 vibration [vai'breiʃən]
对称数 symmetry number ['simitri]
量热熵 calorimetry entropy [,kælə'rimitri]
光谱熵 spectral entropy ['spektrəl]

Statistical thermodynamics was the first foundational physical theory in which probabilistic concepts and probabilistic explanation played a fundamental role. For the philosopher it provides a crucial test case in which to compare the philosophers' ideas about the meaning of probabilistic assertions and the role of probability in explanation with what actually goes on when probability enters a foundational physical theory.
统计热力学是第一门以概率为主要依据的基础物理理论科学。统计热力学为科学家提供了一种重要的测试方案,通过该方案科学家能将其形成的有关概率判断和作用的想法与和实际发生的情况做一对比。

第六节 化学动力学基础

Section 6 Fundamentals of Chemical Kinetics

化学动力学 chemical kinetics
化学反应速率 chemical reaction rate
宏观反应动力学 macroscopic reaction kinetics [ˌmækrəu'skɔpik]
速率方程 rate equation
转化速率 rate of conversion
反应速率 rate of reaction
消耗速率 dissipate rate ['disipeit]
增长速率 increase rate
微分(的) differential [ˌdifə'renʃəl]
积分(的) integral ['intigrəl]
反应总级数 overall order of reaction
一级反应 first order reaction
间歇式反应器 batch reactor
连续式反应器 continuous reactor
反应半衰期 half life of reaction
静态法 stop state methods
流动态法 flow state methods
微分反应器 differential reactor
积分反应器 integral reactor
元反应 elementary reaction
指前参量 pre-exponential parameter [pri:ˌekspəu'nenʃəl]
活化能 activation energy [ˌækti'veiʃən]
活化分子 activated molecule
复合反应 combined reaction
对行反应 opposing reaction
可逆反应 reversible reaction
连串反应 consecutive reaction [kən'sekjutiv]
静态 steady state
平衡态 equilibrium state
表现速率常数 apparent rate coefficient
表观活化能 apparent activation energy
链引发 chain initiation [iˌniʃi'eiʃən]
活性中心 active centers
催化剂寿命 life of catalyst
中毒 poison ['pɔizn]
链传递 chain transfer/propagation [træns'fə:]/[ˌprɔpə'geiʃən]
链终止 chain termination [ˌtə:mi'neiʃən]
直链反应 straight chain reaction
支链反应 side chain reaction
热爆炸 heat explosion [iks'pləuʒən]
链爆炸 chain explosion
催化剂 catalyst
催化作用 catalysis [kə'tælisis]
阻化剂 inhibitor [in'hibitə(r)]
自催化作用 autocatalysis [ˌɔ:təkə'tælisis]
均相催化 homogeneous catalysis
多相催化 heterogeneous catalysis
广义酸碱 generalized acid-base
络合催化 complex catalysis ['kɔmpleks]
主催化剂 principal catalyst ['prinsəp(ə)l, -sip-]
助催化剂 promoter [prə'məutə]
再生 reactivation [riˌækti'veiʃən]

The rates of reactions:

Suppose the chemical reaction is of the form

$$A+B \longrightarrow Y+Z$$

and that concentrations of the species A, B, Y and Z are [A], [B], [Y] and [Z] respectively. The rate of the reaction can be expressed as the rate of change of the concentration of any of the species. Thus the rate of formation of product is d [Y]/dt, and the rate of destruction of A is -d [A]/dt, the formula of $-\frac{1}{a}\frac{d[A]}{dt}=-\frac{1}{b}\frac{d[B]}{dt}=\frac{1}{y}\frac{d[Y]}{dt}=\frac{1}{z}\frac{d[Z]}{dt}$ is obtained.

反应速率：

假设A与B反应生成Y、Z,并分别用[A]、[B]、[Y]和[Z]来表示它们的浓度。反应速率可用这几种物质浓度的变化率来表示,Y的生成速率为d[Y]/dt,A的消耗速率为$-$d[A]/dt,并且有$-\frac{1}{a}\frac{d[A]}{dt}=-\frac{1}{b}\frac{d[B]}{dt}=\frac{1}{y}\frac{d[Y]}{dt}=\frac{1}{z}\frac{d[Z]}{dt}$成立。

第七节 界面层的热力学及动力学

Section 7 Thermodynamics and Kinetics of Interface

界面层 interface layer
分散度 dispersity [dis'pə:siti]
表面功 surface work
表面张力 surface tension
高度分散系统 high dispersed system
附加压力 excess pressure
凸面 convex ['kɔn'veks]
凹面 concave ['kɔn'keiv]
毛细管 capillary [kə'piləri]
曲率半径 curvature radius
沾附润湿 adhesion wetting
浸渍润湿 dipping wetting
铺展润湿 spreading wetting
铺展系数 spreading coefficient
接触角 contact angle
过饱和蒸汽 supersaturated vapor
过冷液体 supercooled liquid
亚稳状态 metastable state [ˌmetə'steibl]
分散 dispersion [dis'pə:ʃən]
淬火 quenching ['kwentʃiŋ]
退火 annealing [æ'ni:liŋ]
表面活性剂 surface active agent; surfactant [sə:'fæktənt]
表面相 surface phase
亲水头部 hydrophilic head
疏水尾部 hydrophobic tail
亲液的 lyophilic [ˌlaiə'filik]
憎液的 lyophobic [ˌlaiə'fɔbik]
两亲结构 amphiphilic structure
单分子表面膜 surface film of unimolecular layer
缔合胶体 associated colloid
胶束 micelle [mi'sel, mai-]
表面压 surface pressure
临界胶团浓度 critical micelle concentration(CMC)
乳化 emulsification [iˌmʌlsifi'keiʃən]
吸附 adsorption [æd'sɔ:pʃən]
增溶 solubilization [ˌsɔljubilai'zeiʃən; -li'z-]
发泡 foaming ['fəumiŋ]
消泡 defoaming [di'fəumiŋ]
洗涤 washing ['wɔʃiŋ] / detergency [di'tə:dʒənsi]
吸附质 adsorbate [æd'sɔ:beit]
吸附剂 adsorbent [æd'sɔ:bənt]
物理吸附 physisorption [ˌfizi'sɔ:pʃən]
化学吸附 chemisorption [ˌkemi'sɔ:pʃən]
表面覆盖度 coverage of surface
单分子层吸附理论 theory of adsorption of unimolecular layer [ˌju:niməu'lekjulə]
吸附平衡常数 equilibrium constant of adsorption

Physisorption and chemisorption: Molecules can adhere to surfaces in two ways, physisorption and chemisorption. In physisorption there is a van der Waals interaction between the surface and the adsorbed molecules. In chemisorption the molecules adhere to the surface as the result of the formation of a chemical, and usually a covalent bond.

物理吸附和化学吸附：分子有两种方式粘附于物质的表面，一种是物理吸附，另一种是化学吸附。物理吸附通过范德华力使分子吸附于物质表面，而化学吸附中，被吸附的分子与物质表面形成了化学键，常为共价键。

第八节 电解质溶液

Section 8 Electrolytic Solution

电解质溶液 electrolyte solution
潜在电解质 potential electrolyte
解离 dissociate [di'səuʃieit]
电解 electrolyze [i'lektrəlaiz]
电离度 ionization degree
离子平均活度 ionic mean activity
电解质活度 activity of electrolyte
正离子活度 activity of positive ion
负离子活度 activity of negative ion
电解质质量摩尔浓度 molality of electrolyte
离子强度 ionic strength
离子氛 ionic atmosphere
电子导体 electronic conductor
离子导体 ionic conductor

电导池 conductance cell
电导率 conductivity [ˌkɔndʌkˈtiviti]
摩尔电导率 molar conductivity
漂移速率 drift rate
无限稀薄摩尔电导率 limiting molar conductivity
电迁移率 electric mobility [məuˈbiliti]
离子独立运动定律 law of the independent migration of ion

1 The simplest way of studying the motion of ions in solution is through their conductivity, their ability to conduct electricity, which can be determined by measuring their electrical resistance.
研究溶液中离子的运动最简单的方法是通过电导率,电导率就是一种物质导电的能力。要确定溶液的电导率,首先要测定其电阻。

2 An ion in solution is subject to variety of forces. If there are concentration gradients, the ions will spread from the regions of high concentration into those of low. The ions also respond to an electrical force. This motion can be interpreted as the result of an electrical force operating on the ions, the magnitude of the force being proportional to the potential gradient.
溶液中的离子会受到很多力的作用,如果溶液存在浓度梯度,离子要从高浓度区域向低浓度区域扩散。在电流的作用下,离子也要运动,因其受到了电场力的作用,该力的大小与电势梯度成正比。

第九节 电化学系统的热力学及动力学

Section 9 Thermodynamics and Kinetics of Electrochemical System

电化学系统 electrochemical system [iˌlektrəuˈkemikəl]
相间电势 potential between two phases
接触电势 contact potential
液接电势 liquid-junction potential
扩散电势 diffusion potential [diˈfjuːʒən]
池图式 cell diagram
电化电池 electrochemical cell
电解池 electrolytic cell [iˌlektrəuˈlitik]
可逆电池 reversible cell
不可逆电池 irreversible cell
干电池 dry battery, dry cell
蓄电池 storage cell, secondary cell
燃料电池 fuel cell
电极 electrode [iˈlektrəud]
甘汞电极 calomel electrode [ˈkæləmel]
电动势 electromotive force
能斯特方程 Nernst equation
标准电极电势 standard electrode potential
温度系数 temperature coefficient
极化 polarization [ˌpəuləraiˈzeiʃən]
超电势 overpotential [ˈəuvəpəˈtenʃəl]
电催化 electrocatalysis [iˈlektrəukəˈtælisis]
分解电压 decomposition voltage
腐蚀 erosion [iˈrəuʒən]
防护 protection [prəˈtekʃən]
电化学传感器 electrochemical sensor
氧化还原反应 redox reaction
生物电化学 Bioelectrochemistry [ˌbaiəuˌiˌlektrəuˈkemistri]
量子电化学 quantum electrochemistry

1 An electrochemical cell consists of two electrodes dipping into an electrolyte solution. If the two electrodes use a common electrolyte solution, the solution potential is common to both electrodes. Such an arrangement is called a cell without a liquid junction. When two electrodes are immersed in different electrolytes the potentials of the solutions differ.
两个电极浸在一种电解质中便构成了一个电化学电池,如果这两个电极共用一种电解质,液接电势对两个电极来说是相同的,这样的电池也叫无液接电池。当两个电极浸于不同的电解质时,溶液液接电势不同。

2 The potential of a single electrode can't be measured, but one electrode may arbitrarily be assigned the value zero and all others referred to it. The electrode chosen to have zero potential is the standard hydrogen electrode. The standard potential is reached when the hydrogen ion activity is unity, the hydrogen gas is at unit fugacity.
我们无法测量单个电极的电势,但可以假定某一个电极电势为零,其他电极都以它为参照。标准氢电极就是我们假定电势为零的电极，其标准的工作条件为氢离子活度为1，氢气的逸度为1。

3 In the general sense of the term all electrodes involve reduction (electronation) and oxidation (de-electronation), but the term oxidation-reduction electrode, or redox electrode, is normally confined to the case where a species exists in the solution in two oxidation states.

从广义角度讲，所有的电极或者为氧化(失电子)电极或者为还原电极(得电子)，但这里所说的氧化-还原电极只局限于某物质在溶液中以两种氧化态的形式存在的情形。

第十节 胶体分散系统及粗分散系统
Section 10 Colloidal and Coarse Dispersed System

分散系统 dispersed system
分散质 dispersed matter
均相分散系统 homogeneous dispersed system
非均相分散系统 heterogeneous dispersed system
胶体，溶胶 colloid ['kɔlɔid]
胶体分散系统 colloid dispersed system
缔合胶束溶液 associated micelle solution
胶体电解质 colloidal electrolyte
大分子溶液 macromolecule solution
憎液胶体 lyophobic colloid
亲液胶体 lyophilic colloid
粗分散系统 coarse dispersed system
乳状液 emulsion [i'mʌlʃən]
泡沫 foam [fəum]
悬浮液 suspension [səs'penʃən]
反射 reflect [ri'flekt]
折射 refract [ri'frækt]
衍射 diffract [di'frækt]
散射现象 scattering phenomena ['skætəriŋ]
扩散系数 diffusion coefficient
运动性质 movement properties
显微镜 microscope ['maikrəskəup]
望远镜 telescope ['teliskəup]
超显微镜 ultramicroscope [ʌltrə'maikrəskəup]
流体性质 fluid properties
流变性 rheology [ri'ɔlədʒi]
相对黏度 relative viscosity
特性黏度 intrinsic viscosity
层流 laminar flow ['læminə(r)]
层速梯度 laminar speed gradient
双电层 double electric layer
斯特恩层 stern layer
紧密层 closed layer
扩散层 diffuse layer
斯特恩面 stern section
滑动面 movable section
热力学电势 thermodynamic potential
斯特恩电势 stern potential
动电电势 moving potential
胶粒 colloidal particle
胶核 colloidal nucleus
盐析 salting out
电动现象 electrokinetic phenomena [i,lektrəukai'netik]
电泳 electrophoresis [i,lektrəfə'ri:sis]
电渗 electroosmosis [i,lektrəuɔz'məusis]
流动电势 flow potential
沉降电势 sedimentation potential
聚沉 coagulation [kəu,ægju 'leiʃən]
絮凝 flocculation [flɔkju'leiʃən]
沉降 sedimentation [,sedimen'teiʃən]
乳析 creaming ['kri:miŋ]
聚沉值 coagulation value [kəu,ægju 'leiʃən]
空间稳定理论 steric stabilization theory
空缺稳定理论 depletion stabilization theory
增溶 solubilization [,sɔljubilai'zeiʃən]
乳液聚合 emulsion polymerization [,pɔlimərai'zeiʃən]
强化采油 enhanced oil recovery
胶束驱油 micelle flooding
洗涤 detergency [di'tə:dʒənsi]
洗涤剂 detergent [di'tə:dʒənt]
胶束催化 micelle catalysis
渗透压 osmotic pressure [ɔz'mɔtik]
膜平衡 membrane equilibrium
唐南效应 Donnan effect
凝胶 gelation [dʒi'leiʃən]
油包水乳液 water in oil emulsion
水包油乳液 oil in water emulsion
乳化 emulsify [i'mʌlsifai]
乳化剂 emulsifying agent

乳化　emulsification [iˌmʌlsifiˈkeiʃən]　|　**破乳**　demulsification [diiˈmʌlsifiˈkeiʃən]

When an aqueous solution containing Cl^- ion is added to one containing Ag^+ ion, under certain conditions the solid AgCl precipitate may form as extremely tiny crystals that remain suspended in the liquid instead of settling out as a filterable precipitate. This is an example of a colloidal system.

在一定的条件下，将一种含有氯离子的溶液加到另一种含有银离子的溶液中，生成的氯化银以微小颗粒的形式分散在溶液中，而不形成沉淀。这样就得到了氯化银胶体。

第四章 化工原理
Chapter 4 Principles of Chemical Engineering

第一节 绪论
Section 1 Introduction

★ 动量传递 momentum [məu'mentəm] transfer
★ 热量传递（传热） heat transfer
★ 质量传递（传质） mass transfer
经验模型 empirical [em'pirikəl] model
半经验模型 semi-empirical model

1 Chemical engineering is a group of industrial processes in which raw materials are changed or separated into useful products.
化学工程是将原料转变成或分离成有用产品的一系列工业过程。

2 Chemical engineering:
- to design the most optimal technology for production of a specified substance from raw materials
- to develop and discover new technological applications for materials

化学工程：
- 由原料生产某种物质的最优技术的设计
- 研究和发展新技术的应用

第二节 流体流动
Section 2 Flow of Fluids

流体动力学 fluid dynamics
★ 质量流率(质量流量） mass flow rate
★ 质量流速 mass velocity [vi'lɔsiti]
质量通量 mass flux
★ 体积流率 volumetric [vɔlju'metrik] flow rate
★ 可压缩流体 compressible [kəm'presəbl] fluid
不可压缩流体 incompressible fluid
★ 牛顿流体 Newtonian fluid
剪应力 shear stress
摩擦力 friction force
动力黏度 dynamic viscosity
运动黏度 kinematic [ˌkaini'mætik] viscosity
★ 伯努利方程 Bernoulli equation
★ 连续性方程 continuity equation
流型(流动型态） flow pattern
湍流 turbulent ['tə:bjulənt] flow
过渡流 transition flow
层流(滞流)区 laminar region
湍流区 turbulent region
过渡流区 transition region
雷诺数 Reynolds ['renəldz] number
无因次数群 dimensionless group
涡旋 vortex ['vɔ:teks]
边界层 boundary layer
层流底层 laminar ['læminə(r)] sub-layer
充分发展流 fully developed flow
粗糙度 roughness
摩擦因子 friction factor
摩擦损失 friction loss
直管阻力损失 head lost due to friction
局部阻力损失 losses in fitting
压降 pressure drop
★ 动压头 kinetic head/ velocity head
★ 静压头 static head
★ 位能 potential energy
速度分布 velocity distribution
当量长度 equivalent length
水力半径 hydraulic radius [hai'drɔ:lik]['reidjəs]
文丘里管 Venturi [ven'tuəri] tube
皮托管 Pitot ['pitəu] tube
★ 文丘里流量计 Venturi meter
★ 孔板流量计 orifice ['ɔrifis] meter
★ 转子流量计 rotameter
流量系数 discharge coefficient

1 The Continuity Equation: So long as the flow q_v is continuous, the continuity equation, as applied to one-dimensional flows, states that the flow passing two points (1 and 2) in a stream is equal at each point, $A_1u_1=A_2u_2$.

$q_v=Au$, $q_m=\rho q_v=\rho Au$, where

q_v=volumetric flow rate,

q_m=mass flow rate,

A=cross － sectional area of flow,

u=average flow velocity, and

ρ= the fluid density.

For steady, one-dimensional flow, q_m is a constant. If, in addition, the density is constant, then q_v is constant.

连续性方程　只要流动是连续的，对一维流动的连续性方程，通过 1、2 两点的流动状态(体积流量)相等，即：$A_1u_1=A_2u_2$，而 $q_v=Au$，$q_m=\rho q_v=\rho Au$。式中：q_v 为体积流速，q_m 为质量流速，A 为流动的横截面积，u 为平均流速，ρ 为流体密度。

对稳定的一维流动，q_m 为常数，若密度为常数，则 q_v 也为常数。

2 The field equation is derived when the energy equation is applied to one-dimensional flows.

Assuming no friction losses and that no pump or turbine exist between sections 1 and 2 in the system,

$\frac{p_1}{\gamma}+\frac{u_1^2}{2g}+z_1=\frac{p_2}{\gamma}+\frac{u_1^2}{2g}+z_2$, where

p_1, p_2=pressure at sections 1 and 2,

u_1, u_2=average velocity of the fluid at the sections,

z_1, z_2=the vertical distance from a datum to the sections (the potential energy),

γ= the specific weight of the fluid, and

g = the acceleration of gravity.

当能量守恒方程应用到一维流动时，得到截面间方程。

假设流动系统中截面 1 和截面 2 间无摩擦损失和泵或叶轮机存在，则

$$\frac{p_1}{\gamma}+\frac{u_1^2}{2g}+z_1=\frac{p_2}{\gamma}+\frac{u_1^2}{2g}+z_2$$

式中：p_1、p_2 分别为截面 1 和截面 2 的压力；u_1，u_2 为流体在两截面上的平均流速；z_1、z_2 是从基准面到两截面的垂直距离(势能)；γ 为流体比重；g 为重力加速度。

3 Flow of A Real Fluid

$$\frac{p_1}{\gamma}+\frac{u_1^2}{2g}+z_1=\frac{p_2}{\gamma}+\frac{u_1^2}{2g}+z_2+h_f$$

The pressure drop as fluid flows through a pipe of constant cross-section and which is held at a fixed elevation is $h_f=(p_1-p_2)/\gamma$, where h_f=the head loss, considered a friction effect, and all remaining terms are defined above.

实际流体的流动

$$\frac{p_1}{\gamma}+\frac{u_1^2}{2g}+z_1=\frac{p_2}{\gamma}+\frac{u_1^2}{2g}+z_2+h_f$$

流体流经恒定横截面积且高度不变时的压力降为：$h_f=(p_1-p_2)/\gamma$，式中：h_f 为压头损失，由摩擦所引起，其他术语同上。

第三节　流体输送机械

Section 3　Pumping Equipment for Fluids

★ 离心泵　centrifugal [sen'trifjugəl] pump

容积式泵　positive displacement pump

活塞泵　piston pump

往复泵　reciprocating [ri'siprəkeitiŋ] pump

叶轮 impeller [im'pelə]
蜗壳 volute [və'lju:t]
★特性曲线 characteristic curve
★工作曲线 operating line
泵效率 pump efficiency
比转速 specific speed
★扬程 height/ head/ lift
扬量 capacity [kə'pæsiti]
汽蚀 cavitation [ˌkævi'teiʃən]
★汽蚀余量 net positive suction head(NPSH)
排风机 fan

1 The first part of the section deals with the transportation of fluids, both liquids and gases. The second part of the section discusses common methods of measuring the flow rate.
本节第一部分讨论的是流体的输送，包括液体和气体；第二部分讨论的是测定流速的常用方法。

2 The pump efficiency is the ratio of fluid power to the total power consumed.
泵的效率是液体所获得的功率与所消耗的总功率之比。

3 PUMPS This section deals with the transportation of liquids through pipes and channels. Liquid are sometimes moved by gravity from elevated tanks, or from a "blowcase"(a storage vessel pressurized from an external source of compressed gas), but by far the most common devices for the purpose are pumps.

Pumps increase the mechanical energy of the liquid, increasing its velocity, pressure, or elevation — or all three. The two major classes are positive-displacement pumps and centrifugal pumps. Positive-displacement units apply pressure directly to the liquid by a reciprocating piston or by rotating members which form chambers alternately filled by and emptied of the liquid. Centrifugal pumps generate high rotational velocities, then convert the resulting kinetic energy of the liquid to pressure energy.

In pumps the density of liquid does not change appreciably and may be considered constant.

泵 本节讨论的是流体通过管道的输送。液体有时靠重力从高位槽或从一个"压力容器"(靠外界气体加压的储罐)中流出，但到目前为止，输送流体最普通的设备是泵。

泵能提高流体的机械能，从而增加流体的流速、压力或高度，或者三者均增加。正位移泵(往复泵)和离心泵是两种主要类型的泵。正位移泵是通过往复塞或活柱交替地在泵缸内吸入和压出液体，直接把压力作用到液体上。离心泵产生高的转速，将液体产生的动能转化成压头能。

对泵来说，流体密度没有明显改变，可认为是常数。

第四节 液体的搅拌

Section 4 Agitation of Liquids

桨式搅拌器 paddle agitator
涡轮搅拌器 turbine agitator
叶轮搅拌器 impeller agitator
框式搅拌器 grid agitator
板片搅拌器 blade agitator
乳液 emulsion [i'mʌlʃən]
搅浆机 change-can mixer

Liquid agitation is an application that is used to keep particles in a liquid bath in suspension. In many cases the use of an air source is the easiest method to accomplish this. Other methods used are primarily mechanical, such as propellers.

液体搅拌是为了保持颗粒在液体中处于悬浮状态。多数情况下，完成这一操作以空气作气源是最容易的方法，其他方法主要是机械法，如螺旋桨搅拌。

第五节 流体通过颗粒层的流动

Section 5 Flow Through Beds of Solids

颗粒群 particle swarm
粒径 particle diameter
颗粒分布 particle size distribution
筛孔直径 sieve [siv] diameter

★ 固液分离 solid-liquid separation
★ 悬浮液 suspension [səs'penʃən]
★ 过滤介质 filtration medium/ filter medium
恒压过滤 constant-pressure filtration
恒速过滤 constant-rate filtration
深层过滤 clarifying filter
助滤器 filter aid
砂滤器 sand-bed filter
★ 真空过滤机 vacuum filter
★ 板框压滤机 plate-and-frame filter press
加压叶滤机 pressure leaf filter
★ 转筒真空过滤机 rotary vacuum drum filter

A filter press contains a set of plates designed to provide a series of chambers or compartments in which solids may collect. The plates are covered with a filter medium such as canvas. Slurry is admitted to each compartment under pressure; liquor passes through the canvas and out a discharge pipe, leaving a wet cake of solids behind.

板框压滤机由一系列滤板和滤框组成，固体在滤框中被收集。过滤介质如帆布覆盖在板上。在一定压力下，滤浆进入到滤框中，滤液穿过帆布从排出管流出，滤饼留在框中。

第六节 颗粒的沉降和流态化
Section 6 Particulate Sedimentation and Fluidization

★ 沉降 sedimentation [ˌsedimen'teiʃən]
曳力 drag force
表面曳力 wall drag
形体曳力 form drag
曳力系数 drag coefficient
★ 机械分离 mechanical separation
★ 重力分离 gravity separation
★ 离心分离 centrifugal separation
惯性分离 inertial [i'nə:ʃəl] separation
重力沉降 gravity setting
★ 旋风分离器 cyclone ['saikləun]
袋滤器 bag filter
静电沉降器 electrostatic precipitator [i'lektrəu'stætik][pri'sipiteitə]
★ 流态化 fluidization [ˌflu:idai'zeiʃən]
散式流态化 particulate fluidization
聚式流态化 aggregative fluidization
鼓泡流态化 bubbling fluidization
流化速度 fluidizing velocity [flu:'idaiziŋ][vi'lɔsiti]
自由沉降速度 free falling velocity
终端速度 terminal velocity
夹带速度 entrainment velocity

A plot of fluid pressure loss across the bed versus superficial fluid velocity through the bed would have the appearance of Figure 4-1.

流体通过床层压力损失与空塔气速的关系见图 4-1。

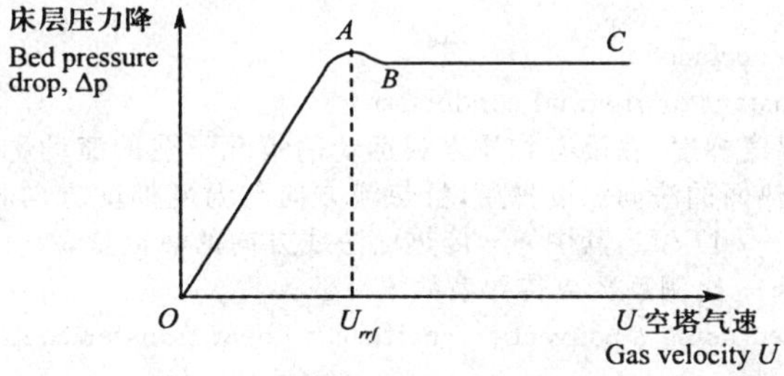

图 4-1 固定床和流化床压力降与流体空塔速度的关系

Figure 4-1 Pressure drop versus fluid velocity for packed and fluidized beds

第七节 传热
Section 7 Heat Transfer

★ 自然对流 natural convection ['nætʃərəl][kən'vekʃən]
★ 强制对流 forced convection
★ 辐射 radiation [ˌreidi'eiʃən]
★ 温度梯度 temperature gradient
★ 传热速率 heat transfer rate

★热流量 heat flow
★热通量(热流密度) heat flux
★总传热系数 overall heat transfer coefficient
传热膜系数 film heat transfer coefficient
★温差 temperature difference
★对数平均温差 logarithmic mean temperature difference
算术平均温差 arithmetic [əˈriθmətik] mean temperature difference
★并流(同向流) co-current flow
★逆流 countercurrent [ˈkauntəˌkʌrənt] flow
★错流 cross flow
热阻 thermal conductivity
污垢(结垢) scale/ fouling [ˈfauliŋ]
★传热单元数 number of heat transfer units
★传热单元高度 height of heat transfer units
导热系数(导热率) thermal conductivity
定态传热 steady state heat transfer
非定态传热 unsteady state heat transfer
温度分布 temperature distribution
吸收率 absorptive [əbˈsɔːptiv]
透射率 transmissivity [trænzmiˈsiviti]
反射率 reflectivity [ˌriːflekˈtiviti]
发射率 emissivity [ˌemiˈsiviti]
换热 heat exchange
★换热器(热交换器) heat exchanger
★管壳换热器(列管换热器) shell-and-tube heat exchanger
套管换热器 double-pipe heat exchanger
板式换热器 plate (type) heat exchanger
浮头换热器 floating head heat exchanger
U形管换热器 U-tube heat exchanger
★管程 tube(side)pass
★壳程 shell(side)pass
夹套 jacket
冷却器 cooler/ chiller
★冷凝器 condenser

1 There are three modes of heat transfer: conduction, convection and radiation. Boiling and condensation are classified as convection.

传热有三种方式:传导、对流和辐射。沸腾给热与冷凝给热属于对流换热。

2 If a temperature gradient exists in a continuous substance, heat can flow unaccompanied by any observable motion of matter. Heat flow of this kind is called conduction, and according to Fourier's law, the heat flux is proportional to the temperature gradient and opposite to it in sign. For one-dimensional heat flow, Fourier's law is

$$dq/dA = -k\, dT/dx$$

where q = rate of heat flow in direction normal to surface

A = surface area

T = temperature

x = distance normal to surface

k = proportionality constant or thermal conductivity

若一连续物质中存在温度梯度,在没有物质宏观运动的情况下热的流动称之为传导。根据傅立叶定律,热流密度正比于传热面的法向温度梯度,且热流方向与温度梯度方向相反。对于一维热传导,傅里叶定律为:$dq/dA = -k dT/dx$,其中 q— 传热面法线方向的热流量,A— 表面积,T— 温度,x— 传热面法线方向的距离,k— 比例系数或传导系数。

3 Convection is determined using a convection coefficient (heat transfer coefficient) α.

$Q = \alpha A\ (T_w - T_\infty)$, where

A = the heat transfer area,

T_w = wall temperature, and

T_∞ = bulk fluid temperature.

Resistance due to convection is given by $R = 1/(\alpha A)$.

对流给热的热流量由对流给热系数 α 来确定。$Q = \alpha A(T_w - T_\infty)$,式中:$A$ 为传热面积;T_w 为壁温度;T_∞ 流体的主体温度。

由对流引起的传热阻力为:$R = 1/(\alpha A)$。

第八节 气体吸收

Section 8 Gas Absorption

★ 物理吸收 physical absorption
化学吸收 chemical absorption
吸收等温线 absorption isotherm [ˈaisəuθəːm]
★ 吸收因子 absorption factor
解吸因子 desorption factor/ stripping factor
★ 传质速率 mass transfer rate
分子扩散系数 molecular diffusivity
涡流扩散系数 eddy diffusivity
★ 菲克定律 Fick's law [fiks][lɔː]
★ 双膜理论 two-film theory
★ 传质单元数 number of mass transfer units
★ 传质单元高度 height of mass transfer units
气相控制(气膜控制) gas phase control
液相控制(液膜控制) liquid phase control
气相传质系数 gas phase mass transfer coefficient
液相传质系数 liquid phase mass transfer coefficient
总传质系数 overall mass transfer coefficient
总传质单元数 number of overall transfer units
★ 总传质单元高度 height of overall transfer unit
气液传质设备 gas-liquid mass transfer equipment
★ 填料塔 packed column
★ 板式塔 tray column
填料支撑板 supporting plate
分布板 distribution plate
散装填料 dumped packing
整装填料 structured packing
拉西环 Raschig ring
鲍尔环 Pall ring
阶梯环 cascade ring

1 This section deals with the mass-transfer operations known as gas absorption and stripping, or desorption. In gas absorption a soluble vapor is absorbed from its mixture with an inert gas by means of a liquid in which the solute gas is more or less soluble. A major application of absorption technology is the removal of CO_2 and H_2S from natural gas or synthesis gas by absorption in solutions of amines or alkaline salts. Another example is the washing of ammonia from a mixture of ammonia and air by means of liquid water.

本节讨论气体吸收、气提或解吸的质量传递操作。在气体吸收中,通过某种液体吸收混合气体(含有惰性组分)中的可溶性气体,可溶性气体在这种液体中多少是溶解的。气体吸收的主要应用是采用铵盐或碱盐溶液吸收天然气或合成气中的 CO_2 和 H_2S。另一个例子是用水从氨-空气混合气中洗涤氨。

2 A plant design calls for an absorber that is to recover 95 percent of the acetone in an air stream, using water as the absorbing liquid. The entering air contains 14 mole percent acetone. The absorber has cooling and operates at 80℉ and 1 atm and is to produce a product containing 5.0 mole percent acetone. The water fed to the tower contains 0.02 mole percent acetone. The tower is to be designed to operate at 50 percent of the flooding velocity.

(a) How many pounds per hour of water must be fed to the tower if the gas rate is 500 ft^3/min, measured at 1 atm and 32℉?

(b) How many transfer units are needed, based on the overall gas - phase driving force?

(c) If the tower is packed with 1-in. Raschig rings, what should be packed height?

For equilibrium. Assume that $P_A = P_A^* \gamma_A x$, where $\ln\gamma_A = 1.95(1-x)^2$. The vapor pressure of acetone at 80℉ is 0.33 atm.

设计一吸收塔吸收空气中的丙酮,用水作吸收剂,吸收在 80℉ 和 1 atm 下进行。要求丙酮吸收率为 95%,出塔气体作为产品,其丙酮的摩尔分数为 5.0%。已知进塔混合气含丙酮摩尔分数为 14%,入塔的水中含丙酮摩尔分数为 0.02%,气体速度为泛点气速的 50%。

(a) 若气体流率为 500 ft^3/min(1 atm,32℉),每小时需提供多少磅水进入吸收塔?

(b) 需要多少传质单元数(以总气相推动力计)?

(c) 若填料为 1 英寸的拉西环,填料高度为多少?

对平衡:假设 $P_A = P_A^* \gamma_A x$,其中 $\ln\gamma_A = 1.95(1-x)^2$,丙酮在 80℉ 时的蒸气压为0.33 atm。

第九节 蒸馏
Section 9 Distillation

★ 连续蒸馏 continuous distillation
★ 间歇蒸馏 batch distillation
★ 馏出液 distillate
★ 残液 residue ['rezidju:]
道尔顿定律 Dalton's law
相对挥发度 relative volatility [ˌvɔlə'tiliti]
微分蒸馏 differential distillation
平衡蒸馏 equilibrium [ˌi:kwi'libriəm] distillation
★ 精馏 rectification [ˌrektifi'keiʃən]
分馏 fractionation [ˌfrækʃə'neiʃən]
★ 提馏 stripping
★ 精馏段 rectification section
★ 提馏段 stripping section
塔板 plate/ tray
★ 理论(塔)板，理论级 theoretical stage
★ 实际(塔)板 actual plate
再沸器 reboiler [ri 'bɔilə]
分凝器 partial condenser
操作线 operating line
★ 回流比 reflux ratio
全回流 total reflux
最小回流比 minimum reflux ratio
芬斯克方程 Fenske's equation
二元混合物 binary mixture
泡罩板 bubble cap tray
筛板 sieve tray
浮阀板 floating valve tray
塔板间距 tray spacing
降液管液柱高度 downcomer backup
溢流堰 overflow weir
堰高 weir height

1 A continuous fractionating column is to be designed to separate 30000 kg/h of a mixture of 40% benzene and 60% toluene into an overhead product containing 97% benzene and a bottom product containing 98% toluene. These percentages are by weight. A reflux ratio of 3.5 mol to 1 mol of product is to be used. The molal latent heats of benzene and toluene are 7360 and 7960 cal/g mol, respectively. Benzene and toluene form a nearly ideal system with a relative volatility of about 2.5; the equilibrium curve is shown in Fig. 21.16. The feed has a boiling point of 95℃ at a pressure of 1 atm. (a) Calculate the moles of overhead product and bottom product per hour. (b) Determine the number of ideal plates and the position of the feed plate (i) if the feed is liquid and at its boiling point; (ii) if the feed is liquid and at 20℃ (specific heat 0.44cal/g. ℃); (iii) if the feed is a mixture of two-thirds vapor and one-third liquid.

设计一连续精馏塔，用来分离流量为 30000 kg/h、含苯 40% 的苯 — 甲苯混合液，塔顶馏出液含苯 97%，釜液含甲苯为 98%(均为质量分数)，回流比为 3.5，苯和甲苯的摩尔汽化热分别为 7360 和 7960 cal/g mol，苯 — 甲苯近似为理想物系，其相对挥发度为 2.5。1atm 下进料的泡点为 95℃。(a) 计算塔顶和塔底的摩尔流率。(b) 确定理论塔板数和进料位置(i) 进料为泡点进料；(ii) 进料为 20℃ 液体进料(平均热容为 0.44cal/g. ℃)；(iii) 气液混合进料，气体占 2/3，液体占 1/3。

2 (a) The molecular weight of benzene is 78, and that of toluene is 92. The concentrations of feed, overhead, and bottoms in mole fraction of benzene are

$$x_F = \frac{\frac{40}{78}}{\frac{40}{78}+\frac{60}{92}} = 0.440 \quad x_D = \frac{\frac{97}{78}}{\frac{97}{78}+\frac{3}{92}} = 0.974 \quad x_B = \frac{\frac{2}{78}}{\frac{2}{78}+\frac{98}{92}} = 0.0235$$

The average molecular weight of the feed is

$$\frac{100}{\frac{40}{78}+\frac{60}{92}} = 85.8$$

The average heat of vaporization of the feed is

$$\lambda = 0.44 \times 7360 + 0.56 \times 7960 = 7696 \text{ cal/g mol}$$

The feed rate F is 30000/85.5 = 350 kg mol/h. By an overall benzene balance, using Eq. (21.5),

$$D = 350\left(\frac{0.440 - 0.0235}{0.974 - 0.0235}\right) = 153.4 \text{ kg mol/h}$$

$$W = 350 - 153.4 = 196.6 \text{ kg mol/h}$$

(b) Next we determine the number of ideal plates and position of the feed plate.

(i) The first step is to plot the equilibrium diagram and on it erect verticals at x_D, x_F, and x_W. These should be extended to the diagonal of the diagram. The second step is to draw the feed line. Here $q = 0$, and the feed line is vertical and is a continuation of line $x = x_F$. The third step is to plot the operating lines. The intercept of the rectifying line on the y axis is, from Eq. (21.19), $0.974/(3.5+1) = 0.216$. This point is connect with the point x_D on the yx reference line. From the intersection of the rectifying operating line and the feed line, the stripping line is drawn. The fourth step is to draw the rectangular steps between the two operating lines and the equilibrium curve. In drawing the steps, the transfer from the rectifying line to the stripping line is at the seventh step. By counting steps it is found that, besides the reboiler, 11 ideal plates are needed and feed should be introduced on the seventh plate from the top.

(ii) The latent heat of vaporization of the feed λ is $7696/85.8 = 89.7$ cal/g. Substitution in Eq. (21.24) gives $q = 1 + 0.44 \times (95 - 20)/89.7 = 1.37$.

From Eq. (21.31) the slope of the feed line is $-1.37/(1-1.37) = 3.70$. when steps are drawn for this case, it is found that a reboiler and 10 ideal plates are needed and that the feed should be introduced on the sixth plate.

(iii) From the definition of q it follows that for this case $q = 1/3$ and the slope of the feed line is -0.5. It calls for a reboiler and 12 plates, with the feed entering on the seventh plate.

(a) 苯和甲苯的分子量分别为 78 和 92;进料、塔顶及塔底苯的摩尔分数分别为 0.440,0.974 和 0.0235;进料的平均分子量为 85.8;进料的汽化热为 7,696 cal/g mol;进料的摩尔流率为 350 kg mol/h。据方程式(21.5),对全塔作苯的物料衡算,分别得到塔顶和塔底的摩尔流率 $D = 153.4$ kg mol/h 和 $W = 196.6$ kg mol/h。

(b) 确定理论塔板数和进料位置。

(i) 第一步画平衡图,并在平衡图上标注 x_D,x_F 和 x_W,过 x_D、x_F 和 x_W 点向上作垂线与对角线相交;第二步画进料线,因为 $q = 0$,故进料线为 $x = x_F$ 的垂线;第三步画操作线,精馏段操作线在 y 轴的截距是 0.216(据方程式 21.19),该点与 x_D 在对角线上的交点相连。从精馏段操作线与进料线的交点,画出提馏段操作线;第四步在操作线和平衡线之间画梯级,在第 7 级精馏段操作线转换到提馏段操作线。通过数梯级可知:需 11 块理论板(除再沸器外),第 7 块为进料板(从塔顶计算)。

(ii) 进料的汽化热为 89.7 cal/g,代入公式(21.24)得到 $q = 1.37$,由方程式(21.31)得到进料线的斜率为 3.70,在该条件下画梯级,需 10 块理论板和 1 个再沸器,第 7 块为进料板。

(iii) 据 q 定义,$q = 1/3$,进料线的斜率为 -0.5,通过画梯级可知,需要一个再沸器和 12 块理论板,在第 7 块板进料。

第十节 固体干燥
Section 10 Drying of Solids

★ 干燥 drying
增湿 humidification [hjuːˌmidifiˈkeiʃən]
减湿 dehumidification
★ 相对湿度 relative humidity
★ 露点 dew point
★ 泡点 bubble point
★ 湿球温度 wet bulb temperature
★ 绝热饱和温度 adiabatic saturation temperature
焓湿图 enthalpy-humidity chart
湿物料 moist material
湿含量 moisture content
临界湿含量 critical moisture content
游离水分 free moisture
结合水分 bound moisture
★ 干燥速率 drying rate
★ 气流干燥器 pneumatic dryer

[nju(:)'mætik]['draiə]

离心干燥器 centrifugal dryer

★ 流化床干燥器 fluidized ['flu:idaizd] bed dryer

★ 喷雾干燥器 spray dryer

1 In general, drying a solid means the removal of relatively small amounts of water or other liquid from the solid material to reduce the content of residual liquid to an acceptably low value. Drying is usually the final step in a series of operations, and the product from a dryer is often ready for final packaging.

一般来说，固体干燥是指从固体物料中除去相对少量的水分或其他液体，以期把湿分含量降至规定的范围内。干燥往往是一系列操作的最后工序，从干燥器出来的产品准备最后的包装。

2 H is the mass of vapor carried by a unit mass of vapor-free gas. So defined, humidity depends only on the partial pressure of the vapor in the mixture when the total pressure is fixed. If the partial pressure of the vapor is P_A atm, the molal ratio of vapor to gas at 1 atm is $P_A/(P-P_A)$. The humidity is therefore

$$H=\frac{M_A p_A}{M_B(P-p_A)}$$

where M_A and M_B are the molecular weights of components A and B, respectively.

The humidity is related to the mole fraction in the gas phase by the equation

$$y=\frac{H/M_A}{1/M_B+H/M_A}$$

Since H/M_A is usually small compared with $1/M_B$, often y may be considered to be directly proportional to H.

H是单位质量的干空气所带有的水汽质量。如此定义它，是因为当总气压一定时，湿度只取决于混合气中水汽的分压。若水汽分压是 p_A(atm)，则在 1atm 下水汽与干空气的摩尔比为 $p_A/(P-p_A)$。即湿度为 $H=M_A p_A/[M_B(P-p_A)]$，式中 M_A 和 M_B 分别为组分 A 和 B 的摩尔质量。

据方程 $y=(H/M_A)/(1/M_B+H/M_A)$，湿度与气相中的摩尔分数有关，因为 H/M_A 相对 $1/M_B$ 通常很小，故可简单地认为 y 正比于 H。

第十一节 其他单元操作
Section 11 Other Unit Operations

液液萃取 liquid-liquid extraction

萃余液 raffinate ['ræfineit]

溶剂萃取 solvent extraction

脉冲筛板塔 pulsed sieve plate column

往复板萃取塔 reciprocating [ri'siprəkeitiŋ] plate column/Karr column

转盘塔 rotating disc contactor (RDC)

浸取(液-固萃取) leaching/liquid-solid extraction

★ 浸滤 lixiviation [lik,siv'eiʃən]

浸渍(浸泡) dipping/ infusion

洗涤(水洗) washing/ scrubbing

超临界萃取 supercritical fluid extraction

自然循环蒸发器 natural circulation evaporator

水平列管蒸发器 evaporator with horizontal tubes

中央循环管蒸发器 calandria [kə'lændriə] type evaporator

浓缩 concentration [ˌkɔnsen'treiʃən]

多效蒸发 multiple-effect evaporation

★ 结晶 crystallization ['kristəlai'zeiʃən]

过饱和 supersaturation

过冷 supercooling

过热 superheating

成核 nucleation [ˌnju:kli'eiʃən]

晶核 crystal nucleus

晶种 seed crystal

亚稳区 metastable region

单晶 single crystal

母液 mother liquor

晶浆 magma

晶体生长 crystal growth

结晶速率 crystallization ['kristəlai'zeiʃən] rate

蒸发结晶 evaporative crystallization

蒸发结晶器 evaporative crystallizer

★ 吸附 adsorption

吸附剂 adsorbent [æd'sɔ:bənt]

吸附质 adsorbate [æd'sɔ:beit]
吸附平衡 adsorption equilibrium [ˌi:kwi'libriəm]
吸附器 adsorber [æd'sɔ:bə]
移动床吸附器 moving bed adsorber
流化床吸附器 fluidized bed adsorber
渗透物 permeate ['pə:mieit]
渗余物 retentate [ri'tenteit]
渗透通量 permeation flux
膜组件 membrane module
管式组件 tubular module
板框组件 plate-and-frame module
中空纤维组件 hollow-fiber module

Many processes for separation of gaseous or liquid mixtures use semipermeable membranes that allow one or more constituents of the mixture to pass through more readily than the others. The membranes may be then layers of a rigid material such as porous glass or sintered metal, but more often they are flexible films of synthetic polymers prepared to have a high permeability for certain types of molecules.
分离气体或液体混合物的许多过程均使用半透性膜,这种膜允许一种或更多种组分更快的通过(与其他组分相比)。它可以是硬质材料如多孔玻璃、烧结金属,但大多数它们是精制的聚合物柔韧薄膜,对某一类型的分子具有高的渗透性。

第五章　生物化学
Chapter 5　Biochemistry

第一节　绪论
Section 1　Introduction

★ 进展，进化　evolution [ˌiːvəˈluːʃən]
多细胞的　multicellular [ˌmʌltiˈseljulə]
★ 前景　perspective [pəˈspektiv]
★ 最新的　updated [ʌpˈdeitid]
说明　elucidation [iljuːsiˈdeiʃən]

1 Biochemistry is a vast subject that occurs at the intersection of a number of disciplines, from chemistry and physics to medicine and psychology.
生物化学是从化学和物理学到药学和心理学等学科相交叉的庞大学科。

2 Bioengineering is a relatively new discipline that combines many aspects of traditional engineering fields such as chemical, electrical and mechanical engineering.
生物工程学是一个相对新的学科，由许多传统工程领域如化学工程学、电子工程学和机械工程学等组成。

第二节　糖类
Section 2　Carbohydrates

★ 糖　carbohydrate [ˈkɑːbəuˈhaidreit] /saccharide [ˈsækəraid] /sugar [ˈʃugə]
单糖　monosaccharide [ˌmɔnəuˈsækəraid]
★ 葡萄糖　glucose [ˈgluːkəus]
★ 果糖　fructose [ˈfrʌktəus]
醛糖　aldose [ˈældəus]
酮糖　ketose [ˈkiːtəus]
丁糖　tetrose [ˈtetrəus]
戊糖　pentose [ˈpentəus]
★ 己糖　hexose [ˈheksəus]
庚糖　heptose [ˈheptəus]
★ 甘油醛　glyceraldehyde [ˌglisəˈrældihaid]
旋光度　optical rotation [ˈɔptikəl][rəuˈteiʃən]
★ 比旋光度　specific rotation [spiˈsifik][rəuˈteiʃən]
★ 变旋　mutarotation [ˌmjuːtərəuˈteiʃən]
哈沃斯投影式　Haworth projection [ˈhɔːəθ][prəˈdʒekʃən]
★ 差向异构体　epimer [ˈepimə]
★ 糖苷　glycoside [ˈglaikəˌsaid]
二糖　disaccharide [daiˈsækəraid]
低聚糖，寡糖　oligosaccharide [ˌɔligəuˈsækəraid]
★ 多糖，聚糖　polysaccharide [pɔliˈsækəraid]
麦芽糖　maltose [ˈmɔːltəus]
★ 乳糖　lactose [ˈlæktəus]
纤维二糖　cellobiose [ˌseləˈbaiəus]
同多糖　homopolysaccharide [ˌhəuməuˌpɔliˈsækəraid]
杂多糖　heteropolysaccharide [ˌhetərəˌpɔliˈsækəraid]
★ 淀粉　starch [stɑːtʃ]
★ 淀粉酶　amylase [ˈæməˌleis]
支链淀粉　amylopectin [ˌæmiləuˈpektin]
★ 糖元　glycogen [ˈglaikəudʒen]
复合糖　glycoconjugate [ˌglaikə ˈkɔndʒugit]

1 All carbohydrates contain at least one asymmetrical (chiral) carbon and are, therefore, optically active.
所有的糖分子至少含有一个不对称(手性)碳原子，因此具有光学活性。

2 Carbohydrates also can combine with lipid to form glycolipids or with protein to form glycoproteins.
糖分子也能与脂质结合形成糖脂或与蛋白质结合形成糖蛋白。

第三节 脂类
Section 3 Lipids

脂肪酸 fatty acid ['fæti][æsid]
★ 硬脂酸 stearic [sti'ærik] acid
油酸 oleic [əu'li:ik] acid
单纯脂 simple lipid ['simpl]['lipid, 'laipid]
蜡 wax [wæks]
复合脂 compound lipid, complex lipid
★ 脂蛋白 lipoprotein [ˌlipə'prəuti:n]
饱和的 saturated ['sætʃəreitid]
不饱和的 unsaturated
★ 皂 soap
★ 甘油 glycerol ['glisəˌrɔl]
甘油三酯 triglyceride [trai'glisəˌraid]
★ 三酰甘油 triacylglycerol [trai'æsilglisəˌrɔl]
两性分子的 amphiphilic [ˌæmfi'filik]
单分子层 monolayer ['mɔnəu'leiə]
★ 去垢剂 detergent [di'tə:dʒənt]
★ 双层结构 bilayer structure
成分，要素 constituent [kən'stitjuənt]
碘值 iodine ['aiədi:n] number
皂化值 saponification[səpɔnifi'keiʃən] number
★ 鞘氨醇 sphingosine ['sfiŋgəusi(:)n]
★ 鞘磷脂 sphingomyelin [ˌsfiŋgəu'maiəlin]
乙醇胺 ethanolamine [əˌeθə'nɔləmi:n]
胆碱 choline ['kəuli:n]
丝氨酸 serine ['seri:n, 'siər-]
萜 terpene ['tə:pi:n]
类固醇 steroid ['stiərɔid]
★ 胆固醇 cholesterol [kə'lestərəul]
外周膜蛋白 peripheral [pə'rifərəl] membrane protein
整合膜蛋白 integral membrane protein
液态镶嵌模型 fluid-mosaic model
★ 脂质体 liposome ['lipəusəum]

1 The sphingomyelins are also membrane lipids. They are the major component of the myelin sheath surrounding nerve fibers.
鞘磷脂也是膜脂类，它们是围绕神经纤维的髓磷脂鞘的主要组分。

2 Liposomes can be used to investigate membrane permeability.
脂质体可用于研究细胞膜通透性。

第四节 蛋白质
Section 4 Proteins

★ 肽 peptide ['peptaid]
氨基酸 amino acid
★ 抗体 antibody ['æntiˌbɔdi]
★ 疫苗 vaccine ['væksi:n]
球蛋白 globulin ['glɔbjulin]/ globular protein ['glɔbjulin]['glɔbjulə]['prəuti:n]
清蛋白 albumin [æl'bjumin]
色蛋白 chromoprotein [ˌkrəumə'prəuti:n]
★ 抗生素 antibiotics [ˌæntibai'ɔtiks]
纤维蛋白 fibrous ['faibrəs] protein
单纯蛋白质 simple protein
结合蛋白质 conjugated ['kɔndʒugitid] protein
蛋白质缺乏症 protein deficiency [di'fiʃənsi]
代谢过程 metabolic process
α-氨基酸 α-amino acid
甘氨酸 glycine ['glaisi:n]
丙氨酸 alanine ['æləˌni:n]
缬氨酸 valine ['væli:n]
亮氨酸 leucine ['lu:si:n, -sin]
异亮氨酸 isoleucine [ˌaisəu'lu:si:n]
苯丙氨酸 phenylalanine [ˌfenəl'æləni:n]
脯氨酸 proline ['prəuli:n]
苏氨酸 threonine ['θri:əni:n]
酪氨酸 tyrosine ['tirəsi:n]
半胱氨酸 cysteine ['sistin]
蛋氨酸 methionine [me'θaiəni:n]
赖氨酸 lysine ['laisi:n]
精氨酸 arginine ['ɑ:dʒini(:)n]
组氨酸 histidine ['histidi:n]
色氨酸 tryptophan ['triptəfæn]
天冬氨酸 aspartic [ə'spɑ:tik] acid
谷氨酸 glutamic [glu:'tæmik] acid
天冬酰胺 asparagine [əs'pærədʒi:n]
谷氨酰胺 glutamine ['glu:təmi:n]
含氮杂环 nitrogen heterocyclic ring
★ 两性离子 zwitterion ['tsvitəraiən]

两性电解质 ampholyte [æmfəlait]
偶极离子 dipolar ion
解离常数 dissociation constant [di,səuʃi'eiʃən]['kɔnstənt]
★ 等电点 isoelectric [,aisəui'lektrik] point (pI)
完全质子化 fully protonated ['prəutəneitid] state
丹磺酰氯 dansyl chloride [dɑ:ŋ'sil]['klɔ:raid]
茚三酮 ninhydrin [nin'haidrin]
★ 巯基乙醇 mercaptoethanol [mə,kæptəu'eθənɔl]
过甲酸 performic acid
★ 中和 neutralization [,nju:trəlai'zeiʃən]
★ 电泳 electrophoresis [i,lektrəfə'ri:sis]
电场 electric field
静电荷 net charge
带正电荷基团 positively charged group
带负电荷基团 negatively charged group
胃蛋白酶 pepsin ['pepsin]
纸色谱 paper chromatography [,krəumə'tɔgrəfi]
★ 层析柱 chromatographic column
肽键 peptide bond
天然肽 natural peptide
生物活性肽 active peptide
★ 胰岛素 insulin ['insjulin]
★ 二硫键 disulfide [dai'sʌlfaid] link
多肽的一级结构 primary structure of peptide
蛋白质测序 protein sequencing
氨肽酶 amino peptidase ['peptideis]
羧肽酶 carboxypeptidase [kɑ:,bɔksi'peptideis]
二级结构 secondary structure
三级结构 tertiary structure
四级结构 quarternary ['kwɔ:tənəri] structure
肽键的平面化 planarity of peptide bond
蛋白质酰胺平面 protein amide plane
共振相互作用 resonance ['rezənəns] interaction
★ 二面角 dihedral [dai'hedrəl] angle
氨基酸残基 amino acid residue
★ α-螺旋 α-helix ['hi:liks]
★ β-折叠 β-pleated sheet
★ β-转角 β-bend/β-turn
★ 无规卷曲 random coil
★ 发夹结构 hairpin ['hɛəpin] structure
★ 反平行 antiparallel [,ænti'pærəlel]
★ 变性 denaturation [di:neitʃə'reiʃən]
★ 复性 renaturation ['ri:,neitʃə'reiʃən]
自我装配 selfassembly [self ə'sembli]
沉降 sedimentation [,sedimen'teiʃən]
★ 沉淀 precipitate [pri'sipiteit]
★ 等离子点 isoionic point
构建进化树 construction of evolutionary tree
★ 极端条件 extreme condition
紫外辐射 ultraviolet ['ʌltrə'vaiəlit] radiation
免疫球蛋白 immunoglobulin [i'mju:nəu'glɔbjulin]
可变区 variable region/variable domain
不变区 constant region/ constant domain
多克隆抗体生产 production of polyclonal antibodies
单克隆抗体生产 production of monoclonal antibodies
邻位基团的立体位阻 steric crowding of neighboring groups
带电基团间的排斥与吸引 repulsion and attraction of charged groups
绿色荧光蛋白标记 green fluorescent protein as a marker
自动多肽固相合成 automated solid-phase peptide synthesis
凝胶过滤色谱 gel-filtration chromatography
离子交换色谱 ion-exchange chromatography
亲和色谱 affinity chromatography
高压液相色谱 high-pressure liquid chromatography
★ 凝胶电泳 gel electrophoresis [i,lektrəfə'ri:sis]
★ 等电聚焦 isoelectric [,aisəui'lektrik] focusing
二维电泳 two-dimensional electrophoresis
★ 超速离心 ultracentrifugation ['ʌltrəsen,trifju'geiʃən]
多肽质量指纹识别 peptide mass fingerprinting
蛋白质纯化 protein purification
★ 盐析 salting out
★ 透析 dialysis [dai'ælisis]

1 The substituent groups have hydrophilic and hydrophobic characters.
取代基具有亲水性和疏水性特征。

2 This orientation of H-bonding produces a helical coiling of the peptide backbone such that the R-groups lie on the exterior of the helix and perpendicular to its axis.
这种氢键的取向使肽链骨架产生了螺旋形卷曲，使 R 基团处于螺旋的外侧并与螺旋轴垂直。

3 Insulin secretion from cells is principally regulated by plasma glucose levels, but the precise

mechanism which the glucose signal is transduced remains unclear.
细胞分泌胰岛素主要由血糖浓度调节，但是血糖信号转换的确切机理尚不清楚。

第五节 酶
Section 5 Enzymes

生物催化剂 biocatalyst [ˌbaiəuˈkætəlist]
★ 加速 accelerate [ækˈseləreit]
★ 绝对专一性 absolute specificity
立体专一性 stereospecificity [ˌstiəriəuˌspesiˈfisiti]
★ 辅助因子 cofactor [kəuˈfæktə]
★ 辅酶 coenzyme [kəuˈenzaim]
★ 辅基 prosthetic [prɔsˈθetik] group
全酶 holoenzyme [ˌhɔləuˈenzaim]
单体酶 monomeric enzyme
寡聚酶 oligomeric [ˈɔligəumərik] enzyme
★ 多酶复合体 multienzyme complex
★ 氧化还原酶类 oxido-reductase [ɔksid][riˈdʌkteis]
★ 转移酶类 transferase [ˈtrænsfəˌreis]
★ 水解酶类 hydrolase [ˈhaidrəleis]
★ 裂合酶类 lyase [ˈlaiəs]
★ 异构酶类 isomerase [aiˈsɔməreis]
★ 连接酶类 ligase [liˈgeis]
★ 底物 substrate [ˈsʌbstreit]
★ 活性中心 active center
★ 诱导契合 induced fit
★ 转换数 turnover [ˈtə:nˌəuvə] number
★ 酶活力 enzyme activity
★ 比活力 specific activity
自我剪接 self-splicing
抗体酶 abzyme [ˈæbzaim]
★ 酶-底物复合物 enzyme-substrate (ES) complex
★ 稳态 steady state
米氏方程 Michaelis-Menten equation
米氏常数 Michaelis constant(KM)
★ 最大反应速度 maximal rate/ Vmax
序列反应 sequential reaction
有序反应 ordered reaction
随机反应 random reaction
乒乓反应 ping-pong reaction
★ 别构酶 allosteric [ˌæləˈsterik] enzyme
★ 协同作用 concerted [ˈkɔnsətid] action
★ 亚基 subunit [sʌbˈju:nit]
代谢途径 metabolic pathway
★ 抑制 inhibition [ˌinhiˈbiʃən]
不可逆抑制 irreversible [ˌiriˈvə:səbl] inhibition
可逆抑制 reversible inhibition
竞争性抑制 competitive inhibition
非竞争性抑制 noncompetitive inhibition
反竞争性抑制 uncompetitive inhibition
★ 最适温度 optimum temperature
★ 激活剂 activator
失活作用 inactivation
★ 修饰 modify [ˈmɔdifai]
磷酸化 phosphorylation [fɔˌsfɔriˈleiʃən]
共价催化 covalent catalysis
酸碱催化 general acid-base catalysis
★ 靠近 approximation [əˌprɔksiˈmeiʃən]
★ 定向效应 orientation [ˌɔ(:)rienˈteiʃən]
★ 形变 distortion [disˈtɔ:ʃən]
★ 协同性 cooperativity [kəuˌɔpərəˈtivəti]
★ 松弛型 relaxed state
★ 紧张型 tensed state
★ 同工酶 isoenzyme [ˌaisəuˈenˌzaim]
★ 激酶 kinase [ˈkineis]
★ 固定化酶 immobilized enzyme
★ 人工酶 artificial enzyme
酶联免疫测定 enzyme-linked immunosorbent assay (ELISA)

1 Enzymes facilitate the formation of the transition state.
酶使过渡态易于形成。

2 The active site is a three-dimensional cleft formed by groups that come from different parts of the amino acid sequence.
活性中心是由来自氨基酸序列不同部位的基团形成的三维凹陷。

3 The nonpolar microenvironment exists in the active site of the enzyme.
非极性微环境存在于酶的活性中心。

4 The specificity of binding depends on the precisely defined arrangement of atoms in an active site.
结合的专一性决定于活性中心上原子的精确排布。

5 Substrates are bound to enzymes by multiple weak attractions.
底物通过多重弱吸引作用与酶结合。

6 The existence of isozymes permits the fine-tuning of metabolism to meet the particular needs of a given tissue or developmental stage.
同工酶的存在使代谢的精密调控成为可能，以满足某个组织或发育阶段的特殊需要。

第六节 核酸
Section 6 Nucleic Acids

★ 基因 gene
常染色体，正染色体 autosome [ˈɔːtəsəum]
性染色体 sex chromosome
★ 遗传的 hereditary [hiˈreditəri]
反义 RNA antisense RNA [ˌæntisens]
核糖核酸 RNA ribonucleic [raibənjuːˈkliːik] acid
嘌呤衍生物 derivative of purine [ˈpjuəriːn]
嘧啶衍生物 derivative of pyrimidine [ˌpaiəˈrimidiːn]
★ 脱氧核糖核酸 deoxyribonucleic [diːˈɔksiˌraibəuˈnjuːkliːik] acid (DNA)
★ 脱氧核糖 deoxyribose [diːˌɔksiˈrɑibəus]
★ 脱氧腺苷 deoxyadenosine [diˈɔksiəˈdenəsiːn]
脱氧鸟苷 deoxyguanosine [diˈɔksiˈgwɑːnəsin]
脱氧胞苷 deoxycytidine [diˈɔksi ˈsaitədin]
胸苷 thymidine [ˈθaimiˌdiːn]
★ 磷酸二酯键 phosphodiester [ˌfɔsfədaiˈestə] bridge
骨架 backbone [ˈbækbəun]
★ 核苷 nucleoside [ˈnjuːkliəsaid]
N-糖苷键 N-glycosidic [ˌglaikəuˈsidik] linkage
★ 碱基配对 base pairing
★ 双螺旋 double helix
进化 evolutionary
★ 基因表达 gene expression
DNA 聚合酶 DNA polymerase
★ 模板 template [ˈtemplit]
★ 遗传密码 genetic code
初生的 nascent [ˈnæsnt]
成熟的 mature [məˈtjuə]
互补的 complementary [kɔmpləˈmentəri]
疏水相互作用 hydrophobic [ˌhaidrəuˈfəubik] interaction
螺旋直径 the diameter of the helix
★ 大沟 major groove
★ 小沟 minor groove
沃森-克里克模型 Watson-Crick model
假设 hypothesis [haiˈpɔθisis]
平行碱基对 parallel base pair
碱基堆垛 base stacking
螺旋桨状扭曲 propeller twisting [prəˈpelə][ˈtwistiŋ]
子代 DNA daughter DNA molecule
父代 DNA parent DNA molecule
★ 回文结构 palindrome sequence [ˈpælindrəum][ˈsiːkwəns]
★ 超螺旋 superhelical form
拓扑学 topology [təˈpɔlədʒi]
^{15}N 氮的重同位素 ^{15}N heavy isotope of nitrogen
密度梯度平衡沉降 density-gradient equilibrium sedimentation
离心机 centrifuge [ˈsentrifjuːdʒ]
氯化铯 cesium chloride [ˈsiːzjəm][ˈklɔːraid]
★ 探测 detect [diˈtekt]
紫外吸收 absorption of ultraviolet light
增色效应 hyperchromic [ˌhaipəˈkrəumik] effect
减色效应 hypochromic effect
杂交 DNA hybrid DNA
★ 复制 duplication [ˌdjuːpliˈkeiʃən]
★ 退火 annealing [æˈniːliŋ]
逆转录酶病毒 retrovirus [ˌretrəuˈvaiərəs]
★ 逆转录酶 reverse transcriptase [trænsˈkripteis]
★ 突变 mutation [mju(ː)ˈteiʃən]
★ 编码 encode [inˈkəud]
★ 切除 excise
★ 错配 mismatch [ˈmisˈmætʃ]
★ 发夹 hairpin [ˈhɛəpin]
自身互补序列 self-complementary sequence
★ 起始 initiation [iˌniʃiˈeiʃən]
★ 终止 termination [ˌtəːmiˈneiʃən]
Western 印迹法 western blotting
限制性酶分析 restriction-enzyme analysis
X-射线结晶学 X-ray crystallography [kristəˈlɔgrəfi]
Sanger 双脱氧 DNA 序列分析法 Sanger dideoxy method of DNA sequencing

核酸固相分析 solid-phase analysis of nucleic acid

聚合酶链反应 polymerase chain reaction (PCR)

转基因动物 transgenic animal

★ 位点特异性突变 site-specific mutagenesis [ˌmjutəˈdʒenisis]

琼脂糖 agarose [ɑ:gərəus]

聚丙烯酰胺 polyacrylamide [ˌpɔləˈkriləmaid]

1 A unit consisting of a base bonded to a sugar is referred to as a nucleoside.
一个碱基与一分子糖结合组成的单位称作核苷。

2 Genetic information is stored in the sequence of bases along a nucleic acid chain.
遗传信息储存在核酸链的碱基序列中。

3 A helical structure consists of two strands.
螺旋结构由两条链组成。

4 These base pairs provide a mechanism for copying the genetic information in an existing nucleic acid chain to form a new chain.
这些碱基对提供了复制已有的核酸链的遗传信息以形成新链的机制。

5 The melting temperature (Tm) is defined as the temperature at which half the helical structure is lost.
熔解温度(Tm)定义为失去双螺旋一半时的温度。

第七节 维生素
Section 7 Vitamins

维生素 vitamin

维生素缺乏症 avitaminosis [eiˌvaitəmiˈnəusis]

脚气病 beriberi [ˈberiˈberi]

视黄醇 retinol [ˈretinɔl]

视黄醛 retinal [retinəl]

β-胡萝卜素 β-carotene [ˈkærəti:n]

视紫红质 rhodopin [ˈrɔdəpin]

生育酚 tocopherol [təuˈkɔfərəul]

凝血酶原 prothrombin [prəuˈθrɔmbin]

硫胺素 thiamine [ˈθaiəmi:n]

烟酸 nicotinic [ˌnikəˈtinik] acid

烟酰胺 nicotinamide [ˌnikəˈtinəmaid]

核黄素 riboflavin [ˌraibəuˈfleivin]

泛酸 pantothenic [ˌpæntəˈθenik] acid

磷酸吡哆醛 pyridoxal [ˌpiriˈdɔksæl] phosphate

四氢叶酸 tetrahydrofolate [ˌtetrəˌhaidrəˈfəuleit]

钴胺素 cyanocobalamin [ˌsaiənəukəˈbæləmin]

血卟啉环 heme porphyrin [ˈpɔ:fərin] ring

生物素 biotin [ˈbaiətin]

活动羧基载体 mobile carboxy group carrier

硫辛酸 lipoic acid

抗坏血酸 ascorbic [əsˈkɔ:bik] acid

1 Vitamins are organic molecules that function in a wide variety of capacities within the body.
维生素是体内功能广泛的有机分子。

2 Riboflavin deficiencies are rare in the world due to the presence of adequate amounts of the vitamin in eggs, milk, meat and cereals.
核黄素缺乏在世界上是少见的,因为其在蛋、奶、肉及谷物中存在丰富。

第八节 代谢概论
Section 8 Metabolism: Basic Concepts and Design

代谢中间物 metabolic intermediate

主要代谢途径 central metabolic pathway

两用代谢途径 amphibolic [ˌæmfiˈbɔlik] pathway

能量代谢 energetic metabolism

基团转移 group transfer

供体 donor [ˈdəunə]

肌收缩 muscle contraction

★ 主动运输 active transport

单向运输 uniport transport

光合生物 photosynthetic organism

开尔文循环 Calvin cycle (dark reaction)

自养生物 autotroph [ˈɔ:təutrɔf]

异养生物 heterotroph [ˈhetərəutrɔf]

细胞色素氧化酶体系 cytochrome [ˈsaitəukrəum] oxidase system

★ 电子载体 electron carrier

铁硫蛋白 iron-sulfur protein
最终电子受体 ultimate electron acceptor
高势能电子 high potential electron
电化学势能 electrochemical [i,lektrəu'kemikəl] potential
氧化还原电势测量 measurement of redox potential
★ 低能分子 energy-poor molecule
★ 高能分子 energy-rich molecule
★ 高能键 high-energy bond
★ 氧化磷酸化 oxidative phosphorylation ['ɔksideitiv][fɔ,sfɔri'leiʃən]
化学渗透假说 chemiosmotic hypothesis [,kemiɔz'mɔtik][hai'pɔθisis]
★ 离子梯度 ion gradient
中间代谢 intermediary metabolism
途径 pathway
相互依赖的 interdependent [,intə(:)di'pendənt]
分解代谢的 catabolic [,kætə'bɔlik]
合成代谢的 anabolic [ə'næbəlik]
★ 自发地 spontaneously [spɔn'teinjəsli]
★ 三磷酸腺苷 adenosine triphosphate (ATP) [ə'denəsi:n][trai'fɔsfeit]
磷酸酐键 phosphoanhydride [,fɔsfə'æn'haidraid] bond
正磷酸 orthophosphate (Pi) [,ɔ:θəu'fɔsfeit]
焦磷酸 pyrophosphate (PPi) [,pairəu'fɔsfeit]
生物合成的 biosynthetic [,baiəsin'θetic]
有营养的 nutrient ['nju:triənt]
★ 通用货币 universal currency
钠钾泵 sodium-potassium ['səudjəm][pə'tæsjəm] pump
水解反应 hydrolytic reaction
去磷酸化 dephosphorylate [di:,'fɔsfərəleit]
水合作用 hydration [hai'dreiʃən]
★ 周转 turnover ['tə:n,əuvə]
★ 再生 regenerate [ri'dʒenərit]
放能的 exergonic [,eksə'gɔnik]
吸能的 endergonic [,endə'gɔnik]
捕获的 trapped [træpt]
★ 三羧酸循环 tricarboxylic [trai,kɑ: bɔk'silik] acid (TCA) cycle/ citric acid cycle
吡啶核苷 pyridine nucleotide
★ 黄素 flavin ['fleivin]
肌氨酸 creatine ['kri:ətin]
磷酸肌酸 phosphocreatine [fɔsfəu'kri:əti:n]
降级 degrade
乙酰辅酶 A acetyl CoA
★ 前体 precursor [pri(:)'kə:sə]
还原剂 reductant [ri'dʌktənt]
标准自由能 standard free energy
★ 能荷 energy charge

1 The role of ATP in energy metabolism is paramount.
ATP 在能量代谢中的作用是非常重要的。

2 Redox reactions involve the transfer of electrons from one chemical species to another.
氧化还原反应包含着电子从一种化学物质向另一种化学物质的转移。

第九节 糖代谢

Section 9 Glycolysis and Gluconeogenesis

★ 糖酵解 glycolysis [glai'kɔlisis]
厌氧的 anaerobic [,æneiə'rəubik]
乳酸发酵 lactic acid fermentation
乙醇发酵 alcoholic fermentation
★ 磷酸烯醇丙酮酸 phosphoenolpyruvate ['fɔsfəuə,nɔ:lpai'ru:veit]
2-磷酸甘油酸 2-phosphoglycerate [,fɔsfə'glisəreit]
★ 3-磷酸甘油醛 glyceraldehyde [,glisə'rældihaid] 3-phosphate
★ 反馈抑制 feedback inhibition
★ 肾上腺素 epinephrine [,epi'nefrin]
触发信号转换 trigger signal transduction
★ 梯瀑 cascade [kæs'keid]
高度放大变化 highly amplified change
细胞间的 intercellular [,intə(:)'seljulə]
细胞内的 intracellular [,intrə'seljulə]
★ 已糖激酶 hexokinase [,heksəu'kaineis]
磷酰基 phosphoryl [,fɔs'fɔ:ril] group
二价金属离子 divalent ['dai,veilənt] metal ion
构象变化 conformational [,kɔnfɔ: 'meiʃənəl] change
果糖的直链状形式 the open-chain form of fructose
酮基 keto group
裂开 splitting ['splitiŋ]

醛缩酶 aldolase [ˈældəˌleis]
羟醛缩合 aldol condensation
磷酸二羟丙酮 dihydroxyacetone [daiˌhaidrəksiˈæsitəun] phosphate
★ 限速步 committed step
载体 transporter [trænˈspɔːtə]
★ 调整 mediate [ˈmiːdiit]
分泌 secretion [siˈkriːʃən]
饥饿 starvation [stɑːˈveiʃən]
骨骼肌 skeletal muscle
羧化作用 carboxylation [kɑːˈbɔksileiʃən]
参与 participate [pɑːˈtisipeit]
动态平衡 homeostasis [ˌhəumiəuˈsteisis]
激发 stimulate
生存能力 viability [ˌvaiəˈbiliti]
★ 双功能酶 bifunctional enzyme
硫酯中间体 thioester intermediate [ˌθaiəuˈestə][ˌintəˈmiːdjət]
柠檬酸合成酶 citrate synthase
α-酮戊二酸脱氢酶 α-ketoglutarate dehydrogenase [ˌkiːtəuˈgluːtəreit][diːˈhaidrədʒəˌneis]
蜕变中间物质 metabolon [meˈtæbəlɔn]
乙醛酸循环 glyoxylate [glaiˈɔksiˌleit] cycle
异柠檬酸裂解酶 isocitrate lyase
苹果酸合成酶 malate synthase
乙醛酸循环体 glyoxysome [ˌgliˈɔksəm]
多功能性 versatility [ˌvəːsəˈtiləti]
构件 building block
琥珀酸盐 succinate [ˈsʌksiˌneit]
硫氧还蛋白 thioredoxin [ˌθaiəuˈriːdɔksin]
质子化作用 protonation [ˌprəutəˈneiʃən]
磷酸戊糖途径 pentose phosphate pathway
★ 转酮酶 transketolase [trænsˈkiːtəuleis]
★ 转醛酶 transaldolase [ˌtrænzˈældəleis]
谷胱甘肽 glutathione [ˌgluːtəˈθaiəun]

1 Glycolysis is the sequence of reactions that metabolizes one molecule of glucose to two molecules of pyruvate with the concomitant net production of two molecules of ATP.
糖酵解是一分子葡萄糖代谢生成两分子丙酮酸伴随 ATP 的净生成的反应序列。

2 The prefix bis- in bisphosphate means that two separate monophosphate groups are present, whereas the prefix di- in diphosphate (as in adenosine diphosphate) means that two phosphate groups are present and are connected by an anhydride bond.
前缀 bis- 在 bisphosphate 中表示存在两个分离的单磷酸基，而前缀 di- 在 diphosphate 中表示存在两个以酐键相连的磷酸基(如在腺苷二磷酸中)。

第十节 脂代谢
Section 10 Lipids Metabolism

肠 intestine [inˈtestin]
食物脂 dietary [ˈdaiətəri] lipid
★ 乳化作用 emulsification [iˌmʌlsifiˈkeiʃən]
★ 胆酸盐 bile salt
胆囊 gallbladder [ˈgɔːlblædə]
★ 摄取 ingestion [inˈdʒestʃən]
酯酶 esterase [ˈestəreis]
胰脂酶 pancreatic [ˌpæŋkriˈætik] lipase
磷脂酶 phospholipase [ˌfɔsfəuˈlaipeis]
二酯酰甘油 diacylglycerol [daiæsilˈglisərɔl]
2-酯酰甘油 2-acylglycerol [ˌæsilˈglisərəul]
溶血磷脂 lysophospholipid [ˌlaisəuˌfɔsfəˈlipeid]
上皮细胞 epithelial [ˌepiˈθiːljəl] cell
脂-蛋白复合体 lipid-protein complex
★ 乳糜微粒 chylomicron [ˌkailəuˈmaikrɔn]
黏膜 mucosa [mjuːˈkəusə]
支配的 predominant [priˈdɔminənt]
淋巴系统 lymphatic [limˈfætik] system
血流 bloodstream
血浆 plasma
毛细管 capillary [kəˈpiləri]
★ 脂肪组织 adipose tissue
残余的 remnant [ˈremnənt]
肝的 hepatic [hiˈpætik]
★ 利用 utilize [juːˈtilaiz]
丙二酸单酰辅酶 malonyl-CoA
乙酰辅酶 A 羧化酶 acetyl-CoA carboxylase
★ 延伸 elongate [ˈiːlɔŋgeit]
酰基载体蛋白 acyl carrier protein(ACP)
乙酰辅酶 A 转酰基酶 acetyl-CoA transacylase
丙二酸单酰辅酶 A 转酰基酶 malonyl-CoA transacylase
β-酮脂酰 ACP 合成酶 β-keto-ACP synthase
β-酮脂酰 ACP 还原酶 β-keto-ACP reductase

3-羟脂酰 ACP 脱水酶 3-OH acyl-ACP dehydratase [di:'haidrəteiz]
烯脂酰辅酶 A 还原酶 enoyl-CoA reductase
不饱和化 unsaturation
三羧酸转运系统 tricarboxylate transport system
精巧调控 exquisitely regulate ['ekskwizit]['regjuleit]
激素敏感型脂酶 hormone-sensitive lipase
★ 酮体 ketone body
脂酰辅酶 A 去饱和酶 fatty acyl-CoA desaturase
★ 细胞色素 cytochrome ['saitəukrəum]
亚油酸盐 linoleate [li'nəuli,eit]
亚麻酸盐 linolenate [,linə'li:neit]
★ 必需脂肪酸 essential fatty acid
β-氧化途径 β-oxidation pathway
渗透性屏障 permeability [,pə:miə'biliti] barrier
★ 肉碱 carnitine ['kɑ:niti:n]
磷脂合成 phospholipid [,fɔsfəu'lipid] synthesis
★ 卵磷脂 lecithin ['lesiθin]
肺表面活性剂 pulmonary surfactant ['pʌlmənəri][sə:'fæktənt]
乙醇胺 ethanolamine ['eθənɔ:l'æmi:n]
心磷脂 cardiolipin [kɑ:diə'lipin]
★ 神经酰胺 ceramide ['serəmaid]
神经细胞 nerve cell
★ 紊乱 disorder [dis'ɔ:də]
★ 缺陷 defect [di'fekt]

1 The pathway for fatty acid synthesis occurs in the cytoplasm, whereas, oxidation occurs in the mitochondria.
脂肪酸合成途径发生在细胞质中,而氧化则发生在线粒体中。

2 The synthesis of fatty acids from acetyl-CoA and malonyl-CoA is carried out by fatty acid synthase, FAS. The active enzyme is a dimer of identical subunits.
由乙酰-CoA和丙二酸单酰-CoA合成脂肪酸是通过脂肪酸合成酶(FAS)进行的,活性酶是同种亚基的二聚体。

第十一节 蛋白质代谢
Section 11 Protein Metabolism

固氮 nitrogen fixation
固氮酶复合体 nitrogenase ['naitrədʒə,neis] complex
细菌固氮酶 bacterial nitrogenase
生物圈 biosphere
谷氨酸脱氢酶 glutamate dehydrogenase ['glu:təmeit][di:'haidrədʒə,neis]
★ 氨 ammonia
酰胺基 amide group
★ 转氨作用 transamination [træn,sæmi'neiʃən]
谷草转氨酶 glutamate-oxaloacetate aminotransferase (GOT) ['glu:təmeit][,ɔksələu'æsiteit][ə,mi:nəu'trænsfə,reis]
谷丙转氨酶 glutamate-pyruvate aminotransferase (GPT) ['glu:təmeit][pai'ru:veit][ə,mi:nəu'trænsfə,reis]
外围组织 peripheral [pə'rifərəl] tissues
★ 尿素 urea ['juəriə]
尿素循环 urea cycle
★ 尿 urine
★ 鸟氨酸 ornithine ['ɔ:niθi:n]
氨甲酰磷酸 carbamoyl [kɑ:'bæməil] phosphate
瓜氨酸 citrulline [si'trʌli:n]
延胡索酸盐 fumarate ['fju:məreit]
精氨琥珀酸盐 argininosuccinate
精氨酸酶 arginase ['ɑ:dʒineis]
毒害神经的 neurotoxic [,njuərəu'tɔksik]
碱毒症 alkalosis [,ælkə'ləusis]
酸毒症 acidosis [,æsi'dəusis]
酸性碳酸盐 bicarbonate [bai'kɑ:bənit]
★ 生酮氨基酸 ketogenic [,ki:təu'dʒenik] amino acid
★ 生糖氨基酸 glucogenic [glu:kə'dʒenik] amino acid
胃蛋白酶原 pepsinogen [pep'sinədʒən]
★ 自身催化地 autocatalytically
蛋白质水解 proteolysis [,prəuti'ɔlisis]
十二指肠 duodenum [,dju(:)əu'di:nəm]
胰腺 pancreas ['pænkriəs]
肽链内切酶 endopeptidase [,endəu'peptideis]
肽链外切酶 exopeptidase
二肽 dipeptide [di'peptaid]

三肽 tripeptide [trai'peptaid]
迂回的 circuitous [sə(:)'kju(:)itəs]
出现 appearance [ə'piərəns]
循环入口 portal circulation
★ 易化扩散 facilitated [fə'siliteitid] diffusion
★ 普遍存在的 ubiquitous [ju:'bikwitəs]
肾脏的 renal ['ri:nl]
氮平衡 nitrogen balance
L-氨基酸氧化酶 L-amino acid oxidase
★ 死亡率 mortality [mɔ:'tæliti] rate
★ 成人期 adulthood ['ædʌlt,hud]
★ 反密码 anticodon ['ænti'kəudɔn]
操纵子模型 operon model
氨酰-tRNA 合成酶 aminoacyl-tRNA synthetase
多核糖体 polysome ['pɔlisəum]
肽酰基转移酶 peptidyl transferase
★ 起始因子 initiation factor
★ 延伸因子 elongation factor
释放因子 release factor

1 That all 64 possible combinations of the 4 nucleotides code for amino acids, i. e. , the code is degenerate since there are only 20 amino acids.
编码氨基酸的 4 种核苷有 64 种可能结合形式,也就是说密码是兼并的,因为氨基酸只有 20 种。

2 DNA transfers information to mRNA in the form of a code.
DNA 将信息以密码的形式转录到 mRNA。

第十二节 核酸代谢
Section 12 Nucleic Acids Metabolism

同源碱基 cognate base
★ 核酸内切酶 endonuclease [,endə'nju:kli,eis]
核苷磷酸化酶 nucleoside phosphorylase ['nju:kliəsaid]['fɔsfəri,leis]
寡核苷酸 oligonucleotide [,ɔligəu'nju:kliəutaid]
向内 inward
核糖-1-磷酸 ribose-1-phosphate
脲酸 uric acid
★ 重新合成 de novo synthesis
★ 补救合成 salvage synthesis
磷酸核糖焦磷酸 5-phosphoribosyl-1-pyrophosphate, PRPP [fɔs'fɔribəsail][,pairəu'fɔsfeit]
★ 消耗 consume [kən'sju:m]
次黄苷,肌苷 inosine ['inəsi:n]
★ 次黄嘌呤 hypoxanthine [,haipəu'zænθi:n]
★ 甲酸盐 formate
★ S-腺苷蛋氨酸 S-adenosylmethionine (SAM) [ə,denəsilmə'θaiəni:n]
二分之一,半族 moiety ['mɔiəti]
超额 excess ['ekses]
嘌呤核苷酸循环 purine nucleotide cycle
★ 分解代谢 catabolism
抗代谢物 antimetabolite ['æntime'tæbə,lait]
免疫的 immune
无能 inability [inə'biliti]
★ 致命的 fatal ['feitl]
★ 幼年 infancy ['infənsi]
给予保护的 protective [prə'tektiv]
★ 自杀性底物 suicide ['sjuisaid] substrate
5-氟尿嘧啶 5-fluorouracil [,flu:ərəu'jurəsil]
叶酸 folic acid
胸腺嘧啶核苷酸合成酶 thymidylate ['θaimi,dileit] synthase
★ 重组 recombination
★ 修复 repair [ri'pεə]
直接修复 direct repair
碱基切除修复 base-excision repair
核苷切除修复 nucleotide-excision repair
端粒 telomere ['teləmiə]
端粒酶 telomerase [teləməreis]
重组酶 recombinase ['ri:kɔmbi'neis]
★ DNA 连接酶 DNA ligase
★ 复制的起点 origin of replication
★ 引物酶 primase ['pri:meis]
★ 复制叉 replication fork
★ 岗崎片段 Okazaki fragment
★ 滞后链 lagging strand
★ 前导链 leading strand
★ 引物 primer
核酸外切酶 exonuclease [,eksəu'nju:klieis]
螺旋酶 helicase ['helikeis]
连环数 linking number (Lk)
拓扑异构体 topoisomer ['tɔpɔ'aisəumə]
★ 泡 bubble
5'帽子 5'cap
多聚腺苷酸尾巴 poly(A) tail

1 Catabolism of the purine nucleotides leads ultimately to the production of uric acid which is insoluble and is excreted in the urine as sodium urate crystals.
嘌呤核苷的分解代谢最终生成尿酸，尿酸不溶而以尿酸钠结晶形式分泌到尿中。

2 Deoxyadenosine and deoxyguanosine are also substrates for deoxycytidine kinase, although the Km for these substrates is much higher than for deoxycytidine.
脱氧腺苷和脱氧鸟苷也是脱氧胞苷激酶的底物，虽然这些底物的 Km 值比脱氧胞苷的高得多。

第十三节　细胞代谢调控
Section 13　Metabolic Regulation in Cell

相反单向反应　opposing unidirectional reaction
★ 前馈　feedforward [ˌfiːdˈfɔːwəd]
★ 反馈　feedback
单价反馈　monovalent feedback
二价或多价反馈　divalent or multivalent feedback
质量作用比　mass-action ratio
★ 腺苷酸化　adenylylation [ædənileiʃən]
脱腺苷酸化　deadenylylation [diædənileiʃən]
腺苷酰转移酶　adenylytransferase [ædənilˈtrænsfəˌreis]
光面内质网　smooth endoplasmic reticulum
★ 运输泡　transport vesicle
质膜体　mesosome [ˈmesəsəum]
★ 双关酶　ambiguous [ˌæmˈbigjuəs] enzyme
★ 信号假说　signal hypothesis
★ 凋亡　apoptosis [ˌæpəˈtousis]
适应条件　adaptive regulation
动作电位　action potential
神经脉冲　nerve impulse
电位门控离子通道　voltage-gated ion channel
配体门控离子通道　ligand-gated ion channel
神经递质　neurotransmitter [ˌnjuərətrænsˈmitə]
旁分泌　paracrine [ˈpærəkrin]

1 Hormones are normally present in the plasma and interstitial tissue at concentrations in the range of 10^{-7} M to 10^{-10} M.
激素通常存在于血浆及间质组织中，其浓度范围为 10^{-7} M ～ 10^{-10} M。

2 There are relatively few metabolic pathways, but each of these can be broken down into many individual, enzyme-specific catalysed steps.
只有相对较少的代谢途径，但是每个代谢途径可以被分解为许多单独的，酶专一性催化的步骤。

第六章 生物学
Chapter 6 Biology

第一节 绪论
Section 1 Introduction

★ 新陈代谢 metabolism
★ 同化作用 assimilation [ə,simi'leiʃən]
★ 异化作用 dissimilation
★ 生长 growth
★ 发育 development
★ 生殖 reproduction
★ 遗传 heredity [hi'rediti]
★ 变异 variation [ˌvɛəri'eiʃən]
★ 进化 evolution [ˌi:və'lu:ʃən]
感应性 irritability [ˌiritə'biliti]
★ 运动 movement
趋性 taxis
感受器 receptor [ri'septə]
效应器 effector
内环境稳定 homeostasis [ˌhəumiəu'steisis]
★ 植物生物学 plant biology
★ 动物生物学 animal biology
★ 微生物学 microbiology [maikrəubai'ɔlədʒi]
人类学 anthropology [ˌænθrə'pɔlədʒi]
★ 分类学 taxonomy [tæk'sɔnəmi]
★ 生理学 physiology [ˌfizi'ɔlədʒi]
★ 生态学 ecology [i(:)'kɔlədʒi]
★ 遗传学 genetics [dʒi'netiks]
★ 生物化学 biochemistry ['baiəu'kemistri]
生物物理学 biophysics ['baiəu'fiziks]
仿生学 bionics [bai'ɔniks]
★ 细胞生物学 cell biology
★ 分子生物学 molecular biology
★ 种群生物学 population biology

1 All living things are organized; the cell is the smallest unit of life.
所有生物都是有组织的，细胞是生命的最小单位。

2 In many organisms, cells form tissues, tissues form organs, and organs form systems.
在许多生物体中，细胞构成组织，组织构成器官，器官构成系统。

3 Organisms are adapted to various types of environments and have different ways of acquiring energy and reproducing.
生物可以适应各种各样的环境，并且具有不同的能量获得途径和生殖方式。

第二节 生命的物质基础
Section 2 Material Basis of Life

★ 元素分析 elementary analysis
★ 基本粒子 elementary particle
★ 水分平衡 water balance
★ 亲水性 water affinity [ə'finiti]
★ 含水量 water content
★ 水势 water potential
★ 水分利用效率 water use efficiency [i'fiʃənsi]
水分胁迫 water stress
蔗糖酶 sucrase ['su:kreis]
糖醇 sugar alcohol
糖-蛋白质相互作用 sugar-protein interaction
纤维素层 cellulose layer
蛋白酶 proteinase ['prəutineis]
蛋白质构象 protein conformation
★ 蛋白质变性 protein denaturation [di:neitʃə'reiʃən]
蛋白质加工 protein processing
模体 motif [məu'ti:f]
结构域 domain
核质 nuclein
核蛋白 nucleoprotein [ˌnju:kliəu'prəuti:n]
核苷酸 nucleotide ['nju:kliətaid]
★ 腺嘌呤 adenine ['ædəni:n]
★ 鸟嘌呤 guanine ['gwɑ:ni:n]
★ 胞嘧啶 cytosine ['saitəsi:n]
★ 胸腺嘧啶 thymine ['θaimi:n]
尿嘧啶 uracil ['juərəsil]
核糖体 RNA ribosomal [ˌraibə'səuməl] RNA

转运 RNA　transfer RNA
信使 RNA　messenger RNA
核内小分子 RNA　small nuclear RNA
酶活化剂　enzyme activator
★ 酶活性　enzyme activity
★ 酶固定化　enzyme immobilization
酶作用机理　enzyme mechanism
酶促合成　enzymic synthesis
酶促反应机理　enzyme reaction mechanism

1 Living things are composed of matter consisting of elements. C、H、N、O、P、S recalls the common elements found in living things.
生物体是由元素组成的物质所构成。C、H、N、O、P、S 被认为是生物体内的共有元素。

2 Each type of small molecule in living things—simple sugars, fatty acids, amino acids, and nucleotides—have a carbon backbone and are characterized by the presence of functional groups.
生物体内每种类型的小分子,如单糖、脂肪酸、氨基酸和核苷酸,都具有一个碳主链,并通过官能团来表现其特性。

3 Although the general mechanism of DNA replication is conceptually simple, the actual process is complex and requires the cooperation of more than a dozen enzymes and other proteins.
虽然 DNA 复制的普通机制简单,但是实际过程是复杂的而且需要更多的酶和其他蛋白的协作。

4 The enzymes that actually make covalent bonds between the nucleotides of a new DNA strand are called DNA polymerases.
在新 DNA 链的核苷之间形成共价键的酶称作 DNA 聚合酶。

5 The relationship between genes and proteins was first proposed in 1909, when English physician Archibald Garrod suggested that genes dictate phenotypes through enzymes, the proteins that catalyze chemical processes.
1909 年,当英国内科医生 Archibald Garrod 提出基因通过酶,即催化化学反应的蛋白,表达显性的时候,基因与蛋白之间的关系被首次提出。

6 When cell or a whole organism reproduces, a complete set of genetic instructions must pass from one generation to the next.
当细胞或一个完整有机体繁殖的时候,一整套遗传指令必然从一代传给下一代。

7 Stating that genetic information in DNA is transcribed into RNA and then translated into polypeptides does not explain how these processes occur.
DNA 所包含的遗传信息转录为 RNA 并翻译成多肽这种说法并不能确切地解释这些过程是如何发生的。

8 Transcription and translation are linguistic terms, and it is useful to think of nucleic acids and polypeptides as having languages, too.
转录和翻译是语言上的术语,这也有益于把核酸和多肽想象成是有语言的。

9 To understand how genetic information passes from genotype to phenotype, we need to see how the chemical language of DNA is translated into the different chemical language of polypeptides.
为了了解遗传信息怎样从基因型传递到表型,我们需要弄清楚 DNA 的化学语言如何翻译成多肽的不同化学语言。

第三节　细胞的显微结构
Section 3　Microstructure of Cell

★ 细胞学说　cell theory
★ 细胞成分　cell component
细胞密度　cell density
★ 细胞发育　cell development
★ 细胞液　cell sap
细胞代谢　cell metabolism
★ 体细胞　somatic [səu'mætik] cell
★ 原核细胞　prokaryotic [prəu,kæri'əutik] cell
★ 真核细胞　eukaryotic [ju'kæriəutik] cell
细胞间通讯　cell-cell communication
★ 细胞培养　cell culture
细胞杂交　cell hybridization

★ 细胞表面　cell surface
★ 细胞全能性　cell totipotency [təu'tipətənsi]
线粒体　mitochondria [ˌmaitə'kɔndriə]
★ 外膜　outer membrane
★ 内膜　inner membrane
膜间隙　inter membrane space
基质　matrix ['meitriks]
嵴　cristae ['kristə]
★ 基粒　elementary particle [ˌeli'mentəri]['pɑːtikl]
ATP 合成酶　ATP synthetase
呼吸链　respiratory chain
★ 叶绿体　chloroplast ['klɔ(ː)rəplɑːst]
类囊体　thylakoid ['θailəˌkɔid]
转运体　translocator
★ 光合作用　photosynthesis [ˌfəutəu'sinθəsis]
★ 光反应　photoreaction [ˌfəutəri'ækʃən]
★ 原初反应　primary reaction
聚光色素　light-harvesting pigment
★ 作用中心色素　reaction center pigment
★ 光合作用单位　photosynthetic unit
PSⅡ复合物　photosystem Ⅱ complex (PSⅡ)
PSⅠ复合物　photosystem Ⅰ complex (PSⅠ)
捕光复合物　light-harvesting complex (LHC)
★ 光合磷酸化　photophosphorylation [ˌfəutəufɔsfəri'leiʃən]
非循环式光合磷酸化　non-cyclic photophosphorylation
循环式光合磷酸化　cyclic photophosphorylation
★ 暗反应　dark reaction
★ 生殖细胞　germ cell
★ 内质网　endoplasmic reticulum(ER) ['endəuˌplæzmik][ri'tikjuləm]
★ 光面内质网　smooth endoplasmic reticulum (sER)
粗面内质网　rough endoplasmic reticulum(rER)
微粒体　microsome ['maikrəusəum]
★ 结合蛋白　binding protein (BP)
★ 高尔基体　Golgi body
高尔基器　Golgi apparatus
小囊泡状　vesicle ['vesikl]
大囊泡　vacuole ['vækjuəul]
分泌泡　condensing vesicle
溶酶体　lysosome ['laisəsəum]
自噬　autophagy [ɔː'tɔfədʒi]
★ 微体　microbody
过氧化物酶体　peroxisome [pə'rɔksisəum]
★ 细胞骨架　cytoskeleton [saitəu'skelitən]
★ 微管　microtubule [ˌmaikrəu'tjuːbjuːl]
自我组装　self-assembly [selfə'sembli]
踏车现象　treadmilling ['tredmiliŋ]
★ 纤毛　cilium
★ 鞭毛　flagellum [flə'dʒeləm]
微丝　microfilament [ˌmaikrəu'filəmənt]
细胞质分裂　cytokinesis [ˌsaitəukai'niːsis]
★ 细胞核　cell nucleus
★ 核膜　nuclear membrane
核包被　nuclear envelope
★ 核外膜　outer nuclear membrane
★ 核孔　nuclear pore
★ 染色质　chromatin ['krəumətin]
★ 染色体　chromosome
★ 组蛋白　histone ['histəun]
非组蛋白　nonhistone
★ 基因组　genome ['dʒiːnəum]
★ 核小体　nucleosome ['njuːkliəsəum]
★ 核型　karyotype ['kæriətaip]
★ 带型　banding pattern
★ 核仁　nucleolus [njuː'kliːələs]

1 Plant cells have a freely permeable cell wall whose main component is cellulose.
植物细胞具有可自由通透的细胞壁，细胞壁的主要成分为纤维素。

2 Different pigments absorb light of different wavelengths, and chloroplasts contain several kinds of pigments.
不同的色素吸收不同波长的光，叶绿体中包含几种色素。

3 The membrane-bound nucleus contains the bulk of the cellular DNA in multiple chromosomes.
在多染色体的细胞中，有膜包围的细胞核包含了大量的细胞 DNA。

4 Transcription of this DNA and processing of the RNA occurs in the nucleus.
DNA 的转录和 RNA 的加工均发生在细胞核内。

5 Nucleoli are contained within the nucleus.
细胞核内含有核仁。

6 To extend our factory metaphor of the cell, the endoplasmic reticulum (ER) is one of the main

manufacturing facilities.
类似于工厂的细胞要扩大规模,细胞中内质网是主要生产设备之一。

7 Endoplasmic reticulum is a membranous labyrinth of tubes and sacs running throughout the cytoplasm.
内质网是细胞质中由膜围成的管状或囊状贯穿细胞质的系统。

8 The ER membrane separates its internal compartment from the surrounding cytosol.
内质网膜将其与周围的胞液隔开。

9 There are two distinct types of ER: rough ER and smooth ER. These two ER components are physically connected, but they differ in structure and function.
内质网膜有两种明显的类型:粗面内质网和滑面内质网。这两种内质网相连,但是在构造和功能上却不同。

10 The wall protects the cell, maintains its shape, and restrains the cell from absorbing so much water that it would pop.
细胞壁保护着细胞,维持其形状并且阻止其因吸收过多的水分而胀裂。

11 Cell walls do not totally isolate plant cells from each other.
细胞壁并不是完全将细胞间互相隔离开。

12 In plants, channels called plasmodesmata pass through the cell walls, connecting the cytoplasm of each cell to its neighbors. These pores allow water and other small molecules to move between cells, integrating the activities of a tissue.
在植物中,称为胞间连丝的管道穿透细胞壁连接到邻近其他细胞的胞质中。这些小孔允许水和其他一些小分子物质在两个细胞间移动,以此实现组织的活跃性。

13 The lysosomes contain degradative, hydrolytic enzymes.
溶酶体含有降解性水解酶。

14 The peroxisomes contain enzymes which destroy certain potentially dangerous free radicals and hydrogen peroxide.
过氧化物酶体含有破坏某些有潜在危害的自由基和过氧化氢的酶。

第四节 细胞膜与细胞内蛋白质合成

Section 4 Cell Membrane and Protein Synthesis in the Cell

★ 细胞膜 cell membrane
★ 质膜 plasma membrane
★ 生物膜 biomembrane [ˌbaiəu'membrein]
★ 磷脂 phospholipid [ˌfɔsfəu'lipid]
糖脂 glycolipid [ˌglaikə'lipid]
★ 外在蛋白 extrinsic [eks'trinsik] protein
内在蛋白 intrinsic [in'trinsik] protein
脂锚定蛋白 lipid-anchored protein
周边蛋白 peripheral [pə'rifərəl] protein
★ 整合蛋白 integral protein
跨膜蛋白 transmembrane protein
★ 单位膜模型 unit membrane model
★ 流动镶嵌模型 fluid mosaic model
质膜的不对称性 plasma membrane asymmetry [æ'simətri]
侧向扩散 lateral diffusion ['lætərəl][di'fjuːʒən]
旋转运动 rotation [rəu'teiʃən]
摆动 flexion ['flekʃən]
★ 翻转 flip-flop
荧光素 fluorescin [fluə'resin]
★ 被动运输 passive transport
★ 自由扩散 free diffusion
★ 简单扩散 simple diffusion
帮助扩散 facilitated [fə'siliteitid] diffusion
★ 运输蛋白 transport protein
★ 载体蛋白 carrier protein
★ 通道蛋白 channel protein
★ 主动运输 active transport
离子泵 ion pump ['aiən][pʌmp]
★ 质子泵 proton pump
伴随运输 cotransport [kəu'trænspɔːt]
同向协同 symport ['simpɔːt]
反向协同 antiport ['æntipɔːt]
★ 核糖体 ribosome ['raibəsəum]

★ 外膜蛋白　extrinsic membrane protein
转运肽　transit peptide
★ 信号肽　signal sequence
★ 转运因子　transitional element
胞吐作用　exocytosis [ˌeksəusai'təusis]
★ 胞吞作用　endocytosis [ˌendəusai'təusis]
吞饮作用　pinocytosis [ˌpainəusai'təusis]
★ 胞饮作用　cellular drinking
吞噬作用　phagocytosis [fægəˌsai'təusis]
★ 前溶酶体　prelysosome
内吞溶酶体　endolysosome
★ 分子伴娘　molecular chaperone ['tʃæpərəun]
★ DNA 结合的结构域　DNA binding domain

1 There are two components of the membrane, lipids and proteins.
膜有两种组分:脂类和蛋白质。

2 In the lipid bilayer, phospholipids are arranged with their polar heads at the surfaces and their hydrophobic tails in the interior.
在脂质双分子层中,磷脂的极性头部位于膜的表面,疏水尾部位于内部。

3 The lipid bilayer is fluid, but acts as a barrier to the entrance and exit of most biological molecules.
脂质双分子层是流动的,是大多数生物分子进出细胞的屏障。

4 During facilitated transport, a protein carrier assists the movement of a molecule down its concentration gradient. No energy is required.
在易化扩散过程中,载体蛋白协助分子顺着浓度梯度进行运输,不需要能量。

5 The plasma membrane and other membranes of the cell are selectively permeable.
质膜和其他细胞膜具有选择透过性。

6 The plasma membrane allows some substances to cross more easily than others and blocks passage of some substances altogether.
那就意味着这种膜允许一些物质比另一些更容易通过,同时阻止一些物质的通过。

第五节　细胞通讯
Section 5　Cell Communication

★ 细胞连接　cell junction
★ 封闭连接　occluding [ɔ'klu:diŋ] junction
不通透连接　impermeable [im'pə:mjəbl] junction
★ 紧密连接　tight junction
★ 锚定连接　anchoring junction
★ 细胞内附着蛋白　intracellular attachment protein
黏合带　adhesion [əd'hi:ʒən] belt
★ 桥粒　desmosome ['desməsəm]
★ 通讯连接　communication junction
间隙连接　gap junction
电耦合　electrical coupling
代谢耦合　metabolic coupling
★ 胞间连丝　plasmodesma [ˌplæzməu'dezmə]
细胞识别　cell recognition
★ 细胞通讯　cell communication
★ 信息分子　signal molecule
内分泌　endocrine ['endəukrain]
★ 第二信使　second messenger
受体　receptor [ri'septə]
G-蛋白耦连受体　G-protein linked receptor
酶连受体　enzyme linked receptor
★ 催化受体　catalytic receptor
cAMP 信号通路　cAMP signal pathway
★ 激活型受体　stimulate receptor
★ 抑制型受体　inhibit [in'hibit] receptor

1 Some of the proteins in the plasma membrane are carriers that transport biologically useful molecules into and out of the cell.
质膜上的一些蛋白质是生物有用分子进出细胞的载体。

2 During facilitated transport (no energy required), small molecules follow their concentration gradients.
在易化扩散过程中(不需要能量),小分子沿着浓度梯度进行运输。

3 During active transport (energy required), small molecules go against their concentration gradients.
在主动运输过程中(需要能量),小分子逆着浓度梯度进行运输。

第六节 细胞增殖与分化

Section 6 Cell Reproduction and Differentiation

★ 细胞增殖 cell reproduction
★ 细胞分化 cell differentiation [ˌdifəˌrenʃi'eiʃən]
★ 细胞周期 cell cycle
★ DNA 合成期 DNA synthetic phase
促成熟因子 maturation promoting factor
★ 分裂期 dividing phase
★ 无丝分裂 amitosis [ˌæmi'təusis]
★ 有丝分裂 mitosis
★ 减数分裂 meiosis [mai'əusis]
核分裂 karyomitosis [ˌkæriəumai'təusis]
间接分裂 indirect division
★ 前期 prophase ['prəufeiz]
★ 早中期 prometaphase [prəu'metəˌfeiz]
★ 中期 metaphase ['metəfeiz]
★ 后期 anaphase ['ænəˌfeiz]
★ 末期 telophase ['teləfeiz]
收缩环 contractile [kən'træktail] ring
★ 细胞板 cell plate
★ 细胞周期的控制系统 cell-cycle control system
★ 二倍体 diploid ['diplɔid]
同源染色体 homologous [hɔ'mɔləgəs] chromosome
★ 单倍体 haploid ['hæplɔid]
★ 染色单体 chromatid ['krəumətid]
★ 受精 fertilization [ˌfə:tilai'zeiʃən]
★ 顶体反应 acrosomal ['ækrəu 'səuməl] reaction
皮层反应 cortical ['kɔ:tikəl] reaction
原核融合 pronuclei fusion
卵裂 cleavage ['kli:vidʒ]
★ 全能性 totipotency [təu'tipətənsi]
★ 看家基因 housekeeping ['hauski:piŋ] gene

Meiosis, the process that produces haploid gametes in diploid organisms, resembles mitosis, but with two special features. The first is the halving of the number of chromosomes. The second special feature of meiosis is an exchange of genetic material—pieces of chromosomes—between homologous chromosomes.

减数分裂,是双倍体生物产生单倍体配子的过程,类似于有丝分裂,但有两个独有的特征。第一个是染色体数量减半。减数分裂的第二个特征是基因材料的互换,即同源染色体片段的交换。

第七节 原核生物

Section 7 Prokaryote

★ 原核生物 prokaryote [prəuˌkæriəut]
★ 细菌 bacteria [bæk'tiəriə]
★ 球菌 cocci ['kɔksai]
★ 杆菌 bacilli [bə'silai]
螺旋菌 spirilla ['spaiərilə]
螺旋体 spirochaete ['spaiərəki:t]
★ 菌落 colony ['kɔləni]
菌苔 lawn [lɔ:n]
★ 细胞壁 cell wall
★ 原生质体 protoplast ['prəutəplɑ:st]
脂多糖 lipopolysaccharide ['lipəuˌpɔli'sækəraid]
间体 mesosome ['mesəsəum]
★ 细胞质 cytoplasm ['saitəuplæzm]
★ 质粒 plasmid
类核 nucleoid ['nju:kliɔid]
★ 细菌染色体 bacterial chromosome
光能自养型 photoautotroph ['fəutəu'ɔ:tətrɔf]
化能自养型 chemoautotroph [ˌkiməu'ɔtətrəuf]
化能异养型 chemoheterotroph [kemə'hetərətrəf]
★ 有氧呼吸 aerobic respiration [ˌeiə'rəubik][ˌrespi'reiʃən]
★ 无氧呼吸 anaerobic respiration [ˌæneiə'rəubik][ˌrespi'reiʃən]
★ 发酵 fermentation [ˌfə:men'teiʃən]
延滞期 lag phase [læg][feiz]
★ 对数期 logarithmic phase [ˌlɔgə'riθmik, -iðm-][feiz]
稳定期 stationary phase ['steiʃ(ə)nəri][feiz]
衰亡期 death phase [deθ][feiz]
★ 基因突变 gene mutation
★ 基因重组 genetic recombination
★ 转化 transformation [ˌtrænsfə'meiʃən]
★ 转导 transduction [træns'dʌkʃən]
古细菌 archaeo ['ɑ:kiəu] bacteria
螺旋体 spirochaeta [ˌspaiərəu'ki:tə]
衣原体 chlamydia [klə'midiə]

支原体 mycoplasma [ˌmaikəuˈplæzmə]
粘细菌 myxobacteria [miksəˌbækˈtiəriə]
放线菌 actinomycete [ˌæktinəumaiˈsiːt]
病毒粒子 virion [ˈvaiəriɔn]
病毒颗粒 virus particle
噬细菌体 bacteriophage [bækˈtiəriəfeidʒ]
刺突 spike [spaik]
★ 增殖 multiplication [ˌmʌltipliˈkeiʃən]
★ 生物合成 biosynthesis
★ 装配 assembly [əˈsembli]
★ 释放 release [riˈliːs]
AIDS acquired immune deficiency syndrome
HIV human immunodeficiency [iˈmjunəudiˈfiʃənsi] virus
真病毒 euvirus
亚病毒 subvirus
类病毒 viroid [ˈvaiərɔid]
朊病毒 prion
病毒卫星 satellite [ˈsætəlait]
疯牛病 bovine spongiform encephalopathy [ˈbəuvain][ˈspʌŋdʒifɔːm][enˌsefəˈlɔpəθi]
辅助病毒 helper virus

1 Bacteria have three basic shapes: rod (bacillus), round (coccus), and spiral (spirillum).
细菌有三种基本形状：杆状(杆状菌)、球状(球菌)和螺旋状(螺旋菌)。

2 Viruses reproduce inside living cells and therefore they are obligate parasites. Since they are parasites they cause diseases in plants and animal.
病毒在活细胞内进行繁殖，因此它们是专性寄生物。因为病毒是寄生物，故可以导致植物和动物产生疾病。

3 All viruses have at least two parts: an outer coat of protein and inner core of nucleic acid.
所有病毒至少有两部分：蛋白质外壳和核酸内核。

4 Most bacteria can be divided up in various ways. One way is to consider which are gram positive and which are gram negative.
多数细菌可以按照多种方法进行区分。一种方式就是划分为革兰氏阳性菌和革兰氏阴性菌。

5 Prokaryotic cells have a plasma membrane, usually enclosed in a rigid cell wall, but no intracellular compartments. They have a single, major circular chromosome.
原核细胞都有质膜，通常其膜外包裹一层刚性的细胞壁，其内没有分隔小室，有一条主要的环状染色体。

6 There are many more differences in cellular structure and in gene between prokaryotes and eukaryotes.
真核生物和原核生物的细胞结构和基因上存在明显差异。

第八节 植物
Section 8 Plant

★ 维管束 vascular bundle
★ 植物性生长 vegetative growth
★ 质体 plastid
★ 液泡 vacuole [ˈvækjuəul]
胞间层 intercellular layer
初生壁 primary wall
次生壁 secondary wall
★ 木质素 lignin [ˈlignin]
木质栓 suberic [sjuːˈberik]
角质 cutin [ˈkjuːtin]
成熟组织 mature tissue
★ 分生组织 meristematic [ˌməristəˈmætik] tissue
★ 原分生组织 promeristem [prəuˈmerəˌstem] tissue
★ 初生分生组织 primary meristem tissue
★ 次生分生组织 secondary meristem tissue
居间分生组织 intercalary [inˈtəːkələri] meristem tissue
★ 薄壁组织 parenchyma [pəˈreŋkimə] tissue
★ 保护组织 protective [prəˈtektiv] tissue
★ 周皮 periderm [ˈperidəːm]
筛管 sieve tube
★ 筛胞 sieve cell
★ 导管 vessel
★ 管胞 tracheid [ˈtreikiid]
机械组织 mechanical tissue
石细胞 stone cell
分泌组织 secretory [siˈkriːtəri] tissue
维管组织 vascular [ˈvæskjulə] tissue
维管系统 vascular system

★韧皮部 phloem ['fləuem]
★木质部 xylem ['zailem]
形成层 cambium ['kæmbiəm]
主根 main root
不定根 adventitious root
根尖 root tip
根冠 root cap
伸长区 elongation [ˌi:lɔŋ'geiʃən] zone
分生区 meristematic zone
★初生生长 primary growth
★初生组织 primary tissue
初生结构 primary structure
中柱鞘 pericycle ['perisaikl]
髓 pith [piθ]
★次生生长 secondary growth
★次生结构 secondary structure
束中形成层 fasciculate cambium [fə'sikjulit]['kæmbiəm]
束间形成层 interfascicular [intəfə'sikjulə] cambium
早材 early wood
晚材 late wood
★年轮 annual ring
★叶片 blade [bleid]
叶柄 petiole ['petiəul]
托叶 stipule ['stipju:l]
栅栏组织 palisade [ˌpæli'seid] tissue
海绵组织 spongy ['spʌndʒi] tissue
叶原基 leaf primordium [prai'mɔ:diəm]
★根 root
★茎 shoot
运转蛋白 transport protein
★主动转运 active transport
扩散 diffusion [di'fju:ʒən]
渗透 osmosis [ɔz'məusis]
束缚能 bond energy
★自由能 free energy
渗透势 osmosis potential [ɔz'məusis][pə'tenʃ(ə)l]
溶质势 solute potential
★压力势 pressure potential
★质壁分离 plasmolysis [plæz'mɔlisis]
短距离运输 short distance transport
侧向运输 lateral ['lætərəl] transport
内聚力学说 cohesion theory
蒸腾-内聚力-张力学说 transpiration-cohesion-tension theory
★蒸腾作用 transpiration [ˌtrænspi'reiʃən]
皮孔蒸腾 lenticular [len'tikjulə] transpiration
★韧皮部装载 phloem loading
★韧皮部卸出 phloem unloading
★矿质元素 mineral element
★必需元素 essential element
★大量元素 major element
有益元素 beneficial element
氨化作用 ammonification [əˌməunifi'keiʃən]
硝化作用 nitrification [ˌnaitrifi'keiʃən]
反硝化作用 denitrification [di:naitrifi'keiʃən]
根瘤 root nodule ['nɔdju:l]
脱分化 differentiation
再分化 redifferentiation
营养繁殖 vegetative propagation [ˌprɔpə'geiʃən]
★无性繁殖 asexual reproduction
★有性繁殖 sexual reproduction
花柄 pedicel ['pedisəl]
花托 receptacle [ri'septəkl]
★花被 perianth ['periænθ]
★雄蕊 androecium [æn'dri:ʃiəm]
★雌蕊 gynoecium [dʒi'ni:siəm]
★胚 embryo ['embriəu]
★胚乳 endosperm ['endəuspə:m]
★种皮 seed coat
★传粉 pollination [pɔl:'niʃən]
★受精 fertilization [ˌfə:tilai'zeiʃən]
★萌发 germination [ˌdʒə:mi'neiʃən]
★种子休眠 seed dormancy ['dɔ:mənsi]
植物激素 plant hormone
激动素 kinetin ['kainətin]
细胞分裂素 cytokinin ['saitəu'kinin]
脱落酸 abscisic [æb'saizik] acid
★光周期 photoperiod [ˌfəutəu'piəriəd]
短日植物 short-day plant
长日植物 long-day plant
日中性植物 day-neutral plant
短-长日植物 short-long-day plant
中日性植物 intermediate day plant
光敏素 phytochrome ['faitəukrəum]
★春化作用 vernalization
苔藓植物 bryophyte ['braiəˌfait]
精子器 antheridium [ˌænθə'ridiəm]
颈卵器 archegonium [ˌɑ:ki'gəuniəm]
卵细胞 egg cell
孢子囊 sporangium [spə'rændʒiəm]

孢子母细胞 spore mother cell
雄器苞 perigonium
★ 维管植物 vascular plant
低等维管植物 lower vascular plant
★ 蕨类植物 fern plant
营养叶 sterile frond ['sterail][frɔnd]
★ 有花植物 flowering plant

1 Plants are photosynthetic organisms adapted to a land existence. Among the various adaptations, all plants protect the developing embryo from desiccation.
植物是适应陆地环境的光合生物体。在不同的适应中,所有植物都会保护发育胚胎避免干燥。

2 All plants have the alternation of generations life cycle. Some types of plants have a dominant gametophyte generation, others have a dominant sporophyte generation.
所有植物都具有生活周期的世代交替。一些植物具有显著的配子体世代,一些植物为显著的孢子体世代。

3 The body of a flowering plant is divided into the root system and the shoot system, which contains stems and leaves. Both systems contain three types of tissues: dermal, ground, and vascular tissue.
显花植物体分为根系和茎系,茎系包括茎干和叶片。根系和茎系包括三种组织:皮组织、基本组织和维管组织。

第九节 动物
Section 9 Animal

★ 神经元 neuron ['njuərɔn]
神经胶质细胞 neuroglial [nju'rɔgliə] cell
★ 肌球蛋白 myosin ['maiəsin]
肌动蛋白 actin
★ 鞭毛 flagella [flə'dʒelə]
肌纤维 muscle fiber
★ 肌节 sarcomere ['sɑːkəmiə]
三联体 triad ['traiəd]
单收缩 twitch
★ 消化 digestion [di'dʒestʃən]
内消化 intracellular digestion
植物性神经系统 vegetative nervous system
★ 自主神经系统 autonomic [ˌɔː təu'nɔmik] nervous system
集团运动 mass movement
内因子 intrinsic [in'trinsik] factor
血液循环 blood circulation
心电图 electrocardiogram [ilektrəu'kɑːdiəugræm]
★ 心动周期 cardiac ['kɑːdiæk] cycle
★ 体液 body fluid
细胞内液 intracellular fluid
细胞外液 extracellular fluid
★ 血浆 plasma
★ 凝集 agglutination [əˌgluːti'neiʃən]
凝集原 agglutinogen [ˌæglu'tinədʒən]
凝集素 agglutinin [ə'gluːtinin]
★ 内呼吸 interal respiration
★ 外呼吸 external respiration
血红素 haem [hiːm]
★ 渗透调节 osmoregulation [ˌɔzməuregju'leiʃən]
★ 排泄 excretion [eks'kriːʃən]
肾孔 nephridiopore [ne'fridiəpɔː]
原管肾 protonephridium [ˌprəutəuni'fridiəm]
扁形动物 platyhelminth [ˌplæti'helminθ]
纽形动物 nemertinean [ˌnemə'tiniən]
轮形动物 rotifer ['rəutifə]
马式小管 malpighian tubule [mæl'pigiən]['tjuːbjuːl]
肾小体 renal corpuscle ['riːnl]['kɔːpəskl]
肾小球 renal glomerulus ['riːnl][gləu'merjuləs]
腔肠动物 protosome coelomate ['prəutəsəum]['siːləmeit]
★ 炎症反应 inflammatory [in'flæmətəri] response
★ 自然杀伤细胞 natural killer cell
补体 complement ['kɔmplimənt]
前列腺素 prostaglandin [ˌprɔstə'glændin]
★ 抗原 antigen ['æntidʒən]
★ 免疫应答 immune [i'mjuːn] response
细胞免疫 cellular immunity
★ 体液免疫 humoral ['hjuːmərəl] immunity
胸腺 thymus ['θaiməs]
★ 骨髓 bone marrow
★ 人工自动免疫 artificial active immunization [ɑːti'fiʃəl]['æktiv][ˌimjuːnai'zeiʃən]
★ 人工被动免疫 artificial passive immunization

★ 生物制品 biological product
★ 死疫苗 dead vaccine ['væksi:n]
★ 活疫苗 live vaccine
★ 类毒素 toxoid ['tɔksɔid]
★ 化学信使 chemical messenger
激素 hormone
雄激素 androgen ['ændrədʒən]
雌激素 estrogen ['estrədʒən]
★ 信息素 pheromone ['ferəməun]
性诱剂 sex attractant
蜕皮激素 ecdysone ['ekdisəun]
★ 树突 dendrite ['dendrait]
轴浆 axoplasm ['æksə,plæzəm]
膜电位 membrane potential
极化 polarization [,pəulərai'zeiʃən]
去极化 depolarization
复极化 repolarization
★ 突触 synapse [si'næps]
突触前细胞 presynaptic [prisi'næptik] cell
突触后细胞 postsynaptic cell
突触小泡 synaptic vesicle [si'næptik][vesikl]
突触小体 synaptic knob
★ 受体蛋白 receptor protein
递质 transmitter substance
抑制性突触电位 inhibitory postsynaptic potential
超极化 hyperpolarization [,haipə,pɔlərai'zeiʃən]
★ 新皮层 neopallium [,ni:əu'pæliəm]
裂 fissure ['fiʃə]
沟 sulcus ['sʌlkəs]
回 gyrus ['dʒaiərəs]
昼夜节律 circadian rhythm [sə:'keidiən]['riðəm]
★ 脑电图 electroencephalogram
[ilektrəuen'sefələugræm]
短期记忆 short-term memory
外感受器 exteroceptor [,ekstərəu'septə]
内感受器 interoceptor [,intərəu'septə]
★ 距离感受器 teleceptor ['telə,septə]
本体感受器 proprioceptor [,prəupriə'septə]
适宜刺激 adequate stimulus
['ædikwit]['stimjuləs]
★ 适应 adaptation
折光体 refractive [ri'fræktiv] body
视紫质 prophyropsin [,prɔ:fi'rɔpsin]
光视蛋白 photopsin ['fəutəpsin]
色弱 color anomaly [ə'nɔməli]
色盲 color blindness

神经丘 neuromast ['njuərəu,mæst]
雌雄同体 sequential hermaphroditism
[si'kwinʃəl][hə:'mæfrədaitizəm]
原始细胞 archaeocyte ['ʌ:kiəsait]
卵泡期 follicular [fə'likjulə] phase
卵泡细胞 follicle ['fɔlikl] cell
★ 动物极 animal pole
★ 植物极 vegetal pole
囊胚 blastula ['blæstjulə]
原肠胚 gastrula ['gæstrulə]
原肠 archenteron [ɑ:'kentərɔn]
胚孔 blastopore ['blæstəupɔ:]
★ 外胚层 ectoderm ['ektəudə:m]
★ 中胚层 mesoderm ['mesədə:m]
★ 内胚层 endoderm ['endəudə:m]
★ 神经管 neural ['njuərəl] tube
★ 脊索 notochord ['nəutəkɔ:d]
羊膜 amnion ['æmniən]
胚盘 blastodisc ['blæstəudisk]
不全卵裂 meroblastic cleavage
[,merəu'blæstik]['kli:vidʒ]
完全卵裂 holoblastic cleavage
[,hɔləu'blæstik]['kli:vidʒ]
上胚层 epiblast ['epiblɑ:st]
下胚层 hypoblast ['haipəublɑ:st]
卵黄囊 yolk sac
绒毛膜 chorion ['kɔ:riɔn]
尿囊 allantois [æ'læntəuis]
内细胞团 inner cell mass
细胞质决定子 cytoplasmic determinant
全能性 totipotent [təu'tipətənt]
★ 行为生物学 ethology [i(:)'θɔlədʒi]
攻击行为 aggressive behavior
防御行为 defensive behavior
繁殖行为 breeding behavior
★ 季节节律 seasonal rhythm
动性 kinesis [ki'ni:sis]
趋光性 phototaxis [,fəutəu'tæksis]
趋湿性 hygrotaxis [,haigrə'tæksis]
趋化性 chemotaxis [,kemə'tæksis]
趋地性 geotaxis [,dʒi:əu'tæksis]
★ 反射 reflection
本能行为 instinctive [in'stiŋktiv] behavior
★ 系统树 phylogenetic [,failəudʒi'netik] tree
★ 适应辐射 adaptive eradiation [i,reidi'eiʃən]
中胶层 mesoglea [,mesəu'gli:ə]

1 Vertebrates include fishes, amphibians, reptiles, birds, and mammals.
脊椎动物包括鱼、两栖动物、爬行动物、鸟和哺乳动物。

2 There are two groups of coelomate animals. In the protostomes, the blastopore becomes the mouth, and in the deuterostomes, it becomes the anus.
体腔动物可以分为两类。原口动物的原口发育成为嘴,而后口动物的原口则发育成为肛门。

第十节 生物的遗传
Section 10 Biologic Heredity

★ 遗传 heredity [hi'rediti]
★ 变异 variation [ˌvɛəri'eiʃən]
★ 回交 backcross ['bækˌkrɔs]
★ 测交 testcross ['testkrɔs;-krɔ:s]
亲组合 parental combination
重组合 recombination
★ 多效性 pleiotropy [plai'ɔtrəpi]
多基因遗传 polygenic inheritance
★ 质量性状 qualitative ['kwɔlitətiv] character
★ 单基因的 monogenic
★ 主基因 major gene
共显性 codominance [ˌkəu'dɔminəns]
★ 上位性 epistasis [i'pistəsis]
★ 数量性状 quantitative ['kwɔntitətiv] character
多基因 polygene
细胞遗传学 cytogenetics [ˌsaitəudʒi'netiks]
X 染色体连锁基因 X-linked gene
★ 连锁 linkage
连锁群 linkage group
★ 连锁图 linkage map
遗传图 genetic map
★ 图距单位 map unit
★ 常染色体 autosome ['ɔ:təsəum]
性染色体 sex chromosome
同配性别 homogametic [ˌhəuməugə'metik] sex
异配性别 heterogametic sex
X-连锁显形遗传 X-linked dominant inheritance
X-连锁隐性遗传 X-linked recessive inheritance
基因组印迹 genomic imprinting
非整倍体 aneuploid ['ænjuplɔid]
★ 单体 monosomic [ˌmɔnəu'səumik]
★ 三体 trisomic [trai'səumik]
四体 tetrasomic ['tetrəˌsəumɔik]
双三体 double trisomic
★ 缺体 nullisomic [ˌnʌli'sɔmik]
同源多倍体 autopolyploid ['ɔ:təu 'pɔliˌplɔid]
异源多倍体 allopolyploid ['ælə'pɔliˌplɔid]
★ 多倍体 polyploidy ['pɔliˌplɔidi]
整倍体 euploid ['ju:plɔid]
★ 染色体畸变 chromosomal aberration
重排 rearrangement [ˌri:ə'reindʒmənt]
重新愈合 reunion [ri:'ju:njən]
重建 restitution [ˌresti'tju:ʃən]
★ 缺失 deletion
★ 重复 duplication [ˌdju:pli'keiʃən]
★ 易位 translocation [ˌtrænsləu'keiʃən]
★ 倒位 inversion [in'və:ʃən]
★ 位置效应 position effect
人类染色体病 chromosomal disease in Human
★ 转化 transformation
★ 碱基对 base pair
复制起始点 origin of replication
复制子 replicon ['replikɔn]
半保留复制 semiconservative replication
复制叉 replication fork
★ 引物 primer ['praimə]
引物酶 primase
误读校正 proof reading
DNA 修复 DNA repair
切除修复 excision [ek'siʒən] repair
营养缺陷型 auxotroph ['ɔ:ksətrɔf]
一个基因一个酶假设 one gene-one enzyme hypothesis
★ 转录 transcription [træns'kripʃən]
信使 RNA messenger RNA
模板链 template strand
★ 遗传密码 genetic code
三联密码 triplet code
★ 翻译 translation
★ 编码链 coding strand
简并密码 degenerate code
同义密码 synonymous code
RNA 聚合酶 RNA polymerase
转录的起始 initiation of transcription
★ 启动子 promoter [prə'məutə]
RNA 链的延长 elongation of the RNA strand

转录的终止 termination of transcription
★ 外显子 extron ['ekstrɔn]
★ 内含子 intron ['intrɔn]
转移 RNA transfer RNA/tRNA
★ 反密码子 anticodon ['ænti'kəudɔn]
核糖体 ribosome
核糖体 RNA ribosomal RNA/rRNA
核酶 ribozyme
★ 起始 initiation [iˌniʃi'eiʃən]
起始因子 initiation factor
延长因子 elongation factor
终止 termination [ˌtə:mi'neiʃən]
多聚核糖体 polyribosome [ˌpɔli'raibəsəum]
★ 自我组装 self-assembly
分子伴侣 molecular chaperone ['ʃæpərəun]
★ 突变 mutation [mju(:)'teiʃən]
点突变 point mutation
转换 transition [træn'ziʒən]
颠换 transversion [trænz'və:ʒən]
错义突变 missense ['misˌsens] mutation
无义突变 nonsense mutation
移码突变 frameshift ['freimʃift] mutation
突变形成 mutagenesis [ˌmjutə'dʒenisis]
★ 诱发突变 induced mutation
诱变剂 mutagens ['mju:tədʒən]
★ 转座子 transposon [træns'pəusɔn]
★ 报告基因 reporter gene
增强子 enhancer [in'hɑ:nsə]
干细胞 embryonic ['embri'ɔnik] stem cell/EScell
★ 基因治疗 gene therapy ['θerəpi]
基因座 gene locus
★ 位点 site
操纵子 operon ['ɔpəˌrɔn]
★ 结构基因 structural gene
★ 操纵基因 operator gene
调节基因 regulatory gene
阻遏物 repressor [ri'presə(r)]
诱导物 inducer
★ 单一序列 unique sequence
重复序列 repetitive sequence
卫星 DNA satellite DNA
插入序列 insertion sequence(IS)
★ 基因簇 gene cluster
★ 假基因 pseudogene ['psju:dəudʒi:n]
基因的可塑性 genome plasticity [plæs'tisiti]
★ 基因扩增 gene amplification
DNA 甲基化 DNA methylation ['meθileiʃən]
基因组内的重排 rearrangement in the genome
交配型 mating type ['meitiŋ]['taip]
免疫球蛋白 immunoglobulin(Ig) [i'mju:nəu'glɔbjulin]
克隆动物 clonal animal
★ 遗传程序 genetic program

1 The nucleic acid sequence is the sequence of bases A, C, G, T/U in the DNA or RNA chain. The sequence is conventionally written from the free 5'- to the free 3'- end of the molecule, for example 5'-ATAAGCTC-3' (DNA) or 5'-AUAGCUUGA-3' (RNA).
核酸序列是指 DNA 或 RNA 链中的碱基 A、C、G、T 或 U 排列顺序，习惯上由分子游离的 5' 端写至 3' 端，例如，5'-TATTGCTC-3'(DNA) 或 5'-AUAGCUUGA-3'(RNA)。

2 With the advent of the PCR technique in molecular biology, DNA can now be extracted and analyzed from ancient remains. Ancient DNA studies hold great potential for anthropologists and archaeologists to address many important issues that cannot be well dealt with in conventional ways. However, the field is still full of technical and interpretive challenges. The most difficult is proving that amplified DNA is authentic ancient DNA.
现代分子生物学中 PCR 技术的发展使得直接提取和分析古代遗骸中的 DNA 成为可能，这为考古学家和人类学家解释许多用传统方法不能解释的论点提供了一种新的研究手段。但该领域仍然充满着技术和解释方面的挑战。如何甄别所获得的 DNA 是真实的古代 DNA 是最大的困难。

第十一节 生态学
Section 11 Ecology

★ 生态学 ecology [i(:)'kɔlədʒi]
★ 种群密度 population density [ˌpɔpju'leiʃən]['densiti]
迁出 emigration [ˌemi'greiʃən]
迁入 immigration [ˌimi'greiʃən]
★ 年龄结构 age structure

性比 sexual ratio
生命表 life table
环境容纳量 carrying capacity
★生物群落 biotic [bai'ɔtik] community
种间竞争 interspecific [ˌintə(:)spi'sifik] competition
★生态分离 ecological separation
互惠共生 mutualistic symbiosis ['mju:tjuəlistik][simbai'əusis]
生态位 ecological niche [nitʃ]
优势种 dominant species
热带雨林 tropical rainforest
真叶林 coniferous [kəu'niferəs] forest
★温带草原 temperate grassland
荒漠 desert [di'zə:t]
★水生群落 aquatic [ə'kwætik] community
★群落演替 community succession [sək'seʃən]
★顶级群落 climax ['klaimæks] community
★初级演替 primary succession
★次级演替 secondary succession
★生产者 producer [prə'dju:sə]
★消费者 consumer [kən'sju:mə]
草食动物 herbivore ['hə:bivɔ:]
食肉动物 carnivore ['kɑ:nivɔ:]
杂食动物 omnivore ['ɔmni,vɔ:]
还原者 decomposer [ˌdi:kəm'pəuzə(r)]
★食物链 food chain
★生物量 biomass ['baiəumæs]
★初级生产 primary production
★次级生产 secondary production
★生物多样性 biodiversity
★可持续发展 sustainable [sə'steinəbl] development

1 Ecosystems contain both abiotic and biotic components. The latter are either producers or consumers. Consumers may be herbivores, carnivores, omnivores, or decomposers.
生态系统包括非生物和生物两个部分。后者要么是生产者,要么是消费者。消费者可能是食草动物,食肉动物,杂食动物或分解者。

2 Energy flow and chemical cycling are two phenomena observed in ecosystems.
能量流动和化学循环是生态系统中所观察到的两个现象。

第七章　微生物学
Chapter 7　Microbiology

第一节　绪论
Section 1　Introduction

★ 微生物学　microbiology
★ 微生物　microorganism
扫描电子显微镜　scanning electron microscope
★ 原核生物　prokaryote
★ 细菌　bacterium/ bacteria(复数)
★ 放线菌　actinomyces [ˌæktinəu'maisiːz]
★ 蓝细菌　blue algae
★ 支原体　mycoplasma [ˌmaikəu'plæzmə]
★ 立克次氏体　rickettsia [ri'ketsiə]
★ 衣原体　chlamydia [klə'midiə]
★ 古生菌　archaea
★ 真核微生物　eukaryotic microorganism
★ 真菌　fungus ['fʌŋgəs] / fungi(复数)
★ 酵母菌　yeast
霉菌　mold [məuld]
★ 蕈菌　mushroom
藻类　algae
非细胞微生物　acellular microorganism
★ 病毒　virus
★ 亚病毒　subvirus
★ 类病毒　viroid
★ 拟病毒　virusoid/ viroid-like
★ 朊病毒　prion
人类免疫缺陷病毒　human immunodeficiency virus (HIV)
艾滋病　acquired immunodeficiency syndrome(AIDS)
★ 毒素　toxin
鼠疫　plague [pleig]
结核　tuberculosis [tjuˌbəːkju'ləusis]
流感　flu/ influenza ['influ'enzə]
生物工程学　biotechnology
模式生物　model organism
生物转化　bioconversion/ biotransformation
嗜极菌　extremophile
普通微生物学　general microbiology
应用微生物学　applied microbiology
真菌学　mycology
细菌学　bacteriology
病毒学　virology

1 The word microbe is used to describe an organism that is too small to be seen clearly by the unaided eye — that is microorganisms cannot be seen without the use of a microscope.
微生物通常是指一切若不借助显微镜,用肉眼看不清的微小生物。

2 Life on earth could not exist without bacteria because bacteria make possible many of the essential functions of ecosystems, including the capture of nitrogen from the atmosphere, decomposition of organic matter and photosynthesis in many aquatic communities.
没有细菌,地球上就不存在生命,因为细菌使很多重要的生态功能成为可能,包括大气中氮的固定,有机物的降解以及许多水生群落中的光合作用。

第二节　原核生物的形态、构造和功能
Section 2　Morpha, Structure and Function of Prokaryotes

核区　nuclear region
杆菌　bacillus [bə'siləs]
螺菌　spirillum [spai'riləm]
弧菌　vibrio
纳米细菌　nano-bacteria
★ 革兰氏染色法　Gram stain
★ 革兰氏阳性细菌　Gram positive bacteria (G+)
★ 革兰氏阴性细菌　Gram negative bacteria (G−)
★ 肽聚糖　peptidoglycan [ˌpeptidəu'glaikæn]
N-乙酰葡糖胺　N-acetylglucosamine [ˌəsetilˌglukə'sæmin]
N-乙酰胞壁酸　N- acetylmuramic acid

β-1,4一糖苷键 β-1,4一glucosidic bond
双糖单位 disaccharide [dai'sækəraid] unit
溶菌酶 lysozyme ['laisəzaim]
四肽尾 tetrapeptide side chain
肽桥 peptide interbridge
★ 磷壁酸 teichoic acid
★ 脂多糖 lipopolysaccharide (LPS) [lipəu,pɔli'sækəraid]
类脂 A lipid A
核心多糖 core polysaccharide
O-特异侧链 O-specific side chain
周质空间 periplasmic ['periplæzəmik] space
★ 假肽聚糖 pseudo-peptidoglycan
N-乙酰塔罗糖胺糖醛酸 N-acetyltalosaminouronic ['æsitil 'tæləus'æminəurɔnik] acid
★ 缺壁细菌 cell wall deficient bacteria
L 型细菌 L-form of bacteria
磷脂 phosphatide ['fɔsfətaid]
phospholipid [fɔsfəu'lipid]
聚-β-羟丁酸 poly-β-hydroxy-butyrate (PHB) ['pɔli'haidrɔksi 'bju:tireit]
聚羟链烷酸 polyhydroxy-alkanoate(PHA)
异染粒 metachromatic granule [,metəkrəu'mætik]['grænju:l]
★ 糖被 glycocalyx ['glaikəu'kæliks]
★ 荚膜 capsule
黏液层 slime layer
菌胶团 zoogloea [,zəuə'gli:ə]
★ 鞭毛 flagellum [flə'dʒeləm]
"拴菌"试验 tethered-cell experiment
单端鞭毛菌 monotricha [mə'nɔtrikə]
端生丛毛菌 lophotricha
两端鞭毛菌 amphitricha
周毛菌 peritricha [pe'ritrikə]
★ 菌毛 fimbria/ fimbriae(复数)
★ 性菌毛 sex-pili/ F-pili
★ 芽孢 endospore
伴孢晶体 parasporal crystal
二分裂 binary fission
★ 芽殖 budding
★ 孢子丝 spore-bearing mycelium
★ 分生孢子 conidiospore [kəu'nidiə spɔ:]
水华 water bloom
类支原体 mycoplasma-like organism(MLO)
植原体 phytoplasma ['faitə,plæzəm]
异形胞 heterocyst ['hetərəusist]
静息孢子 akinete
赤潮 red tide
★ 克隆 clone
菌膜 scum [skʌm]
★ 基内菌丝 substrate mycelium
★ 气生菌丝 aerial mycelium
类立克次氏体 Rickettsia-like bacteria, RLB

1 There are fundamentally different types of cells — prokaryotic cell and eukaryotic cell. Bacteria belong to prokaryote, and fungi belong to eukaryote.
有不同细胞类型 —— 原核细胞和真核细胞。细菌属于原核生物，真菌属于真核生物。

2 Prokaryotic cells have a much simpler morphology than eucaryotic cells. Prokaryotes are small, single-celled microorganisms whose chromosome is not contained within a nuclear membrane, and which lack many of the organelles and internal structures found in eukaryotic cells.
原核细胞在形态上比真核细胞更加简单。原核生物是一类微小的单细胞微生物，其染色体没有核膜包围，还缺少许多在真核细胞中发现的细胞器和内部结构。

3 The prokaryotes can be split into two groups, the archaebacteria and the eubacteria.
原核生物分为两个类群，古细菌和真细菌。

4 Eukaryotic cells have a membrane-bound nucleus containing chromosomes made up of both nucleic acids and proteins.
真核细胞具有膜包围的细胞核，细胞核内含有由核酸和蛋白构成的染色体。

5 Bacteria are single-celled prokaryote and do not have membrane-bound nuclei, nor do they have chromosomes of the kind present in eukaryote. Instead, bacterial DNA consists of a single chromosome. The DNA is circular, tightly supercoiled. Their naked circular DNA is localized in a zone of the cytoplasm called the nucleoid.

细菌是单细胞原核生物，既没有膜包围的细胞核，也没有真核生物那样的染色体。细菌的 DNA 由一条染色体构成，环状、致密螺旋状。裸露的环状 DNA 位于细胞质中被称之为拟核的区域。

第三节 真核微生物的形态、构造和功能

Section 3 Morpha, Structure and Function of Eukaryotes

菌物界 mycetalia/ fungi
真菌 eumycetes [juːˈmaisiːts]
★黏菌 myxomycetes [ˌmiksəumaiˈsiːts] fungi-like protozoa
★假菌 pseudofungi
原生动物 protozoon/ protozoa(复数) [prəutəuˈzəuə]
★甘露聚糖 mannan [mænæn]
★葡聚糖 glucan [ˈgluːkən]
★几丁质 chitin [ˈkaitin]
纤毛 cilium [ˈsiliə]
几丁质酶体 chitosome
氢化酶体 hydrogenosome [ˌhaidrəˈdʒenəsəm]
★蜗牛酶 helicase [ˈhelisiːn]
★麦角甾醇 ergosterol [əːˈgɔstərɔl]
工程菌 engineering bacteria
假酵母(拟酵母) pseudo-yeast
★有性繁殖 sexual multiplication
裂殖 fission
★掷孢子 ballistospore [bəˈlistəˌspɔː]
★厚垣孢子 chlamydospore [kləˈmidəspɔː]
★子囊 ascus [ˈæskəs]
★子囊孢子 ascospore [ˈæskəuspɔː]
★质配 plasmogamy [plæzˈmɔgəmi]
★核配 caryogamy [ˈkæriəu][ˈgeimi]
生命周期 life cycle
单倍体 haploid [ˈhæplɔid]
★丝状真菌 filamentous fungi
★菌丝 hypha/ hyphae(复数)
★菌丝体 mycelium/ mycelia(复数)
★营养菌丝 vegetative mycelium
★气生菌丝 aerial mycelium
假根 rhizoid [raizɔid]
匍匐菌丝 stolon [ˈstəulɔn]
吸器 haustorium [hɔːsˈtɔriəm]
附着胞 adhesive cell
附着枝 adhesive branch
菌核 sclerotium [skliˈrəuʃiəm]
★子实体 fruiting body/ sporocarp/ fructification
★分生孢子头(分生孢子穗) conidia head
★孢子囊 sporangium [spəˈrændʒiəm]
★有性孢子 sexual spore
★无性孢子 asexual spore
★分生孢子器 pycnidium
★分生孢子座 sporodochium [ˌspɔːrəˈdəukiəm]
★分生孢子盘 acervulus
★分生孢子梗 conidiophore [kəuˈnidiəfɔː]
★子囊果 ascocarp [ˈæskəukɑːp]
闭囊壳 cleistothecium [ˌklaistəˈθiːsiəm]
子囊壳 perithecium [ˌperiˈθiʃiəm]
子囊盘 apothecium [ˌæpəˈθisiəm]
★游动孢子 zoospore [ˈzəuəspɔː]
★孢囊孢子 sporangiospore [spəˈrændʒiəspɔː]
★分生孢子 conidiospore [kəuˈnidiəspɔː]
节孢子 arthrospore [ˈɑːθrəspɔː]
★卵孢子 oospore [ˈəuəspɔː]
★接合孢子 zygospore [ˈzaigəˌspɔː]
★担孢子 basidiospore [bəˈsidiəuspɔː]
★食用菌 edible mushroom
★锁状联合 clamp connection
菌盖 cap/ pileus
菌褶 gills/ lamella
菌环 collar/ ring
菌蕾 button
菌托 volva [ˈvɔlvə]

1 The main features of eukaryotes are the presence of a nucleus, organelles, such as mitochondria and chloroplasts, and complex internal membranes. In a word, eukaryotes have complex, membrane-bound, nucleus and subcellular organelles which compartmentalize cell functions.
真核生物的主要特征是有细胞核，诸如线粒体、叶绿体的细胞器及复杂的内膜系统。总之，真核生物具有复杂的、有膜包被的细胞核和亚细胞器，使细胞功能区域化。

2 Fungi are heterotrophic eukaryotes with a filamentous, tubular structure which is called a hypha. A network of hyphae is called a mycelium.

真菌是异养的真核生物，具有被称为菌丝的丝状或管状结构。菌丝形成的网络状结构称为菌丝体。

3 Hyphae are bound by firm, chitin-containing cell walls and contain most eukaryotic cell organelles.

菌丝由坚硬的含几丁质的细胞壁包被，内含大量真核生物的细胞器。

第四节 病毒和亚病毒
Section 4 Virus and Subvirus

★ 寄生物 parasite ['pærəsait]
★ 专性寄生物 obligate parasite
潜伏性侵染 latent infection
★ 衣壳 capsid ['kæpsid]
★ 衣壳粒 capsomere ['kæpsəmə]
核衣壳 nucleocapsid [nju:kliə'kæpsid]
★ 包膜 envelope
包涵体 inclusion body
★ 噬菌斑 plaque [plɑ:k]
枯斑 lesion ['li:ʒən]
空斑 plaque [plɑ:k]
★ T 偶数噬菌体 Even type phage
★ 噬菌体 phage
★ 烈性噬菌体 virulent phage
★ 裂解性周期 lytic cycle
增殖性周期 productive cycle
★ 吸附 adsorption/ attachment
★ 侵入 penetration, injection
★ 释放 release
★ 效价 titre/ titer ['taitə]
★ 敏感宿主 sensitive host
★ 噬菌斑形成单位 plaque-forming unit(pfu)
感染中心数 infective centre
★ 双层平板法 two layer plating method
★ 一步生长曲线 one-step growth curve
★ 潜伏期 latent phase
★ 隐晦期 eclipse phase
胞内累计期 intracellular accumulation phase
★ 裂解期 rise phase
裂解量 burst size
★ 平稳期 plateau ['plætəu, plæ'təu]
★ 溶原性 lysogeny [lai'sɔdʒini]
★ 溶源菌 lysogenic bacteria
★ 温和噬菌体 temperate phage
前噬菌体 prophage
★ 裂解性循环 lytic cycle
★ 溶源性循环 lysogenic cycle
★ 脱壳 encoating/ uncoating
人畜共患病 zoonosis [ˌzəuə'nəusis]
病毒卫星 satellite
蛋白侵染子 protein infection
杀菌剂 germicide ['dʒə:misaid]
免疫原性 immunogenicity [iˌmju:nəu'dʒenisiti]
发酵 fermentation

1 Viruses are obligate intracellular parasites and vary from 20-200nm in size.

病毒是细胞内专性寄生物，其大小为 20 ～ 200 nm。

2 Virus contains only RNA or DNA. The intact particle is termed 'virion' which consists of a capsid.

病毒只含有 RNA 或 DNA。完整的颗粒被称为“病毒粒子”，它由一个衣壳构成。

3 Viruses are resistant to antibiotics.

病毒抗抗生素。

4 As obligate intercellular parasites, virus must enter and replicate in living cells in order to 'reproduce' themselves.

病毒作为细胞内专性寄生物，必须进入活细胞，并在其内进行复制来完成自我“繁殖”的过程。

5 This 'growth cycle' involves specific attachment of virus, penetration and uncoating, nucleic acid transcription, protein synthesis, maturation and assembly of the virus and their subsequent release from the cell by budding or lysis.

这种“生长周期”包括病毒特异性的吸附、侵入及脱壳、核酸转录、蛋白质合成、病毒粒子的成熟和装配，随后通过出芽或裂解的方式从细胞中释放出来。

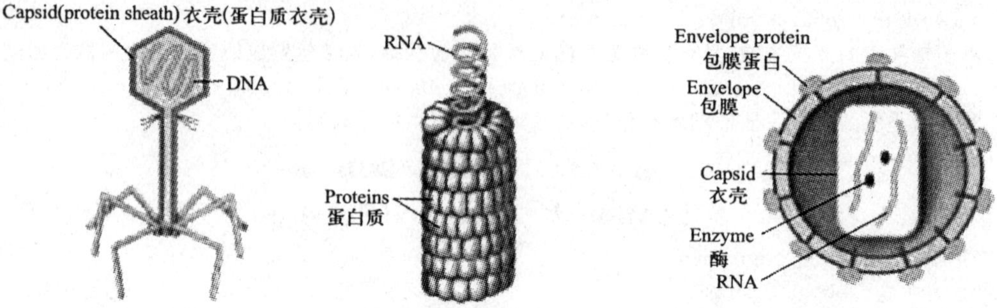

(a) Bacteriophage
(a) 噬菌体

(b) Tobacoo mosaic virus (TMV)
(b) 烟草花叶病毒

(c) Human immunodeficiency virus (HIV)
(c) 人类免疫缺陷病毒

第五节 微生物的营养和培养基
Section 5 Nutrition and Medium of Microbe

营养 nutrition [nju:'triʃən]
★ 营养物 nutrient
★ 培养基 medium
★ 营养要素 nutrient key element
碳源 carbon source
大量营养物 macronutrient
★ 碳源谱 spectrum of carbon sources
★ 异养微生物 heterotrophic microorganism
★ 自养微生物 autotrophic microorganism
有机碳源 organic carbon source
无机碳源 inorganic carbon source
有机酸 organic acid
明胶 gelatin ['dʒelətin, 'dʒelə'ti:n]
琼脂 agar
★ 双功能营养物 difunctional nutrient
氮源 nitrogen source
★ 氮源谱 spectrum of nitrogen sources
牛肉膏 beef extract
蛋白胨 peptone ['peptəun]
酵母膏 yeast extract
玉米浆 corn steep liquor
肝浸液 liver infusion
麦芽汁 malt extract
尿素 urea
carbamide ['kɑ:bəmaid]
硝酸盐 nitrate ['naitreit]
铵盐 ammonium cuprate [ə'məunjəm]['kju:preit]
★ 氨基酸自养型生物 amino acid autotrophs
★ 氨基酸异养型生物 amino acid heterotrophs
菌体蛋白 single cell protein(SCP)
★ 能源 energy source
★ 生长因子自养型微生物 auxoautotrophs [ˌɔ:ksəu'ɔ:təuˌtrɔf]
★ 生长因子异养型微生物 auxoheterotrophs
维生素 vitamin
无机盐 mineral salts
微量元素 microelements
★ 光能有机营养型 organotrophic [ˌɔ:gə'nɔtrəfik]
★ 无机化能营养型 chemolithotrophy [keməli'θɔtrəfi]
★ 有机化能营养型 organolithotrophy
★ 单纯扩散 simple diffusion
★ 被动运送 passive transport
★ 促进扩散 facilitated diffusion
渗透酶 permease ['pɜ:mieis]
移位酶 translocase [trænsləu'keiz]
★ 移位蛋白 translocator protein
★ 主动运送 active transport
离子泵 ion pump
贫养菌(寡养菌) oligophyte
★ 基团移位 group translocation
难养菌 fastidious [fæs'tidiəs] microorganism
水活度 water activity
相对湿度值 equilibrium [ˌi:kwi'libriəm] relative humidity(ERH)
★ 琼脂平板 agar plate
摇瓶培养 shake culture
发酵罐 fermenter
★ 天然培养基 natural medium
牛肉膏蛋白胨培养基 beef extract peptone medium
麦芽汁培养基 malt extract medium

★ 组合培养基　chemical defined medium
★ 半组合培养基　semi-defined medium
★ 半合成培养基　semi-synthetic medium
马铃薯蔗糖培养基　potato sucrose medium
★ 液体培养基　liquid medium
★ 固体培养基　solid medium
★ 固化培养基　solidified medium
凝固剂　gelling agent
硅胶　silica gel
滤膜　membrane filter
★ 半固体培养基　semi-solid medium
★ 脱水培养基　dehydrated culture medium
脱水商品培养基　dehydrated commercial medium
★ 选择性培养基　selected medium
★ 加富性选择培养基　enriched selected medium
富集　enrichment
★ 抑制性选择培养基　inhibited selected medium
★ 鉴别性培养基　differential medium
伊红美蓝乳糖培养基　Eosin Methylene Blue ['i:əsin]['meθili:n][blu:]

1 A culture medium is a solid or liquid preparation used to grow, transport, and store microorganisms.
培养基是用于微生物生长、移植和保存的固体或液体制品。

2 The effective medium must contain all the nutrient the microorganisms requires for growth.
有效的培养基必须含有微生物生长所需要的全部营养物。

3 Specialized media are essential in the isolation and identification of microorganisms, the testing of antibiotics sensitivities, water and food analysis.
特定的培养基对微生物分离、鉴定、抗生素敏感性试验、水和食品分析都很重要。

4 Although all microorganisms need sources of energy, carbon, nitrogen, phosphorus, sulfur, and various minerals, the precise composition of a satisfactory medium will depend on the species one is trying to cultivate because nutritional requirements vary so greatly.
尽管所有的微生物都需要能源、碳、氮、磷、硫和各种无机物，但令人满意的培养基的精确组成取决于你要培养的种类，因为不同种类的营养需求有很大差异。

5 Each kind of microbes has an optimum temperature, oxygen concentration, pH and water activity for growth.
每一种微生物都有一个生长适合温度、氧的浓度、pH 和水的活度。

6 The metabolic type of microorganisms are placed four nutritional classes based on their primary sources of carbon, energy, and electrons: photolithotrophic autotroph, photoorganotrophic heterotroph, chemolithotrophic autotroph, and chemoorganotrophic heterotroph.
根据微生物的碳、能量和电子最初来源不同，微生物的代谢类型分为 4 种营养类型：光能无机自养型、光能有机异养型、化能无机自养型和化能有机异养型。

第六节　微生物的新陈代谢
Section 6　Metabolism of Microbes

★ 腺苷三磷酸　adenosine 5′-triphosphate(ATP)
还原力　reducing power
同化　assimilation [ə,simi'leiʃən]
最初能源　primary energy source
通用能源　universal energy source
★ 生物氧化　biological oxidation
★ EMP 途径　Embden-Meyerhof-Parnas pathway
★ 糖酵解途径　glycolysis [glai'kɔlisis] pathway
★ 己糖二磷酸途径　hexose diphosphate pathway
★ HMP 途径　hexose monophosphate pathway
★ 戊糖磷酸途径　pentose phosphate pathway
★ 氧化磷酸化　oxidative phosphorylation ['ɔksideitiv][fɔ,sfɔri'leiʃən]
硝酸盐呼吸　nitrate respiration
反硝化作用　denitrification [di:naitrifi'keiʃən]
同化性硝酸盐还原作用　assimilative nitrate reduction
异化性硝酸盐还原作用　dissimilative nitrate reduction
硫酸盐呼吸　sulfate respiration
硫呼吸　sulphur respiration
铁呼吸　iron respiration
碳酸盐呼吸　carbonate respiration
延胡索酸呼吸　fumarate ['fju:məreit] respiration
同型酒精发酵　homo-alcoholic fermentation

同型乳酸发酵 homolactic fermentation
V. P. 试验 Vogos-Prouskauer test
异型乳酸发酵 heterolactic fermentation
★ 化能自养微生物 chemoautotroph [ˌkiməuˈɔtətrəuf]
生长速率 growth rate
硝化细菌 nitrifying bacteria
亚硝酸氧化酶 nitrite oxidase
★ 光能营养细菌 phototroph
★ 光合细菌 photosynthetic bacteria
菌绿素 bacteriochlorophyll(Bchl) [bæktiəriəˈklɔ(:)rəfil]
不产氧光合作用 anoxygenic photosynthesis
★ 嗜盐菌 halophile [ˈhæləfail]
紫膜蛋白 purple membrane protein
光介导 ATP 合成 light-mediated ATP synthesis
细菌视紫红质 bacteriorhodopsin [bækˌtiəriərəuˈdɔpsin]
★ 紫膜光合磷酸化 photophosphorylation by purple membrane
两用代谢途径 amphibolic pathway
★ 葡糖异生作用 gluconeogenesis [ˌgluːkəˌniːəuˈdʒenisis]
★ 乙醛酸循环 glyoxylate cycle
★ 生物固氮 biological nitrogen fixation
★ 固氮微生物 nitrogen-fixing organism
自生固氮菌 free-living nitrogen-fixer
共生固氮菌 symbiotic nitrogen-fixer
固氮酶 nitrogenase [ˈnaitrədʒəˌneis]
放氧性光合生物 oxygenic phototroph
超氧化物歧化酶 superoxide dismutase (SOD)
选择毒力 selectic toxicity
自溶素 autolysin [ˌɔːtəuˈlaisin]
次生代谢物 secondary metabolite
生物药物素 biopharmaceutin
代谢调节 regulation of metabolism
诱导酶 induced enzyme
组成酶 constitutive enzyme

1 Metabolism is divided into catabolism and anabolism.
新陈代谢可分为分解代谢和合成代谢。

2 In catabolism larger and more complex molecules are broken down into smaller, simpler molecules with the release of energy.
在分解代谢中随同能量的释放,较大的和较复杂的分子被分解成较小的、较简单的分子。

3 Anabolism is the synthesis of complex molecules from simpler ones with the input of energy.
合成代谢是从较简单的分子伴随能量的吸收而合成复杂的分子。

4 EMP (Embden-Meyerhof-Parnas) named Glycolysis pathway, is undoubtedly the most common pathway for glucose degradation to pyruvate in stage two of catabolism.
EMP 途径又称糖酵解途径。在分解代谢第二阶段葡萄糖降解成丙酮酸时无疑是最常见的途径。

5 Most microbes utilize glycolytic pathway for the catabolism of carbohydrates such as glucose and fructose. The products of this pathway are pyruvate.
大多数微生物都利用糖酵解途径对碳水化合物如葡萄糖、果糖进行分解代谢。该途径的产物是丙酮酸。

6 Glycolysis pathway is located in the cytoplasm of microbes and can function in the presence or absence of oxygen.
糖酵解途径定位在微生物细胞质中,有氧和无氧条件都能进行。

7 Among bacteria, *Pseudomonas*, *Rhizobium* and *Agrobacter* substitute the Entner-Doudoroff pathway for the glycolytic pathway.
细菌中的假单胞菌、根瘤菌、土壤杆菌可以利用 ED 途径,替代糖酵解途径。

8 A second pathway, Hexose Monophosphate pathway also named pentose phosphate pathway may be used at the same time as the glycolytic pathway or the Entner-Doudoroff sequence.
在利用糖酵解或 ED 途径的同时,微生物也利用第二条途径即己糖单磷酸又称戊糖磷酸途径。

9 The pentose phosphate pathway can operate either aerobically or anaerobically and is important in biosynthesis as well as catabolism.
戊糖磷酸途径在有氧和无氧条件下都能进行,在合成和分解代谢中也一样重要。

10 Tricarboxylic acid cycle can be named citric acid cycle. The metabolism of pyruvate to CO_2 by the citric cycle is the major mechanism of ATP generation in the cell and is also an important source of carbon skeletons for biosynthesis.

柠檬酸循环又称为三羧酸循环。丙酮酸经过三羧酸循环代谢为 CO_2，是细胞中 ATP 产生的主要机制，也是合成碳素骨架的重要来源。

第七节 微生物的生长及其控制
Section 7 Growth and Control of Microbes

繁殖 reproduction
★ 培养物 culture
★ 高密度培养 high cell-density culture(HCDC)
★ 菌落计数法 colony-counting method
平板浇注 pour plate
涂布平板 spread plate
★ 菌落形成单位 colony forming unit(cfu)
★ 同步培养 synchronous [ˈsiŋkrənəs] culture
★ 同步生长 synchronous [ˈsiŋkrənəs] growth
★ 生长曲线 growth curve
生长速率常数 growth rate constant
★ 接种物 inoculum [iˈnɔkjuləm] inocula(复数) [iˈnɔkjulə]
★ 指数期 exponential [ˌekspəuˈnenʃəl] phase
★ 对数期 logarithmic [ˌlɔgəˈriθmik] phase
代时 generation time
倍增时间 doubling time
平衡生长 balanced growth
生长限制因子 growth-limited factor
生长得率 growth yield
初生代谢物 primary metabolite
稳定期产物 idiolite
代谢产物合成期 idiophase [ˈidiəfeiz]
生物测定 bioassay [ˌbaiəuˈæsei]
★ 衰亡期 decline phase, death phase
★ 自溶 autolysis [ɔːˈtɔlisis]
★ 连续培养 continuous culture
★ 开放培养 open culture
★ 单批培养 batch culture
★ 密闭培养 closed culture
★ 连续生长 continuous growth
恒浊器 turbidostat [ˈtəːbidəˌstæt]
恒化器 chemostat [ˈkeməstæt]
单级连续培养器 one-step continuous fermentor
多级连续培养器 multi-step continuous fermentor
★ 菌体生长期 trophophase
★ 产物合成期 idiophase [ˈidiəfeiz]
稀释率 dilution rate
连续发酵 continuous fermentation
高密度培养 high cell-density culture(HCDC)
下游工程 down-stream processing
生长温度三基点 three cardinal points
最适生长温度 optimum growth temperature
★ 好氧微生物 aerobe [ˈɛərəub]
★ 厌氧微生物 anaerobe [æˈneiərəub]
★ 专性好氧菌 obligate or stricaerobe [strikˈɛərəub]
★ 兼性厌氧菌 facultative anaerobe
★ 兼性好氧菌 facultative aerobe
★ 耐氧菌 aerotolerant anaerobe
★ 专性厌氧菌 strict-anaerobe
产甲烷菌 methanogen [ˈməθənəˌdʒen]
过氧化物酶 peroxidase [pəˈrɔksideis]
试管斜面 test-tube slant
琼脂平板 agar plate [ˈeigɑː][pleit]
克氏扁瓶 kolle flask
★ 高层琼脂柱 high agar column
韦荣氏管 Veillon tube
★ 厌氧培养皿 anaerobic petri dish
★ 恒盖特滚管技术 Hungate roll-tube technique
★ 厌氧罐 anaerobic jar
厌氧手套箱 anaerobic glove box
台式发酵罐 bench top fermentor
传感器 sensor
固体曲 solid mouldy bran
浅盘培养 shallow pan cultivation
★ 生物反应器 bioreactor
固定化细胞 immobilized cell
污染 contamination [kənˌtæmiˈneiʃən]
★ 灭菌 sterilization [sterilaiˈzeiʃən]
★ 杀菌 bacteriocidation
★ 溶菌 bacteriolysis [bækˌtiəriˈɔləsis]
★ 消毒 disinfection [ˌdisinˈfekʃən]
★ 防腐 antisepsis [ˌæntiˈsepsis]
★ 防腐剂 antiseptic [ˌæntiˈseptik]
★ 化疗 chemotherapy [ˌkeməuˈθerəpi]
选择毒力 selective toxicity
★ 干热灭菌法 dry heat sterilization
灼烧 incineration[inˌsinəˈreiʃən]/ combustion
★ 湿热灭菌法 moist heat sterilization
★ 巴氏消毒法 pasteurization [ˌpæstəraiˈzeiʃən]

★ 低温维持法 low temperature holding method (LTH)
★ 高温瞬时法 high temperature holding method (HTS)
★ 间歇灭菌法 fractional sterilization/ tyndallization
加压蒸汽灭菌法 normal autoclaving [ɔ:təu'kleiviŋ]
连续加压蒸汽灭菌法 continuous autoclaving
热死时间 thermal death time
热死温度 thermal death point
滤器 filter
最低抑制浓度 minimum inhibitory concentration(MIC)
★ 半致死剂量 50% lethal dose(LD50) ['li:θəl][dəus]
★ 最低致死剂量 minimum lethal dose(MLD)
表面消毒剂 surface disinfectant [disin'fekt(ə)nt]
石炭酸系数 phenol coefficient(P.C.) [kəui'fiʃənt]
★ 供试菌 test organism
抗代谢药物 antimetabolite ['æntime'tæbəlait]
磺胺药 sulphonamide [ˌsʌl'fɔnəmaids]
sulfa drug
竞争性拮抗作用 competitive antagonism
牛津单位 oxford unit
广谱型抗生素 broad-spectrum antibiotic
半合成抗生素 semi-synthetic antibiotic
★ 生物药物素 biopharmaceutin
抗药性 antibiotic resistance

1 Growth of single-celled microbes in liquid medium follows a typical pattern often called growth curve.
单细胞微生物在液体培养基中的生长遵循典型模式规律称之为生长曲线。

2 When microorganisms are introduced into fresh culture medium, usually no immediate increase in cell number occurs and therefore this period is called the lag phase.
当微生物被接种在新鲜培养基时，通常其数量并不立即增加，因此这个阶段称之为延滞期。

3 Exponential (or log) growth is seen as a straight line on a plot of $\log_{10}$ viable count of cells against time.
对数生长期是以活细胞数的对数对时间所做的曲线，在一定区域内的可视为一段直线。

4 During the exponential or log phase, microorganisms are growing and dividing at the maximal rate possible given their genetic potential, the nature of the medium, and the conditions under which they are growing.
在对数期或指数期，微生物在一定的培养基和生长条件下尽其可能的遗传潜力以最快的速度进行生长和分裂。

5 Growth rate of microorganisms is constant during the exponential phase.
在对数生长期，微生物的生长速度不变。

6 Once nutrient becomes exhausted or metabolites accumulate, the net increase in cell numbers stops. This is called the stationary phase, which may be followed after a period of time by the death phase in which the number of bacteria decreases due to cell lysis.
一旦营养耗尽或有代谢物积累，细胞数目的净生长停止，称之为稳定期。再经过一段时间后，紧接着是衰亡期，在这一阶段由于细胞自溶，细菌数目下降。

7 Continuous microbes growth can be maintained in systems, such as the chemostat, where fresh medium is constantly fed into a culture vessel. Removal of excess liquid by an overflow mechanism ensures that the culture volume is kept constant.
在诸如恒化器的系统中，微生物可保持连续生长。在恒化器中，新鲜培养基恒速补料进培养容器，通过溢流机制除去过量的液体，以保证培养液体积恒定。

第八节 微生物的遗传变异和育种

Section 8 Heredity Variation and Breeding of Microbes

★ 变异 variation [ˌvεəri'eiʃən]
★ 遗传型 genotype ['dʒenətaip]
表型 phenotype ['fi:nətaip]
饰变 modification [ˌmɔdifi'keiʃən]
致病菌 pathogenic bacteria
千碱基对 kilo base pair

兆碱基对 mega base pair
操纵子 operon [ˈɔpəˌrɔn]
结构基因 structure gene
操纵基因 operator [ˈɔpəreitə]
启动基因 promoter [prəˈməutə]
阻遏物 repressor [riˈpresə(r)]
★密码子 codon [ˈkəudən]
松弛性复制控制 relaxed replication control
严紧型复制控制 stringent replication control
整合 integration
克隆载体 cloning vector
聚丙烯凝胶电泳 polyacrylamide gel electrophoresis (PAGE) [ˌpɔləˈkriləmaid][dʒel][ilektrəfəˈri:sis]
密度梯度离心法 density gradient centrifugation
致育因子 fertility factor
性因子 sex factor
转座因子 transposable element
Col 质粒 Colicin plasmid/ Col plasmid
细菌素 bacteriocin [bækˈtiəriəsin]
Ti 质粒 tumor inducing plasmid
Ri 质粒 root inducing plasmid
Mega 质粒 megaplasmid
★基因突变 gene mutation
★野生型菌株 wild type strain
★突变株 mutant
★选择性突变株 selectable mutant
非选择性突变株 non-selectable mutant
★营养缺陷型 auxotroph [ˈɔ:ksətrɔf]
★基本培养基 minimum medium, MM
★抗性突变型 resistant mutant
★条件致死突变型 conditional lethal mutant
★温度敏感突变株 temperature sensitive mutant/ Ts mutant
★形态突变型 morphological mutant
★抗原突变型 antigenic mutant
★产量突变型 metabolite quantitative mutant
突变率 mutation rate
★回复突变株 back mutant/reverse mutant
★抗药性突变株 resistance mutant
正突变 plus-mutant [plʌsˈmju:tənt]
负突变 minus-mutant [ˈmainəsˈmju:tənt]
★影印平板培养法 replica plating
★点突变 point mutation
★诱发突变 induced mutation
★置换 substitution [ˌsʌbstiˈtju:ʃən]
★转换 transition [trænˈziʒən]
★颠换 transversion [trænzˈvə:ʒən]
★移码突变 fram-shift mutation/ phase-shift mutation
移码突变株 frame-shift mutant
烷化剂 alkylating [ˈælkileitiŋ] agent
★染色体畸变 chromosomal aberration
★缺失 deletion
★重复 duplication
★插入 insertion
★易位 translocation
★倒位 inversion
★转座 transposition
转座子 transposon, Tn
插入序列 insertion sequence(IS)
跳跃基因 jumping gene
正向重复 direct repeat
反向重复 inverted repeat
转座酶基因 transposase gene
靶序列 target sequence
★自发突变 spontaneous mutation
紫外线 ultraviolet ray(UV)
嘧啶二聚体 pyrimidine dimmer
★光复活作用 photoreactivation [ˌfəutəuˌriæktiˈveiʃən] / photorestoration
单体 monomer [ˈmɔnəmə]
光解酶 photolyase [fəuˈtəulaiəs]
★切除修复 excision repair
★暗修复 dark repair
★诱变育种 breeding by induced mutation
离子束 ion beam
★出发菌株 original strain
★表型延迟 phenotype lag
★协同效应 synergism [ˈsinədʒizm]
选择性遗传标记 selective genetic marker
★梯度平板法 gradient plate
代谢拮抗物 antimetabolite [ˈæntimeˈtæbəˌlait]
★琼脂块培养法 agar block method
★完全培养基 complete medium(CM)
★补充培养基 supplemental medium(SM)
★野生型 wild type/ wild strain
★原养型 prototroph [ˈprəutətrɔf]
★夹层培养法 layer plating method
生长谱法 auxanography [ˌɔ:ksəˈnɔgrəfi]
★基因重组 gene recombination
★遗传重组 genetic recombination
★受体菌 recipient cell [riˈsipiənt][sel] / re-

ceptor
★ 供体菌　donor cell
球状体　spheroplast
感受态　competence [ˈkɔmpətəns]
转化因子　transforming principle
★ 转染　transfection
★ 缺陷噬菌体　defective phage
★ 转导子　transductant
完全转导(普遍转导)　complete transduction
转导颗粒　transducing particle
★ 流产转导　abortive transduction
★ 局限转导　specialized transduction/restricted transduction
误切　abnormal excision
★ 低频转导　low frequency transduction, LFT
低频转导裂解物　LFT lysate [ˈlaiseit]
★ 高频转导　high frequency transduction(HFT)
高频转导裂解物　HFT lysate
★ 双重溶源菌　double lysogen [ˈlaisədʒen]
★ 溶源转变　lysogenic conversion
★ 接合　conjugation/ mating
接合子　conjugant
★ 高频重组菌株　high frequency recombination strain
性导　sexduction [ˌseksˈdʌkʃən]
F 质粒媒介转导　F-mediated transduction
★ 原生质体融合　protoplast fusion
亲本菌株　parent strain
★ 聚乙二醇　polyethylene glycol [ˌpɔliˈeθiliːn][ˈglaikɔl]
★ 有性杂交　sexual hybridization
★ 准性杂交　parasexual hybridization
半知菌类　fungi imperfecti
核融合　nuclear fusion
体细胞交换　somatic crossing-over
有丝分裂交换　mitotic crossing-over
准性杂交育种　breeding by parasexuality [ˌpærəˌsekjuˈæliti]
遗传工程　genetic engineering
限制性核酸内切酶　restriction endonuclease
黏性末端　cohesive end
连接酶　ligase [liˈgeis]
嵌合体　chimaera [kaiˈmiərə]
干扰素　interferon [ˌintə(ː)ˈfiərɔn]
白细胞介素　interleukin
转基因动物　transgenic animal
转基因植物　transgenic plant
★ 衰退　degeneration [didʒenəˈreiʃ(ə)n]
★ 纯种分离法　pure culture isolation
离异　dissociation [diˌsəuʃiˈeiʃən]
★ 菌种保藏　preservation conservation maintenance [ˈmeintinəns]
★ 冷冻干燥保藏法　lyophilization [laiˌɔfilaiˈzeiʃən]/ freeze-drying method
典型培养物　type culture

1 The genotype of an organism is the information stored in its genetic material: the sequence of the bases in its DNA.
生物的遗传型就是贮存在遗传物质中的信息,也即 DNA 的碱基序列。

2 The phenotype of an organism is the expression of that genotype as it can be observed or measured.
生物的表型是遗传型的表达,它能被观察或测量。

3 Any change to the sequence of bases in the DNA of an organism is a mutation.
突变是生物体 DNA 中碱基序列的任何改变。

4 The change may be a single-base change called a point mutation or involve a much larger sequence of DNA in which case it is called a multi-site mutation.
一个碱基的改变,称为点突变。大片段 DNA 碱基序列的改变,称多位点突变。

5 Replacement of a purine with another purine (or a pyrimidine with another pyrimidine) is called a transition.
一个嘌呤被另一个嘌呤(或一个嘧啶被另一个嘧啶)所取代称为转换。

6 The change of a purine to a pyrimidine or vice verse is a transversion.
一个嘌呤被另一个嘧啶(或一个嘧啶被另一个嘌呤)所取代称为颠换。

7 Plasmids are extra-chromosomal molecules of DNA that replicate independently of the bacterial chromosome. They are normally covalently closed, circular, supercoiled molecules.
质粒是染色体外的 DNA 分子,独立于细菌染色体复制。它们通常是共价、闭合、环状、超螺旋分子。

8 Conjugation is the mechanism by which genetic material is transferred between two bacterial cells by plasmids.
接合是同通过质粒使遗传物质在两个细菌细胞间转移的机制。

9 Transduction is the movement of host DNA from one bacterial cell (donor) to another cell (recipient) by a bacteriophage.
通过噬菌体把一个细菌细胞(供体)的宿主 DNA 转移到另一个细胞(受体)的过程称为转导。

第九节 微生物的生态
Section 9 Ecology of Microbes

★ 生态系统 ecosystem [i:kə'sistəm]
★ 群落 community
★ 种群 population
个体 individual [ˌindi'vidjuəl]
器官 organ
组织 tissue
细胞器 organelle [ˌɔ:gə'nel]
★ 贫营养细菌 oligotrophic [ˌɔligəu'trɔfik] bacteria
自净作用 self cleaning
★ 滤膜培养法 filter membrane culture method
材料劣化 material deterioration
霉变 mildew ['mildju:] / mouldness [məuldnis]
腐朽 decay [di'kei]
腐烂 putrefaction [ˌpju:tri'fækʃən] / rot
腐蚀 corrosion [kə'rəuʒən]
真菌毒素 mycotoxin [ˌmaikəu'tɔksin]
★ 嗜热微生物 thermophile ['θə:məfail]
★ 耐热菌 thermotolerant bacteria
兼性嗜热菌 facultative thermophile ['fækəltətiv]['θə:məfail]
专性嗜热菌 obligately thermophile
★ 极端嗜热菌 extremothermophile
超嗜热菌 hyperthermophile
嗜极酶 extreme enzyme
★ 嗜冷微生物 psychrophile [ˌsaikrəu'fail]
耐冷微生物 psychrotolerant [saikrəu'tɔlərənt]
★ 嗜酸微生物 acidophile [ə'sidəfail]
耐酸微生物 acidtolerant
★ 嗜碱微生物 alkalinophile
洗涤剂 detergent [di'tə:dʒənt]
★ 嗜盐微生物 halophile ['hæləfail]
耐盐微生物 halotolerant ['heiləu'tɔlərənt]
嗜压微生物 barophile ['bærəfail]
★ 正常菌群 normal flora
微生态学 microecology
★ 条件致病菌 opportunist pathogen
★ 内源感染 endogenous infection
益生菌剂 probiotics [prəˌbai'ɔtiks]
★ 无菌动物 germ-animal
★ 悉生生物 gnotobiota [ˌnəutəubai'əutə]
悉生生物学 gnotobiology [ˌnəutəubai'ɔlədʒi]
★ 共生 symbiosis [simbai'əusis]
★ 根际微生物 rhizosphere microorganism
★ 附生微生物 epibiotic [ˌepibai'ɔtik] microorganism
★ 寄生 parasitism ['pærəsaitizəm]
★ 捕食 predation [pri'deiʃən]
★ 拮抗 antagonism [æn'tægənizəm]
★ 互生 metabiosis [ˌmetəbai'əusis]
偏害共栖 amensalism
混合培养 mixed cultivation
混合发酵 mixed fermentation
地衣 lichen ['laikən]
菌根 mycorhiza [ˌmaikɔ:'raizə]
菌套 mantle ['mæntl]
★ 寄生物 parasitism ['pærəsaitizəm]
★ 宿主 host
★ 兼性寄生物 facultative parasite ['fækəltətiv]['pærəsait]
生物农药 bio-pesticide
食物链 food chain
分解者 decomposer [ˌdi:kəm'pəuzə]
碳素循环 carbon cycle
氮素循环 nitrogen cycle
★ 氨化作用 ammonification [əˌməunifi'keiʃən]
铵盐同化作用 assimilation of ammonium
★ 反硝化作用 denitrification [di:naitrifi'keiʃən]
★ 硫化作用 sulfur oxidation
★ 细菌沥滤 bacterial leaching
磷素循环 phosphorus cycle
植酸盐 phytate ['faiteit]
★ 富营养化 eutrophication [ju:ˌtrɔfi'keiʃən]
赤潮 red tide
★ 生化需氧量 biochemical oxygen demand

★ 化学需氧量 chemical oxygen demand
共代谢 co-metabolism
★ 活性污泥 activated sludge ['æktiveitid][slʌdʒ]
有机垃圾好氧生物反应器 aero-bioreactor of organic garbage
可再生资源 renewable resource
★ 沼气 marsh gas/ swamp gas
生物气 biogas ['baiəu,gæs]
甲烷形成(沼气发酵) methanogenesis [,meθənəu'dʒenisis]
发光细菌 luminescent [,lu:mi'nesnt] bacteria
自诱导 autoinduction ['ɔ:təuin'dʌkʃən]
自诱导物 autoinducer ['ɔ:təuin'dju:sə]

1 Microbes are found in environments ranging from hydrothermal vents where temperatures reach 100℃ to polar regions.
微生物存在于各种环境之中,从温度高达100℃的温泉口,直到两极地区。

2 Microbes perform many important tasks such as nutrient recycling (carbon cycle, nitrogen cycle, sulfur cycle) and many form important symbiotic associations with plants, increasing soil fertility and plant growth.
微生物可以从事很多重要工作,例如营养再循环(碳循环,氮循环,硫循环等),和植物形成的许多重要的生物共生关系,提高土壤肥力,促进植物的生长。

第十节 传染与免疫
Section 10 Infection and Immunity

病原微生物 pathogenic microorganism
★ 传染 infection
★ 传染病 infectious disease/ communicable disease/ contagious disease
★ 毒力 virulence ['virulәns]
★ 致病力 pathogenicity [,pæθədʒi'nisiti]
★ 侵袭力 invasiveness
★ 外毒素 exotoxin
★ 内毒素 endotoxin
★ 免疫力 immunity [i'mju:niti]
★ 免疫防御 immunologic defence
★ 免疫稳定 immunologic homeostasis [,həumiəu'steisis]
★ 免疫监视 immunologic surveillance
★ 隐性传染 inapparent infection
带菌者 carrier
★ 显性传染 apparent infection
★ 急性传染 acute infection
★ 慢性传染 chronic infection
局部感染 local infection
全身感染 systemic infection
败血症 septicemia [,septi'si:miə]
★ 非特异性免疫 nonspecific immunity
★ 先天免疫 innate immunity
★ 自然免疫 natural immunity
★ 屏障结构 barrier structure
★ 血脑屏障 blood-brain barrier
白细胞 leukocyte ['lju:kəsait]
红细胞 erythrocyte [i'riθrəusait]
血小板 blood platelet ['pleitlit]
多形核白细胞 polymorphonuclear leukocyte (PMN) [,pɔli,mɔ:fə'nju:kljə]['lju:kəsait]
粒细胞 granulocyte ['grænjuləsait]
★ 巨噬细胞 macrophage ['mækrəfeidʒ]
骨髓干细胞 stem cell
白细胞介素 inter leukin-1(IL-1)
★ 炎症 inflammation [,inflə'meiʃən]
抗病毒蛋白 antiviral protein
★ 转移抑制蛋白 translation inhibitory protein
★ 特异性免疫 specific immunity
★ 获得性免疫 acquired immunity
★ 体液免疫系统 humoral immunity
★ 细胞免疫系统 cellular immunity
★ 细胞介导免疫系统 cell-mediated immunity
免疫细胞 immunocyte [,imjunə'sait]
免疫活性细胞 immunologically competent cell
★ T淋巴细胞 T lymphocyte ['limfəsait]
免疫原 immunogen [i'mju:nədʒən]
★ 抗原性 antigenicity [,æntidʒə'nisiti]
★ 完全抗原 complete antigen
★ 不完全抗原 incomplete antigen
★ 抗原决定簇 antigenic determinant
表面抗原 surface antigen
★ 交叉反应 cross-reaction antigen
★ 多克隆抗体 polyclonal antibody [,pɔli'kləunəl]['ænti,bɔdi]

多发性骨髓瘤 multiple myeloma [ˈmʌltipl][ˌmaiəˈləumə]
重链 heavy chain
轻链 light chain
铰链区 hinge region
初次免疫应答 primary immune response
再次免疫应答 secondary immune response
滴度 titre [taitə]
杂交瘤 hybridoma [ˌhaibriˈdəumə]
★ 血清学反应 serological [ˌsiərəuˈlɔdʒikəl] reaction
免疫诊断学 immunodiagnostic [ˌimjunəuˌdaiægˈnəustik]
★ 凝集反应 agglutination [əˌglu:tiˈneiʃən]
★ 沉淀反应 precipitation
★ 单向琼脂扩散法 simple agar diffusion
★ 双向琼脂扩散法 double agar diffusion
★ 对流免疫电泳法 counter immuno-eletrophoresis(CIE)
中和试验 neutralization [ˌnju:trəlaiˈzeiʃən] test
免疫标记技术 immunolabelling technique
免疫定位分析 immunolocalization
免疫亲和层析 immunoaffinity chromatography [imjunəuəˈfiniti][ˌkrəuməˈtɔgrəfi]
免疫生物传感器技术 immunobiosensor technique
★ 免疫酶技术 immunoenzymatic technique
酶免疫测定法 enzyme immunoassay
胶体金 colloidal gold
★ 酶联免疫吸附法 enzyme linked immunosorbant assay(ELISA)
免疫印迹 immunoblot
★ 免疫调节剂 immunoregulative preparation
★ 活疫苗 live vaccine
★ 减毒活疫苗 attenuated [əˈtenjueitid] vaccine
★ 死疫苗 dead vaccine
亚单位疫苗 subunit vaccine
抗菌血清 antibacterial serum

第十一节 微生物的分类和鉴定
Section 11 Classification and Identification of Microbes

微生物系统学 microbial systematics [maiˈkrəubiəl][ˌsistiˈmætiks]
★ 分类 classification
★ 鉴定 identification
★ 命名 nomenclature
分类单元 taxon/ taxa(复数)/ category
界 kingdom
门 phylum [ˈfailəm]
纲 class
目 order
科 family
★ 属 genus
★ 种 species [ˈspi:ʃiz]
典型菌株 type strain
模式种 type species
微生物基因组学 microbial genomics
新种 species nova/ sp. nov/ nov sp.
域 domain
★ 双名法 binominal nomenclature [baiˈnɔminl][nəuˈmenklətʃə]
属名 generic name
变种 variety(var) [vəˈraiəti]
型 form
★ 纯培养物 pure culture
菌物界 mycetalia
动物界 animalia
植物界 plantae
原生生物界 protista
真菌界 fungi
原核生物界 monera
病毒界 archetista
★ 三域学说 Tree Domain Theory
★ 细菌域 bacteria
★ 真核生物域 eukarya
★ 化学分类法 chemotaxonomy [keməuˈtæksənəmi]/ chemotaxosystematics
★ 数值分类法 numerical taxonomy
解链温度 melting temperature(Tm)
★ 核酸分子杂交法 hybridization of nucleic acid
★ 统计分类法 taxonometrics
操作分类单元 operational taxonomic units
简单匹配相关系数 simple matching coefficient
相似度矩阵 similarity matrices [ˌsimiˈlæriti][ˈmeitrisi:z]
树状谱 dendrogram [ˈdendrəgræm]

1 The objective of current taxonomic schemes is to creat monophyletic groups of microorganisms which are assumed to have a single ancestor. This is achieved by studying characters of an individual and comparing them with phylogenetic trees can be constructed that can indicate probable evolutionary sequences of protistan microbes.

现行分类体系的目标是建立一个微生物的单源群，假定它们具有单一的祖先。这一结论是通过研究个体特征以及与其系统进化树比较而得出的。系统进化树可以推断出原生生物大概的进化顺序。

2 Living organisms are grouped into three general categories called domains. One of the domains, the eukaryotes, is subdivided into four kingdoms: protists, fungi, plants, and animals.

生物分为三大类，即三个域，其中真核生物域又分为四个界，分别为原生生物界，真菌界，植物界和动物界。

第八章　细胞生物学
Chapter 8　Cell Biology

第一节　绪论
Section 1　Introduction

★ 细胞学　cytology
★ 细胞生物学　cell biology
★ 细胞外被　cell coat
★ 细胞凋亡　cell apoptosis
细胞融合　cell fusion
细胞系　cell line
细胞通讯　cell communication
细胞液　cell sap
细胞衰老　cell senescence/ cellular aging
细胞株　cell strain
★ 细胞表面受体　cell surface receptor
细胞分化　cell differentiation
细胞周期　cell cycle
细胞分裂　cell division
★ 细胞工程　cell engineering
非细胞体系　cell-free system
生物膜　biomembrane

1 Cell theory: all living beings — animals, plants or protozoan — are composed of cells and cell products.
细胞学说：一切生物 —— 动物、植物或原生动物，都是由细胞及其产物组成的。

2 The cell is the fundamental unit of all life on earth.
细胞是地球上所有生命的基本单位。

3 Cell biology defines the very center of all efforts to understand all aspects of biology and human disease.
细胞生物学介绍了生物学和人类疾病所有方面取得成就的核心内容。

4 Cell biology elucidates the molecular basis of how cells are constructed, how the thousands of cell types accomplish their individual tasks, and finally how these different cells cooperate to form tissues, systems, and organisms.
细胞生物学是以阐明细胞结构的分子基础，研究上千种类型的细胞如何完成各自的任务，并且最终这些不同的细胞是如何协作形成了组织、系统和器官的。

第二节　细胞的基本知识
Section 2　Basic Knowledge about the Cell

★ 细胞增殖　cell proliferation [prəuˌlifəˈreiʃən]
★ 细胞识别　cell recognition
★ 细胞繁殖周期　cell reproductive cycle
★ 古核生物　archaeon
★ 细胞癌基因　cellular oncogene [ˈɔnkəˌdʒiːn]
病毒癌基因　viral-oncogene
真细菌　eubacteria
★ 常染色质　euchromatin [juːˈkrəumətin]
★ 真核细胞　eukaryotic cell
★ 无丝分裂　amitosis [ˌæmiˈtəusis]
鞭毛蛋白　flagellin [ˈflædʒilin]
盐细菌　halobacteria [hæləubækˈtɪərɪə]
中膜体　mesosome
有丝分裂器　mitotic apparatus
核壳体　nucleocapsid
★（细胞）质膜　plasma membrane
★ 原核细胞　procaryotic cell
★ 原核生物　prokaryote
★ 复制子　replicon

1 All organisms are composed of one or more cells.
所有生物体都是由一个或多个细胞组成的。

2 The cell is the structural unit of life.
细胞是生命的结构单位。

3 Cells are highly complex and organized.

细胞是高度复杂和有序的。

4 Cells possess a genetic program and the means to use it.
细胞拥有一套遗传程序和使用它的方法。

5 Cell can acquire and utilize energy.
细胞能获得和利用能量。

6 Cells carry out a variety of chemical reactions.
细胞执行各种化学反应。

第三节 细胞生物学研究方法
Section 3 The Research Method of Cell Biology

★ 电子显微镜 electron microscope
★ 光学显微镜技术 light microscopy
★ 荧光显微镜技术 fluorescence [fluə'resns] microscopy
★ 绿色荧光蛋白 green fluorescent protein (GFP)
★ 分辨率 axial resolution
★ 录像增差显微镜技术 video-enhance microscopy
★ 激光共焦点扫描显微镜 laser scanning confocal microscope
★ 微分干涉显微镜 differential-interference microscope
★ 冷冻蚀刻 freeze-etching
生物信息学 bioinformatics
★ 核体 karyoplast
★ 低温电镜技术 cryoelectron microscopy
★ 差速离心 differential centrifugation
★ 电融合技术 electron fusion method
成纤维样细胞 fibroblast ['faibrəublɑ:st] like cell
★ 流式细胞仪 flow cytometry(FCM)
★ 异核体，异核融合细胞 heterokaryon
★ 同核体，同核融合细胞 homo karyon
★ 免疫荧光技术 immunofluorescence technique
★ 原位杂交 *in situ* hybridization
★ 显微分光光度测定技术 microspectrophotometry
★ 微量注射 microinjection
★ 显微操作技术 micromanipulation
★ 单克隆抗体 monoclonal antibody
★ 突变体 mutant
★ 蛋白质组学 proteomics
★ 原生质体 protoplast
★ 快速冷冻深度蚀刻 quick freeze deep etch
★ 放射免疫沉淀 radioimmuno-precipitation ['reidiəu,imjunəu][pri,sipi'teiʃən]
★ 扫描电子显微镜 scanning electron microscope (SEM)
★ 扫描隧道显微镜 scanning tunneling microscope(STM)
传代(培养)细胞 subculture cell
融合核细胞 synkaryon [sin'kæri,ɔn]
★ 超薄切片 ultra thin section
★ 免疫印迹法 western blotting
★ 液泡 vacuole ['vækjuəul]

1 Electron microscope as a tool for the biologist brought about a complete reappraisal of the micro-anatomy of biological tissues, organisms and cells.
电子显微镜作为一种工具为生物学家带来了对生物组织、器官和细胞的显微解剖学上的完全重新认识。

2 The two most common types of electron microscopes are the transmission electron microscope (TEM) and the scanning electron microscope (SEM).
两种最常用的电子显微镜是透射电子显微镜和扫描电子显微镜。

第四节 细胞膜与细胞表面
Section 4 Cell Membrane and Cell Surface

★ 心磷脂 cadiolipin
黏着带 adhesion belt
附着蛋白 attachment protein
细菌紫膜质 bacteriorhodopsin [bæk,tɪərɪərəʊ'dɔpsɪn]
带状桥粒 belt desmosome ['desməsəm]
钙粘素 cadherin
★ 胶原 collagen ['kɔlə,dʒən]

连接子 connexion [kə'nekʃən]
★ 去垢剂 detergent [di'tə:dʒənt]
电偶联 electrical coupling
电突触 electronic junction
★ 细胞外基质 extracellular matrix
细胞外小页断裂面 extrocytoplasmic face(EF)
★ 细胞外表面 extrocytoplasmic surface(ES)
纤连蛋白 fibronectin
光脱色恢复技术 fluorescence recovery after photobleaching(FRAP)
黏着斑 focal adhesion
★ 功能基因组学 functional genomics
★ 间隙连接 gap junction
半桥粒 hemidesmosome
★ 膜内在蛋白 integral membrane protein
整联蛋白 integrin
层粘连蛋白 laminin
★ 凝集素 lectin ['lektin]
本质素 lignin ['lignin]
脂筏模型 lipid rafts model
★ 脂质体 liposome
运动蛋白 movement protein
膜骨架 membrane associated cytoskeleton
★ 封闭连接 occluding junctions
成斑现象 patching
植物抗毒素 phytoalexin [ˌfaɪtəuə'leksɪn]
★ 胞间连丝 plasmodesmata
★ 孔蛋白 porin
原代(培养)细胞 primary culture cell
★ 前胶原 procollagen [prəu'kɔlədʒən]
蛋白聚糖 proteoglycan [prəutiə'glaikæn]
★ 原生质表面 protoplasmic surface(PS)
红膜肽，血影蛋白 spectrin [spektrɪn]
★ 突触 synapse
★ 紧密连接 tight junction
★ 跨膜蛋白 transmembrane protein
单位膜模型 unit membrane model
★ 选择素 selectin

1 Cell membrane is the permeability barrier of the cell, separating the cytoplasm from the environment.
细胞膜是细胞的通透性屏障，把细胞质和外界环境分开。

2 While the plant cell has a rigid cell wall, an animal cell membrane is a flexible lipid bilayer.
植物细胞有坚硬的细胞壁，而动物细胞膜是柔韧的质脂双分子层。

3 The lipid molecules (mostly phospholipids) that make up the membrane have a polar, hydrophilic head and two hydrophobic hydrocarbon tails.
构成细胞膜的质脂分子(大部分是磷脂)有极性，包括一个亲水性的头和两个非亲水性的碳氢化合物尾。

第五节 物质的跨膜运输与信号传递

Section 5 Transmembrane Transport of the Material and Signals Transmission

★ 通讯连接 communicating junction
动作电位 active potential
★ 主动运输 active transport
接触性依赖通讯 contact-dependent signaling
★ 接触抑制 contact inhibition
★ 共转移 cotranslocation
★ 协同运输 cotransport
组成型胞吐途径 constitutive exocytosis [ˌeksəusai'təusis] pathway
离子通道偶联受体 ion-channel-linked receptor
★ 接头蛋白 adaptor protein
接合素蛋白 adaptin [ə'dæptin]
★ 自分泌 autocrine
批量运输 bulk transport
★ 钙调蛋白 calmodulin (CaM)
钙调蛋白激酶 CaM - kinase
★ 载体蛋白 carrier protein
催化性受体 catalytic receptor
★ 网格蛋白 clathrin
有被小窝 coated pit
★ 有被小泡 coated vesicle
去极化 depolarization [di:ˌpəulərai'zeiʃən]
二聚化 dimerization [ˌdaimərі'zeiʃən]
双信使系统 double messenger system
★ 下降调节 down regulation
★ 弹性蛋白 elastin
电化学梯度 electrochemical gradient
胞吞泡 endocytic vesicle
★ 胞吞作用 endocytosis
★ 胞吐作用 exocytosis [ˌeksəusai'təusis]
伸展蛋白 extensin [iks'tensin]
★ 协助扩散 facilitated diffusion

融合复合物 fusion complex
★ 融合蛋白 fusion protein
鸟苷酸环化酶 G-cyclase(GC)
G 蛋白偶联的受体 G-protein linked receptor
细胞内受体 intracellular receptor
★ 配体门离子通道 ligand-gated channel
★ 低密度脂蛋白 low-density lipoprotein (LDL)
★ 膜转运蛋白 membrane transport protein
★ 分子开关 molecular switches
神经信号 neuronal signaling
★ 渗透压 osmotic pressure
乌本苷 ouabain [wɑːˈbeiin]
★ 旁分泌 paracrine
★ 被动运输 passive transport
★ 肽信号 peptide signal
★ 通透酶 permease
★ 吞噬作用 phagocytosis [fægəˌsaiˈtəusis]
(激酶)磷酸化级联反应 phosphorylation cascade [fɔˌsfɔriˈleiʃən][kæsˈkeid]
★ 胞饮作用 pinocytosis [ˌpainəusaiˈtəusis]
多向性效应 pleiotropic effect
★ 受体 receptor
★ 受体下行调节 receptor down-regulation
★ 受体介导的胞吞作用 receptor mediated endocytosis
★ 调节型胞吐途径 regulated exocytosis pathway
静息电位 resting potential
★ 第二信使 second messenger
第二信使学说 second messenger theory
★ 信号分子 signal molecule
★ 信号网络系统 signal network system
★ 信号转导 signal transduction
★ 细胞信号通路 signaling pathway
★ 简单扩散 simple diffusion
压力激活通道 stress-activated channel
超极化 super polarization
★ 开关蛋白 switch protein
★ 共运输 symport
跨细胞的转运 transcytosis
★ 递质门(离子)通道 transmitter-gated channel
★ 膜泡运输 vesicular transport
★ 电压门通道 voltage-gated channel

1 Virtually all of the basic sciences are affected by fundamental research into how cells communicate.
实际上所有基础科学都受到对细胞如何通讯的基础研究的影响。

2 Cell communication, or signal transduction, is simply the means by which cells in the body respond to signals coming from outside those cells.
细胞通讯或信号转导,简单地说就是体内细胞对胞外信号做出反应。

第六节 细胞质基质与细胞内膜系统

Section 6 Cytoplasmic Matrix and Endomembrane System

★ 内吞溶酶体 endolysosome
★ 内膜系统 endomembrane system
ER 内质网 endoplasmic reticulum
★ 嵴 cristae
★ 细胞质基质 cytoplasmic matrix, cytomatrix
★ 协助装配 aided-assembly
★ 装配 assembly
原子力显微镜 atomic force microscope
★ 高尔基体顺面网状结构 cis Golgi network (CGN)
★ 胞质溶胶 cytosol [ˈsaitəusɔl]
★ 停泊蛋白 docking protein(DP)
★ 逃逸蛋白 escaped protein
★ 微体 microbody
★ 门控转运 gated transport
★ 高尔基体(器) Golgi body/ Golgi apparatus
★ 高尔基复合体 Colgi complex
★ 微粒体 microsome
★ 单核细胞 monocyte
粘蛋白 mucin
N-连接糖基化 N-linked glycosylation
★ 神经多肽 neuropeptide
★ 过氧化物酶体 peroxisome
★ 吞噬细胞 phagocyte [ˈfægəusait]
★ 异噬溶酶体 phagolysosome
磷酸葡萄糖苷酶 phosphoglycosidase
磷脂转换蛋白 phospholipid exchange protein (PEP)
磷脂转位因子 phospholipid translocator
后转移 post translation
★ 初级溶酶体 primary lysosome
★ 蛋白质分选 protein sorting
★ 蛋白质定向转运 protein targeting
★ 蛋白酶体 protosome

★ 残余小体　residual [ri'zidjuəl] body
★ 回收信号(序列)　retrieval [ri'tri:vəl] signal
逆行转运　retrograde transport
★ 粗面内质网　rough endoplasmic reticulum(rER)
肌质网　sarcoplasmic reticulum
★ 次级溶酶体　secondary lysosome
★ 分泌溶酶体　secretory lysosome
★ 信号假说　signal hypothesis
★ 信号区　signal patch
★ 信号识别颗粒　signal recognition particle (SRP)
★ 信号序列，信号肽　signal sequence/ signal peptide
起(开)始转移序列　start transfer sequence
★ 停止转移序列　stop transfer sequence
★ 自我装配　self-assembly
高尔基反面网状结构　trans Golgi network (TGN)
★ 易位子　translocon

1 Golgi body (also called the Golgi apparatus or golgi complex) is a flattened, layered, sac-like organelle and is located near the nucleus.
高尔基体(也叫高尔基器或高尔基复合体)是扁平的、多层的、囊状细胞器，位于核附近。

2 Lysosome (also called cell vesicles) is round organelles surrounded by a membrane containing digestive enzymes.
溶酶体(也叫细胞囊泡)是被膜包成的含有消化酶的圆形细胞器。

第七节　细胞的能量转换—线粒体和叶绿体
Section 7　Energy Conversion of the Cell-Mitochondria and Chloroplast

★ 叶绿体　chloroplast
★ 叶绿体膜　chloroplast membrane
腺苷酸环化酶　A-cyclase
★ 叶绿素　chlorophyll ['klɔ:rəfil]
叶绿素 a/ b 结合蛋白　chlorophyll a/ b binding protein
★ 自磷酸化　autophosphorylation
★ ATP 合成酶　ATP synthetase
顶体精子　acrosome
★ 肌动蛋白　actin ['æktin]
卡尔文循环　calvin cycle
二氧化碳同化　CO_2 assimilation [ə,sɪmi'leɪʃən]
★ 碰撞假说　collision hypothesis [kə'liʒən] [hai'pɔθisis]
循环式光合磷酸化　cyclic photophosphorylation
★ 细胞色素　cytochrome ['saitəukrəum]
★ 暗反应　dark reaction
★ 电子传递链　electron-transport chain
铁氧还蛋白　ferredoxin(Fd) [fe're,dɔksin]
基粒片层　grana lamella [lə'melə]
基粒类囊体　granum thylakoid ['θailə,kɔid]
热休克蛋白　heat shock protein
膜间隙　inter membrane space
脊内间隙　intracristal space lamin
★ 导肽，前导肽　leader peptide
白色体　leucoplast [lju:kəuplæst]
捕光色素　light harvesting pigmen
★ 光反应　light reaction
★ 基质　matrix
★ 分子伴侣　molecular chaperone
非循环式光合磷酸化　noncyclic photophosphorylation
★ (细)胞核　nucleus
★ 光合磷酸化　photophosphorylation [,fəutəufɔsfəri'leiʃən]
★ 光合作用　photosynthesis
★ 光合作用单位　photosynthetic unit
光系统Ⅰ、Ⅱ　photosystem Ⅰ，Ⅱ
★ 原初反应　primary reaction
前片层体　prolamellar body
★ 前质体　proplastid [prəu'plæstid]
质子动力势　proton motive force
旋转催化　rotational catalysis
★ 半自主性细胞器　semiautonomous ['semiɔ:'tɔnəməs] organelle
定子　stator ['steitə]
基质片层　stroma lamella
基质类囊体　stroma thylakoid
转运肽，转运序列　transit peptide/ transit sequence

1 Mitochondrion is a spherical to rod-shaped organelles with a double membrane. The inner membrane is infolded many times, forming a series of projections (called cristae).
线粒体为圆形至棒状的具双层膜的细胞器。内层膜多次内折，形成系列突起的嵴。

2 The mitochondrion converts the energy stored in glucose into ATP (adenosine triphosphate) for the cell.
线粒体能把储藏在葡萄糖中的能量转换成 ATP (三磷酸腺苷) 为细胞所需。

第八节　细胞核与染色体
Section 8　The Nucleus and Chromosomes

★ 染色单体连接蛋白　chromatid linking protein
★ 染色中心　chromocenter
染色粒　chromomere
★ 中央颗粒　central granule(CG)
★ 中央栓　central plug
★ 中心粒　centriole [ˈsentriəul]
★ 中心体　centrosome [ˈsentrəsəum]
中心体(或中心粒)周期　centrosome cycle/centriole cycle
有色体　chromoplast
染色体重排　chromosomal rearrangement
染色体列队　chromosome alignment
★ 染色体　chromosome
染色体支架　chromosomal scaffold
★ 收缩环　contractile [kənˈtræktail] ring
★ 通道蛋白　channel protein
活性染色质　active chromatin
★ 中央结构域　central domain
锚定连接　anchoring junction
★ 锚蛋白　ankyrin
★ 对向运输　antiport
自主复制 DNA 序列　autonomously replicating DNA sequence
AR 着丝粒 DNA 序列　centromere DNA sequence (CEN)
顺式作用元件　Cis-acting element
★ 亲核蛋白　karyophilic protein
★ 核型　karyotype [ˈkæriətaip]
★ 关键序列　key sequence
★ 动粒结构域　kinetochore domain
★ 螺旋体　coiled body(CB)
★ 共有序列　consensus sequence
组成型异染色质　constitutive heterochromatin
胞质滞留因子　cytoplasmic retention factor
★ 胞质环　cytoplasmic ring
双标记免疫荧光显微术　double label microscopy
二重对称轴　dyad axis
二分对称性　dyad symmetry
蜕皮激素　ecdysone [ˈekdisəun]
★ 核内有丝分裂　endomitosis
★ 必需基因　essential gene
★ 纤维中心　fibrillar center(FC)
功能元件　functional element
★ 遗传图谱　genetic map
★ 基因组　genome [ˈdʒi:nəum]
巨大染色体　giant chromosome
颗粒组分　granular component(GC)
★ 异染色质　heterochromatin
螺旋-环-螺旋结构模式　helix-loop-helix motif
螺旋-转角-a 螺旋模式　helix-turn-a-helix motif
激素应答元件　hormone response element
★ 非活性染色质　inactive chromatin
★ 抑制蛋白　inhibitor protein
★ 核型模式图　ideogram [ˈidiəgræm]
★ 蛋白质印迹法/免疫印迹法　immunoblotting
★ 奢侈基因　luxury [ˈlʌkʃəri] gene
★ 连接 DNA　linker DNA [ˈliŋkə]
腔内亚单位　luminal [ˈljuminæl] subunit
★ 中着丝染色体　metacentric chromosome
微带　miniband
微小染色体　minichromosome
多级螺旋模型　multiple coiling model
★ 负生长调节子　negative growth regulator
★ 非必需基因　nonessential gene
★ 非组蛋白　nonhistone
★ 核篮　nuclear basket
★ 核被膜　nuclear envelope
★ 核输出信号　nuclear export signal(NES)
★ 核纤层　nuclear lamina [ˈlæminə]
★ 核定位信号　nuclear localization signal(NLS)
★ 核基质　nuclear matrix
★ 核孔复合体　nuclear pore complex(NPC)
★ 核质环　nuclear ring
★ 核骨架　nuclear skeleton
★ 核仁周期　nucleolar cycle
★ 核仁组织区　nucleolar organizing region(NOR)
★ 核仁蛋白,核仁素　nuclein
★ 核质蛋白　nucleoplasmin
★ 核孔蛋白　nucleoporin(Nup)
★ 核小体　nucleosome [ˈnju:kliəsəum]

★ 核小体组蛋白 nucleosomal histone
核小体犁 nucleosome plow
★ 外核膜 outer nuclear membrane
★ 配对结构域 pairing domain
核周池 perinuclear cisternae
核周间隙 perinuclear [ˌperiˈnjukliə] space
★ 孔膜区 pore membrane domain
★ 前核仁体 prenucleolar body(PNB)
★ 主缢痕 primary constriction
胀泡 puff [pʌf]
反带 reverse band
★ 核糖体，核糖核蛋白体 ribosome
★ 核酶 ribozyme
★ 随体 satellite [ˈsætəlait]
★ 次缢痕 secondary constriction
★ 小分子核仁核糖核蛋白 small nucleolar ribonucleo protein(snoRNP)
★ 剪接体 spliceosome [ˈsplaisəsʌm]
螺线管 solenoid
辐 spoke
近中着丝粒染色体 submetacentric [ˈsʌbˌmetəˈsentrik] chromosome
应急因子 stringent [ˈstrindʒənt] factor
端着丝粒染色体 telocentric [ˌtiː ləˈsentrik] chromosome
★ 端粒酶 telomerase
★ 端粒 DNA 序列 telomere DNA sequence
四聚体 tetramer
转录辅阻抑制物 transcriptional co-repressor
★ 上升调节 upregulation
★ 锌指模式 zinc finger motif

1 Nuclear is a membane — enclosed structure in cells of eukaryon that contains the genetic material — chromosomes.
细胞核是在真核细胞中被膜包裹的结构，其中含有遗传物质 —— 染色体。

2 Chromosomal DNA is packaged into a compact structure with the help of specialized proteins called histones. The complex DNA plus histones in eukaryotic cells is called chromatin.
染色体 DNA 是由叫组蛋白的特殊的蛋白包裹形成的紧密结构。在真核细胞中 DNA 复合体加上组蛋白被称之为染色质。

3 Nucleosome is a subunit of chromatin composed of a short length of DNA wrapped around a core of histone proteins.
核小体是染色质的亚单位，由一段短链的 DNA 盘绕核心组蛋白构成。

第九节 细胞骨架
Section 9 Cytoskeleton

★ 细胞骨架 cytoskeleton
★ 基粒 basal body
★ 基膜 basal lamina
★ 生物发生 biogenesis [ˌbaiəuˈdʒenisis]
★ 肌动蛋白纤维 actin filament
秋水仙素 colchicine [kɔltʃisiːn]
胞质环流 cyclosis [ˌsaiˈkləusis]
DNA 复制复合体 DNA replication complex
动力蛋白臂 dynein arm
表染色质 epichromatin [ˌepiˈkrəumətin]
★ 纤维形肌动蛋白 F-actin
★ 球形肌动蛋白 G-actin
★ 基因缺失 gene deletion
★ 基因剔除 gene knockout
GTP 酶活化蛋白 GTPase activating protein (GAP)
★ GTP 结合调节蛋白质，G 蛋白 GTP binding regulatory protein
促旋酶，旋转酶 gyrase [ˈdʒaiəreis]
异二聚体 heterodimer
同源岛 homology island
★ 中间丝，中间纤维 intermediate filament [ˈfiləmənt]
★ 中间纤维结合蛋白 intermediate filament associated protein(IFAP)
★ 微丝 microfilament
★ 肌纤维 myofiber
★ 肌原纤维 myofibrils
★ 肌球蛋白 myosin
★ 神经元纤维 neurofilament
★ 神经胶质纤维 neuroglial filament
★ 非组蛋白骨架 nonhistone protein scaffold
★ 原丝 protofibril [ˌprəutəˈfaibril]
★ 原纤维 protofilament

★ 放射环模型 radial loop model
放射辐条 radial spoke
★ 复制灶 replication foci
★ 复制叉 replication fork
★ 复制位点 replication site
★ 应力纤维 stress fiber
紫杉酚 taxol [tæk'sɔl]
★ 转基因 transgene
★ 踏车现象 tread milling
★ 原肌球蛋白 tropomyosin (Tm) [ˌtrɔpəu'maiəsin]
★ 肌钙蛋白 troponin (Tn) ['trɔpənin]
★ 微管蛋白 tubulin [tju:bjulin]
★ 波形纤维 vimentin filament

1 The cytoskeleton is composed of three well-defined filamentous structures — microtubules, microfilaments, and intermediate filaments — which together form an elaborate interactive network.
细胞骨架是由三种丝状结构 —— 微管、微丝和中间纤维共同形成的精细的交互的网状结构。

2 Cytoskeleton is associated with the changes of cell shape, function, signaling and so on.
细胞骨架与细胞形状的变化、功能和信号传递等都有关系。

第十节 细胞增殖及其调控
Section 10 Cell Proliferation and the Control

★ 后期 anaphase ['ænəˌfeiz]
星体 aster
★ 细胞分裂周期 cell division cycle
★ 检验点 checkpoint ['tʃekpɔint]
化学偶联假说 chemical coupling hypothesis
化学突触 chemical synapse
化学渗透偶联假说 chemiosmotic coupling hypothesis
交叉 chiasma/crossover
★ 着丝粒 centromere ['sentrəˌmiə]
着丝粒-动粒复合体 centromere-kinetochore complex
★ 驱动蛋白 kinesin
★ 动粒 kinetochore
★ 凝集期 condensation stage
中板聚合 congression
★ 细胞周期蛋白 cyclin ['saikliŋ]
周期蛋白框 cyclin box
★ 周期中细胞 cycling cell
破坏框 destruction box
★ 终变期 diakinesis [ˌdaiəkai'ni:sis]
DNA 复制执照因子学说 DNA replicationlicence factor theory
★ DNA 合成期，S 期 DNA synthesis phase
★ 双线期 diplotene/diplonema
赤道面 equatorial [ˌekwə'tɔ:riəl] plane
赤道板 equatorial plate
★ 分裂沟 furrow ['fʌrəu]
★ 同源染色体 homologous chromosome
★ 分裂间期 interkinesis [ˌintəkai'ni:sis]
细线期 leptotene [leptəuti:n]
★ 成熟促进因子 maturation-promoting factor (MPF)
★ 微丝结合蛋白 microfilament associated protein
★ 微管相关蛋白 microtubule associated protein (MAP)
★ 微管组织中心 microtubule organizing center (MTOC)
微绒毛 microvili [ˌmaikrəu'vilai]
★ 中间体 midbody
★ 细胞促分裂因子 mitosis promoting factor
★ 复制起始点识别复合体 origin recognition complex(ORC)
粗线期 pachytene['pækiti:n]/ pachynema
中心粒外基质 pericentriolar matrix
原生质团，变形体 plasmodia
★ 极性微管 polar microtubule
★ 减数分裂前间期 premeiotic interphase [ˌpri:mai'ɔtik]['intə(:)feiz]
★ 前期 prophase
★ 静止期细胞 quiescent cell
★ 限制点 restriction point
纺锤体装配检验点 spindle assembly checkpoint
纺锤体极体 spindle polar body
★ 标准细胞周期 standard cell cycle
监控机制 surveillance [sə:'veiləns] mechanism
联会复合体 synaptonemal complex(SC)
★ 端粒 telomere ['teləmiə]
★ 末期 telophase
★ 温度敏感突变体 temperature-sensitive mutant
★ 端化 terminalization
四分体 tetrad ['tetræd]

泛素(肽)　ubiquitin

偶线期　zygotene [ˈzaigəˌtiːn]/ zygonema

1 Cell division is the process in which a cell splits to form two new cells that are just like the original. The original cell often is called the parent cell. The two new cells produced by cell division are called daughter cells.
细胞分裂是一个细胞分裂成两个和原始细胞一样的新细胞的过程。原始细胞通常叫做母细胞，两个分裂出来的新细胞叫子细胞。

2 Every cell in human body came from another cell. Most cells can divide over and over again. But cells aren't necessarily dividing all the time. They go through a cell cycle.
人体的每个细胞都来自于其他的细胞。多数细胞能多次分裂。但是细胞并不是一直在分裂，他们需要经过一个细胞周期。

第十一节　细胞分化与基因表达调控
Section 11　Cell Differentiation and the Control of Gene Expression

癌　cancer
纤毛　cilia [ˈsiliə]
核仁内染色质　intranucleolar chromatin
★ 内含子　intron [intrɔn]
组成型剪接　constitutive splicing
去分化，脱分化　dedifferentiation
决定　determination [ditəːmiˈneiʃən]
决定子　determinant [diˈtəːminənt]
DNA 足迹技术　DNA foot printing
DNA 甲基化　DNA methylation
定向干细胞　directional stem cell
★ 胚胎干细胞　embryo stem cell
胚胎诱导　embryonic induction
增强子　enhancer
成纤维细胞生长因子　fibroblast [ˈfaibrəublɑːst] growth factor (FGF)
通用转录因子　general transcription factor
基因组印记　genomic imprinting
生殖质　germ plasm
★ 生长因子　growth factor
★ 生长因子受体　growth factor receptor
同源异型基因　homeotic selector gene
★ 管家基因　house-keeping gene
★ 恶性肿瘤　malignancy
★ 恶性转化　malignant transformation
隐蔽 mRNA　masked mRNA
转移灶　metastases
甲基转移酶　methyl transferase
单能干细胞　monopotential cell
多级调控系统　multistage regulation system
★ 癌基因　oncogene [ˈɔnkəˌdʒiːn]
多潜能性　pluripotency
★ 位置效应　position effect
前起始复合体　preinitiation complex
★ 加工水平的调控　processing-level control
★ 启动子　promoter
蛋白质异形体　protein isoform
原癌基因　proto-oncogene
假原质团　pseudoplasmodium [ˌ(p)sjuːdəplæzˈməudjəm]
★ 再生　regeneration
★ 调节基因　regulated gene
剪接增强子　splicing enhancer
★ 干细胞　stem cell
★ 对称序列　symmetrical sequence
★ 终末分化　terminal differentiation
组织特异性基因　tissue-specific gene
★ 全能性　totipotency [təuˈtipətənsi]
反式作用因子　trans acting factor
★ 转录因子　transcription factor
转录激活子　transcriptional activator
★ 转录水平的调控　transcriptional level control
★ 转录阻抑　transcriptional repression
★ 转分化　transdifferentiation
★ 翻译水平的调控　translational level control
★ 肿瘤细胞　tumor cell
★ 抑癌基因　suppressor gene
甲基化不足　under methylation
非翻译区　untranslated region(UTR)

1 The germ cell differentiated into the three primary germ lines that make up the body — endoderm, ectoderm and mesoderm — and subsequently into arrays of tissue cells such as cartilage, bone, muscle, neural and gut cells.

受精卵分化成构成人体的三个最初胚层——内胚层、外胚层和中胚层。然后分化成各种组织细胞像软骨、骨、肌肉、神经和内脏细胞。

2 All the information needed to trigger your cells to divide was in your genes.
所有引发细胞分裂的信息都储存在基因里。

第十二节 细胞衰老与凋亡
Section 12 Cell Senescence and Apoptosis

选择性剪接 alternate splicing
★ 衰老小体 apoptic [ˌæpəupˌtik] body
★ 细胞凋亡 apoptosis
★ 凋亡小体 apoptosis body
癌细胞 cancer cell
凋亡蛋白酶活化因子 apoptosis protease-activating factor(Apaf)
铁调蛋白 iron regulatory protein(IRP)
铁应答元件 iron-response element(IRE)
彗星电泳法 comet assay
黄色素 yellow pigment
★ 死亡效应子结构域 death effector domain (DED)
死亡诱导信号复合物 death-inducing signaling complex(DISC)
致密体 dense body
★ DNA 梯状条带 DNA ladders
染色体外 RDNA 环 extrachromosomal rDNA circle(ERC)
铁蛋白 ferritin [ˈferitən]
★ Hayflick 界限 Hayflick limitation
异性活化 hetero activation
同性活化 homo activation
★ 坏死 necrosis
★ 细胞编程性死亡 programmed cell death
活性氧基团或分子 reactive oxygen species (ROS)

1 Cancer cells can be distinguished from normal cells by their motility in culture.
癌细胞能根据游动性培养与正常细胞加以区别。

2 Apoptosis is both positively and negatively regulated. In addition to the numerous factors involved in the induction of apoptosis, many other factors have been identified that prevent cell death.
细胞凋亡既有正向的又有负向的调节作用。除了一些诱导细胞凋亡的因子以外，能阻止细胞死亡的其他因子也已经被确认。

第九章　分子生物学
Chapter 9　Molecular Biology

第一节　DNA 的结构
Section 1　Structure of DNA

★ 分子生物学　molecular biology
★ 一级结构　primary structure
★ 超螺旋 DNA　superhelical DNA [ˌsjuːpəˈhelikəl]
遗传信息　genetic message
共线性　co/ linearity [kəuˌlainˈæriti]
选择性表达　selective expression
★ 主链　main chain
小沟　minor groove
螺距　pitch
疏水键　hydrophobic [ˌhaidrəuˈfəubik] bond
★ 反向重复序列　inverted repeats
空间构象　spatial conformation [ˈspeiʃəl][ˌkɔnfɔːˈmeiʃən]
★ 复性　renaturation/ annealing
拓扑异构酶　topo isomerase
碱基配对　base pairing
左手螺旋　left-handed helix
右手螺旋　right-handed helix
DNA 酶 Ⅰ　DNA-ase Ⅰ
螺线管　solenoid [ˈsəulinɔid]
复制型　replication form(RF)
★ 脱氧核糖核苷酸　deoxyribonucleic [diːˈɔksiˌraibəuˈnjuːklek] acid(DNA)
自我复制　self-replication
滚环型　rolling circle
D-环型　D-loop
解链酶　helicase [ˈhelikeis]
引发体　primosome [ˈpriːməusəm]
引发前体　preprimosome [priːˈpriːmuˈsm]
★ 引发酶　primase [ˈpraimeiz]
转座(移位)　transposition
可移动因子　transposable [trænsˈpəuzəbl] element
转座酶　transposase
解离酶　resolvase [riˈsɔlveitz]
杂种不育　hybrid dysgenesis [ˈhaibrid][disˈdʒenisis]
细胞质因子　cytotype

1 The basic building block of DNA was known to be a nucleotide consisting of the five-carbon sugar deoxyribose to which one phosphate is esterified at the 5′ position of the sugar ring and one nitrogenous base is attached at the 1′ site. There are two nitrogenous bases present in a nucleic acid, pyrimidines, which contain a single ring, and purines, which contain two rings. DNA contains two different pyrimidines, thymine (T) and cytosine(C), and two different purines, guanine (G) and adenine(A). The nucleotides were known to be covalently linked to one another to form a linear polymer, or strand, with a backbone composed of alternating sugar and phosphate groups joined by 3′, 5′-phosphodiester bonds. The bases attached to each sugar were thought to project from the backbone like a column of stacked shelves.

DNA 的基本结构单元是核苷酸，核苷酸由五碳糖脱氧核糖和以酯键结合在其糖环 5′ 位置上的一个磷酸以及一个结合在其 1′ 位置上的含氮碱基组成。在核酸中有两种含氮碱基，含有单环的嘧啶和含有双环的嘌呤。DNA 含有两种不同的嘧啶，胸腺嘧啶(T) 和胞嘧啶(C) 以及两种不同的嘌呤，鸟嘌呤和腺嘌呤。核苷酸通过共价结合形成线性多聚体或链，其骨架由通过 3′,5′- 磷酸二酯键连接在一起的糖和磷酸的交替重复组成。结合在每一核糖上的碱基被认为像梯子的横杆一样从骨架上伸出。

2 A nucleotide has a polarized structure: one end, where the phosphate is located, is called the 5′ end (pronounced “five-prime end”), while the other end is the 3′ end. Because each of the stacked nucleotides in a strand faces the same direction, the entire strand has a direction. One end is the 3′ end, the other is the 5′ end. X-ray diffraction analysis indicated that the distance between the nucleotides of the stack was 3.4Å(0.34 nm) and suggested the presence of a large structural repeat every 3.4 nm.

核苷酸有一个极性结构：带有磷酸基团一端，称为 5′ 端(突出的"5-primer 末端")，另一端是 3′ 端。由于核苷酸链中的每个核苷酸面向相同的方向，因而整条链就有方向性。一端是 3′ 端，另一端是 5′ 端。X射线衍射分析指出，堆叠在一起的核苷酸之间的距离是 3.4Å，并且提出每隔 3.4 nm存在一个大的结构重复。

第二节 基因的组织与结构
Section 2 Organization and Structure of Gene

编码蛋白质 coding protein
调控单位 regulatory unit
重叠序列 overlap
蛋白质复合物 protein complex
调控蛋白 regulatory protein
编码序列 coding sequence
重复频度 repetitive frequency, repetition frequency
非重复序列 nonrepetitive sequence
同源性 homology [hɔ'mɔlədʒi]
浮力密度 buoyant ['bɔiənt] density/ buoyancy ['bɔiənsi] density
梯度离心 gradient centrifugation
★ 基因家族 gene family
重复结构单元 constitutional repeating unit
不连续基因 discontinuous [ˌdiskən'tinjuəs] gene
蛋白质结构域 protein structure domain
假基因 pseudogene ['sju:dəudʒi:n]

1 The two strands of a DNA molecule are held together by weak, noncovalent bonds. When DNA is dissolved in saline solution and the solution is slowly warmed. Within a few degrees, the process is generally complete and the solution contains single-stranded molecules that are completely separated from their original partners. The progress of thermal denaturation (or DNA melting) is usually monitored by following the increase in absorbance of the dissolved DNA. The nitrogenous bases of a nucleic acid molecule absorb ultraviolet radiation with an absorbance maximum near 260 nm. Once the two DNA strands have separated, the hydrophobic interactions that result from base stacking are greatly decreased, which changes the electronic nature of the bases and increases their absorbance of ultraviolet radiation. The temperature at which the shift in absorbance is half completed is termed the melting temperature (Tm). The higher the GC content (%G + %C) of the DNA, the higher the Tm. This reflects increased stability of GC-containing DNA compared with AT pairs.

DNA 分子的两条链由弱的非共价键结合在一起。当把 DNA 溶解在盐溶液中并被缓慢加热。到某一温度时两条链开始解离。通常在几摄氏度以内完成这一过程，溶液中含有与其互补链完全分离开的单链分子。热变性(或 DNA 熔解) 的进程一般通过追踪溶解的 DNA 的吸收增加进行监控。核酸分子中的含氮碱基吸收紫外辐射，最大吸收值在 260nm 附近。一旦 DNA 两条链分离开，由碱基堆积产生的疏水作用大大降低，从而改变碱基的带电性并增加它们对紫外辐射的吸收。当一半的 DNA 分子解离为单链时的温度称为解链温度(Tm)。DNA 中 GC 含量越高，解链温度越高，反映了与 AT 碱基对相比，富含 GC 碱基对的 DNA 稳定性增加。

2 An operon is a unit of prokaryotic gene expression which includes coordinately regulated (structural) genes and control elements which are recognized by regulatory gene products.

操纵子是原核基因表达单位。包括协同调节基因和能被调控基因产物识别的控制元件。

3 A large proportion of the DNA in eukaryotic cells does not code for protein; most genes have large intron sequences, and genes or gene clusters are separated by large stretches of sequence of unknown function. Much of this DNA consists of multiple repeats, sometimes thousands or hundreds of thousands of copies, of a few relatively short sequence elements.

真核生物细胞的大部分 DNA 不编码蛋白质。大部分基因含有大段的内含子序列，基因或基因簇被大段未知功能的序列所分割。这样的 DNA 由许多序列元件的重复序列组成，有时重复次数高达数百至上万个。

第三节 蛋白质与 DNA 的相互作用
Section 3 Reciprocity of Protein and DNA

非特异性结合 non-specific conjunction
特异性结合 specific conjunction
★ 纯化 purification/depuration [ˌpjuərifiˈkeiʃən][ˈdepjureiʃən]
★ 聚丙烯酰胺凝胶电泳 polyacrylamide gel electrophoresis [ˌpɔləˈkriləmaid] [dʒel] [iˌlektrəfəˈri:sis]
★ 提取 extraction/retrieval [iksˈtrækʃən][riˈtri:vəl]
★ 核酸酶 nuclease [ˈnju:kliˌeis]
修饰碱基 modifier base/modifying gene
基本结构 infrastructure [ˈinfrəˈstrʌktʃə]
结合蛋白 conjugated [ˈkɔndʒugeitid] protein
拉链结构 zipper structure
放射自显影 radioautography [ˌreidiəuɔ:ˈtɔgrəfi] / autoradiography [ˈɔ:təuˌreidiˈɔgrəfi]
★ 探针 probe [prəub]
★ 残基 residue [ˈrezidju:]
转移 RNA transfer RNA/transfer ribonucleic acid(tRNA)
★ 编码链 coding strand
反义链 anticoding strand
简并 degeneracy [diˈdʒenərəsi]
同义密码子 synonymous codon [siˈnɔniməs][ˈkəudən]
单顺反子 mRNA monocistronic [ˈmɔnɔsistrɔnik] mRNA
多顺反子 mRNA polycistronic [ˌpɔlisisˈtrɔnik] mRNA
★ 信使核糖核酸 messenger RNA (mRNA)
可译框架 open reading frame(ORF)
★ 前导肽，引肽 leader peptide
脂质小泡 vesicle [ˈvesikl]

1 There is an intermediate between a gene and its polypeptide; the intermediate is messenger RNA (mRNA).

在基因和多肽之间存在中间体，这种中间体就是信使 RNA。

2 A messenger RNA is assembled as complementary copy of one of the two DNA strands that make up a gene. The synthesis of an RNA from a DNA template is called transcription.

mRNA 是组成基因的 DNA 双链中某一条的互补拷贝。这种以 DNA 为模板合成 RNA 的过程称为转录。

3 Because the nucleotide sequence of messenger RNA is complementary to that of the gene from which it is transcribed, the mRNA retains the same information as the gene itself. The use of messenger RNA allows the cell to separate information storage from information utilization.

因为 mRNA 的核苷酸序列与被转基因的核苷酸序列互补，所以 mRNA 保留了与基因本身相同的信息。mRNA 使得细胞对信息储存和信息利用在空间上进行隔离。

4 While the gene remains stored away in the nucleus as part of a huge, immobile DNA molecule, its information can be imparted to a much smaller, mobile RNA that passes into the cytoplasm. Once in the cytoplasm, the mRNA can serve as a template to direct the incorporation of amino acids in a particular order encoded by the nucleotide sequence of the DNA and mRNA. The use of a messenger RNA also allows a cell to greatly amplify its synthetic output. One DNA molecule can serve as the template in the formation of many mRNA molecules, each of which can be used in the formation of a large number of polypeptide chains.

虽然基因保持储存在细胞核中巨大的、不动的 DNA 分子中，但它的信息能转变成许多小的，可以移动到细胞质中的 RNA 上。一旦进入细胞质之后，以 mRNA 作为模板，按照 DNA 和 mRNA 的核苷酸序列编码的特定顺序指导氨基酸的结合。mRNA 的作用也可以使细胞的合成量大大增强，即一个 DNA 分子可以作为模板合成许多 mRNA 分子，而每个 mRNA 分子又可以指导合成大量肽链。

第四节 DNA的复制
Section 4 Replication of DNA

基因调节 gene regulation
★ 引物酶 primase [praiˈmeiz]
RNA 聚合酶 ribonucleic acid polymerase
解螺旋酶 unwinding enzyme/untwisting enzyme/unwindase/helicase
旋转酶 gyrase [ˈdʒaiəreis]
★ 模板链 template strand
★ 冈歧片段 Okazaki fragment/Okazaki piece
DNA 损伤 deoxyribonucleic acid damage/deoxyribonucleic acid injury/deoxyribonucleic acid impairment
复制的方向 direction of replication
★ 双向复制 bidirectional replication
★ 模板 template/templet
★ 引物 primer
半不连续复制 semidiscontinuous replication
DNA 复制的调控 DNA replication regulatory
★ 染色体复制 chromosome replication/chromosome reduplication

1 DNA replication is accomplished by an enzymatic polymerization reaction in which a DNA strand is used as a template to synthesize a copy with a base sequence that is complementary to the template.
DNA 复制是通过酶催化的聚合反应完成的，反应过程中，DNA 链作为模板合成一个拷贝，这个拷贝的碱基序列与模板完全互补。

2 Replication is semiconservative; each parent single strand is present in one of the double-stranded progeny molecules. The growth of a DNA strand is catalyzed by a DNA polymerase enzyme. Strand growth always moves in the 5′-to-3′direction. Since double-stranded DNA is antiparallel, only one strand (the leading strand) grows in the direction of movement of the replication fork. The second daughter strand (the lagging strand) is synthesized in the opposite direction as short, precursor fragments that are subsequently joined together.
复制采取半保留的方式：每条亲代单链存在于其中一个双链子代分子中。DNA 链的伸长是在 DNA 聚合酶催化下进行。链的伸长总是沿 5′→3′ 方向移动。因为双链 DNA 是反向平行的，所以只有一条链（前导链）沿复制叉移动的方向生长。另一条子链（后随链）沿相反的方向合成短的前体片段，最后这些片段连接在一起。

3 DNA polymerase cannot initiate synthesis, so a primer is always needed. In most replication systems the primer is a short RNA oligonucleotide synthesized either by RNA polymerase or by DNA primase.
DNA 聚合酶不能起始合成，所以需要引物。在大多数复制系统中，引物是短的 RNA 寡聚核苷酸，由 RNA 聚合酶或 DNA 引物酶合成。

4 In E. coli two enzymes, polymerase I and polymerase III, carry out nucleotide addition. Pol I is used in the completion of precursor fragments and also in repair processes; pol III is responsible for overall chain growth. The primer is removed at later stages of replication by the 5′-3′ exonuclease activity of DNA polymerase I. Pol I and pol III both possess a 3′- 5′ exonuclease activity that corrects incorporation errors.
在 E. coli 中有两种酶，DNA 聚合酶 I 和 DNA 聚合酶 III，促使核苷酸的添加，聚合酶 I 用于前体片段的完成，也用于修复过程；聚合酶 III 负责总体链的伸长。在复制后期，通过 DNA 聚合酶 I 的 5′→3′ 核酸外切酶活性去掉引物。聚合酶 I 和聚合酶 III 都具有 3′→5′ 核酸外切酶活性，可以纠正错误的掺入。

5 Replication of both linear and circular DNA molecules is usually bidirectional; the replication loops enlarge by movement of replication forks at both ends of the loop. A DNA molecule of a prokaryote is usually initiated at a single site, and has a single replication loop; in contrast, eukaryotic DNA molecules have multiple initiation sites, and hence many loops.

线性和环状DNA分子的复制通常都是双向的；通过复制叉的移动在环的两个末端使其扩大。原核细胞的DNA分子通常在单一部位起始，具有单一的复制环；相反的，真核DNA分子具有多个起始部位，因而具有许多环。

第五节　基因的转录和转录后加工
Section 5　Transcription and Post-transcriptional Processing of Gene

★ 转录　transcription [træns'kripʃən]
★ 中心法则　central dogma
★ 基因转录　genetic transcription
编码区　coding region
双链超螺旋形式　double-stranded supercoiling
基因活性　gene activity
★ 起始　initiation [i,niʃi'eiʃən]
★ 延伸　extension
底 物　zymolyte ['zaiməlait] / substratum ['sʌb'strɑːtəm] / substrate
★ 分子量　molecular weight
★ 阻遏蛋白　repressor protein
起始复合物　initiation complex
调控区域　regulatory domain
相互作用　interaction [,intər'ækʃən]
元件　component [kəm'pəunənt]
增强子　enhancer/enhanser
加工　processing [prəu'sesiŋ]
识别　recognition/recognize
剪接　splicing ['splaisiŋ]
编辑　editing ['editiŋ]
选择性加工　selective processing

1 RNA polymerase recognizes the terminator which causes no further ribonucleotides to be incorporated. This sequence is commonly a hairpin structure. Some terminators require an accessory factor called rho for termination RNA.

RNA聚合酶通过识别终止子来结束核糖核酸的合成。这个序列一般是发夹结构。一些终止子需要一个叫做rho的附件因子终止RNA合成。

2 The central dogma is the original proposal that DNA makes RNA makes protein, which had happened *via* the processes of transcription and translation respectively. Retroviruses reverse transcribe RNA into DNA, a number of viruses are able to replicate RNA directly into an RNA copy, and a number of organisms can edit a messenger RNA sequence so that the protein coding sequence is not directly specified by DNA sequence.

中心法则是"DNA生成RNA生成蛋白质"，分别通过转录与翻译两个过程进行的。反转录病毒能将RNA反转录为DNA，很多病毒能直接以RNA为模板复制RNA，还有很多生物能编辑信使RNA序列，所以编码蛋白质的序列不仅仅直接由特异的DNA序列决定。

3 Transcription is the synthesis of a single-stranded RNA from a double-stranded DNA template. RNA synthesis occurs in the $5' \rightarrow 3'$ direction and its sequence corresponds to that of the DNA strand which is known as the sense strand.

转录是以一条双链DNA片段为模板合成一条单链RNA。RNA合成是沿着$5' \rightarrow 3'$方向进行的，其序列与DNA有义链相一致。

4 RNA polymerase is the enzyme responsible for transcription. It binds to specific DNA sequences called promoters to initiate RNA synthesis. These sequences are upstream (to the 5 -end) of the region that codes for protein, and they contain short, conserved DNA sequences which are common to different promoters. The RNA polymerase binds to the dsDNA at a promoter sequence, resulting in local DNA unwinding. The position of the first synthesized base of the RNA is called the start site.

RNA聚合酶是负责转录的酶，它结合到称之为启动子的特定DNA序列上起始RNA的合成。这些序列位于蛋白质编码区的上游(5′末端)，并含有一些短的保守序列，在不同启动子中这些序列是相同的。RNA聚合酶结合到双链DNA的启动子序列上，引起DNA结合部位解旋。合成的第一个RNA碱基的位置称为起始位点。

第六节 蛋白质的生物合成
Section 6 Biosynthesis of Protein

共同序列 consensuce sequence [kən'sensəs]['siːkwəns]
帽子结构 cap structure
★ 开放阅读框架 open reading frame
★ 发夹结构 hairpin ['hɛəpin] structure
氨基酸活化 amino acid activation
★ 起始密码子 start codon/initiation codon
终止密码子 terminator codon/termination codon/stop codon
核糖体 ribosome ['raibəsəum]
可溶性蛋白质因子 soluble protein factor
★ 肽链 peptide chain
肽键 peptide bond/peptide linkage
转位 transposition [ˌtrænspə'ziʃən]
结构性调控 structural regulatory ['strʌktʃərəl]['regjulətəri]
翻译区域 translation region
非翻译区域 non-translated region/non-translational region/untranslated region
质膜 plasma membrane/plasmalemma [ˌplæzmə'lemə]/plasmolemma
跨膜运输 transmembrane transport
多顺反子 polycistron [ˌpɔli'sistrɔn]/ multicistron
反密码子 anticodon ['ænti'kəudɔn]/ nodoc

1 Three eukaryotic polymerases transcribe different sets of genes. Their activities are distinguished by their different sensitivities to the fungal toxin α-amanitin.

- RNA polymerase Ⅰ is located in the nucleoli. It is responsible for the synthesis of the precursors of most rRNAs.
- RNA polymerase Ⅱ is located in the nucleoplasm and is responsible for the synthesis of mRNA precursors and some small nuclear RNAs.
- RNA polymerase Ⅲ is located in the nucleoplasm. It is responsible for the synthesis of the precursors of 5SrRNA, tRNAs and other small nuclear and cytosolic RNAs.

三种真核聚合酶转录不同种的基因。它们的活性依据它们对于真菌毒素 α-鹅膏蕈碱的敏感性来区分。

- RNA 聚合酶 Ⅰ 位于核仁内，负责大多数 rRNA 的前体合成。
- RNA 聚合酶 Ⅱ 位于核质内，主要负责 mRNA 前体和一些小的核 RNA 的合成。
- RNA 聚合酶 Ⅲ 位于核质内，负责 5SrRNA，tRNA 和其他的小的核 RNA 及胞质 RNA 前体的合成。

2 Ribosomal RNA genes

The pre-RNA transcription units contain three sequences that encode the 18S, 5.8S, and 28S rRNAs.

核糖体 RNA 基因

前 RNA 转录单元包括三种编码 18S、5.8S、和 28S rRNA 的序列。

3 Proteins are synthesized in the cytoplasm by a complex process called translation. Translation requires the participation of dozens of different components, including ribosomes.

在细胞质中蛋白质是通过复杂的翻译过程合成的，这个过程要包括核糖体在内的许多成分的参与。

4 Ribosomes are nonspecific components of the translation machinery. These complex, cytoplasmic "machines" can be programmed, like a computer, to translate the information encoded by any mRNAs.

核糖体是非特异性的翻译机器。这些复杂细胞质内的机器，可以像计算机一样，翻译任何 mRNA 编码的信息。

5 A functioning ribosome consists of a large and a small subunit. The ribosome is assembled from these subunits at the time it begins to synthesize a particular polypeptide and then is disassembled when synthesis has been completed. Ribosomes consist of RNA and protein.

核糖体由大亚基和小亚基组成，在开始合成特定多肽时，亚基组装成核糖体，合成结束时，核糖体又解离成亚基。核糖体由 RNA 和蛋白质组成。

6 The RNAs of a ribosome are called ribosomal RNAs (or rRNAs), and like mRNAs, each is transcribed from one of the DNA strands of a gene. Rather than functioning in an informational capacity, rRNAs are adapted to recognize and bind other molecules, to provide structural support, and to catalyze the chemical reaction in which amino acid are covalently linked to one another.

组成核糖体的 RNA 称为核糖体 RNA 或 rRNAs,像 mRNA 一样,rRNA 也由基因的某一条 DNA 链转录而成。rRNA 不是信息载体而适于作为识别和结合其他分子,支持核糖体的结构成分起作用,并且催化氨基酸通过共价键结合在一起。

第七节 原核生物基因表达调控
Section 7 Regulation of Procaryotic Gene Expression

诱导反应 induced reaction
阻遏作用 repression/block
正调节 positive regulation/positive control/up regulation
负调控 negative regulation/down regulation/negative control
结构基因 structural gene/constitutive gene
控制模式 control pattern
★操纵基因 operator gene/operator
乳糖操纵子 lactose operon/lac operon
转录调控 transcriptional control
乳糖代谢 metabolism of lactose
调控模型 regulatory model
顺式作用元件 cis-acting element
活化蛋白 activating protein
衰减子 attenuator [ə'tenjueitə(r)]
前导序列 leader sequence/targeting sequencing
序列分析 sequence analysis
诱导物 inducer [in'dju:sə]
特异性元件 specific element
正调控模型 positive regulatory model
负调控模型 negative regulatory model
永久型 constitutive ['kɔnstitju:tiv]
适应型 adaptive
调节型 regulated
★基因表达 gene expression
基因表达调控 gene regulation/gene control
mRNA 加工成熟水平上的调控 differential processing of RNA transcription
★转录产物 primary transcript
转录单元 transcription unit
★终止子 terminator ['tə:mineitə]
起点 start point
上游 upstream
下游 downstream
★增强子(强化子) enhancer
★启动子 promoter [prə'məutə]
★操纵区 operator
代谢物阻遏效应 catabolite repression
色氨酸操纵子 tryptophane ['triptəfæn] operon
★半乳糖操纵子 galactose [gə'læktəus] operon
异构酶 UDP-galactose-4-epimerase(gale) [gə'læktəus]['epiməreis]
半乳糖激酶 galactose kinase(galk)
自我调节 autogenous [ɔ:'tɔdʒinəs] regulation
自控活性 autoregulating activity
结合子诱导 zygotic induction

1 Transcription of a single gene or operon starts at the promoter, ends at the terminator and produces a monocistronic or polycistronic messenger RNA. The coding regions of the message are translated by the ribosome from the start codon (close to the ribosome binding site) to the stop codon. Transfer RNAs deliver the appropriate amino acid, according to the genetic code, to the growing protein chain.

单个基因或操纵子的转录从启动子开始,结束于终止子,产生单顺反子或多顺反子的信使 RNA。核糖体从起始密码子(靠近核糖体结合位点)到终止密码子的信使 RNA 的编码区被核糖体翻译。转运 RNA 根据遗传密码把合适的氨基酸递送到延长的蛋白链上。

2 The synthesis of particular gene products is controlled by mechanisms collectively called gene regulation. Gene regulation generally occurs at the level of transcription or translation.

特定基因产物的合成由基因调节机制来控制,基因调节通常发生在转录水平或者翻译水平。

3 Transcriptional regulation can be characterized as positive or negative. In negative regulation an inhibitor (repressor) bound to the DNA must be removed before transcription can occur. In positive regulation an activator molecule must bind to the DNA to allow transcription. The E. coli lactose system was first explained by the operon model, and is an example of both negative and positive regulation. In this system, lactose is an inducer that alters the shape of a repressor (lacI) so the repressor can no longer bind to its operator. Transcription of the lactose operon, however, also requires the presence of the cAMP-CRP activator complex.

转录调节可以是正调节或者负调节。在负调节中，在转录发生前必须先去除结合于DNA上的抑制剂(阻抑物)。在正调节中，激活剂分子一定要结合DNA才能转录。*E. coli* 的乳糖系统是第一个用操纵子模型解释的系统，也是负调节和正调节的一个例子。在这个系统中，乳糖是一种诱导物，它改变阻抑物(lacI)的形状，于是阻抑物不再结合到它的操纵基因上。然而，乳糖操纵子的转录还需要有cAMP-CRP激活复合物的参与。

第八节 真核生物基因表达
Section 8 Eucaryotic Gene Expression

结构蛋白 structural protein/constitutive ['kɔnstitju:tiv] protein
常染色质 euchromatin [ju:'krəumətin]
蛋白质修饰 protein modification
遗传操作 genetic manipulation
★靶序列 target sequence
衰减作用 deamplification/extenuation deamplification
持家基因 housekeeping gene
时序基因 temporal ['tempərəl] gene
激活 activation
顺式作用 cis-acting
真核基因转录调控区 transcriptional control region of eukaryotic gene
启动子上游调控元件 upstream regulatory element of promoter
转录调控区 transcriptional control region
反式调控因子 trans-regulatory factor
激活结构域 structural domain of activation
糖基化 glycosylation [ˌglaikəsi'leiʃən]
★磷酸化 phosphorylation [fɔˌsfɔri'leiʃən]
转录活化结构域 structural domain of transcription activation
转录后水平调控 post transcriptional regulation
翻译水平调控 translational regulation
蛋白质加工水平的调控 protein processing regulation
成熟和剪接 maturation and splicing
高度保守序列 consensus sequence [kən'sensəs]['si:kwəns]
卵母细胞 oocyte ['əuəsait]
营养细胞 nurse cell
营养遗传型 genotrophic
★传感基因 sensor gene
★受体基因 receptor gene
超敏感位点 hypersensitive [ˌhaipə(:)'sensitiv] site
★终止区 terminator ['tə:mineitə]
印迹法 Northern blotting
★沉默子 silencer(S)
应答元件 response element
热休克蛋白 heat shock protein
热激因子 heat shock factor(HSF)
胞质受体 cytosol receptor
磷脂酰肌醇 phosphatidyl inositol [ˌfɔsfə'taidil][i'nəusitəul]
胰高血糖素 glucagons ['glu:kəgɔns]
胰泌素 secretin [si'kri:tin]
外周蛋白(表蛋白) extrinsic protein
内在蛋白(整合蛋白) intrinsic protein/integral protein
组成型剪接 constitutive splicing
帽子结合蛋白 cap binding protein(CBP)

1 In most cases monocistronic messenger RNAs are transcribed from a gene, initiated at a promoter. The resulting pre-messenger RNA is capped at the 5′-end and has a poly(A) tail added to the 3′-end. Intron are removed by splicing before the mature mRNA is exported from the nucleus to be translated by ribosomes in the cytoplasm.

大多数情况下转录起始于启动子，单顺反子信使 RNA 从一个基因转录而来。产生出的前信使 RNA 在 5′ 端有帽子结构，在 3′ 端有多聚 A 尾巴。在成熟 mRNA 从核中释放到细胞质中被核糖体翻译之前，其中的内含子通过剪接被去除。

2 Disassembling chromatin to copy the DNA slows the movement of eukaryotic replication forks and gives rise to the short size of eukaryotic lagging strand fragments. Leading and lagging strands are initiated by the primase and polymerase activities of DNA polymerase α. DNA polymerase δ then takes over elongation of both strands while DNA polymerase ε may complete the lagging strand fragments.

染色质解组装以复制 DNA 减慢真核生物复制叉的移动，并且在后随链上产生小片段。引发酶及 DNA 聚合酶 α 的聚合酶活性可起始前导链与后随链的复制，然后 DNA 聚合酶 δ 负责前导链和后随链的延伸，而 DNA 聚合酶 ε 完成后随链的复制。

3 Regulation of genetic systems in eukaryotes is accomplished in a variety of ways, most of which are quite different from the mechanisms observed in prokaryotes. In eukaryotes, three different RNA polymerases fulfill transcriptional tasks which in prokaryotes are carried out by a single RNA polymerase. It is at this level of transcription where decisions are often made with regard to gene expression.

真核细胞中基因系统的调节可以通过各种途径实现，其中大多数途径与原核细胞中发现的机制大不相同。在真核细胞中，3 种不同的 RNA 聚合酶完成转录任务，而在原核细胞中只有一种 RNA 聚合酶。通常在这一转录水平决定了有关基因的表达。

4 Sequences upstream of eukaryotic genes known as promoters are required for initiation of transcription. Activation of transcription is carried out by binding of transcription factors to these sites. Additional regulatory sequences, known as enhancers, are often associated with genes. They also stimulate transcription, but in contrast to promoters, they often are located several kilobases away from the coding region, upstream or downstream. Once mRNA is transcribed, further regulation can occur during splicing events that remove eukaryotic introns and create mRNA sequences that can be translated into functional proteins.

转录的起始需要启动子，它是真核基因的上游序列。转录的激活由转录因子结合这些序列所开始。其他调节序列如增强子，常与基因有联系。它们也能够刺激转录，但是与启动子相比，它们通常位于距离编码区的上游或下游数千个碱基的位置。一旦转录了 mRNA，剪接反应去除真核内含子，产生能够翻译功能蛋白的 mRNA 序列，在这些反应中能够进行进一步的调节。

第十章 生化分离工程
Chapter 10 Biochemical Separation Engineering

第一节 绪论
Section 1 Introduction

★ 生物技术 biotechnology
★ 下游加工过程 downstream processing ['daunstri:m][prəu'sesiŋ]
回收率 recovery [ri'kʌvəri]
★ 纯化因子 purification factor
选择性 selectivity [silek'tiviti]
★ 终产品 final product
过程评价 process evaluation [prəu'ses][i,vælju'eiʃən]
★ 生物加工过程 bioprocess [,baiəu'prəuses]
浓缩率 concentration factor [,kɔnsen'treiʃən]['fæktə]
★ 分离因子 separation factor [sepə'reiʃən]['fæktə]

1 Bioseparation processes are defined as the manufacturing steps undertaken to purify products derived from animals, tissues, cells, or parts of cells.
生物分离过程是指用于纯化动物、组织、细胞或细胞成分产品所经过的各生产步骤。

2 Bioseparation processes encompass processes that yield such divergent final products as dairy products and monoclonal antibodies, or citric acid and polio vaccines.
生物分离过程包括生产不同的终产品如奶制品与单克隆抗体，或柠檬酸与脊髓灰质炎疫苗等过程。

3 The processes that are involved in making these seemingly disparate products can be treated as a single group because the end products, many of which are labile, are created in dilute aqueous solutions that can contain myriad other components. Because of these shared characteristics, a framework can be developed to study biological product purification in a systematic way — hence the term bioseparation process science.
生产这些似乎是不同产品的各过程可以按单一的一类来处理，因为终产品(很多都是不稳定的)，是从含有大量其他组分的稀水溶液中生产出来的。正是因为这些共同特征，可以构建这样一个平台以系统研究生物产品的纯化 —— 这才是生物分离过程科学这一术语的含义。

第二节 发酵液的预处理和固液分离
Section 2 Pretreatment of Fermentation Fluid and Solid-liquid Separation

★ 凝聚 coagulation
★ 絮凝 flocculation [flɔkju'leiʃən]
★ 离心 centrifugation [sen,trifju'geiʃən]
★ 沉降系数 sedimentation coefficient
★ 差速离心 differential centrifugation
★ 区带离心 zonal ['zəunl] centrifugation
★ 超离心机 ultracentrifuge [,ʌltrə'sentrifju:dʒ]
管式离心机 tubular bowl centrifuge
★ 碟片式离心机 disc-type bowl centrifuge
助滤剂 filter aid
★ 细胞破碎 cell disruption
★ 化学渗透 chemical permeation
★ 高压匀浆 high-pressure homogenization [,həumədʒənai'zeiʃən]
珠磨机 bead mill
★ 超声波破碎法 ultrasonication [,ʌltrə,sɔni'keiʃən]
酶溶法 enzymatic lysis
★ 渗透压冲击法 osmotic shock
包涵体 inclusion body

1 When a cell or cell population is the desired product, the separation of solids and liquids may be the entire goal of the bioseparation processing steps.
当细胞或细胞群既为目标产品时，固液分离则是生物分离加工过程各步骤的全部目的。

2 On the one hand, recovering the product without prior cell removal is an attractive option because it reduces both capital costs and the number of processing steps needed. On the other hand, solid/liquid separation is often decoupled from product recovery in commercial processes.

一方面，不预先除去细胞既能回收产品是一种吸引人的方法，因为既减少了资金投入，又减少了加工步骤。另一方面，商业上固液分离常与产品回收相脱节。

3 If the product is excreted from cells that are cultured in a bioreactor during the manufacturing process, then removing the cells in the first bioseparation step may prove helpful.

如果产品是在生产过程中从生物反应器中所培养的细胞中分泌出来的，则在生物分离的第一步除去细胞会是很有益的。

4 Cell removal minimizes enzymatic product degradation and eliminates many of the components that can interfere with later purification steps and introduce occupational and product safety hazards.

去除细胞会减少酶对产物的降解，并消除可干扰后续纯化步骤及产生职业危害与产品安全危险的多种成分。

第三节 萃取
Section 3 Extraction

★ 萃取剂 extractant [iks'træktənt]
★ 反萃取 back extraction
稀释剂 diluent ['diljuənt]
★ 分配系数 distribution coefficient
★ 澄清器 settler ['setlə]
重相 heavy phase
轻相 light phase
萃取因子 extraction factor
★ 超临界流体 supercritical fluid
★ 超临界二氧化碳 supercritical carbon dioxide
★ 临界点 critical point
助表面活性剂 cosurfactant
★ 夹带剂 entrainer [in'treinə]
改性剂 modifier
★ 共溶剂 cosolvent [kəu'sɔlvənt]
★ 相图 phase diagram
不相溶性 incompatibility ['inkəm,pætə'biliti]
聚乙二醇 polyethylene glycol [,pɔli'eθili:n]['glaikɔl]
★ 葡聚糖 dextran ['dekstrən]
疏水性因子 hydrophobic factor
交错分配法 cross partitioning
★ 双水相萃取 two-phase aqueous extraction
★ 反胶束 reversed micelle
★ 临界胶束浓度 critical micelle concentration
★ 表面活性剂 surfactant [sə:'fæktənt]
水化半径 hydration radius [hai'dreiʃən]['reidjəs]

1 Organic solvents that form a second phase when added to water have been used to extract low-molecular-weight biochemicals, such as carboxylic acids, amino acids, alcohols, and antibiotics.

加入到水相可产生第二相的有机溶剂已被用于萃取低分子量的生化物质，如羧酸、氨基酸、醇类及抗生素。

2 In general, organic solvents denature proteins, nucleic acids, and cells. As a consequence, they are rarely used for purifying these biomolecules.

有机溶剂常可使蛋白质、核酸及细胞变性。结果，其很少用于纯化这些生物分子。

3 Water plays a complex role in the organic phase. Solutes in the organic phase that form complexes with water can improve the distribution coefficient and control solubility to manage crystallization from the predominantly organic phase.

水在有机相中起多种作用。有机相中的溶质与水形成复合体后可改变分配系数并控制溶解性以利于从有机相中进行结晶。

4 The old chemistry adage that "like dissolves like" applies to organic solvent extraction.

古老的化学格言"相似者相溶"应用于有机溶剂萃取。

5 Above the critical point, the fluid is considered supercritical.

临界点以上的流体被认为是超临界流体。

6 As researchers have found, the fluid density (i. e., the number of molecules per unit volume) can vary near the critical point.

因为研究者已发现，流体密度（即每单位体积的分子数）在临界点附近变化较明显。

7 Supercritical fluid extraction(SFE) with carbon dioxide, has found application in the recovery of relatively non-polar, non-ionic compounds having therapeutic value, such as alkaloids.
二氧化碳超临界流体萃取已应用于具有治疗价值的相对非极性、非离子化合物（如生物碱）的回收。

8 The addition of one or two water-soluble polymers along a mineral salt can cause two distinct phases to form.
一种或两种水溶性聚合物与一种矿物盐加到一起可生成不同的两相。

9 If the system includes a higher concentration of ionic buffer salts, we can modify the protein partition coefficient by changing the pH, ionic strength, or the type of electrolyte in the solution.
如果系统含有一种高浓度离子缓冲盐，我们可通过改变溶液 pH 值、离子强度或电解质类型而调整蛋白质分配系数。

10 Reverse-micelle formation is promoted by several factors. For example surfactant molecules tend to stay at the organic solvent-water interface, as they have both a nonpolar and a hydrophilic section. In addition, water molecules can take advantage of the greater number of opportunities to form hydrogen bonds.
有几种因素可促进反胶团生成。例如，表面活性剂分子倾向于处在有机溶剂 —— 水界面上，因为其既有非极性又有亲水性部分。另外，水分子可利用大量机会形成氢键。

12 The two most important effects that have been studied are pH and salt type and concentration. Reverse-micelles will solubilize and extract proteins that have a net charge opposite to that of the polar group of the surfactant. This property can be exploited to separate proteins based on their pI values.
已研究的两个最重要的影响因素是 pH 值与盐的类型及浓度。反胶团将溶解并萃取带有与表面活性剂极性基团具有相反净电荷的蛋白质。这一种特性可用于根据其等电点分离蛋白质。

第四节 膜分离

Section 4 Membrane Separation

★ 微滤 microfiltration
★ 超滤 ultrafiltration [ˌʌltrəfil'treiʃən]
★ 反渗透 reverse osmosis
★ 透析 dialysis [dai'ælisis]
★ 电渗析 electrodialysis [ɪˌlektrəudaɪ'ælɪsɪs]
渗透汽化 pervaporation [pəˌveipə'reiʃən]
对称膜 symmetric membrane
★ 不对称膜 asymmetric membrane
动态膜 dynamic membrane
核孔微滤膜 nuclepore membrane
泡点法 bubble point method
★ 膜组件 membrane module
★ 浓差极化 concentration polarization [ˌkɔnsen'treiʃən][ˌpəulərаɪ'zeiʃən]
★ 凝胶极化 gel polarization
截留率 rejection coefficient
体积浓度 bulk concentration
★ 截留分子量 MWCO, molecular weight cut-off
终端过滤 dead-end filtration
★ 错流过滤 cross-flow filtration
连续搅拌釜式反应器 continuous stirred tank reactor
洗滤 diafiltration
★ 膜污染 membrane fouling ['fauliŋ]
反洗 back flushing
热原 pyrogen ['paiərəudʒin]
★ 膜生物反应器 membrane bioreactor

1 The separation of molecules through the use of a barrier is commonly known as membrane filtration. The composition of this barrier which can be either an insoluble liquid phase or a solid phase, is a critical factor in membrane separations. Commonly used membranes consist of cellulose, Teflon, or nylon polymers that have been formed into sheets or tubes.
通过使用一种屏障而进行的分子分离常称为膜分离。该屏障（可以是不溶的液相或固相）的组成在膜分离中是一个关键因素。常用的膜是由已用于生产纸张、试管的纤维素、聚四氟乙烯或尼龙聚合物制成。

2 Three fundamental driving forces are commonly used for membrane separations: pressure, concentration, and electrical potential.
常用于膜分离的有三种基本驱动力:压力、浓度和电位。

3 Pressure-driven membrane processes can be classified into microfiltration(MF), ultrafiltration(UF), and reverse osmosis(RO) separations. MF membranes reject particulates of different sizes depending on the membrane pore size which can range between 0.02 and 10 μm. UF membranes reject macromolecules with molecular-weight values ranging between 1000 and 1 000 000. The manufacturer typically specifies the molecular weight cut-off(MWCO) of an UF membrane. RO membrane prevent the passage of low molecular-weight solutes such as salts, and their performance is evaluated in terms of the percentage of salt rejected.
压力驱动的膜分离过程可分为微滤(MF)、超滤(UF)和反渗透(RO). 微滤膜根据膜孔径大小(0.02 ~10 μm)截留不同大小粒子。超滤膜截留分子量为 1000 ~ 1000000 的大分子。生产厂家会有代表性地详细说明一种超滤膜的截断分子量(MWCO)。反渗透膜可阻止低分子量溶质(如盐)的通过,并且其性能是根据盐的截留百分率来评价的。

第五节 电泳
Section 5 Electrophoresis

★ 凝胶电泳 gel electrophoresis
自由流电泳 free-flow electrophoresis
★ 迁移率 mobility [məu'biliti]
重叠 overlapping
★ 等电点聚焦 isoelectric focusing [ˌaisəui'lektrik]['fəukəsiŋ]
二维电泳 two-dimentional electrophoresis
★ 毛细管电泳 capillary electrophoresis [kə'piləri][iˌlektrəfə'ri:sɪs]
★ 电渗 electroosmosis [iˌlektrəuɔz'məusis]
★ 聚丙烯酰胺凝胶电泳 polyacrylamide [ˌpɔliə'krɪləmaɪd] gel electrophoresis(PAGE)
★ 十二烷基磺酸钠 sodium dodecyl ['dəudəsil] sulfate(SDS)
(热的)对流 convection [kən'vekʃən]
交联度 degree of cross-linking
巯基乙醇 mercaptoethanol [məˌkæptəu'eθənɔl]
荷质比 charge-to-mass ratio
等速电泳 isotachophoresis [ˌaisəˌtækəfə'ri:sis]

1 Electrophoretic methods are similar to chromatographic methods in that fixed barrier phases (ranging from paper to polymer gels) are employed to facilitate separation.
电泳与色谱方法相似,都是使用固定的屏障相(从滤纸到聚合物凝胶)以利于分离。

2 Gel electrophoresis is known as a zonal electrophoresis method. The apparatus separates the sample into well-defined zones. The applied sample migrates through a polymer matrix — usually composed of polyacrylamide, starch, agar, or agarose.
凝胶电泳是一种区带电泳方法,因为电泳装置可将样品分离为明确的区带。所加样品通过一种聚合物基质(通常由聚丙烯酰胺、淀粉、琼脂或琼脂糖组成)而迁移。

3 By far, the most popular laboratory-scale gel electrophoresis method is polyacrylamide gel electrophoresis (PAGE). Acrylamide can be polymerized in place so as to vary the gel pore size. Because polyacrylamide does not carry a charge and produces very small pores (less than 80 nm), this technique minimizes unwanted effects, such as electro-osmosis and Joule heating-related convection. Moreover, polyacrylamide pore size can be controlled by varying the monomer concentration or the degree of cross-linking.
目前,最受欢迎的实验室规模的凝胶电泳方法是聚丙烯酰胺凝胶电泳(PAGE)。丙烯酰胺可适度聚合以形成不同大小的凝胶孔径。因为聚丙烯酰胺不带电荷且可产生很小的孔径(小于 80 nm),这使得不希望产生的效应(如电渗与焦耳热对流)减到最小。而且,聚丙烯酰胺孔径大小可通过改变单体浓度或交联度而加以控制。

4 Capillary electrophoresis (CE) does not require the use of a supporting medium(gel or paper) and can take place in free aqueous solution. The most commonly used capillary material is fused silica, although glass and Teflon have been used for this purpose in the past.
毛细管电泳不需要使用支持介质(凝胶或滤纸)并且可在游离的水溶液中进行。最常用的毛细管材料为熔合硅,尽管过去使用过玻璃及聚四氟乙烯。

第六节 吸附与离子交换
Section 6 Adsorption and Ion Exchange

★ 吸附 adsorption
★ 吸附剂 adsorbent
脱附 desorption
★ 离子交换剂 ion exchanger
可离子化基团 ionizable group
★ 阳离子交换剂 cation exchanger
★ 阴离子交换剂 anion exchanger
★ 羧甲基 carboxymethyl(CM) [kɑːˈbɔksiˈmeθil]
★ 二乙胺乙基 diethyl [daiˈeθil] aminoethyl (DEAE)
★ 吸附等温线 adsorption isotherm [ˈaisəuθəːm]
反离子 counterion [ˈkauntəaiən]
流化床 fluidized [ˈfluːidaizd] bed
移动床 moving bed
搅拌釜吸附 stirred tank adsorption

1 In adsorption, molecules in a fluid phase concentrate on a solid surface without any chemical change. This phenomenon occurs because physical surface forces from the solid phase attract and hold certain molecules — namely the substrate — from the fluid surrounding the surface. Silica, alumina, activated carbon, and macroporous non-ionic polymers frequently serve as adsorbents. Low-molecular-weight organic solvents are the most common eluents.
在吸附作用中,流动相中的分子浓集在固相表面而不发生任何化学变化。该现象能发生,是因为固相物理表面力可从其表面周围的流动相吸引并保持住某些分子 —— 即底物。二氧化硅、氧化铝、活性炭与大孔非离子聚合物常常作为吸附剂。低分子有机溶剂是最常用的洗脱剂。

2 In ion-exchange chromatography, true heteropolar chemical bonds are reversibly formed between ionic components in the mobile and stationary phases. The ion-exchange beds may have either hydrophobic or hydrophilic matrices that contain ionizable groups.
在离子交换色谱中,真正的有极化学键在流动相与固定相的离子成分之间可逆形成。离子交换床可以是含有可离子化基团的疏水性或亲水性基质。

3 Because proteins are biological polymers of zwitterion amino acids, some amino acid residues can have a net positive, net negative, or neutral charge depending on the pH.
因为蛋白质是兼性离子氨基酸的生物聚合物,根据 pH 的不同,一些氨基酸残基可带净正电荷、净负电荷或中性电荷。

第七节 色谱
Section 7 Chromatography

★ 色谱,层析 chromatography [ˌkrəuməˈtɔgrəfi]
★ 固定相 stationary [ˈsteiʃ(ə)nəri] phase
★ 流动相 mobile phase
★ 洗脱展开 elution development
迎头分析 frontal analysis
★ 顶替展开 displacement development
顶替剂 displacer
★ 洗脱时间 elution time
★ 保留时间 retention time
洗脱体积 elution volume
★ 理论板当量高度 height equivalent to a theoretical plate
★ 分离度,分辨率 resolution [ˌrezəˈljuːʃən]
容量因子 capacity factor
★ 吸附色谱 adsorption chromatography
★ 活性炭 activated charcoal [ˈæktiveitid][ˈtʃɑːkəul]
金属螯合作用色谱 metal chelation [kiːˈleiʃən] chromatography
★ 共价作用色谱 covalent interaction chromatog-

raphy
★ 分配色谱 distribution chromatography
★ 正相色谱 normal phase chromatography
★ 羟基磷灰石 hydroxyapatite [hai.drɔksi'æpətait]
★ 色谱柱 chromatographic column
★ 洗脱液 eluant ['eljuənt]
洗脱物 eluate ['eljuit]
★ 硅胶 silica gel ['silikə][dʒel]
★ 氧化铝 aluminum oxide [ə'lju:minəm]['ɔksɪd]
★ 纸层析 paper chromatography
★ 薄层层析 thin layer chromatography(TLC)
可电离基团 ionogen [aiə'nɔdʒin]
离子化基团 ionogenic group
★ 凝胶过滤色谱 gel filtration chromatography (GFC)
★ 恒定洗脱法 isocratic elution [ˌaɪsəu'krætɪk][i'lju:ʃən]
★ 排阻极限 exclusion limit
★ 分级范围 fractionation range
床体积 bed volume
★ 空隙体积 void volume
★ 亲和作用 affinity [ə'finiti]
★ 生物亲和作用 bioaffinity [ˌbaiəuə'finiti]
★ 间隔臂 spacer
★ 非共价相互作用 noncovalent interactions
非特异性相互作用 nonspecific interactions
★ 分子识别 molecular recognition
★ 受体-配基复合体 receptor-ligand complex
★ 抗生物素蛋白-生物素 avidin-biotin ['ævidin]['baiətin]
★ 疏水性相互作用色谱 hydrophobic [ˌhaɪdrəu'fəubɪk] interaction chromatography
★ 反相色谱 reverse phase chromatography
离液离子 chaotropic [keiɔ'trɔpik] ion
反离液离子 antichaotropic ion
★ 线性梯度洗脱法 linear gradient elution
★ 逐次洗脱法 stepwise elution
★ 十八烷基键合硅胶 octadecyl silica(ODS)

1 Chromatography is a useful and popular technique that can provide a high degree of resolution for a wide range of bioseparations. A chromatographic system includes four basic elements: a stationary phase, a mobile phase, a pumping system, and an on-line detector. Using these core elements, many different physical, chemical, and biological interactions can be exploited to separate biochemicals and biopolymers

色谱是一项实用且受欢迎的技术,其可为一系列生物分离提供高度分辨率。一个色谱系统包括四个基本构成要素:固定相、流动相、泵系统和在线检测器。使用这些核心部分可以利用多种不同的物理学、化学及生物学相互作用分离生化物质和生物聚合物。

2 The solution known as the mobile phase is pumped through a column containing the stationary phase. The stationary phase typically consists of a spherical insoluble polymer, ranging in size from 5 to 300 μm. While the mobile phase flows at a constant rate through the column packed with the stationary phase, the sample is introduced at a specified time. At the end of the column, a detector tracks the concentrations of the now-separated components of the sample. After detection, the column effluent can be collected as separate fractions for further analysis or processing.

称为流动相的溶液被泵至含有固定相的柱内。固定相一般是由球形不溶性聚合物构成,其大小为5～300 μm。当流动相以恒定速率流经装填有固定相的色谱柱时,样品在某一指定时间加入。在色谱柱的末端,检测器跟踪样品中正被分离组分的浓度。检测后,可收集柱流出物为不同的洗脱部位以进一步分析或加工处理。

3 Most stationary phases include two major components: 1) the composition of the polymer matrix, which serves as the structural component, dictating the stationary phase's size and porosity, and 2) the chemical or biopolymer functional component, which confers a specific ability for interactions with the particular solute.

大部分固定相包括两种主要成分:1) 聚合物基质的组成,其作为结构成分,决定固定相颗粒大小及孔隙度,和 2) 化学或生物聚合物功能成分,其具有与某种溶质产生特异性相互作用的能力。

4 Smaller solutes can freely pass in and out of its pores; larger solutes, on the other hand, have more restricted pore movement — some are even totally excluded from the fluid in its pores. Thus the

molecular weight cut off point, which provides a general guide as to the largest solute that can enter the pores, is an important design variable. Smaller solutes spend a longer time in the column because they can freely enter the pore volume, while larger solutes pass quickly through the column.
较小的溶质可自由进出其孔隙,另一方面,较大的溶质在孔内的活动受到较多的限制 —— 一些甚至全部被流体排阻在孔外。因此,截断分子量(对可进入空隙的最大溶质能提供一般的指南)是一个重要的设计变量。较小的溶质在柱内停留较长时间,因为它们能自由进入孔容积,而较大的溶质则快速流出柱。

5 The term "bioaffinity" is used to describe the physical and chemical interactions between various biomolecules—namely, ligands and receptors. A ligand will bind to its receptor counterpart through noncovalent means such as ionic bonds, hydrogen bonds, hydrophobic interactions, and van der Waals forces. These binding relationships can be mathematically described using either the thermodynamic approach or the equilibrium approach. In specific interactions, antibodies with antigens, proteins with DNA, ligands with cell receptors, enzymes with substrates, biotin with avidin or streptavidin, and lectins with carbohydrates. These specific interactions can be exploited in bioseparation design, as in affinity chromatography.
术语"生物亲和作用"用于描述不同生物分子之间 —— 即配基与受体之间的物理和化学相互作用。配基将与其受体的相应部位通过非共价键(如离子键、氢键、疏水性相互作用及范德华力)结合。这些结合关系可以用热力学方法或平衡方法进行数学描述。在特异性相互作用中,抗体将与抗原、蛋白质与DNA、配基与细胞受体、酶与底物、生物素与抗生物素蛋白或抗生物素蛋白链菌素、凝集素与糖发生作用。在生物分离设计方面,可以利用这些特异性相互作用,例如亲和色谱。

第八节 离心
Section 8 Centrifugation

★ 离心 centrifugation
离心机 centrifuge
相对离心力 relative centrifugal force
重力沉降 gravitational sedimentation [ˌgrævi'teiʃənəl][ˌsedimen'teiʃən]
沉降分离 settlement separation
密度梯度离心 density gradient centrifugation
★ 等密度离心 isopycnic [ˌaisə'piknik] centrifugation
预形成梯度 performed gradient
自形成梯度 self-formed gradient
浮力 buoyancy ['bɔiənsi] force
角速度 angular velocity
★ 垂直转子 vertical rotor

1 Centrifuges are used primarily in solid-liquid or liquid-liquid separation processes, where the separation is based on density differences between the solid or liquid and the surrounding fluid.
离心机最初用于固-液或液-液分离过程,其分离是基于固体或液体与周围流体之间的密度差。

2 The higher-density phase will settle through the lower-density phase under the influence of the applied force created through centrifugal acceleration.
在离心加速所产生力的影响下,高密度相将会穿越低密度相而沉降。

3 Currently, centrifugation is used to separate such bioproducts as DNA, macromolecules, mammalian cells, and intracellular components.
目前,离心用于分离DNA、大分子、哺乳动物细胞及胞内组分等生物产品。

第九节 结晶
Section 9 Crystallization

★ 吸藏 occlusion [ə'klu:ʒən]
★ 包藏 inclusion [in'klu:ʒən]
★ 成核 nucleation [ˌnju:kli'eiʃən]
★ 不稳区 liable zone
★ 初级成核 primary nucleation
★ 二次成核 secondary nucleation
均相成核 homogeneous nucleation
★ 晶习 crystal habit

★ 过饱和　supersaturation
★ 母液　mother liquor

Crystallization — the formation of solid particles of defined shape and size from a homogeneous liquid phase — is the oldest and most widely used type of purification. It occurs when the liquid phase becomes supersaturated because of cooling, solvent evaporation, or the addition of a nonsolvating diluent. Most bulk pharmaceutical and organic chemicals are marketed as crystalline products because crystals of exceptional purity can be manufactured. In addition, the use of a crystalline-form product facilitates the finishing steps in the production process, such as filtering and drying. Apart from conferring these advantages, crystallization also improves the product's appearance — an important aspect of consumer acceptance.

结晶 —— 从均相液体中生成一定形状和大小的固体粒子 —— 是最古老和使用最广泛的纯化方式。当因冷却、溶剂蒸发或加入非溶剂化的稀释剂使液相成为过饱和溶液时就会发生结晶现象。多数批量药品和有机化学品是以晶形产品进入市场的，因为可以生产出异常纯度的晶体。另外，晶形产品的使用有利于产品加工过程中的最后步骤，例如过滤和干燥。除这些优势外，结晶也会改善产品的外观 —— 消费者接受的一个重要方面。

第十节　蒸发与干燥
Section 10　Evaporation and Drying

★ 蒸发　evaporation
★ 相对湿度　relative humidity
非结合水　unbound water
盘架传导干燥　tray conduction drying
★ 恒速干燥阶段　constant rate drying
★ 降速干燥阶段　falling rate drying
盘架干燥器　tray dryer [trei]['draiə]
★ 冷冻干燥器　freeze dryer
传送带式干燥器　belt dryer
转筒干燥器　tumble dryer
★ 喷雾干燥器　spray dryer
★ 升膜式蒸发器　rising-film evaporator
★ 降膜式蒸发器　falling-film evaporator
★ 流化床干燥器　fluidized bed dryer
转鼓式干燥机　drum dryer

Evaporation and drying are popular techniques for solvent removal. Evaporation can take place in a semibatch process such as Rayleigh distillation, though it is more commonly performed in a continuous fashion. Flash evaporators, short-and long-tube vertical evaporators, and agitated-film evaporators are widely used in industry. Drying can take place via fluidized bed dryers, flash dryers, spray dryers, drum dryers, and freeze dryers. All solvent removal processes must take care to limit the exposure of thermolabile compounds to excessive temperatures.

蒸发与干燥是受欢迎的除去溶剂技术。蒸发可以以半连续方式(如 Rayleigh 蒸馏)进行，尽管其更常以连续方式进行。闪烁蒸发器、短管和长管垂直蒸发器及搅拌膜式蒸发器在工业上广泛使用。干燥可通过流化床干燥器、闪烁干燥器、喷雾干燥器、转鼓式干燥器和冷冻干燥器而进行。所有除溶剂过程必须注意限制热不稳定化合物暴露于过高温度。

第十一章　生化设备

Chapter 11　Biochemical Equipment

第一节　生物反应器

Section 1　Bioreactor

★ 连续发酵　continuous fermentation
间歇发酵　batch fermentation
深层发酵　deep submerged fermentation
固态发酵　solid state fermentation
★ 发酵罐　fermenter
发酵周期　fermentation period
★ 发酵罐搅拌　fermenter agitation
★ 发酵醪　fermentation mash
★ 发酵液　fermented broth
消泡剂　antifoam ['æntifəum]
★ 通风发酵设备　ventilation [venti'leiʃən] fermentative plant
好氧微生物　aerobic microbe
★ 搅拌式发酵罐　stirred tank fermenter
★ 机械搅拌通风发酵罐　mechanical agitation ventilation fermenter
气升式发酵　air-lift fermentation
体积氧传递系数　volumetric oxygen transfer coefficient
★ 酒精发酵罐　alcohol fermenter
酒度　degree of spirit
酒化酶　zymase ['zaimeis]
★ 酒精酵母　alcohol yeast
★ 酒精精馏　alcohol rectification
酒精前发酵期　before fermentation phase of alcohol
(罐)批酒精发酵　batch alcoholic fermentation
啤酒花　hop
啤酒酵母　brewers yeast
啤酒酿造　brewing of beer
★ 啤酒发酵设备　beer fermentative plant
★ 啤酒发酵　beer fermentation
酒母　yeast mash
半连续培养　semi-continuous culture
好氧培养　aerobic culture
厌氧培养　anaerobic culture
★ 培养基　culture medium
★ 植物细胞培养反应器　plant cell culture reactor
★ (细胞)悬浮培养反应器　cell suspension culture reactor
★ 固定化细胞反应器　immobilized [i'məubilaiz] cell reactor
★ 动物细胞培养反应器　zooblast ['zəuəblɑ:st] cell culture reactor

1

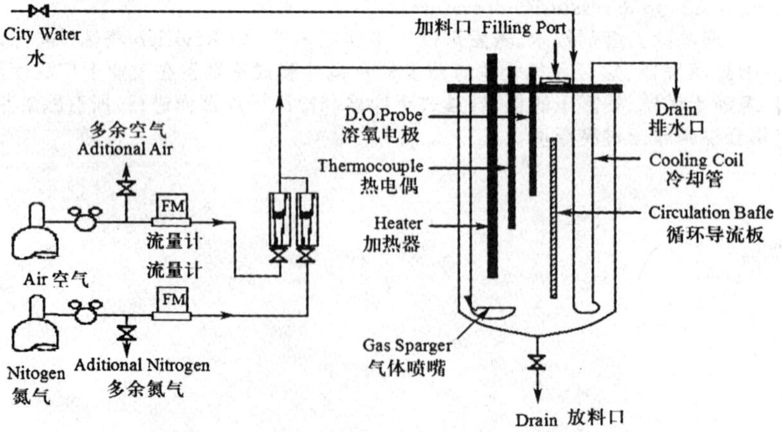

Fig 11-1　Schematic of the Air-lift Fermenter Apparatus
图 11-1　气升式发酵装置示意图

2 Air-Lift Bioreactor A schematic diagram of the air-lift bioreactor is depicted in Figure 11-1 and consists of the following equipment and instrumentation:

(1) A cylindrical glass vessel approximately 15 inches long and 3 inches diameter, with an internal divider separating the cylinder into two unequal volumes (about 2 : 1). The smaller side is called the downcomer and the larger side is the riser.

(2) The top of the bioreactor is fitted with ports for the following: thermocouple, gas sparger, dissolved O_2 electrode, filling port, cooling coil, and heater.

(3) Two mass-flowmetes for air and nitrogen flow into the fermenter.

(4) Two rotameters for visual adjustment of the air and nitrogen flows to the fermenter.

(5) Dissolved oxygen probe (Ocean Optics Inc. Model FOXY 2) connected to a SF2000 spectrophotometer.

(6) Two additional gas flow control valves for control of air and nitrogen to other containers.

(7) A PC computer equipped with the data acquisition program (LABTECH CONTROL) where oxygen concentration, flow, temperature, and time data are collected, averaged, and recorded every ten seconds.

气升式发酵罐　图 11-1 为气升式发酵罐示意图，它由下列部件及仪器构成：

(1) 在约有15英寸长、直径3英寸的玻璃圆筒内，有一隔板（导流板）把圆筒体积分成约2∶1的比例。体积较小侧为下降管，体积较大侧为气升管。

(2) 发酵罐顶部设有热电偶插孔、气体喷雾器插孔、溶氧电极插孔、进料口、蛇形冷却管插孔以及加热器插孔。

(3) 两个控制空气和氮气流入发酵罐的质量流量计。

(4) 两个可视调节的转子流量计控制空气和氮气流入发酵罐。

(5) 一个溶氧电极(Ocean Optics Inc. Model FOXY 2 型）与 SF200 型分光光度计相连接。

(6) 两个附加的分流控制阀用来控制到其他容器的空气和氮气。

(7) 一台配有数据采集程序的PC计算机，每10秒钟对氧浓度、流速、温度和时间进行采集、记录并计算其平均值。

3 The bioreactor has a working volume of approximately 2 liters and can be filled by adding solutions through the port on the top of the bioreactor. The culture medium is circulated throughout the bioreactor by an air-lift action of the sparged air. Air is bubbled through the riser lowering the liquid density as compared to the other side, which then causes the aqueous medium in the downcomer to flow into the bottom of the riser. The medium recirculates from the top of the riser into the downcomer. The air flow rate and to a lesser degree, the bioreactor and sparger designs determine the degree of liquid and gas circulation, affecting the oxygen transfer capacity of the bioreactor. Measured amounts of air and nitrogen can be added to the sparger. These gas flows are adjusted manually using the rotameters and the signals from the mass flowmeters are send to the computer. The measured gas flows are combined and sent to the bioreactor.

该发酵罐的工作容积约为 2L，并可通过发酵罐顶部的孔进行补料，发酵液在发酵罐中通过喷射（无菌）空气而被循环，空气在气升管中起泡，使发酵液体的密度比下降管要低并向上运动，导致在下降管中的发酵液流到气升管的底部，而气升管顶部的发酵液再循环到下降管（形成循环流动）。空气流速决定了气体和液体循环的程度，影响发酵罐中氧气的传递能力，发酵罐及（空气）喷嘴的设计也对其有一定的影响。一定比例的空气和氮气被送至喷嘴，其流速通过转子流量计手动调节，并用质量流量计将信号传输至计算机，一定比例的气体（空气和氧气）则被混合送入发酵罐。

第二节　物料的处理与培养基制备
Section 2 Materials Processing & Preparing Culture Media

★ 筛选机　screening machine/sieving machine | ★ 锤式粉碎机　hammer breaker

球磨机 ball mill
★ 连续稀释 serial dilution
★ 蒸煮 cooking
蒸煮锅 cooking kettle
★ 糖化 saccharification
糖蜜 molasses [mə'læsiz]
原料 feedstock ['fi:dstɔk]
糖浆 syrup ['sirəp]
★ 麦芽 malt
★ 麦芽汁 wort
★ 麦汁过滤机 mash filter
★ 麦汁煮沸锅 wort boiling kettle
醪液 mash
★ 培养基灭菌 (culture) medium sterilization
保留时间 holding time
活化能 activation energy
杂菌感染 microbial contamination [mai'krəubiəl][kən,tæmi'neiʃən]
★ 灭活 inactivation
真空干燥器 vacuum dryer
回旋分离器 whirlpool separator ['(h)wə:lpu:l]['sepəreitə]
冷冻干燥机 lyophilizer
★ 喷雾干燥 spray drying
小型试验设备 pilot plant unit

1 Continuous Sterilization Batch sterilization uses steam or direct firing to elevate the temperature, and then cooling water stops the process and brings the material back toward room temperature. Both the heat and the cooling water are spent with no opportunity for energy recovery. Large volumes should be passed continuously through for energy economy with the hot, treated fluid heating the cold, incoming feed.

One method of continuous sterilization injects steam into the medium (no heat exchanger). The medium stays in a loop for a predetermined holding time until the entire medium is sterile.

Better heat economy comes from substituting heat exchangers for direct steam injection. Instead of having a cold water stream to cool the sterile media, the lower temperature unsterile media stream absorbs heat from the warm stream, cooling the sterile media.

A system for continuous sterilization has a holding coil for detention long enough to kill all of the microorganisms. The medium from a make up vessel flows through the exchanger, is held in the coil, and passes back through the heat exchanger, heating more unsterile medium while becoming cool itself, as it is collected in a sterile fermenter.

Heat economy is not important for a small pilot plant unit for continuous sterilization, so direct steam injection is simpler. A heat exchanger is then needed with cooling water to bring the medium back down quickly to a temperature at which it is not over cooked.

Advantages:

- Uniform steam requirements throughout the duration of the sterilization
- Simplified process control
- Shorter sterilization time means less thermal degradation of medium

Disadvantages:

- High demand for steam in a shorter period of time than batch
- Concentration of media becomes dilute due to steam condensation
- Since steam is actually dispersed in media, steam must be clean to avoid contamination

High temperatures for short times are used in preparing nutrient media for industrial fermentations and in pasteurizing milk, because this causes less damage to biochemicals than more prolonged times at lower temperatures. This exploits the temperature effects on activation energies because bacterial killing is affected by a temperature change more than is heat destruction of biochemicals.

连续灭菌　间歇灭菌是采用蒸汽或直接加热的方法使培养基升温，再用水冷却使灭菌后的培养基降至室温，这种方法消耗的热及冷却水能量不能再回收利用。大量培养基的灭菌应连续经过换热器，已灭菌的热流体加热未灭菌的冷料液，以节省能量。

连续灭菌的方法之一是将蒸汽直接喷射到培养基上(无换热器),培养液在维持段管子内停留预定时间,直至全部被灭菌。

用换热器代替直接蒸汽喷射是较好的节能方法。温度较低的待灭菌培养液通过换热器从温度较高的已灭菌培养液中吸收热量,同时冷却该液体,从而代替了用冷却水冷却已灭菌培养基的方法。

一个连续灭菌系统拥有一段维持管,在其滞留的时间足以杀死全部微生物,待灭菌料液从储罐流入换热器,进入维持管,然后返回到换热器,以加热更多的待灭菌培养基,同时自身被冷却,最后被收集在已灭菌的发酵罐中。

小型试验设备的连续灭菌,节能不太重要,因此直接蒸汽加热更为简单,但需要用冷却水把已灭菌的培养基迅速降至使培养基不能过热的温度,这一过程在换热器中进行。

优点:

- 整个灭菌过程蒸汽负荷均衡
- 过程控制简单
- 较短的灭菌时间意味着培养基的热能消耗较少

缺点:

- 与间歇灭菌相比,在较短的时间内蒸汽的需求量大
- 蒸汽的冷凝使培养基的浓度降低
- 由于蒸汽直接分散在培养基中,要求蒸汽必须是无菌的,以免污染

对于工业发酵及生产巴氏灭菌奶,高温短时间的灭菌方法用于制备营养基,因为它比低温长时间的灭菌对生物制剂(营养成分)的破坏要小。这是利用温度对活化能影响的关系,因为细菌灭死程度受温度变化的影响高于生物制剂(营养成分)的热破坏。

2 System for continuous sterilization To sterilize a large volume of medium in a reasonable time, the flow rate through the sterilizer must be high. For example, suppose that you wish to sterilize 100000 liters in 100 minutes. The pumping rate would be 1000 l/m. If the holding time in the coil must be 2 minutes, the coil volume must be 2000 l. Use the applet to calculate the length of coil for various diameters.

连续灭菌系统 为了在合适的时间内完成大量培养基的灭菌,培养基通过灭菌装置的流速必须很大。如要在 100 分钟内灭菌 100000 公升,泵的流速应该是 1000 公升 / 分。若在维持段的保留时间是 2 分钟,则维持段的体积是 2000 公升,用 applet 编程计算不同直径下维持段管子的长度。

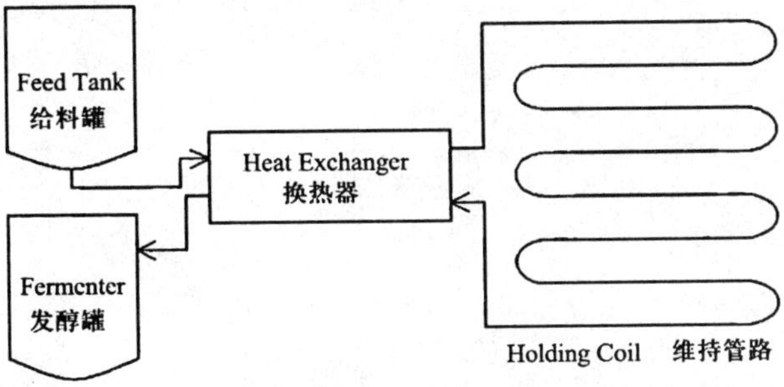

第三节 辅助系统设备

Section 3 Auxiliary System Equipment

★ 空气除菌 air sterilization	空气分配器 air distributor
★ 空气过滤器 air filter	空气冷凝器 air condenser
★ 过滤除菌 sterilization by filtration	空气压缩机 air compressor
空气调节 air conditioning	斗式提升机 bucket elevator

皮带运输机 band conveyor
螺旋输送机 screw conveyor
气力输送机 pneumatic conveyor [nju(:)'mætik][kən'vei]

Air Sterilization The magnitude of the air sterilization problem is seen from the usual needs of a highly aerobic fermentation where roughly 1 volume of air per volume of medium per minute may be used. For a factory with 20 fermenters of 100000 L each, 2 million L/m (70000 ft^3 /m) of air is handled. Very large compressors are used, and at least two are required so that one can be down for maintenance.

空气灭菌 由培养基与空气每分钟的体积比大约为 1∶1 的高耗氧发酵可以知道空气灭菌所需的空气量。对一个拥有 20 个 100000 升发酵罐的工厂，需要处理 2×10^6 升/米(即:70000 英尺^3/米)的空气，并需配备处理量很大的压缩机，至少需要两个，其中一个可在空闲时做维护。

第十二章　现代生化分析

Chapter 12　Contemporary Biochemistry Analysis

第一节　紫外-可见吸收光谱法

Section 1　Ultraviolet and Visible Absorption Spectrometry

吸收光谱法　absorption spectrometry
★ 紫外光　ultraviolet light
★ 可见光　visible light
真空紫外　vacuum ultraviolet
近红外　near infrared
远红外　far infrared
无线电波　radio frequency
★ 普朗克常数　Planck's constant
★ 电子光谱　electronic spectrum
★ 电子跃迁　electron transition
★ 价电子　valence electron
★ 生色团　chromophore [ˈkrəuməfɔ:]
★ 红移　red shift [red][ʃift]
★ 蓝移　blue shift
★ 紫外-可见分光光度计　UV Spectrophotometer [ˌspektrəufəˈtɔmitə]
★ 光源　light source
★ 钨丝灯　tungsten [ˈtʌŋstən] filament lamp
★ 氘灯　deuterium [dju:ˈtiəriəm] lamp
★ 单色器　monochromator [mɔnəuˈkrəumeitə]
色散　dispersive [disˈpə:siv]
光栅　grating
滤光片　filter
★ 狭缝宽度　slit width
固定狭缝宽度　fixed slit width
多通道检测器　multichannel [ˌmʌltiˈtʃænl] detector
转换器　transducer [trænzˈdju:sə]
光强度　light intensity [inˈtensiti]
单色光　monochromatic light
入射光　incident light
比色皿　colorimetric cuvette
★ 样品池　sample cell
吸收池　absorption cell
放大　amplification [ˌæmplifiˈkeiʃən]
★ 空白　blank
被分析物质的浓度　analyte concentration
★ 透光度　transmittance [trænsmittns]
百分透光率，T%　percent transmittance/T%
★ 吸光度　absorbance [əbˈsɔ:bəns]
朗白-比尔定律　Lambert-Beer Law
★ 定量检测　quantitative measurement [ˈkwɔntitətiv][ˈmeʒəmənt]
★ 定性检测　qualitative measurement
定量光谱检测方法　quantitative spectrometric method

1 The wavelength can be changed continuously and smoothly over some wavelength range. This method is called simply scanning the wavelength.
使波长在一定范围内连续、平稳变化的方法简称为扫描波长。

2 Absorbance is abbreviation by an upper case A. Another name, optical density.
吸光度缩写为大写字母 A，也称为光密度。

3 The absorbance of a sample is proportional to the total amount of material that absorbs the incident light. Experimentally, it can be shown that $A = abc$
In other words, the absorbance is directly proportional to the following:
(1) a constant that is a property of the material itself as well as the wavelength of the measurement;
(2) the length of the path through which the light travels in the samples; and
(3) the concentration of the material that absorbs the light.
样品的吸光度与吸收入射光物质的总量成比例关系。经实验，可表示为 $A = abc$
也就是说，吸光度与下列因素成正比关系：
(1) 常数，取决于在检测波长下物质自身性质；
(2) 光通过样品的路径长度；
(3) 吸收光的物质浓度。

4 Equation A = abc describes the Lambert-Beer law. When the concentration c is in (mol L^{-1}) and the path length in cm, then the constant will have the molar extinction coefficient, or molar absorptivity (the term currently preferred). With labels included, we see A(unitless) = ε (L $mol^{-1}cm^{-1}$) b (cm) c (mol L^{-1}) or, simply, A = εbc

公式 A = abc 为朗白-比尔定律。当浓度单位为 mol L^{-1}，光通过样品的路径长度单位为 cm 时，其常数为摩尔消光系数，或称为摩尔吸收率(常用的术语)；将单位代入后，得到 A(无单位) = ε(L $mol^{-1}cm^{-1}$)b(cm)c(mol L^{-1}) 或简写为 A = εbc

第二节 红外吸收光谱法
Section 2 Infrared Absorption Spectrometry

★ 红外光谱法 infrared spectrometry ['infrə'red][spek'trɔmitri]
★ 红外光区 infrared region
★ 波数 wavenumber
★ 红外光谱 infrared spectra
★ 振动光谱 vibrational spectrum
★ 转动光谱 rotational [rəu'teiʃənəl] spectrum
★ 振动频率 vibrational frequencies
★ 特征频率 characteristic frequencies
★ 胡克定律 Hooke's law
★ 对称伸缩 symmetric stretch
★ 不对称伸缩 asymmetric stretch
★ 对称弯曲 symmetric bend
★ 不对称弯曲 asymmetric bend
★ 弯曲振动 bending vibration
力常数 force constant
★ 折合质量 reduced mass
★ 官能团区 functional group region
★ 指纹区 fingerprint region
★ 基团频率 group frequencies
输出信号 output signal
★ 红外分光光度计 infrared spectrophotometer
热电偶 thermocouple ['θə:məu,kʌpl]
光电管 phototube
光电倍增管 photomultiplier [,fəutəu'mʌltiplaiə]
液体池 solution cell
可拆卸液体样品池 demountable liquid sample cell
★ 气体样品池 gas sample cell
★ 固体溴化钾片 solid potassium bromide plate
★ 傅里叶变换仪 Fourier-transform instrument
傅里叶变换红外分光光度计(FTIS) Fourier-transform infrared spectrophotometer(FTIS)
迈克尔逊干涉仪 Michelson interferometer ['maikəlsən][,intəfiə'rɔmitə]
对映体 enantiomer [i'næntiəumə]
光学异构体 optical isomer
芳香族化合物 aromatic compound
卤素 halogen ['hælədʒən]
★ 定性分析 qualitative analysis
★ 定量分析 quantitative analysis
c-c 基团振动 c-c group vibration
玛瑙研钵 agate mortar

1 Infrared spectra can be obtained on samples as small as a few μg. In this manner, IR spectra are used in qualitative analysis of solids, liquids, and gases. With the exception of enantiomers (optical isomers), no two compounds have the same infrared spectrum.

用几微克的样品就可以得到红外光谱。红外光谱可用于固体、液体及气体的定性分析。除对映体(光学异构体)外，不同的化合物，均具有不同的红外光谱。

2 Infrared spectra can be used to identify molecular components of samples by matching the spectrum of an unknown to a library spectrum.

通过未知样品谱图与谱图库的谱图进行对比，用红外光谱可鉴别样品的分子组成。

3 In addition to identification or comparisons of compounds, a second qualitative use of infrared spectra is to obtain structural information. Functional groups can be identified because their absorption bands lie in relatively narrow, characteristic regions of the infrared region.

除鉴别或比较化合物外，红外光谱的另一个用途是获得分子结构信息。由于不同官能团的吸收带位于红外光区相应窄的、特征区域内，为此可用于官能团的鉴定。

4 Spectra of the blanks are obtained and stored, and the blank is subtracted after each sample is run under identical conditions.

得到空白光谱并保存，在同样的条件下测定样品，然后减掉空白。

第三节　荧光光谱法
Section 3　Fluorescence Spectrometry

化学发光　chemiluminescence [ˌkemiˌluːmiˈnesns]
发光源　luminous [ˈljuːminəs] source
激发光源　emitting source
激发光　emitted light
发光　luminescence [ˌluːmiˈnesns]
★ 荧光光谱测定法　fluorescence spectrometry
磷光光谱法　phosphorescence spectrometry
激发光谱　excitation spectrum
磷光发射　phosphorescence [ˌfɔsfəˈresns] emission
激发发射　stimulate emission
★ 无辐射跃迁　nonradiative transitions
★ 单重态　singlet state
★ 三重态　triplet state
★ 荧光效率　fluorescence efficiency
★ 氙灯　xenon [ˈzenɔn] lamp
汞灯　mercury [ˈməːkjuri] vapor lamp
★ 荧光分光光度计　fluorescence spectrometer
背景信号强度　background signal intensity
荧光素　fluorescein [ˌfluəˈresiːin]
稀释　attenuation [əˌtenjuˈeiʃən]/dilute [daiˈljuːt]
空白样品　blank sample
浓度范围　concentration range
★ 自吸收　self-absorption
★ 猝灭剂　quencher [kwentʃ]
光谱数据　spectral data
选择性　selectivity
灵敏度　sensitivity
吸收光谱　absorbance spectrum
已知光谱　known spectra
峰形　spectral shape
半自动　semiautomatic [ˈsemiˌɔːtəˈmætik]
背景检测　background measurement
背景校正　background correction
混合物组分　components of mixtures
分析波长　analytical wavelength
精密度　precision [priˈsiʒən]
化学干扰　chemical interference
重复性　repetition

1 Luminescence can be divided into two types: fluorescence and phosphorescence. These two can be distinguished experimentally by their emission that may persist for an appreciable time after the illumination is removed. This persistent emission is phosphorescence. Fluorescence emission, on the other hand, ceases virtually immediately.
发光可分为两类：荧光和磷光。这两种类型的光可以通过移开激发光后发射光保持的时间的实验加以区分。持续时间长的为磷光，而荧光发射会立即停止。

2 The light is absorbed by the sample and then remitted in all directions at longer wavelengths.
样品吸收光，被激发出波长较长的光，向四外发射。

3 A fluorescence assay was to be tested to see over what range of concentrations it was linear.
用荧光实验来确定呈线性关系的浓度范围。

第四节　色谱法引论
Section 4　Introduction of Chromatography

★ 色谱法　chromatography method
★ 色谱分析　chromatography analysis
★ 色谱柱　chromatography column
有色化合物　colored chemical compound
无色化合物　colorless chemical compound
色谱术语　chromatographic nomenclature [nəuˈmenklətʃə]
★ 色谱图　chromatogram
色谱峰　chromatographic peak
★ 基线　baseline [ˈbeislain]
★ 保留体积　hold-up volume/retention volume
死体积　dead volume
总保留体积　total retention volume
调整保留体积　net retention volume
死时间　dead time
★ 保留时间　retention time
调整保留时间　net retention time
相对保留值　relative retention
分配系数　distribution coefficient/partition coefficient

容量因子 capacity factor
柱效 column efficiency
★ 理论塔板高度 height equivalent to a theoretical plate
★ 范第姆特方程 Van Deemter equation
★ 涡流扩散 eddy diffusion
★ 纵向扩散 longitudinal [lɔndʒi'tju:dinl] diffusion
★ 传质阻力 mass transfer resistance
★ 峰宽 peakwidth
标准偏差 standard deviation
对称峰 symmetric peak
★ 吸附色谱 adsorption chromatography
★ 分配色谱 partition chromatography
★ 离子交换色谱 ion-exchange chromatography
★ 空间排阻色谱 size-exclusion chromatography
吸附剂 adsorbent
吸附 adsorption
柱容量 column capacity
柱层析 column chromatography
柱效率 column efficiency
基线漂移 baseline drift
基线噪声 baseline noise

1 The full width at half maximum is abbreviated as fwhm or $W_{1/2}$.
峰高一半处对应的宽度，简称半峰宽或 $W_{1/2}$。

2 This is a sample composed of mixed components A and B.
这是含有 A 和 B 混合组分的样品。

3 The column is packed with a solid adsorbent.
柱内填充固体吸附剂。

第五节 气相色谱法
Section 5 Gas Chromatography

★ 气相色谱仪 gas chromatography
★ 载气 carrier gas
流量控制器 flow controller
★ 进样口 injection [in'dʒekʃən] ports
★ 填充柱 packed column
★ 毛细管柱 capillary column
★ 程序升温法 temperature programmed
热导池检测器 TCD thermal conductivity detector(TCD)
氢火焰离子化检测器 FID flame ionization detector(FID)
电子捕获检测器 ECD electron capture detector (ECD)
火焰光度检测器 FPD flame photometric detector(FPD)
★ 校正因子 correction factor
★ 归一化 normalization [ˌnɔ:məlai'zeiʃən]
★ 内标 internal standard
★ 外标 external standard
响应因子 response factor
气-固色谱 gas-solid chromatography
气-液色谱 gas-liquid chromatography
高温分解 pyrolysis [ˌpaiə'rɔlisis, ˌpi-]
分解产物 decomposition product

GC instrument differs significantly from that for HPLC: gas chromatography requires precise control of the flows of gases instead of liquids, and the columns are longer and usually narrower. Also, the stationary phases differ. Many, but not all, detectors differ both in construction and in operation from those used for HPLC. Finally, since the carrier gas does not act as a solvent, the temperature of the column is a critical variable in optimizing separations and must be strictly controlled.
气相色谱仪与高效液相色谱仪显著的区别：气相色谱分析需对载气而不是对液体的流速进行准确的控制。气相色谱柱一般较长、较细。固定相也不同。另外大部分气相色谱检测器在结构和使用上与高效液相色谱不同。最后一点，由于流动相是载气而不是溶剂，为优化分离可改变柱温，但柱温必须严格控制。

第六节 高效液相色谱
Section 6 High-performance Liquid Chromatography

液相色谱 liquid chromatography(LC)
气相色谱 gas chromatography(GC)
液体洗脱液 liquid eluent
★ 流速 flow rate

★ 固定液 stationary liquid
流动相 flowing liquid/moving fluid phase
★ 流动相 mobile phase
载体 solid support
固定相 static/stationary phase
进样器 injector
柱长 length of the column
柱箱温度调节装置 column thermostat ['kɔləm]['θə:məstæt]
★ 检测器 detector
废液 waste
等度洗脱 isocratic elution [ˌaisəu'krætik][i'lju:ʃən]
梯度洗脱 gradient elution
填料 packing material
吸附剂 sorbent/adsorbent
★ 正相 normal-phase
★ 反相 reversed-phase
★ 极性固定相 polar stationary phase
★ 分配色谱法 partition chromatography
★ 吸附色谱法 absorption chromatography
★ 离子色谱法 ion chromatography
离子交换 ion exchange
凝胶过滤 gel filtration
排阻 exclusion
凝胶渗透 gel-permeation
抑制 suppression
抑制物 suppressor
亲和色谱法 affinity chromatography
阳离子交换树脂 cation-exchange resin
阴离子交换树脂 anion-exchange resin
凝胶渗透色谱法 gel-permeation chromatography
超临界液相色谱法 supercritical liquid chromatography(SFC)
紫外-可见吸收检测器 UV-visible absorption detector
示差折光率检测器 differential refractive index detector
电导检测器 conductivity detector
蒸发光散射检测器 evaporative light scattering ['skætəriŋ] detector
安培检测器 amperometric [æmpi:rə'metrik] detector
脱气 degas

1 In reversed-phase chromatography, the stationary phase is less polar than the mobile phase.
在反相色谱中,固定相的极性小于流动相的极性。

2 For partition chromatography, a solid support was coated with an organic solvent such as n-butanol, benzene, or chloroform, and the mobile phase was water and/or a polar organic solvent. Partition chromatography is, thus, truly a liquid-liquid extraction chromatography, but the unbonded, organic coating tends to wash off over time.
在分配色谱中,固体载体被涂上一层如正丁醇、苯或氯仿的有机溶剂,而流动相为水和(或)一种极性有机溶剂。分配色谱实质上是液-液提取色谱法,但由于没有键合,随着时间的推移,载体上覆盖的有机溶剂会被冲刷掉。

第七节 薄层色谱法
Section 7 Thin Layer Chromatography

薄层色谱 TLC thin layer chromatography (TLC)
★ 平面色谱 planar chromatography
★ 平板色谱 flat-bed chromatography
纸色谱 paper chromatography
柱色谱 columnar chromatography
上行色谱法 ascending chromatography
下行色谱法 descending chromatography
★ 展开 development
★ 展开剂 developing solvent
★ 单向展开 single development
★ 双向展开 developed in two dimensions
颗粒 particle
★ 高效薄层色谱法 high performance thin-layer chromatography
薄层色谱扫描仪 thin layer chromatogram scanner
层析分离 chromatographic fractionation

1 Retention factor or retention index is abbreviated as R_f.
保留因子或保留指数,简称 R_f。

2 $R_f = \frac{\text{distance from origin to spot}}{\text{distance from origin to solvent front}}$

$$R_f = \frac{\text{斑点至原点的距离}}{\text{溶剂前沿至原点的距离}}$$

第八节 电泳
Section 8 Electrophoresis

★ 电迁移 electromigration [iˈlektrəumaiˈgreiʃən]
电渗 electroosmosis [iˈlektrəuɔzˈməusis]
★ 电场 electric field
★ 离子迁移率 ion mobility
二价离子 divalent [ˈdaiˌveilənt] ion
两性离子 zwitterion [ˈtsvitəraiən]
★ 琼脂糖凝胶 agarose gel
★ 平板胶 slab gel
★ 管状胶 tube gel
★ 不连续胶 discontinuous gel
★ 梯度凝胶 gradient gel
★ 电聚焦 electrofocusing [iˈlektrəuˈfəukəsiŋ]
等电聚焦电泳 isoelectrofocusing electrophoresis
★ 脉冲电场凝胶电泳 pulsed field electrophoresis(PFGE)
毛细管电泳 capillary electrophoresis [kəˈpiləri][iˌlektrəfəˈri:sis]

1 Negative ions tend to migrate toward the higher (positive) voltage and positive ions tend to migrate toward the lower (negative) voltage whether the ions are in solution or in a gas.
无论离子在溶液还是气体中,负离子总是向较高电压(正极)移动,而正离子是向较低电压(负极)移动。

2 The technique is called capillary electrophoresis (CE) or capillary zone electrophoresis (CZE).
这种技术称为毛细管电泳或毛细管区带电泳。

第九节 DNA 核苷酸序列分析
Section 9 DNA Nucleotides Sequencing

双脱氧核苷酸链终止法 dideoxy-nucleotide chain termination method
★ 聚合酶 polymerase [ˈpɔliməˌreis]
★ 模板 DNA template DNA
★ 标记 tagged
随机链终止 random chain terminations
★ 放射性标记 radioactive labeled
★ 修饰 modify
★ 荧光物质 fluorochrome
数据库 database
化学裂解法 Base-Specific Chemical Cleavage Method
计算机序列分析 computer analysis of sequences

第十节 糖的检测
Section 10 Carbohydrates Analysis

★ 还原糖 reducing sugar
★ 兰-爱农法 Lane-Eynon
★ 斐林试剂 Fehling solution
★ 氧化亚铜沉淀 precipitate [priˈsipiteit] of cuprous oxide
★ 指示剂 indicator
★ 标准溶液 standard solution
★ 样品制备 sample preparation
★ 标准曲线 standard curve
色谱条件 chromatographic conditions
二硝基水杨酸 dinitrosalicylic acid
蒽酮比色法 anthrone colorimetry

1 Spot the hydrolyzed sample on the filter paper with glass capillary, spot a drop of standard monosaccharide mixture, which has known the content right next to it.
用玻璃毛细管将多糖样品水解液点于滤纸上,同时点已知组分的标准单糖混合液。

2 Hang the filter paper inside chromatography tank and develop for 16 ～ 20h, pick out the paper and dry in the air .
将滤纸置于层析缸中层析 16 ～ 20h,将滤纸取出,于空气中干燥。

3 Evaporative light scattering detector is not dependent on the spectral properties of the sample or mobile phase.

蒸发光散射检测器不依赖于样品或流动相的光谱特性。

第十一节 脂的检测
Section 11 Lipids Analysis

★ 磷脂 phospholipid
★ 衍生物 derivative [di'rivətiv]
★ 索氏提取法 Soxhlet Extraction Method
索氏提取器 soxhlet extraction apparatus
★ 酸水解法 acid hydrolyze method
罗兹-哥特里法 Rose Gottlieb Method
肌醇 inositol [i'nəusitəul]
★ 脲 urea
脂肪酸 fatty acid
滤纸 filter paper
必需脂肪酸 essential fatty acid

第十二节 蛋白质与氨基酸的检测
Section 12 Protein and Amino Acid Analysis

★ 蛋白质 protein
★ 凯氏定氮 total Kjeldahl nitrogen
★ 双缩脲试剂 Biuret [ˌbaiju'ret] reagent
★ 双缩脲法 Biuret Method
★ 甲醛 formaldehyde [fɔː'mældiˌhaid]
福林-酚试剂法 Lowry Method
考马斯亮蓝法 Bradford Method
★ 肽键 peptide bond
氨基酸分析仪 amino acid analyzer
氨基酸(排列)序列 amino acid sequence
双缩脲反应 biuret reaction
福林-酚试剂 Folin-Phenol reagent
显色剂 ingrain agent
茚三酮反应 ninhydrin reaction
考马斯亮蓝 Coomassie brilliant blue

1 Molecular weight of protein is assayed by SDS-polyacrylamide Gel Electrophoresis.
利用 SDS- 聚丙烯酰胺凝胶电泳法测定蛋白质的分子量。

2 Polyacrylamide Gel Electrophoresis can separate different macromolecular compound, because these molecule take different charge and are different size.
聚丙烯酰胺凝胶电泳能将不同的大分子化合物分开,是由于这些分子所带电荷的差异和分子大小不同之故。

3 If protein samples are treated so that they have a uniform charge, electrophoretic mobility then will depend primarily on molecular weight.
如果使蛋白质样品电荷带相同的电荷和电量,那么电泳淌度的大小则将主要取决于分子量。

4 SDS is the shot term of sodium dodecyl sulfate.
SDS 是十二烷基硫酸钠的简称。

5 SDS is a kind of anion surface-active agent. It can open the hydrogen bond and hydrophobic bond in the electrophoresis system and combine with the protein molecule.
SDS 是一种阴离子表面活性剂,电泳系统中它能使蛋白质的氢键、疏水键打开,并结合到蛋白质分子上。

第十三节 酶活性检测
Section 13 Enzymes Activity Analysis

★ 酶学 enzymology [ˌenzai'mɔlədʒi]
★ 活力 activity
★ 活力单位 active unit
★ 比活力 specific activity
DEAE-纤维素 diethylaminoethyl-cellulose ['dai'eθil'æminəueθil-'seljuləus]
羧甲基-纤维素 carboxymethyl-cellulose [kɑːbɔksimeθil-'seljuləus]
透析 dialysis [dai'ælisis]
高效性 high efficiency
特异性 high specificity
酶活性测定 determination of enzyme activity

酶活性单位 unit of enzyme activity

酶原激活 activation of zymogen

第十四节 维生素检测

Section 14 Vitamins Analysis

视黄素 retinal ['retinɔl]

胡萝卜素 carotene/carotenoid

硫胺素 thiamine ['θaiəmi:n]

核黄素 riboflavin [ˌraibəu'fleivin]

吡哆醛和吡哆胺 pyridoxal and pyridoxamine

钴氨酸 cobalamine [kəu'bɔləmin]

抗坏血酸 ascorbic acid

脂溶性维生素 fat soluble vitamin

水溶性维生素 water soluble vitamin

第十五节 核酸、RNA、DNA 检测

Section 15 Nucleic acid、RNA and DNA Analysis

溴化乙啶 ethidium bromide [e'θidiəm]['brəumaid]

诱变剂 mutagen ['mju:tədʒən]

退火 anneal [ə'ni:l]

光谱测定的 spectrometric

小玻璃试管 cuvette [kju:'vet]

样品玻璃试管 sample cuvette

样品浓度 sample concentration

一定 pH 值的缓冲溶液 a fixed-pH buffer

工作曲线 working curve

滴定指示剂 titration indicator [tai'treiʃən]['indikeitə]

盐酸 hydrochloric [ˌhaidrəu'klɔ:rik] acid

用溶剂稀释溶液 dilute the solution with solvent

1 Pulsed field gel electrophoresis (PFGE) isolate macro molecular DNA.
脉冲电场凝胶电泳分离检测大分子 DNA。

2 DNA is analyze by agarose gel electrophoresis.
琼脂糖凝胶电泳分离检测 DNA。

第十三章 生物信息学

Chapter 13 Bioinformatics

第一节 绪论

Section 1 Introduction

★ 生物信息学 bioinformatics
★ 在线服务 online service
★ 因特网服务商 internet service provider(ISP)
元搜索引擎 meta-search engine
★ 文件传输协议 file transfer protocal
虚拟图书馆 virtual library
分子生物学电子邮件服务器 molecular biology E-mail server
★ 万维网浏览器 world wide web browser
收录号 accession number

Bioinformatics: genome informatics is a scientific discipline that encompasses all aspects of genome information acquisition, processing, storage, distribution, analysis, and interpretation. The researchers can get the relative information quickly by collecting, organizing and managing the data and then use the information expediently. We can gain the deep-seated biological knowledge by dealing with and digging the data and then enhance our understanding to the biology world. In the applications and researches of biology and iatrology, we can improve the efficiency and scientize the researches and exploitation by using the data and analysed results.

生物信息学:即基因组信息学,是包括基因组信息获得、处理、存储、分类、分析和解释所有方面的一门学科。通过收集、组织、管理(生物分子)数据,使研究人员能够迅速地获得和方便地使用相关信息;通过处理、挖掘(生物分子)数据,得到深层次的生物学知识,加深对生物世界的认识;在生物学、医学的应用和研究中,利用(生物分子)数据及其分析结果,可以大大提高研究和开发的科学性及效率。

第二节 基因组研究

Section 2 The Research of Genome

★ 人类基因组计划 human genome project(HGP)
★ 单核苷酸多态性 single nucleotide polymorphisms(SNP)
★ 缺失 indel
★ 基因组树 genome tree
★ 同型蛋白质家族 homeomorphic protein families
★ 同源物 homolog
★ 同源结构域 homologous domain
★ 同源序列 homologous sequence
同型性 homoplasy [ˈhəuməˌplæsi]
X 光晶体成像技术 X-ray crystallography [kristəˈlɔgrəfi]
★ 蛋白质序列分析 protein sequence analysis
★ 蛋白质结构预测 protein structure prediction
★ 模式分析 pattern analysis
★ 簇 cluster [ˈklʌstə]
★ 基序 motifs
算法 algorithm

1 The Human Genome Project (HGP) began in 1990 as an effort by researchers from around the world to map and sequence the human genome.
人类基因组计划(HGP)开始于 1990 年,是全世界研究者绘制并测定人类基因组序列草图的一种尝试。

2 The Human Genome Project has explored our molecular selves. Our DNA is almost identical with chimpanzee DNA. It is only about 2 percent different. Now people probably aren't surprised that we share our genes with monkeys.
人类基因组计划揭示了我们自身的分子。我们的 DNA 与黑猩猩的 DNA 几乎一致。仅有 2% 的不同。现在人们可能不再惊讶于我们和猴子有共同的基因。

第三节 生物学数据库种类

Section 3 The Kind of Biological Database

★ 美国国立生物技术信息中心 national center for biotechnology information

★ 美国国立医学图书馆 national library of medicine

神经网络 neural networks

★ 基因预测 gene prediction

★ 启动子预测 promoter prediction

★ 蛋白质二级结构预测 protein secondary structure prediction

★ 蛋白突变数据库 protein mutant database (PMD)

★ 核酸突变数据库 nucleic acid mutant database (NMD)

1 When trying to identify papers from the primary literature, you may find it useful to use the National Center for Biotechnology Information (NCBI) website. This site hosts PubMed, an online literature search resource. Many of the titles and abstracts that can be reached through PubMed also have links to online versions (PDFs or full text versions) of the associated papers.

当想从原始文献找出同一篇论文时，你会发现国家生物技术信息中心(National Center for Biotechnology Information-NCBI) 的网站相当好用。这个网站有个在线文献搜寻资源 -PubMed。许多标题与摘要可于 PubMed 找到并可链 / 连接至相关论文的在线版本(PDF 档或全文）。

2 Another online literature search tool is the Web of Science database. This tool is similar in some respects to PubMed, except that it is not so strongly focused on medically-related journals, and includes journals not indexed on PubMed.

另一个在线文献搜寻的工具是科学资料库网(Web of Science database)。这个工具在某些方面与 PubMed 相似，它除了不重点强调医学相关期刊外，但包含没被 PubMed 编入索引的期刊。

3 Web of Science also enables searches of the Science Citation Index(SCI), which allows you to ask the question, "What papers have been written recently that cite a paper I already have?" This type of search is very useful for keeping up-to-date in a research field.

科学资料库网，亦能查询科学引用指数(SCI)，也可以让你问"哪些最近完成的论文曾引用你目前已搜集的文献"这样的问题。这种方式的搜寻对于跟上一个研究领域的最新知识是非常有帮助的。

第四节 GenBank 序列数据库

Section 4 GenBank Sequence Database

★ 美国国立卫生研究院 national institution of health(NIH)

★ 蛋白质信息资源 protein information resource (PIR)

★ 蛋白质数据库 protein database(PDB)

★ 正态分布 normal distribution

NR 数据库 NR databases

★ 核酸序列 nucleic acids sequences

★ 核小体结构 nucleosome structure

核苷酸序列数据库合作组 nucleotide sequence database collaboration [kəˌlæbəˈreiʃən]

染色体重组 chromosomal rearrangement

复合基因预测 composite genes prediction

染色体上的基因簇 gene clusters on chromosome

1 DNA Sequence searches can be carried out using the GenBank search tools available on the NCBI site.

DNA 序列查询可使用位于 NCBI 网站上 GenBank 内的许多查询工具。

2 Most valuable of these is probably the Entrez browser, which can be used to find DNA sequences based upon keyword searches.

其中以 Entrez 网页最有价值，它可依据关键字搜索找出 DNA 序列。

第五节 结构数据库

Section 5 Structural Database

★ 结构与结构的对位排列 structure-by structure alignment

结构分类数据库 structure classification database

★ 结构形成 structure formation

★ 结构建模 structure modeling

核心区兼容性 core compatibility fold

折叠公共性 commonality threading

★ 蛋白质组间比较 between-proteome comparison

★ 直系同源群簇 EST 数据库 clusters of orthologous groups EST database

★ 直系同源物搜索 orthologs search

★ 家族与结构域分析 family and domain analysis

★ 基因与基因组关系 gene and genome relationship

1 Other structure-classification tools like the Structural Classification of Proteins (SCOP) project, classify structural information on evolutionary relationships.

其他的结构分类工具像蛋白质结构的分类项目(SCOP),按在进化关系上的结构信息分类。

2 DDIB — Database of Domain Interactions and Bindings, collects information of domain-domain interactions and domain interactions with biological molecules such as RNA, DNA, peptides, inorganic ions, phospholipids and cholesterols.

DDIB—— 结构域相互作用和结合数据库,收集结构域和结构域相互作用和结构域与生物分子,诸如与 RNA、DNA、多肽、无机离子、磷脂和胆固醇的相互作用的信息。

第六节 数据库在线分析

Section 6 The Analysis of Database Online

★ 全局对位排列 global alignment

★ 局部对位排列 local alignment

★ 显著性估计 significance assessment

★ 多重序列对位排列 multiple sequence alignment

★ 功能域 domain

★ 交互作用 interaction

★ 复合基因 composite genes

★ 点标矩阵 dot matrix

★ 第一个全基因组序列 first complete genome sequence

★ 第一个基因组数据库 first genome database

★ 开放阅读框比较 open reading frames comparison

★ 蛋白质序列比较 protein sequences comparison

★ DNA 微阵列 DNA microarrays

★ 寡聚核苷酸引物 oligonucleotide primer [ˌɔligəu'nju:kliəutaid]['praimə]

综合法 synthesis method

1 An open reading frame (ORF) is a portion of a gene's sequence that contains a sequence of bases, uninterrupted by stop sequences, that could potentially encode a protein.

开放阅读框(ORF)是基因序列的一部分,包含一段可以编码蛋白的碱基序列,不能被终止子打断。

2 ORF identification involves scanning each of the six reading frames and determining which one(s) contains a stretch of DNA sequence bounded by a start and stop codon, yet containing no start or stop codons within it; a sequence meeting these conditions could correspond to the actual single product of the gene.

ORF 识别包括检测这六个阅读框架并决定哪一个包含以启动子和终止子为界限的 DNA 序列,而其内部不包含启动子或终止密码子,符合这些条件的序列有可能对应一个真正的单一的基因产物。

第七节 DNA、蛋白质序列同源分析及进化树构建 BLAST
Section 7 DNA, Protein Sequence Homologous Analysis and the Building of Evolutionary Tree BLAST

★ 系统发育分析 phylogenetic analysis
★ 系统发育预测 phylogenetic prediction
★ 相似性转换 similarity conversion
★ 物种起源 origin of species
★ 系统发育学 cladistics [klæ'distiks]
★ 特性及其状态 characters and character states
最大节约 maximum parsimony(MP)
最大似然 maximum likelihood(ML)
侧线 paralinear
转化 log-det
排除不变位点 exclude constant site
不加权配对组算术方法 unweighted pair group method with arithmetic mean(UPGMA)
相邻连接方法 neighbor joining(NJ)
最小进化方法 minimum evolution(ME)
排列结尾概率 permutation tail probability(PTP)

In some cases, rather than searching for DNA sequences that closely match your own sequence, you may want to identify proteins whose sequences are homologous to those encoded by your DNA element. To do this, use the BLASTx tool. This tool will translate the DNA sequence you enter and will compare it to known protein sequences or translation products of known DNA sequences.

在某些状况下，你并不想查询与你的序列相似的DNA序列，而可能会想找出有哪些蛋白质与你的DNA序列可转译出的蛋白质序列相似。此时，可使用BLASTx工具。这个工具将你输入的DNA序列转译成蛋白质，并与已知的蛋白质序列或已知DNA序列所转译的产物比对。

第八节 生物芯片
Section 8 Biochip

★ 芯片 chip
★ DNA芯片 DNA chip
★ 蛋白质芯片 protein chip
★ 生物芯片 biochip
★ 扫描阵列 scan array
译码器 decoder strategy
再测序 resequencing
★ 杂交测序 sequencing by hybridization(SBH)
★ 界标 landmark
★ 微阵列 microarray

1 Biochip technique is a fire-new microanalysis method.
生物芯片技术是一全新的微量分析技术。

2 Gene chips are silicon slices, glass slide, and plastic plate arranged in particular orders fixed with quantity of gene probes, gene fragments.
基因芯片就是按特定的排列方式，固定有大量基因探针、基因片段的硅片、玻片、塑料片。

第十四章　发酵工程
Chapter 14　Fermentation Engineering

第一节　绪论
Section 1　Introduction

需氧发酵　aerobic fermentation
厌氧发酵　anaerobic fermentation
生物反应器　bioreactor
浅盘发酵　shallow pan fermentation
微生物转化发酵　microbiological conversion fermentation

1 Fermentation comes from the Latin fervere, meaning to boil, describes well the sensation created by the carbon dioxide bubbles when pickled foods are produced.
发酵一词来源于拉丁文“发泡”，意思是沸腾，该词形象地描述了腌渍食品时产生二氧化碳气泡的现象。

2 In 1857 Pasteur decisively proved that fermentation was a physiological process, for he showed that the yeast which produced fermentation was no dead mass, as assumed by Liebig, but consisted of living organisms capable of growth and multiplication.
1857 年，Pasteur 证实了发酵是一种生理过程，他证明了引起发酵的酵母不是 Liebig 假设的死亡体，而是由具有生长繁殖能力的活的生物所组成的。

3 The antagonistic effect of penicillin on bacteria was first observed by the Scottish biologist Sir Alexander Fleming in 1928. Although he recognized the therapeutic potential of penicillin, it was not until 1941 that a group of biologists working in England, including Sir H. W. Florey and E. B. Chain, purified the substance and established its effectiveness against infectious organisms and its lack of toxicity to humans. Despite the development of hundreds of different antibiotics in recent decades, penicillin remains important in antibiotic therapy.
1928 年苏格兰生物学家亚历山大·弗莱明首次观察到青霉素对细菌的拮抗效果，尽管弗莱明认识到青霉素的治疗潜能，但直到 1941 年英国的一批生物学家包括弗洛里和钱恩才纯化了这种物质，并证实了青霉素对传染生物具有防效而对人类无毒。尽管近几十年来，发现了数百种不同的抗生素，但青霉素作为抗生素治疗剂始终保持着其重要性。

第二节　工业微生物的生长与产物的生物合成
Section 2　Growth and Metabolite Synthesis of Industrial Microorganisms

初级代谢产物　primary metabolite
次级代谢产物　secondary metabolite
聚合　polymerization [.pɔlimərai'zeiʃən]
协同反馈抑制　concerted cooperative feedback inhibition
顺序反馈抑制　sequential feedback inhibition
反馈阻遏　feedback repression
反馈调节　feedback regulation
多价阻额　repression
葡萄糖效应　glucose effect
诱导调节　induced regulation
自身反馈抑制　autofeedback inhibition
前体　precursor [pri(:)'kə:sə]

The yield of an amino acid depends on the quantities and qualities of the enzymes. The yield increases if the enzymes involved in the production of the target amino acid are present in large quantities under workable conditions, while it decreases if the enzymes are present in small quantities. Suppose that a microorganism has a metabolic pathway A → (a) → B → (b) → C → (c) → D (a, b, and c are enzymes). In order to produce only amino acid C in large amounts, you have to enhance the actions only of enzymes a and b and to get rid of the action of enzyme c.
氨基酸的产量依赖于酶的数量及质量。如果目标氨基酸产量有关的酶数量多且处于活跃状态，则氨基酸的产量就高，反之则少。假定微生物的代谢途径为 A → (a) → B → (b) → C → (c) → D (a, b, c 为酶)，为了大量生产氨基酸 C，就必须提高酶 a 和 b 的活性，去除酶 c 的活性。

第三节 工业微生物的菌种选育
Section 3 Screening and Breeding of Industrial Microorganisms

自然选育 natural selection
出发菌株 original strain
杂交育种 hybridization
原生质体融合 protoplast fusion
分子育种 molecular breeding

Fermentation technique A laboratory scale stainless steel stirred fermentor of 15 L capacity with working volume of 9 L was used for citric acid fermentation. The fermentation medium consisting; clarified cane molasses 300.0 (sugar 15%), $K_4Fe(CN)_6$ 200 ppm at an appropriate initial pH. The vegetative inoculum was transferred to the production medium at a rate of 5% (v/v) based on total working volume of the fermentation medium. The incubation temperature was varied during different batches. Air supply was maintained at different levels from an aeration pump. Necessary agitation intensity was also maintained. Sterilized silicone oil (Antifoam AE-11, Germany) was used to control the foaming during fermentation.

发酵过程 实验室用 15L 不锈钢蒸汽发酵罐，进行柠檬酸发酵时工作容积为 9L。发酵培养基为：糖蜜澄清液 300.0(含糖 15%)，在适宜的起始 pH 时，$K_4Fe(CN)_6$ 200 ppm，接种营养物为发酵培养基总体积 5%(v/v)，培育温度随不同批次而不同。通气量根据气泵不同而保持不同水平，保持一定强度的搅拌也是必要的，灭菌的硅树脂被用来控制发酵过程中产生的气泡(消沫剂 AE-11，德国)。

第四节 培养基
Section 4 Medium

孢子培养基 spore medium
种子培养基 seed medium
发酵培养基 fermentation medium

第五节 灭菌
Section 5 Sterilization

分批灭菌 batch sterilization
热交换器 heat exchanger
连续灭菌系统 continuous sterilization system
补料系统 fed system
消沫剂系统 defoamer [di:'fəumə(r)] system
肉汤培养基 broth medium
斜面培养基 slant medium

1 Medications that we inject into people should be sterile, removing the health hazard due to pathogenic organisms is called disinfection.

注射进人体的所有药物均应该被灭菌，以去除病原微生物对健康的危险称为消毒。

2 In the past, industrial bioprocess could tolerate a very few foreign organisms relative to the desired organisms. Costs of achieving absolute sterility was not justified for processes such as the penicillin fermentation where the production culture is hardy and will outcompete a few foreign organisms and will usually develop a high titer of product anyway. The situation is different today with cell cultures because they grow slowly and are very easily contaminated.

在过去，工业生物过程能够容忍和目标生物有关的极少数外来生物的存在。要达到绝对无菌的费用无法保证，对于诸如整个生产培养很困难的青霉素发酵过程要和少数外来生物竞争，并通过各种方法以获得产品的高效价。如今，由于细胞培养生长较慢且常易受到污染，形势有所不同。

3 Usually, a fermenter and its medium are sterilized together, and all connecting lines are flushed with steam. A fermenter may be sterilized and then filled with sterile medium. If so, it is common practice to sterilize water and insensitive materials in the fermenter so as to minimize the volume that must be

sterilized by other methods such as filtration or passage through a continuous sterilizer. Among the several factors that influence killing are temperature, pH, osmotic pressure, shear, and concentrations of extraneous substances.

通常,发酵罐和培养基一同灭菌,所有相关的管路采用蒸汽灭菌。发酵罐也可先被灭菌,然后再灌入灭菌的培养基。如果采用这种方式灭菌,通常的经验是在发酵罐内进行水和不敏感培养基的灭菌,以减少必须采用其他方式灭菌的培养基的体积,如采用过滤或连续灭菌的方式。影响灭菌的因素中包括温度、pH、渗透压、剪切力以及外加物质的浓度。

第六节 生产菌种的培养与保藏技术

Section 6 Inoculum development and preservation techniques

培养时间 cultivation time

冷藏时间 cryopreserved [ˌkraɪəupresəːvəd] time

接种 inoculate

接种环 inoculating hood

接种针 inoculating needle

种龄 seed age

菌种保藏 culture preservation

第七节 通气与搅拌

Section 7 Aeration and Agitation

溶解氧浓度 dissolving oxygen concentration

呼吸强度 respiratory intensity

呼吸临界氧浓度 respiratory critical oxygen concentration

培养液 inoculum

搅拌器 agitator/mixer/stirrer

溶解氧的饱和度 saturation degree of dissolving oxygen

传递阻力 transfer resistance

传递速率 transfer rate

比表面积 specific surface area/specific area

液相 liquid phase

氧传递系数 oxygen transfer coefficient

摄氧率 oxygen uptake rate

离子强度 ionic strength

剪切速度 shear rate

流体动力学 fluid dynamics

hydrodynamics [ˈhaidrəudaiˈnæmiks]

牛顿液体 Newton liquid

搅拌速率 stir speed

表观黏度 apparent viscosity [visˈkɔsiti]

泡沫发生器 foam generator

鼓沫器 sparger

溶氧 dissolved oxygen

氧传感器 oxygen sensor

European bioreactors often have agitator shafts entering below the liquid level. While steam can be injected into the seals around the shafts to provide a barrier to the entry of foreign organisms.

欧洲生物反应器常常有直通入液体下面的搅拌轴。当蒸汽被注入密封的发酵罐内,周围的搅拌轴能够起到防止外来生物侵入的屏障作用。

第八节 发酵工艺的控制

Section 8 Control of Fermentation Process

搅拌功率 power of agitator

空气流量 air flow

浊度 cloudiness/turbidity [təːˈbiditi]

基质浓度 substrate concentration

溶解氧浓度 dissolved oxygen concentration

氧化还原电位 oxidation-reduction potential

产物的浓度 product concentration

比消耗速率 specific consumption rate

分批培养 batch culture

补料分批培养(FBC) fed-batch culture

偶联 couple

生长期 growth phase

生产期 production phase

产物合成阶段 product synthesis phase

产物分泌期 product secretory phase

自溶阶段 autodigestion [ˌɔːtəudiˈdʒesʃən] phase

菌体浓度 cell concentration
氧传递速率(OTR) oxygen transfer rate
温度系数(Q_{10}) temperature coefficient
生物热($Q_{生物}$) biothermal [ˈbaɪəʊˈθəːməl]
搅拌热($Q_{搅拌}$) stirred heat
蒸发热($Q_{蒸发}$) evaporation heat
辐射热($Q_{辐射}$) radiant heat
发酵热($Q_{发酵}$) fermentation heat
比生产速率 specific production rate
咕林醇酰胺(B因子) cobinamide [kəuˈbinəmaid]
半连续培养 semi-continuous culture
半连续发酵 semi-continuous fermentation
热致死 thermal death
反馈控制 feedback control
传感器 sensor/transducer [trænzˈdjuːsə]
控制器 controller
驱动器 actuator [ˈæktjueitə]/driver
放料和补料 withdraw and fill
流态泡沫 fluid foam
聚氧乙烯氧丙烯甘油 polyoxyethylene oxypropylene glycerol(GPE)
高碳醇 alcohol
天然油脂 natural oil and fat
氧化丙烯 oxypropylene [ɔksi ˈprəupiliːn]/propene [ˈprəupiːn] oxide/propylene epoxide
环氧乙烷 epoxyethane [eˈpɔksiθein]/ethylene epoxide/ethylene oxide oxacyclopropane
惰性载体 inert carrier
吐温-80 Tween-80/polysorbate 80
乳剂 emulsion [iˈmʌlʃən]

第九节 发酵动力学
Section 9 Fermentation Dynamics

质量平衡法 mass balance method
内源代谢 endogenous metabolism
维持能 preservation energy
生长得率常数 growth yield coefficient
化学计量式 chemical equation
元素平衡 element conservation
物质平衡 matter conservation
饱和常数 saturation constant
生长限制基质浓度 substrate concentration for growth restriction
产物抑制 production inhibition
生长动力学 growth kinetics
细胞比死亡率 cell specific death rate
表观比生长率 apparent specific growth rate
生化模型 biochemical model
基质抑制模型 substrate inhibition model
氧限制模型 oxygen inhibition model
简单分批发酵 simple batch fermentation
恒浊器 turbidostat [ˈtəːbidəˌstæt]
基质比消耗率 substrate specific consumption [kənˈsʌmpʃən] rate
临界稀释率 critical dilution ratio

The dynamics of fungi, yeasts, and lactic acid bacteria during fermentation of four musts were studied during fermentation of wines. Fungi disappeared quickly in the fermenting must. The lactic acid bacteria population diminished during alcoholic fermentation, then they increased and performed malolactic fermentation. Yeasts grew quickly, reaching maximum populations at different times depending on the vinification treatment

在酒的发酵过程中,四种葡萄汁发酵过程中真菌、酵母菌和乳酸菌的动力学必被研究。在发酵葡萄汁中真菌很快消失。在酒精发酵过程中乳酸菌数量锐减,然后有所增加,完成苹果乳酸发酵。依据葡萄酒酿造方法的不同,不同时期内酵母菌增长很快,达到最大密度。

第十节 发酵过程检测与自控
Section 10 Detection and Autocontrol of Fermentation Process

可靠性可通过 good engineering practice (GEP)
离线传感器 off-line sensor
在线传感器 on-line sensor
原位传感器 in situ sensor
力敏元件 force sensor
热敏元件 thermal sensor/thermal element/heat-sensitive eye
光敏元件 light sensor
磁敏元件 magnetic sensor
电化学传感器 electrochemical sensor
电极 electrode [iˈlektrəud]
生物传感器 biosensor

★ 耗氧率 oxygen uptake rate (OUR)
质谱 mass spectra/mass spectrum
★ 校准 calibrate ['kælibreit]
★ 基质消耗率 substrate consumption rate
★ 氧消耗率 oxygen consumption rate
★ 前馈控制 feedforward control
★ 反馈控制 feedback control
★ 开关控制 off-on control
串级反馈控制 cascade feedback control
自适应控制 adaptive control

As the objective of control systems for batch and fed-batch processes is always to ensure high quality of the products.
作为分批发酵和补料分批发酵过程控制系统的目标就是确保高质量的产品。

第十一节 发酵过程的实验室研究、中试和放大
Section 11 Laboratory Study, Pilot and Amplification

均匀设计 uniform design
正交设计 orthogonality [ɔ:θɔgə'næliti] design
中试 pilot
缩小 scale-down
放大 scale-up

1 Orthogonality is a system design which enables the making of complex designs feasible and compact.
正交是使复杂的设计可行化和简洁化的一种系统设计。

2 When the experimenter wants to investigate more factors (for instances, more than 4 factors) in an experiment one cannot expect to obtain a good result without the use of modern experimental designs. For example, suppose that the experimenter wants to consider various reacting temperature, time, pressure, speed and several materials in an experiment. Choose five testing levels for each of the factors. The total of level-combinations is $5^5 = 3125$, If we use the so-called orthogonal design, the number of experiments can be reduced to 25.
当在一个实验中需要调查多个因素(比如超过 4 个因素)时,若没有现代实验设计,要取得理想的结果是不可能的。例如,在一个实验中需要考虑不同影响温度、时间、压力、速度和几种材料时,每个因素选择 5 个试验水平,实验总数为 $5^5 = 3125$,如果采用正交设计,实验数目减少到 25 次。

3 The Uniform design is another such efficient fractional factorial design. It was proposed by Professor Fang Kai-Tai and Professor Wang Yuan in 1980. It has been successfully used in various fields such as chemistry and chemical engineering, pharmaceutics, quality engineering, system engineering, survey design, computer sciences and natural sciences.
均匀设计是另外一种有效的部分因子设计。1980 年由方开泰教授和王云教授提出,已被成功地应用于不同的领域如化学、化学工程、制药学、质量工程、系统工程、勘探设计、计算机科学和自然科学。

第十二节 现代生物技术在发酵工业中的应用
Section 12 Applications of Modernistic Biotechnology in Fermentation Industry

★ 基因工程 genetic engineering
★ 细胞工程 cell engineering
★ 发酵工程 fermentation engineering
★ 生化工程 biomical engineering
抑制素 somatostatin [səumætəu'stætin]
★ 悬浮培养 suspension culture
★ 贴壁培养 anchorage culture
★ 组织培养(愈伤组织培养) callus culture
吸附法 adsorption method
包埋法 embedding method
交联法 crosslinking method
生物反应器 bioreactor

第十三节 发酵工业与环境保护
Section 13 Fermentation Industry and Environmental Protection

★ 好氧生物 aerobic organism
★ 厌氧生物 anaerobic organism
★ 生物处理 biotreatment
★ 生物膜 biofilm/biological film/biomembrane ['baɪəu'membrein]
深井曝气 borehole aeration ['bɔ:həul][ˌeiə'reiʃən]
氧化沟 oxidation channel
氧化塔 oxidation column/oxidation tower

上流式厌氧污泥床反应器 up-flow anaerobic sludge bed reactor(UASB)
★ 厌氧流化床 anaerobic fluidized bed(AFB)
★ 厌氧过滤器 anaerobic filter(AF)
硬性填料 hard packing/hard filler
软性填料 soft packing
★ 厌氧消化 anaerobic digestion

第十四节 发酵过程经济学
Section 14 Fermentation Economics

★ 固定成本 fixed cost
★ 可变成本 variable cost
★ 原料成本 input material cost
★ 公用事业成本 utility cost
生产规模 production capacity
年产出计划 annual outgoing design
日常生产计划 daily production program
日产量 daily output/daily production/current yield
发酵周期 fermentation period
★ 经济效益 economic effectiveness
输出功率 output power
额定功率 rated power
负荷率 load rate
产物浓度 product concentration
发酵批产量 fermentation batch output
★ 发酵产率 fermentation production rate
基质转化率 substrate conversion
★ 单位产量 specific duty
★ 发酵成本 fermentation cost

第十五节 β-内酰胺类抗生素
Section 15 β-lactam Antibiotics

头霉素 cephamycin [ˌsefəˈmaisin]
克拉维酸 clavulanic
噻纳霉素 thienamycin
★ β-内酰胺抗生素 β-lactam antibiotics
单菌胺 monobactam
6-氨基青霉烷酸 6-amino penicillanic acid (6-APA)

1 The mass production of antibiotics began during World War II with streptomycin and penicillin. Now most antibiotics are produced by staged fermentations in which strains of microorganisms producing high yields are grown under optimum conditions in nutrient media in fermentation tanks holding several thousand gallons.

第二次世界大战期间，链霉素和青霉素开始大量生产。如今大多数抗生素通过分批发酵生产，在几千加仑的发酵罐中，配制营养培养基，在合适的条件下，通过高产微生物发酵生产。

2 Antibiotic, any of a variety of substances, usually obtained from microorganisms, that inhibit the growth of or destroy certain other microorganisms.

抗生素通常是通过微生物获得，能够抑制或破坏其他微生物生长的物质。

第十六节 非β-内酰胺类抗生素
Section 16 Non-β-lactam Antibiotics

★ 四环类抗生素 tetracyclines(TCs) [tetrəˈsaiklins]
★ 大环内酯抗生素 macrolides antibiotics
★ 多烯大环内酯抗生素 polyene macrolides antibiotics
麦迪霉素 medemycin/midemycin
交沙霉素 josamycin [dʒɔsəˈmaisin]
制霉菌素 nystatin [ˈnistətin]
两性霉素 amphotericin [ˌæmfəˈterəsin]
曲古霉素 hachimycin
trichomycin [ˌtrikəˈmaisin]
利福霉素 rifamycin [ˌrifəˈmaisin]
★ 蒽环类抗生素 anthracene [ˈænθrəsiːn] ring antibiotics
★ 蒽环糖苷抗生素 anthracene ring glucoside antibiotics
★ 链霉素 streptomycin [streptəˈmaisin]
2-脱氧链霉胺 2-deoxy streptamine (2-DOS) [diˈɔksi ˈstreptəˌmiːn]

新霉素 neomycin [ni(:)əu'maisin] / fradiomycin [ˌfrædiə'maisin] /mycifradin
卡那霉素 mycifradin
庆大霉素 gentamycin [ˌdʒentə'maisin]/gentamicin [ˌdʒentə'maisin]
林可霉素 lincomycin [ˌliŋkə'maisin]
★ 肽类抗生素 peptide antibiotics
短杆菌肽 gramicidins [ˌgræmi'saidins]
多粘菌素 polymyxin [ˌpɔli'miksin]
杆菌肽 bacitracin [ˌbæsi'treisn]

Tetracycline, any of a group of antibiotics produced by bacteria of the genus *Streptomyces*, is effective against a wide range of Gram positive and Gram negative bacteria.
四环素是链霉菌属细菌产生的,对革兰氏阳性和阴性菌具有广谱抑菌效果的一组抗生素。

第十七节 氨基酸和维生素

Section 17 Amino Acid and Vitamine

★ 核黄素 lactochrome ['læktəˌkrəum]
维生素 B_{12} vitamine B_{12}
钴维生素 cobalamin [kəu'bɔləmin]
β-胡萝卜素 β-carotene ['kærəti:n]

1 Currently, the amino acids products are mainly manufactured by the fermentation method using natural materials, similar to yogurt, beer, vinegar, miso(bean paste), soy sauce,etc.
目前,氨基酸产品主要通过使用天然材料发酵方法获得,类似于酸奶、啤酒、醋、日本豆面酱、酱油等。

2 The amino acid fermentation method is a method for the production of amino acids utilizing the phenomenon that microorganisms convert nutrients to various vital components necessary to themselves.
氨基酸发酵法是一种利用微生物将营养物质转化成多种其自身所必需的重要组分的现象来进行氨基酸生产的方法。

第十八节 甾类化合物的微生物转化

Section 18 Microbial Conversion of Steroids

★ 甾类化合物 steroids ['stiərɔids]
★ 胆甾醇 cholesterol [kə'lestərəul]
胆酸 cholic acid
★ 性激素 sex hormone
肾上腺皮质激素 adrenal [ə'dri:nl] cortex hormone
可的松 cortisone ['kɔ:tisəun]
氢化可的松 hydrocortisone

第十九节 核酸类物质及药用酶

Section 19 Nucleic Acid and Medical Enzyme

★ 肌苷 carnine
鸟苷 guanosine
胞苷 cytidine
黄苷 xanthosine
尿苷 uridine
腺苷 adenosine
★ 药用酶 medical enzyme
★ 辅酶 coenzyme/coferment [kəu'enzaim][kəu'fə:mənt]
蛋白酶 Protease
青霉素酰化酶 penicillin acylase [ˌpeni'silin]['æsileis]

第二十节 农牧用抗生素

Section 20 Agriculture Antibiotics

★ 农用抗生素 agriculture antibiotics
春日霉素(春雷霉素) kasugamycin [ˌkɑ:sugə'maisin]
有效霉素 validamycin [vælidə'maisin]
除草霉素(除莠霉素) herbimycin [hə:bimaisin]
赤霉素 gibberellin [ˌdʒibə'relin]
阿维菌素 avermectin [ə'və:məktin]
莫能星 monensin [mɔnənsin]
盐霉素 salinomycin [ˌsælinə'maisin]

第十五章 基因工程
Chapter 15 Gene Engineering

第一节 绪论
Section 1 Introduction

可移动遗传因子 mobile genetic element
★ 基因座 gene locus
★ 基因操作 gene manipulation
★ 基因标签 gene tag
基因寻靶 gene targeting
★ 遗传图谱 genetic map
★ 基因拷贝 gene copy
★ 基因表达系统 gene expression system
基因融合 gene fusion
★ 基因定位 gene localization
生物碱 alkaloid
★ 生物信息学 Bio-Informatics

1 In DNA, there are four heterocyclic bases: adenine (A) and guanine (G) are purines; cytosine (C) and thymine (T) are pyrimidines. In RNA, thymine is replaced by the structurally very similar pyrimidine, uracil (U).

DNA 含有 4 种杂环碱基:嘌呤类有腺嘌呤(A) 和鸟嘌呤(G),嘧啶类有胞嘧啶(C) 和胸腺嘧啶(T);而在 RNA 中尿嘧啶(U) 替代了结构非常相似的胸腺嘧啶。

2 DNA most commonly occurs as a double helix. Two separate and antiparallel chains of DNA are wound around each other in a right-handed helical (coiled) path, with the sugar-phosphate backbones on the outside and the bases, paired by hydrogen bonding and stacked on each other, on the inside. Adenine pairs with thymine; guanine pairs with cytosine. The two chains are complementary; one specifies the sequence of the other.

DNA分子通常以双螺旋形式存在。两条独立的反向平行的DNA链以右手螺旋方式相互缠绕,糖-磷酸骨架在外,靠氢键和堆积力相互配对的碱基在内。腺嘌呤(A) 与胸腺嘧啶(T) 配对,鸟嘌呤(G) 与胞嘧啶(C) 配对。两条链是互补的,一条链决定了另一条链的序列。

第二节 DNA 的提取与纯化
Section 2 Extraction and Purification of DNA

共价闭合环状 DNA covalently closed circular DNA
循环测序法 cycle sequencing
差异杂交 differential hybridization
DNA 指纹图 DNA finger-printing
★ 表达图 expression map
碱基插入 base insertion
★ 碱基堆积 base stacking
假基因 pseudogene
基体 basal body
必需基因 essential gene
畸形 malformation
单体性 monosomy [ˌmɔnəˈsəumi]
小鼠模型 mouse model
功能性克隆表达 functional cloning expression
富 GC 序列 GC-rich sequence
★ 蛋白-蛋白交互作用 protein-protein interaction
基因剔除 gene knock-out
通用转录因子 general transcription factor
★ 基因组印记 genomic imprinting
协助装配 aided-assembly
选择性剪接 alternate splicing
后期促进因子复合物 anaphase-promoting complex

1 The extinction coefficient of nucleic acid bases depends on their environment. The absorbance of isolated nucleotides is greater than that of RNA and singe-stranded DNA, which is in turn greater than that of double-stranded DNA. Double-stranded DNA is hypochromic with respect to single-stranded DNA.

核酸碱基的消光系数与它所处的环境有关。分离的核苷酸的吸收值比RNA和单链DNA的吸收值最大，单链DNA又比双链DNA大。这种双链DNA相对于单链DNA吸收值减少的现象就称为减色效应。

2 The absorbance at 260nm is used to determine the concentration of nucleic acids. At a concentration of 1mg ml-1 and 1 cm pathlength, double-stranded DNA has A260 = 20. RNA and single-stranded DNA have A260 = 25. The values for RNA and single-stranded DNA depend on base composition and secondary structure.

核酸260nm处的光吸收可用于测定其浓度。当浓度为1mg·ml^{-1}，光程为1cm时，双链DNA的A_{260} = 20，RNA和单链DNA的A_{260} = 25。对于RNA和单链DNA，A_{260}的值取决于碱基的组成和二级结构。

3 The A260/A280 ratio of a double-stranded DNA sample can be used to assess its purity. For pure DNA, the value is 1.8. Values above 1.8 suggest RNA contamination and those below 1.8 suggest protein contamination.

A_{260}/A_{280}比值可用于估计双链DNA样品的纯度。对于纯DNA，该比值为1.8；若比值大于1.8，则意味着RNA污染；若比值小于1.8，则有蛋白质污染。

4 Increased temperature can bring about the denaturation of DNA and RNA. RNA denatures gradually on heating, but double-stranded DNA 'melts' cooperatively to give single strands at a defined temperature, T_m which is a function of the G+C content of the DNA. Denaturation may be detected by the change in A260.

升高温度可使DNA和RNA变性。RNA随着温度升高而逐渐变性，双链DNA在某一确定温度"熔解"为单链DNA，这一温度表示为T_m，该值是DNA的(G+C)含量的函数。核酸的变性可以A_{260}的变化来测定。

第三节　基因文库

Section 3　Gene Library

★ mRNA差异显示法　mRNA differential display [ˌdifəˈrenʃəl][dɪˈspleɪ]
★ 基因组文库　genomic library
动物模型　animal model
常染色体显性　autosomal dominant
★ 自主性因子　autonomous element
★ 细菌人工染色体　bacterial artificial chromosome
★ cDNA微矩阵分析　cDNA micro-array analysis
组合库　combinatorial library
★ 感受态细胞　competent cell
★ 文库　library
座位　locus
★ 序列标记位点　sequence-tagged site
光谱核型　spectral karyotype
自杀基因　suicide gene
综合症　syndrome [ˈsindrəum]
先天缺陷　birth-defect
骨髓移植　bone-marrow transplantation
后选基因　candidate gene
★ cDNA文库　cDNA library
★ 细胞遗传图　cytogenetic [ˌsaitəudʒiˈnetik] map

1 Gene libraries made from genomic DNA are called genomic libraries and those made from complementary DNA are known as cDNA libraries. The latter lack nontranscribed genomic sequences (repetitive sequences, etc.). Good gene libraries are representative of the starting material and have not lost certain sequences due to cloning artifacts.

由基因组DNA所制成的基因文库称为基因组文库，而由互补DNA所制成的称为cDNA文库。cDNA文库不包括不能转录的核基因序列(重复序列等)。好的基因文库要覆盖起始材料的全部基因，不能丢失因克隆操作丧失其中的基因或某些序列。

2 A gene library must contain a certain number of recombinants for there to be a high probability of it containing any particular sequence. This value can be calculated if the genome size and the average size of the insert in the vector are known.

基因文库必须包含一定数量的重组体，以便具有包含任何特定序列的可能性。如果知道基因组大小以及插入载体片段的平均大小，则可以计算出所需重组体的数量。

3 For making libraries, genomic DNA, usually prepared by protease digestion and phase extraction, is fragmented randomly by physical shearing or restriction enzyme digestion to give a size range appropriate for the chosen vector. Often combinations of restriction enzymes are used to partially digest the DNA.

为了构建文库，通常用蛋白酶降解及分相抽提制备基因组 DNA，要用物理剪切或限制性酶解来进行随机分割，以形成适合于所用载体大小范围的片段。常用一组限制酶对 DNA 进行部分降解。

第四节 基因扩增

Section Gene Amplification

★ 基因扩增 gene amplification

基因治疗 gene therapy

★ 基因转移 gene transfer

★ 遗传密码 genetic code

遗传咨询 genetic counseling

分子伴侣 chaperone [ˈʃæpərəun]

多重 PCR 扩增系统 multiple PCR application system

前体 mRNA pre-mRNA

技术转让 technology transfer

1 The polymerase chain reaction (PCR) is used to amplify a sequence of DNA using a pair of oligonucleotide primers each complementary to one end of the DNA target sequence. These are extended towards each other by a thermostable DNA polymerase in a reaction cycle of three steps: denaturation, primer annealing and polymerization.

聚合酶链反应(PCR)利用与DNA模板序列的两端互补的一对寡核苷酸引物来扩增一段DNA序列。有一种热稳定的 DNA 聚合酶经三步反应即变性、引物退火和聚合的循环从两个引物来相对延伸。

2 The reaction cycle comprises a 95℃ step to denature the duplex DNA, an annealing step of around 55℃ to allow the primers to bind and a 72℃ polymerization step. Mg2＋ and dNTPs are required in addition to template, primers, buffer and enzyme.

反应循环有三部组成：95℃ 变性双螺旋 DNA，约 55℃ 退火时引物与模板结合，以及 72℃ 聚合反应。除模板、引物、缓冲液和酶外，还需要 Mg^{2+} 和 dNTP。

3 Almost any source that contains one or more intact target DNA molecule can, in theory, be amplified by PCR, providing appropriate primers can be designed.

几乎所有的含一个或多个完整的目的 DNA 分子的 DNA 在理论上均可用作 PCR 扩增的模板，其条件是只要能设计出合适的引物。

第五节 基因重组

Section 5 Gene Recombination

★ 平齐末端 blunt end

★ 黏性末端 cohesive end

同裂酶 isoschizomer [ˌaisəuˌskiˈzəumə]

同尾酶 isocaudamer [ˌaisəuˌkɔːdəːmə]

★ 杂交位点 hybrid site

★ 完全消化作用 complete digestion

★ 局部酶切消化 partial digestion

专利权 patent

谱系 pedigree

★ 定位克隆 positional cloning

接合型质粒 conjugative plasmid

非接合型质粒 non-conjugative plasmid

★ 质粒的不亲和性 plasmid incompatibility [ˈinkəmˌpætəˈbiliti]

不亲和群 incompatibility group

多聚衔接物 poly linker

抗终止因子 anti-terminator

★ 插入型载体 insertion vector

★ 置换型载体 replacement vector

存活噬菌体 viable phage

★ 衔接物 linker

碱性磷酸酶 alkaline phosphatase

★ 微卫星 DNA micro-satellite DNA

1 Genes on the same chromosome (A,a;B,b) recombine by crossing-over and genes on different chromosomes are shuffled as members of chromosome pair reassort.
在同一染色体上的基因(A,a;B,b)由于交换而重新组合;而在不同染色体上的基因则任意地重新分配组成染色体对。

2 Restriction endonucleases are bacterial enzymes which cut (hydrolyze) DNA into defined and reproducible fragments. In bacteria, they form part of the restriction modification defense mechanism against foreign DNA. They are the basic tools of gene cloning.
限制性内切核酸酶来自细菌,可将 DNA 酶切(水解)成特定的、可再生的片段。在细菌中限制性内切核酸酶是抵抗外源 DNA 侵入的限制修饰防卫机制的一部分。限制性内切核酸酶是进行基因克隆的基本工具。

3 Agarose gels separate linear DNA on the basis of size, by the migration of DNA through a matrix under the influence of an electric field. Electrophoresis may be used to determine the gross organization of plasmid molecules.
琼脂糖凝胶电泳可以在电场作用下,根据线性DNA分子的大小及DNA在凝胶基质中的迁移率分离DNA分子。电泳还可用于测定质粒分子的大致构型。

4 T4 DNA ligase repairs breaks in a dsDNA backbone and can covalently rejoin annealed cohesive ends in the reverse of a restriction enzyme reaction, to create new DNA molecules.
T4 DNA 连接酶可以修复双链 DNA 骨架的断裂,催化限制酶酶解反应的逆反应,即使退火的互补黏性末端共价连接在一起,形成新的 DNA 分子。

第六节 重组体检测

Section 6 Examination of Recombinants

微丝结合蛋白 micro filament associated protein
显微分光光度测定法 microspectrophotometry
微管相关蛋白 microtubule associated protein
微管组织中心 microtubule organizing center
★ 微带 mini band
微小染色体 mini chromosome
小卫星染色体 mini satellite DNA
细胞促分裂因子 mitosis promoting factor
有丝分裂器 mitotic apparatus
接收器位点 receptor site
★ 原位杂交 insitu hybridization
缺口转移 nick translation
杂交选择转译 hybrid selected translation
杂交抑制的转译 hybrid arrested translation
无细胞翻译系统 cell free translation system
★ 杂交释放转译 hybrid released translation
鸟枪法 shot gun
沉默子 silencer

1 The use of a restriction enzyme, followed by DNA ligase, can create recombinant plasmids, with a target DNA fragment inserted into a vector plasmid.
限制酶及随后 DNA 连接酶的使用,可以产生目的 DNA 片段插入载体质粒的重组质粒。

2 Bacteria which have taken up a plasmid are selected by growth on a plate containing an antibiotic to which the plasmid vector encodes resistance.
吸收了质粒的细菌可以通过在含某种抗生素的平板上生长的方法进行筛选,该质粒载体上携有编码抗性基因。

3 The efficiency of the transformation step is given by the number of anti-biotic-resistant colonies per microgram of input plasmid DNA.
转化率定义为每毫克转化的质粒 DNA 所产生的具抗生素抗性的克隆数。

4 Single colonies from a transformation plate are grown in liquid medium, maintaining the antibiotic selection for the plasmid, and a portion of the culture is stored for later use as a frozen glycerol stock.
从转化平板上挑取单克隆,接种于含有选择标记抗生素的液体培养基中培养,然后将部分菌液冻存于含一定浓度甘油的保存液中备用。

第七节 克隆基因的表达

Section 7 Expression of Cloning Gene

★ 有意义链 sense strand
★ 反意义链 anti sense strand
抗终止子 antiterminator
★ 衰减子 attenuator [ə'tenjueɪtə]
前导序列 leader sequence
★ 间隔序列 spacer sequence
★ 一致序列 consensus sequence
★ 副作用成分 negatively acting component
重复性基因外回文序列 repetitive extra genic palindromic [ˌpælidrɔmik] sequence
★ 受体质粒 acceptor plasmid
穿梭载体 shuttle vector
中间载体 intermediate vector
高速微型子弹射击法 high-velocity micro projectile
★ 共转染现象 co-transfection
显微注射法 microinjection

1 Determining various properties of a recombinant DNA molecule, such as size, restriction map, orientation of any gene present and nucleotide sequence, constitutes the process of clone characterization. It requires a purified preparation of the cloned DNA.
克隆的鉴定过程包括测定重组 DNA 分子的各种特性如大小、限制图谱、基因的方向以及核苷酸序列，该过程需要纯的已克隆化的 DNA 样品。

2 The nucleic acid in lanes of a gel is transferred to a membrane, bound and then hybridized with a labeled nucleic acid probe. Washing removes nonhybridized probe, and the membrane is then treated to reveal the bands produced. Specific RNA species are detected on Northern blots, whereas the DNA bands on Southern blots could be genes in genomic DNA or parts of cloned genes.
将凝胶泳道中的核酸带转移至膜上，固定之后与标记的核酸探针进行杂交，经洗脱去未杂交的探针，然后对膜进行处理以显示出杂交带。Northern 杂交可检测特定的 RNA 分子，而经 Southern 杂交显示出的 DNA 带应该是 DNA 中基因或已克隆基因的部分。

第八节 酵母的基因工程

Section 8 Gene Engineering of Yeast

★ 自主复制序列 autonomously [ɔː'tɔnəməsli] replicating sequence
整合载体 integrating vector
★ 酵母人工染色体 yeast artificial chromosome
显微操作器 micromanipulator [ˌmaikrəumə'nipjuleitə]
★ 转化基因组 transforming genome

The yeast S. cerevisiae is currently the most popular microbial eukaryote for recombinant protein production. Cloned genes are often placed under the control of the GAL, which is normally upstream of the gene.
酵母 S. cerevisiae 是目前生产重组蛋白的最普遍的真核微生物。克隆的基因一般在 GAL 启动子的控制之下，GAL 启动子可以被半乳糖所诱导。

第九节 植物基因工程

Section 9 Gene Engineering of Plant

★ 体细胞克隆变异 somaclonal ['səuməkləunl] variation
★ 二元系统 binary system
★ 交叉保护 cross-protection
★ 阻抑作用 repression
自发突变体 spontaneous mutant
组成型基因表达 constitutive gene expression
组成型异染色质 constitutive heterochromatin

1 The two main methods of DNA sequencing are the Maxam and Gilbert chemical method in which end-labeled DNA is subjected to base-specific cleavage reactions prior to gel separation, and Sanger's enzymic method. The latter uses dideoxynucleotides as chain terminators to produce a ladder of molecules generated by polymerase extension of a primer.

DNA测序有两种主要方法，一是Maxam和Gilbert的化学法，即在对末端标记的DNA进行碱基特异切割反应后进行凝胶分离的方法；另一是Sanger的酶学法，即用双脱氧核苷酸作为链终止试剂，通过聚合酶的引物延伸产生一系列大小不同的分子后再进行分离的方法。

2 Special computer software is used to search nucleic acid and protein sequences for the presence of patterns (e. g. restriction enzyme sites) or similarities (e. g. to new sequences).

用特定的计算机软件来对核酸及蛋白质的序列进行检索与分析，以找出某图谱(如限制酶切位点)或相似性(如新序列)。

第十节 哺乳动物基因工程
Section 10 Gene Engineering of Mammal

★ 原病毒 provirus
★ 包装细胞 packaging cell
辅助自由 helper free
生殖系 germ line
超级小鼠 super mouse
★ 转基因动物 transgenic animal
★ 转基因小鼠 transgenic mouse
白血病 leukemia [lju:'ki:miə]

1 The various applications of gene cloning include recombinant protein production, genetically modified organisms, DNA fingerprinting, diagnostic kits and gene therapy.

基因克隆技术有多种应用，包括重组蛋白的生产、遗传修饰生物体的构建、DNA指纹分析、诊断试剂盒以及基因治疗。

2 Introducing a foreign gene into an organism which can propagate creates a genetically modified organism. Transgenic sheep have been created to produce foreign proteins in their milk.

向一个生物体引入一个可在其内增殖的外源基因，构成一种遗传修饰生物体。例如已做出的转基因羊能在其奶汁中生产外源蛋白。

3 For proteins with complex and essential glycosylation structures an animal cell might be the only type of host within which the active protein can be synthesized.

对于复杂糖基化结构的蛋白，动物细胞是可以合成有活性蛋白的唯一寄主。

第十一节 医药工业基因工程
Section 11 Gene Engineering of Medicine Industry

★ 瞬时表达 transient ['trænziənt] expression
可溶性受体 soluble receptor
★ 前激素 pre-hormone
血友病 hemophilia [ˌhi:mə'filiə]
弱毒疫苗 attenuated vaccine
★ 亚单位疫苗 subunit vaccine
★ 功能元件 functional element
★ 共价催化 covalent catalyse
★ 共价修饰 covalent modification
效应区域 effector domain
双特异性抗体 bispecific antibody [baispi'sifik]
肿瘤坏死因子 tumor necrose factor

1 The sequence information derived from cloning medically important genes has allowed the design of many diagnostic test kits which can help predict and confirm a wide range of disorders.

可根据已克隆的具有重要医学价值基因的序列信息设计多种诊断试剂盒，来对多种疾病进行预测与确诊。

2 Attempts to correct a genetic disorder by delivering a gene to a patient is described as gene therapy.

通过向病人导入特定基因试图来治疗某种遗传疾病的方法称为基因治疗。

3 However, it must also be appreciated that many important new bio-products will still be synthesized chemically from models derived from existing biological molecules, e. g. , new drugs based on the interferons.

但也应理解，还会有很多重要的新的生物产品仍以化学方法，按现有的生物分子模型进行合成，例如以干扰素为基础的新药。

4 Insulin, synthesized by the β-cell of the Islets of Langerhans in the pancreas, controls the level of glucose in the blood.

胰岛素是由胰腺中的胰岛中的β细胞合成的，可以控制血液中的血糖水平。

第十六章 酶工程

Chapter 16 Enzyme Engineering

第一节 绪论

Section 1 Introduction

★ 酶 enzyme ['enzaim]
脲酶 urease ['juəri,eis]
极端微生物 extremophiles [iks'tri:məfails]

Enzyme engineering (or enzyme technology) is the application of fermentation process using catalytic activity of isolated enzymes, to produce new metabolites or to convert some compounds into another's (biotransformation).
酶工程(酶技术)是指在发酵过程中利用分离出的酶的催化作用,生产新的代谢产物或进行物质转化(生物转化)的技术。

第二节 酶学基础

Section 2 Basic Knowledge of Enzyme

★ 推荐名 recommended name [rekə'mendid][neim][,sisti'mætik][neim]
★ 氧化还原酶 oxidoreductase ['ɔksɪidəuri'dʌkteis]
★ 裂解酶 lyase ['laiəs]
复合酶 conjugated enzyme ['kɔndʒugeitid]['enzaim]
辅因子 cofactor [kəu'fæktə]
★ 辅酶 coenzyme [kəu'enzaim]
辅基 prosthetic group [prɔs'θetik][gru:p]
酶原激活 zymogen activation ['zaimədʒen][,ækti'veiʃən]
过渡态 transition state [træn'ziʒən][steit]
★ 竞争性抑制 competitive inhibition [kəm'petitiv][in'hibitə(r)]
★ 激活剂 activator ['æktiveitə]
★ 比活力 specific activity [spi'sifik][æk'tiviti]
★ 动力学 kinetics [kai'netiks]

1 Enzymes are systematically classified according to the nature of the reaction that they catalyze, such as the transfer of a phosphate group(phosphotransferase) or an oxidation-reduction(oxidoreductase).
酶根据它们催化反应的性质而进行系统分类,如磷酸基团转移(磷酸转移酶)或氧化-还原(氧化还原酶)。

2 In 1913, in a much more widely read paper, Michaelis and Menten expanded on the earlier work of Henri and rederived the enzyme rate equation that today bears their names. The Michaelis-Menten equation, or more correctly the Henri-Michaelis-Menten equation, is a cornerstone of much of the modern analysis of enzyme reaction mechanisms.
1913年,在一篇传播较广的论文中,米歇尔和曼吞扩展了亨利的早期工作,再次推导得到了酶速度方程,也就是现在以他们的名字命名的方程。米歇尔-曼吞方程,或者更准确地说是亨利-米歇尔-曼吞方程,是许多现代酶反应机理分析的基础。

第三节 产酶微生物的分离和选育

Section 3 Separation and Selection of Enzyme-production Microorganism

完全培养基 complete medium [kəm'pli:t]['mi:djəm]
基本培养基 minimal medium ['miniməl]['mi:djəm]
有限培养基 limited medium ['limitid]['mi:djəm]
补充培养基 supplemental medium [,sʌpli'mentl]['mi:djəm]

1 There are a number of properties which may be improved or altered by genetic engineering including

the yield and kinetics of the enzyme, the ease of downstream processing and various safety aspects.
运用基因工程可以改善原有酶的各种性能，包括提高酶的产率和动力学，减轻下游加工和多方面的安全问题。

2 Enzymes from dangerous or unapproved microorganisms and from slow growing or limited plant or animal tissue may be cloned into safe high-production microorganisms.
可以将原来由有害的、未经批准的微生物产生的酶基因，或由生长缓慢的动植物产生的酶基因，克隆到安全高产的微生物体内进行生产。

第四节 酶的生物合成与发酵生产
Section 4 Biosynthesis and Fermentation Production of Enzyme

核糖体 ribosome ['raibəsəum]
★ 悬浮培养 suspension culture [səs'penʃən]['kʌltʃə]
微载体 microcarrier [ˌmaikrəu'kæriə]

End-product repression of enzyme biosynthesis may be prevented by any or all of the following approaches:
(a) limitation of end-product accumulation by adding an inhibitor to the medium;
(b) ensuring that end-products do not occur in the medium supplied;
(c) selection of regulatory mutants (constitutive mutants) which are not repressed by end-products.
可以采用以下任意一种或全部方法防止酶生物合成的终产物阻遏：
(a) 在培养基中加入抑制剂，限制终产物的积累；
(b) 确保添加的培养基中不含终产物；
(c) 不被终产物阻遏突变菌株的选择(组成型突变型)。

第五节 酶的提取、分离与纯化
Section 5 Extraction, Separation and Purification of Enzyme

★ 细胞破碎 cell disruption [sel][dis'rʌpʃən/]
★ 沉淀 precipitation [priˌsipi'teiʃən]
★ 吸附层析 adsorption chromatography [æd'sɔːpʃən][ˌkrəumə'tɔgrəfi]
★ 分配层析 partition chromatography [pɑː'tiʃən][ˌkrəumə'tɔgrəfi]
★ 离子交换层析 ion-exchange chromatography ['aiən][iks'tʃeindʒ][ˌkrəumə'tɔgrəfi]
★ 凝胶层析 gel chromatography [dʒel][ˌkrəumə'tɔgrəfi]
★ 亲和层析 affinity chromatography [ə'finiti][ˌkrəumə'tɔgrəfi/]

Filtration is defined as the process in which particles are separated from a liquid by passing the liquid through a permeable material. The porous filter medium is the permeable material that separates particles from the liquid passing through it and is known as a filter.
过滤可定义为使液体通过可渗透材料从而除去其中微小颗粒的过程。这种可渗透材料是一种多孔物质，称为过滤介质(滤布、滤纸等)，液体通过它后固体颗粒被截留下来。

第六节 酶与细胞的固定化
Section 6 Immobilization of Enzyme and Cell

★ 固定化 immobilization [iˌməubilai'zeiʃən]
原生质体 protoplast ['prəutəplɑːst]
★ 吸附法 adsorption method [æd'sɔːpʃən]['meθəd]
静态法 static process ['stætik][prə'ses]
电沉积法 electrodeposition process [iˌlektrəudi'pɔziʃən][prə'ses]
混合振摇装载法 mixing or shaking bath loading process ['miksiŋ][ɔː]['ʃeikiŋ][bɑːθ]['ləudiŋ][prə'ses]
反应器装载法 reactor loading process [ri(ː)'æktə]['ləudiŋ][prə'ses]

★ 包埋法 entrapment method [in'træpmənt]['meθəd]
脂质体 liposome ['lipəusəum]
★ 结合法 coupling method ['kʌpliŋ]['meθəd]
离子键偶联法 ionic coupling [ai'ɔnik]['kʌpliŋ]
共价键偶联法 covalent coupling [kəu'veilənt]['kʌpliŋ]
★ 交联法 cross linking method [krɔs]['liŋkiŋ]['meθəd/]

1 In practice, immobilization of an enzyme will be achieved by covalent attachment to the surface of a water-insoluble material; by crosslinking with suitable agents to give insoluble particles; by entrapment inside a matrix or gel that is permeable to the enzyme, substrate and products; by encapsulation; and by adsorption of enzymes on solid supports.
实际操作中,酶的固定化方法可以将酶共价结合到水不溶性材料的表面;可以通过合适的试剂交联到某种不溶颗粒上;可以将酶包埋到能使酶、底物、产物透过的基质或凝胶中;可以包埋到微囊中;还可以将酶吸附到固相载体上。

2 High immobilization yields, expression of high activity of the bound enzyme and stabilization against inactivation/denaturation are the principal yardsticks for the measurement of the success of any enzyme immobilization procedure.
固定化酶的高产量、高活力、对失活剂和变性剂的稳定性是酶固定化成功与否的首要评价标准。

第七节 酶反应器
Section 7 Enzyme Reactor

★ 酶反应器 enzyme reactor ['enzaɪm][ri(:)'æktə]
★ 填充床式反应器 packed bed reactor [pækt][bed][ri(:)'æktə]
★ 流化床式反应器 fluidized bed reactor ['flu:idaizd][bed][ri(:)'æktə]
★ 鼓泡式反应器 bubble column reactor ['bʌbl]['kɔləm][ri(:)'æktə]
★ 膜反应器 membrane reactor ['membrein][ri(:)'æktə/]

Enzyme reactor consists of a vessel, containing a reactional medium, used of perform a desired conversion by enzymatic means. Enzymes used in this process are free in the solution or immobilized in particulate, membranous or fibrous support.
酶反应器是装有反应介质的管状容器,用来进行特定的酶转化反应。在这个过程中所使用的酶在溶液中是游离的,或固定在粒状、膜状、纤维状载体上。

第八节 酶传感器
Section 8 Enzyme Sensor

★ 生物传感器 biosensor [ˌbaiəu'sensə]
★ 酶传感器 enzyme sensor ['enzaim]['sensə]
★ 酶电极 enzyme electrode ['enzaim][i'lektrəud]
★ 光纤 fiber optic ['faibə]['ɔptik]
★ 离子敏场效应晶体管 ion sensitive field effect transistor ['aiən]['sensitiv][fi:ld][i'fekt][træn'zistə/]

1 Hundeck et al. (1992) employed a four-channel enzyme thermistor for on-monitoring of glucose, maltose and sucrose during cultivation of *Bacillus licheniformis*.
1992 年,Hundeck 等人应用四通道酶热敏电阻实时监测了地衣芽孢杆菌培养过程中葡萄糖、麦芽糖和蔗糖的含量。

2 The rapid response and reliability of ion-selective electrodes has been combined with the sensitivity and selectivity of enzyme reactions in the production of enzyme probes.
酶探针的制作是离子选择电极的快速反应和可靠性与酶反应的灵敏性和选择性的结合。

第九节 酶的分子修饰
Section 9 Modification of Enzyme Molecule

★ 分子修饰 molecule modification ['mɔlikju:lɜ][ˌmɔdifi'keiʃən]
★ 胃蛋白酶原 pepsinogen [pep'sinədʒən]
★ 胰蛋白酶原 trypsinogen [trip'sinədʒən]
★ 胰凝乳蛋白酶原 chymotrypsinogen [ˌkaiməu'trip'sinədʒən]
★ 定点突变 site directed mutagenesis [sait][di'rektid][ˌmjutə'dʒenisis]
★ 定向进化 directed evolution [di'rektid][ˌi:və'lu:ʃən]
易错 PCR error-prone PCR ['erə][prəun]
DNA 改组 DNA shuffling ['ʃʌfliŋ]
交错延伸 stagger extension ['stægə][iks'tenʃən]
体外随机引发重组 Random ['rændəm] priming *in vitro* recombination ['praimiŋ] ['ri:kɔmbi'neiʃən/]

The use of site-directed mutagenesis (in which one amino acid residue is substituted for another) and deletional mutagenesis (in which sections of the polypeptide chain of a protein are eliminated) have allowed enzymologists to pinpoint the chemical groups that participate in ligand binding and in specific chemical steps during enzyme catalysis.
定点突变(一个氨基酸残基被另外一个替代)和删除突变(蛋白多肽链的一部分被删除)技术的使用令酶学家查明了在酶催化过程中参与配基结合以及参与某一化学反应步骤的化学基团。

第十节 酶的蛋白质工程
Section 10 Protein Engineering of Enzyme

★ 蛋白质工程 protein engineering ['prəuti:n][ˌendʒi'niəriŋ]['sekəndəri]['strʌktʃə]
★ 超二级结构 super secondary structure ['sju:pə]['sekəndəri]['strʌktʃə]
★ 结构域 domain [dəu'mein/]

The deduction of three-dimensional structures from X-ray diffraction or NMR methods depends on knowledge of the arrangement of amino acids along the polypeptide chain of the protein.
利用 X- 射线衍射或核磁共振方法推测蛋白质的三维结构需要首先知道氨基酸在蛋白多肽链上的排列顺序。

第十一节 非水介质中酶的催化作用
Section 11 Enzymatic Catalysis in Non-aqueous Media

★ 非水体系 non-aqueous system ['nɔn'eikwiəs]['sistəm]
★ 超临界流体 supercritical fluids [ˌsju:pə'kritikəl]['flu(:)idz]
★ 低共溶混合体系 eutectic mixture [ju:'tektik]['mikstʃə]
水不互溶有机溶剂单相体系 monophasic organic solution [ˌmɔnə'feizik][ɔ:gænik][sə'lju:ʃən]
水互溶有机溶剂单相体系 monophasic aqueous-organic solution [ˌmɔnə'feizik] ['eikwiəs] [ɔ:gænik][sə'lju:ʃən]
水-有机溶剂两相体系 biphasic aqueous-organic solution [bai'feizik] ['eikwiəs] [ɔ: gænik] [sə'lju:ʃən]
★ 反胶束 reverse micelle [ri'və:s][maisel]
★ 必需水 essential water [i'senʃəl]['wɔ:tə/]

Some enzymes have been shown not only to retain their catalytic function but also to catalyze novel reactions and exhibit remarkable stability in organic solvents.
在有机溶剂中一些酶不但能保留催化功能而且能够催化新反应,稳定性也相当好。

第十七章　生物统计
Chapter 17　BioStatistics

第一节　绪论
Section 1　Introduction

★ 统计学　statistics [stə'tistiks]
★ 统计量　statistic
★ 生物统计　biostatistics, biometry, biometrics [,baiəustə'tistiks][bai'ɔmitri][baiə'metriks]
数理统计　mathematical statistics [,mæθə'mætikəl] [stə'tistiks]
抽样误差　sampling error
系统误差　systematic error [,sisti'mætik]['erə]
测量误差　measurement error
随机事件　random event ['rændəm][i'vent]

1 *Statistical analysis*　Statistical method are mathematical approaches that allow us to arrive at conclusions which are unlikely to be wrong. They also assess how unlikely it is that these conclusions may be wrong.

统计分析　统计方法是能让我们得出正确结论的数学方法。同时也可以估计这些结论出现错误的可能性。

2 *Introducing the Curve Fitting Tool*

You can fit curves and surfaces to data and view plots with the Curve Fitting Tool GUI. This tool lets you:

- Create, plot, and compare multiple fits.
- Use linear or nonlinear regression, interpolation, smoothing, and custom equations.
- View goodness-of-fit statistics, display confidence intervals and residuals, remove outliers and assess fits with validation data.
- Automatically generate code to fit and plot curves and surfaces, or export fits to the workspace for further analysis.

曲线拟合工具箱介绍

该工具箱可以对数据进行曲线拟合和曲面拟合，在图形用户界面(GUI)下观察图形。可以：

- 创建图形、比较多个拟合
- 进行线性和非线性回归、插值和平滑以及自定义方程
- 观察统计的拟合优度，显示置信区间、残差，剔除异常值，用有效数据评估拟合
- 自动生成拟合、绘制曲线和曲面的代码，或为进一步分析输出拟合到工作空间

第二节　描述性统计
Section 2　Descriptive Statistics

★ 总体　population [,pɔpju'leiʃən]
观察单位　element ['elimənt]
★ 样本容量　sample size
样本变异　sample variation
概率　probability [,prɔbə'biləti]
加法原则　addition rule
乘法原则　multiplication rule
二项分布　binomial distribution [bai'nəumiəl][,distri'bju:ʃən]
数据类型　data types
定名测量数据　nominal scale data ['nɔminl][skeil]['deitə]
定序测量数据　ordinal scale data ['ɔ:dinl][skeil]['deitə]
定距测量数据　interval scale data ['intəvəl][skeil]['deitə]
定比测量数据　ratio scale data ['reiʃiəu][skeil]['deitə]
离散型变量　discrete variables [dis'kri:t]['veəriəbəlz]
连续型变量　continuous variables [kən'tinjuəs]['veəriəbəlz]

★ 频数分布 frequency distribution
['fri:kwənsi][,distri'bju:ʃən]
成组频数分布 grouped frequency distribution
相对频数分布 relative frequency distribution
累计频数分布 cumulative frequency distribution
['kju:mjulətiv]['fri:kwənsi][,distri'bju:ʃən]
频率表 frequency table
★ 计数 count
四分位(数)间距 inter-quartile range
[in'tə:- 'kwɔ:tail][reindʒ]
十分位(数)间距 inter-decile range
上四分位数 upper quartile
下四分位数 lower quartile
正态分布 normal distribution
集中趋势 central tendency
['sentrəl]['tendənsi]
均数 mean
中位数 median ['mi:diən]
众数 mode [məud]
分散 dispersion [dis'pə:ʃən]
变化性 variability [,veəriə'biliti]
离散程度描述 measures of dispersion and variability
★ 方差(均方) variance(mean square)
['veəriəns(mi:n/skwεə)]
★ 标准差 standard deviation(SD)
['stændəd][,di:vi'eiʃən]
★ 极差 range [reindʒ]
★ 平方和 sum of squares(SS)
★ 自由度 degree of freedom
★ 显著性 significance [sig'nifikəns]
★ 残差 residuals [ri'zidjuəlz]

1 The mean, or arithmetic average, plays a crucial role. We are interested in two classes of mean value - a population mean and a sample mean.
均值或算术平均值都扮演着重要的角色,我们感兴趣的是两类均值,即:总体均值和样本均值。

2 The mode is simply the most frequently-appearing numerical value in a set of data. In the set [4,7,9,8,8,10], 8 is the mode.
在一组数据中,众数是最大频数出现的值。在[4,7,9,8,8,10]一组数据中,8是众数。

第三节 统计推断
Section 3 Inferential Statistics

统计推断 inferential statistics
[,infə'renʃəl][stə'tistiks]
★ 参数 parameter [pə'ræmitə]
均数的随机抽样分布 random sampling distribution of mean
随机样本的均值 mean of random samples
中心极限定理 central limit theorem
['sentrəl]['limit]['θiərəm]
★ 标准误差 standard error
★ 总体均数的估计 estimating the mean of a population
★ 可信限 confidence limits
['kɔnfidəns]['limits]
★ 精密度 precision [pri'siʒən]
★ 准确度 accuracy ['ækjurəsi]
t 值表 t tables

Inferential statistics involve making inference that go beyond the actual data. They usually involve inductive reasoning(i. e. ,generalizing to a population after having observed).
统计推断是对超出实际范围的数据做出判断,通常是归纳推理(即:从观察的样本来推断总体)。

第四节 假设检验
Section 4 Hypothesis Testing

★ 假设检验 hypothesis test
[hai'pɔθisis][test]
★ 显著性水平 level of significance
['levəl][əv][sig'nifikəns]
★ 置信区间 confidence intervals
['kɔnfidəns]['intəvəlz]
★ 接受域与拒绝域 acceptance and rejection regions
★ 单尾检验 one-sided test(one-tail test)
★ 双尾检验 two-sided test(two-tail test)
★ 无效假设与备择假设 null and alternative hypotheses
[nʌl][ənd][ɔ:l'tə:nətiv][hai'pɔθisi:z]
I 类错误与 II 类错误 type I versus type II errors

★ 单样本检验　one-sample test
★ 双样本检验　two-sample test
★ 方差分析　analysis of variance (ANOVA) [ə'næləsis][əv]['veəriəns]
一维方差分析　one-way analysis of variance
配对样本设计　paired-sample design
★ 独立样本设计　independent-sample design
★ 多重比较　multiple comparisons ['mʌltipəl][kəm'pærisən]
★ 参数检验与非参数检验　parametric and non-parametric tests
★ 两个独立样本比较　comparing two independent samples
★ 两个相关样本比较　comparing two related samples
★ 卡方检验　Chi-square test
列联表　contingency-table method [kən'tindʒənsi]['teibəl]['meθəd]

1 Analysis of variance is used in comparing more than two sample groups.
方差分析用于两个以上样本组的比较。

2 To test a hypothesis about a mean, the steps are as follows:

- State the null and alternative hypotheses, H_0 and H_A.
- Select the decision criterion α (or "level of significance").
- Establish the critical values.
- Draw a random sample from the population, and calculation the mean of that sample.
- Calculate the standard deviation and estimate standard error of the sample.
- Calculate the value of the test statistict that corresponds to the mean of the sample.
- Compare the calculated value of t with the critical values of t, and then accept or reject the null hypothese.

检验一个均数的假设，步骤如下：

- 陈述无效假设与备择假设，H_0 与 H_A；
- 选择决策准则 α(或显著性水平)；
- 建立临界值；
- 从总体中抽取随机样本，并计算该样本的平均数；
- 计算样本的标准差，并估计样本的标准误差；
- 计算与样本均数相对应的 t 检验的值；
- 将 t 的计算值与 t 的临界值进行比较，确定是否接受无效假设。

3 *Confidence Interval Approach* (comparing a standard value to the sample mean)

Step 1. Formulate the hypothesis.

H_0: the sample mean is not different from the standard value (sample = standard)

H_A: the sample mean is different from the standard value (sample $\neq$ standard)

Step 2. Determine α, and select the sample size, n.

Step 3. Select the statistical test to be used.

Step 4. Specify the acceptance / rejection of the alternative hypothesis, also known as the decision rule.

Step 5. Perform the experiment / test.

Step 6. Make the decision based on the computed confidence interval and the standard value.

置信区间方法(标准值与样本均值比较)

(1) 陈述假设，H_0：样品均值 = 标准值；H_A：均值 $\neq$ 标准值。
(2) 确定 α，选择样本容量 n。
(3) 选择一种统计检验。
(4) 根据决策准则，指定备择假设的接受域及拒绝域。
(5) 进行实验及检验。
(6) 根据计算的置信区间及标准值做出判断。

第五节 回归与相关分析

Section 5 Regression and Correlation Analysis

★ 相关 correlation [ˌkɔriˈleiʃən]
偏相关 partial correlation
多重相关 multiple correlation
正相关 positive correlation
负相关 negative correlation
★ 因变量 dependent variable
★ 独立变量 independent variable
相关系数类型 type of correlation coefficient [ˌkɔriˈleiʃən][kəuiˈfiʃənt]
★ 决定系数 coefficient of determination
双变量散点图 bivariate scatter plot [baiˈvɛərieit][ˈskætə][plɔt]
★ 线性回归 linear regression
★ 非线性回归 nonlinear regression
★ 多元回归 multiple regression [ˈmʌltipəl][riˈgreʃən]
最小二乘方程 least squares equation
变异性 variability [ˌvɛəriəˈbiliti]
★ 总平方和 total sum of squares(Total SS)
回归平方和 regression sum of squares(Reg SS)
★ 残差平方和 residual sum of squares(Res SS)
回归平均平方 regression mean squares(Reg MS)
残差平均平方 residual mean squares(Res MS)

1 Correlation, which is used to establish and quantify the strength and direction of the relationship between two variables.
相关是用来建立和量化两个变量之间关系的程度与方向。

2 Regression, which is used to express the functional relationship between two variables, so that the value of one variable can be predicted from knowledge of the other.
回归是用来表达两个变量之间的函数关系,因此一个变量的值可以从另一个变量的信息预测出来。

3 A regression analysis is appropriate when two variables, y and x, are related. y is termed the dependent variable, and x, the independent variable. The dependent variable, y, is the quantitative measured response when the independent variable, x, is set at various levels or points.
当 y 与 x 有联系时,采用回归分析是恰当的。y 称为因变量,而 x 为自变量(独立变量),当自变量 x 做系列变化时,因变量 y 也随之做相应变化。

4 The most widely used value in correlation analysis is the correlation coefficient, r. The r value ranges between -1 and $+1$, negative if the regression slope is negative, and positive if the regression slope is positive.
在相关分析中最广泛使用的是相关系数 r,r 值在 -1 与 $+1$ 之间,若回归的斜率是负值,r 值为负,若回归的斜率是正值,r 值为正。

5 Coefficient of Determination: r^2
The coefficient of determination, r^2, is an expression more useful than the correlation coefficient, r, because it can be interpreted directly. An $r^2 = 0.90$ means that 90% of the data variability can be explained by the regression equation. This is very valuable to know. In the previous example, where $r = -0.96$, $r^2 = 0.92$, meaning only 92% of the variability in the data was explained by the regression equation. When the $xi = 0$ values are omitted from the equation, $r = -0.98$, and $r^2 = 0.96$. This means that 96% of the variability in the data is explained by the regression equation. The r^2 is always a positive value.
决定系数 r^2:
决定系数 r^2 是一个比相关系数 r 更为实用的表示方式,这是由于它能更为直接的表达。$r^2 = 0.90$,说明 90% 的数据变化可以用回归方程表示。这是一个很有价值的信息。在前面的例子中 $r = -0.96$,$r^2 = 0.92$,说明只有 92% 的数据变化可以用回归方程表示。当把 $xi = 0$ 的值从方程中删去后,则变为 $r = -0.98$,$r^2 = 0.96$,这意味着有 96% 的数据变化可以用回归方程表示。r^2 值永远是正值。

第十八章　植物细胞工程
Chapter 18　Plant Cell Engineering

前　言
Foreword

★ 植物细胞工程　plant cell engineering or plant cell biotechnology
★ 胚胎拯救　embryo rescue

第一节　植物细胞工程的理论基础——细胞全能性学说
Section 1　Theoretical Elements of Plant Cell Engineering — Cell Totipotency

★ 细胞全能性学说　cell totipotency [sel][təu'tipətənsi]
★ 滋助培养　nurse culture
★ 胚性细胞　embryogenic cell or embryonic cell
★ 胚胎干细胞　embryonic stem cell
组织干细胞　tissue stem cell
★ 干细胞后裔　progeny of stem cell
子代细胞　daughter cell
干细胞促进途径　stem cell promoting pathway
鱼雷胚时期　torpedo stage
不定胚　adventitious [ˌædven'tiʃəs] embryo
不定芽　adventitious bud
★ 脱分化　dedifferentiation [diːˌdifərenʃi'eiʃən]
★ 再分化　redifferentiation
★ 体细胞胚胎发生　embryogenesis [ˌembriəu'dʒenisis]
同源异型基因　homeobox gene
器官发生　organogenesis [ˌɔːgənəu'dʒenisis]

第二节　细胞培养的设备和基本方法
Section 2　Facility and Cardinal Techniques of Cell Culture

萘乙酸　naphthalene ['næfθəliːn] acetic acid (NAA)
★ 吲哚乙酸　indoleacetic acid(IAA)
★ 吲哚丁酸　indole-3-butyric acid(IBA)
★ 激动素　kinetin(KT) ['kainətin]
★ 玉米素　zeatin(ZT) ['ziːətin]
★ 赤霉素　gibberellins(GA3) [ˌdʒibə'relin]
脱落酸　abscisic acid (ABA)
植物激素　plant hormones
植物生长调节剂　plant growth regulator
★ 生长素　auxin
细胞分裂素　cytokinin

第三节　植物细胞培养脱毒快速繁殖技术
Section 3　Toxin Doffing and Rapid Propagation Techniques of Plant Cell Culture

微繁　micropropagation ['maikrəuˌprɔpə'geiʃən]
★ 快速繁殖　rapid propagation
微嫁接　micrografting
WPM 培养基　wood plant medium
温热疗法　thermotherapy [ˌθəːməu'θerəpi]
冷冻法　cryotherapy ['kraiəuθerəpi]
抗血清　antiserum ['ænti'siərəm]
陷入法　trapping

第四节　胚胎培养和胚乳培养
Section 4　Embryo and Albumen culture

★ 双受精　double fertilization
合子　zygote
原胚　proembryo [ˌprəu'embriəu]
球形胚　globular embryo
心形胚　heart-shaped embryo
鱼雷形胚　torpedo-shaped embryo
拐杖形胚　walking-stick-shaped embryo
倒 U 形胚　inverted U-shaped embryo
成熟胚　mature embryo
★ 胚胎因子　embryo factor

★ 提前萌发 precocious germination

第五节 体细胞胚胎培养和人工种子
Section 5 Somatic Embryo Culture and Artificial Seeds

★ 体细胞胚胎发生 somatic embryogenesis
★ 体细胞胚 somatic embryo
不定胚 adventitious embryo
★ 胚性愈伤组织 embryogenic callus
胚性细胞复合体 embryogenic cell complex
★ 人工种子 artificial seed
体细胞种子 somatic seed

第六节 体细胞无性系变异与植物改良
Section 6 Somaclonal Variation and Plant Upgrading

★ 体细胞无性系变异 somaclonal variation
内复制 endoduplication
核内有丝分裂 endomitosis
逆转座子 retrotransposon [ˌri:trəuˌtræns'pəusɔn]

第七节 单倍体的诱导与单倍体育种
Section 7 Haploid Induction and Breeding

★ 半配合 semigamy ['semi'geimi]
雄核发育 androgenesis ['ændrəu 'dʒenisis]
花粉胚 pollen embryo
★ 花粉愈伤组织 pollen callus
花粉植株 pollen plant
★ 花粉培养 pollen culture
加倍单倍体植株 doubled haploid plant
★ 小孢子培养 microspore culture
游离小孢子培养 isolated microspore culture
散落法 shed pollen method

第八节 原生质体培养和体细胞杂交
Section 8 Protoplast Culture and Somatic Cell Crossing

微滴培养 droplet culture
★ 体细胞杂种 somatic hybrid
★ 异核体 heterokaryon [ˌhetərəu"kæriɔn]
★ 胞质杂种 cybrid ['sibrid]

第九节 药用植物细胞的大量培养
Section 9 Quantitative Culture of Officinal Plant Cells

★ 生长培养基 growth medium
★ 生产培养基 production medium
★ 诱导子 elicitor [i'lisitə]
防御物质 phytoalexin [ˌfaitəuə'leksin]
★ 细胞悬浮培养 cell suspension culture
★ 生物反应器 bioreactor
细胞刮削器 scraper ['skreipə]
转化酶 invertase [in'və:teis]

第十节 植物细胞转化技术
Section 10 Transforming Techniques of Plant Cells

微弹轰击法 microprojectile bombardment ['maikrəuprə'dʒektail][bɔm'bɑ:dmənt]
★ 花粉管通道转化法 pollen tube pathway or pollen transformation

第十一节 无融合生殖与杂种优势的永续利用
Section 11 Apomixis and Eternal Utilization of Heterosis

★ 无融合生殖 apomixis [ˌæpə'miksis]
★ 专性无融合生殖 obligate apomixis
★ 兼性无融合生殖 facultative apomixis
★ 配子体无融合生殖 gametophytic apomixis
★ 孢子体无融合生殖 sporophytic apomixis
★ 孤雌生殖 parthenogenesis [ˌpɑ:θinəu'dʒenisis]
★ 无配子生殖 apogamy [ə'pɔgəmi]
★ 单雄生殖 androgenesis ['ændrəu'dʒenisis]
★ 假受精 pseudogamy [ˌsju:'dɔgəmi]
★ 二倍体无融合生殖 diploid parthenogenesis

第十九章　动物细胞培养
Chapter 19　Animal Cell Culture

第一节　组织细胞培养的基本概念
Section 1　Rudimentary Concepts of Tissue and Cell Culture

★ 组织培养　tissue culture
器官培养　organ culture
复制式细胞培养　replicate cell culture
摇瓶培养　shake-flask culture
锚着依存性细胞　anchorage dependent cells
细胞分裂指数　mitotic index（MI）
含核体　synkaryon
★ 原种　stock cells
★ 细胞融合　cell fusion
★ 细胞杂交　cell hybridization
汇合　confluent [ˈkɔnfluənt]
体外恶性转化　in vitro malignant transformation
集落形成率　plating efficiency
饱和密度　saturation density
贴壁率　attachment efficiency
★ 传代培养　passage culture
贴壁依赖性　anchorage-dependent
★ 细胞一代时间　cell generation time
群体倍增时间　population doubling time
★ 细胞系　cell line
★ 细胞株　cell strain
★ 有限细胞系　finite cell line
★ 连续或无限细胞系　continuous cell line or infinite cell line
获得不死性　immortality [imɔːˈtæləti]
★ 接触抑制　contact inhibition
★ 密度抑制　density inhibition

1 The study of animal cells has helped us gain an insight not only in the structure and function of cells and tissues but also in different physiological, biochemical and immunological processes.
动物细胞的研究能够帮助我们获得许多知识，不仅是细胞和组织的结构和功能方面，也包括生理学方面、生物化学方面和免疫学过程方面的知识。

2 Animal cells are known to produce many chemicals of great value to humans.
众所周知，动物细胞能产生许多对人类有价值的化学制品。

第二节　细胞培养的准备
Section 2　Preparation of Cell Culture

★ 细胞培养箱　cell incubator
★ 超净工作台　clean bench
★ 倒置显微镜　invert microscope
★ 干热灭菌　dry sterilization
★ 湿热灭菌　wet sterilization
★ 无菌操作　sterile optation

1 A cell incubator is an apparatus used to grow and maintain cell cultures. The incubator keeps cultures at an optimal temperature and humidity. Many will also regulate the carbon dioxide(CO_2) and oxygen content of the atmosphere inside, in which case they are also called CO_2 incubators.
细胞培养箱是用来培养和保存细胞培养物的设备。细胞培养箱要保持在一个最适的温度和湿度进行培养。许多细胞培养箱要调控内部空气中氧和二氧化碳的浓度，因此也叫二氧化碳培养箱。

2 Cell incubators are essential for a lot of experimental work in cell biology and molecular biology and are used to culture both bacterial as well as eukaryotic cells.
细胞培养箱对开展细胞生物学和分子生物学的很多实验工作很重要，并且也用于培养真核细胞和细菌。

第三节　细胞培养用液
Section 3　Solution for Cell Culture

★ 平衡盐溶液　balanced salt solution(BSS)
★ 离子交换水　ion exchange water

蒸馏水 distilled water
★ 胰蛋白酶 trypsin [ˈtripsin]
抗霉素 antimycin [ˌæntiˈmaisin]
★ 青霉素 penicillin [ˌpeniˈsilin,pəˈni]
★ 氨苄青霉素 ampicillin [ˌæmpɪˈsɪlɪn]
★ 天然培养基 natural media
药用培养基 pharmic media
★ 基础培养基 minimal medium
★ 胎牛血清 fetal bovine serum(FBS)
★ 小牛血清 calf serum(CS)
★ 合成培养基 synthetic media
转铁蛋白 transferrin
大豆胰酶抑制剂 soybean trypsin inhibitor
纤维连接素 fibronectin
多聚赖氨酸 poly-lysine

1 Trypsin enzyme acts to degrade protein; it is often referred to as a proteolytic enzyme, or proteinase. Trypsin is one of the three principal digestive proteinases, the other two being pepsin and chymotrypsin.
胰蛋白酶的作用是降解蛋白,经常作为蛋白水解酶或蛋白酶。胰蛋白酶是三种主要的消化蛋白酶之一,另外两种是胃蛋白酶和胰凝乳蛋白酶。

2 In the digestive process, trypsin acts with the other proteinases to break down dietary protein molecules to their component peptides and amino acids.
在消化过程中,胰蛋白酶和其他蛋白酶一起将食物中的蛋白分子分解成多肽和氨基酸。

第四节 细胞培养的基本技术
Section 4 Basic Technology of Cell Culture

★ 消化 digestion
★ 分散 discrete
匀浆机 homogenizer [həˈmɔdʒinaizə]
★ 台盼蓝 trypanblau
★ 盖玻片 cover slip
★ 载玻片 slide
★ 核型 karyotype [ˈkæriətaip]

1 Methods of cultivation of animal cells have anchorage dependent culture and suspension culture.
动物细胞培养方法有贴壁依赖式培养和悬浮式培养。

2 Trypanblau is a vital dye. Viable cell counts using trypan blue.
台盼蓝是一种活体染料。活细胞计数使用台盼蓝。

第五节 细胞培养污染的检测与排除
Section 5 Detection and Elimination of Cell Culture Contamination

疱疹病毒 herpesvirus [ˌhəːpiːzˈvaiərəs]
乳多空病毒 papovavirus
逆转录病毒 retrovirus [ˌretrəuˈvaiərəs]
前病毒 provirus [prəuˈvaɪərəs]
★ 污染 contaminate
★ 检验 inspection
★ 检疫 quarantine
★ 排除 eliminate

1 The contamination of cell cultures by mycoplasmas remains a major problem in cell culture.
细胞培养中由支原体引起的污染,是细胞培养的主要问题。

2 Mycoplasmas are resistant to most antibiotics commonly employed in cell cultures.
支原体能抵抗细胞培养中使用的许多抗生素。

第六节 原代细胞培养
Section 6 Primary Cell Culture

★ 原代培养 primary culture
★ 组织块培养法 tissue culture
组织块预消化培养法 tissue predigest [ˈpriːdiˈdʒest] cell culture
气液界面培养法 biphasic cell culture
悬浮细胞培养法 suspension cell culture
微载体细胞培养法 microcarrier cell culture
中空纤维细胞培养法 hollow fibre cell culture
切除 excise
★ 外植体 explant

无钙镁 Hanks CMF-Hanks
单层细胞培养法 monolayer cell culture
成纤维样的 fibroblastic
异质的 heterogeneous [ˌhetərəu'dʒiːniəs]
小鼠 mouse
脾细胞 spleen cell

1 Primary cultures, which are obtained directly from an animal and can keep the differentiated state for a short period (days to weeks).
原代培养是直接从动物体获得细胞并在短时间内保持其特有形态(几天或几星期)。

2 Functionally differentiated primary cell cultures have a limited life span, and although maintenance of the differentiated properties has been improved by additives to the culture medium, components of the extracellular matrix or by different forms of co-culture, cell specific functions will eventually decline.
原代培养细胞的特有功能的保持是有限的,由于加入的培养液,细胞外基质的组成或不同形式的共培养,其保持的特性将被改变,细胞特有的功能最终将被减弱。

第七节 二倍体细胞株的培养
Section 7 The Culture of Diploid Cell Strain

★ 哺乳动物细胞 mammalian cell
细胞群体倍增 population doubling
成纤维细胞 fibroblast cell
★ 二倍体细胞系 diploid cell line

Cells in higher eukaryotic organisms proliferate in the diploid phase to form a multicellular organism, only the gametes being haploid.
高等真核生物的细胞以二倍体形式繁殖形成多细胞生物,只有配偶子是单倍体。

第八节 传代细胞的培养
Section 8 Culture of Subcultured Cell

贴壁依赖性细胞 anchorage-dependent cell
胰酶消化作用 trypsinization [tripsinai'zeiʃən]
胰酶终止液 trypsin inhibition
★ 传代细胞 subcultured cell
★ 有限细胞系 finite cell line
★ 细胞周期时间 time of cell cycle(TC)
★ 人子宫颈癌细胞 Hela cell
★ 传代时间 generation time
非贴壁依赖性细胞 anchorage-independent cell
★ 淋巴细胞 lymphocyte ['limfəsait]

1 Primary cell cultures typically will have a finite life span in culture whereas continuous cell lines are, by definition, abnormal and are often transformed cell lines.
原代细胞培养的特征是:细胞的生命是有限的,而连续细胞系是无限的、非正常的,通常是被转化的细胞系。

2 Liquid N_2 is used to preserve tissue culture cells.
液氮通常用来保存组织培养细胞。

3 The freezing medium typically use 90% serum, 10% DMSO.
冻存液通常用 90% 的血清和 10% 的 DMSO。

第九节 细胞克隆
Section 9 Cell Colony

克隆形成法 colony formation
均一的细胞集团 homogeneous cell population
评价 appraise
★ 培养板 plate
克隆形成率 plating efficiency
非转化细胞 untransformed cell
★ 转化细胞 transformed cell
转化的恶性性质 malignant transformation
琼脂平板培养法 agar plate culture

Colony formation test of the transduced cells was studied by soft-agarose cell colony formation test.
转导细胞克隆形成试验是用软琼脂糖细胞克隆法。

第十节 细胞培养技术的应用

Section 10 Application of Cell Culture Technology

细胞病变效应 cytopathic effect(CPE)
空斑单位 plaque forming unit
Rous 肉瘤病毒 rous sarcoma virus(RSV)
原癌基因 proto-oncogene
标记基因 marker gene
★ 转染 transfection
羟基脲 hydroxyurea [hai,drɔksi'juəriə]
前黄素 proflavine [prəu'fleivi:n]
放线菌酮 cycloheximide [ˌsaikləu'heksimaid]
秋水仙胺 colcemid
骨髓瘤 myeloma [ˌmaiə'ləumə]
饲养 feeder
★ 融合 fusion
胸腺嘧啶核苷 thymidine
★ 细胞因子 cytokines [ˌsaitəukai'ni:sis]
★ 肿瘤坏死因子 tumor necrosis factor, TNF
血凝素 hemagglutinin [hi:mə'glu:tɪn]
人抗鼠抗体 human anti-mouse antibody
再程序化 reprogram
共转化技术 co-transformation
内细胞团 inner cell mass
★ 胚胎干细胞 embryonic stem cell(ES)
胚胎生殖细胞 embryonic germ cell(EG)
树突细胞 dendritic cell(DC)
工程组织 engineered tissues
工程器官 engineered organs
细胞生长支持物 cell growth supplement
人类突变细胞库 human mutant coll repository (HMCR)
衰老细胞库 ageing cultured cell repository(ACCR)
仙台病毒 Sandai virus
电融合 electrofusion
生殖细胞 germ cell
杂交瘤 hybridoma
饲养细胞 feeder cell
嵌合抗体 chimeric antibody
重构抗体 reshaped antibody
人化抗体 humanized antibody

1 The hybrid cell is a hybridoma cell that produces a monoclonal antibody.
杂交细胞是能产生一种单克隆抗体的杂种细胞。

2 Embryonic stem cells (ES cells) could generate all the different cell types in the mammalian body, and that adult stem cells are more restricted in their developmental potential.
胚胎干细胞能分化成哺乳动物体的所有不同类型的细胞，成体干细胞在分化潜能上受到限制。

3 Many researchers have observed that adult neural and haematopoietic stem cells are not limited to forming cells of the central nervous system or blood, respectively, but can also generate liver, intestinal cells, and heart and skeletal muscle.
许多研究者发现成人的神经和造血干细胞是不被限制的，能形成中枢神经系统细胞和血细胞，也能形成肝、肠和心、骨骼肌细胞。

4 The cytokines play an important role in mediating many immune reactions of the body, such as tumor immunity, transplantation immunity, and autoimmunity. The effects of cytokines have a character of multiplication and networking. They can not only act lonely, but also interact with other cytokines by synergism or antagonism.
细胞因子在介导机体多种免疫反应，如肿瘤免疫、移植免疫、自身免疫等过程中发挥重要作用。细胞因子的作用具有多相性、网络性的特点。它们不仅可以单独发挥生物学作用，而且不同的细胞因子之间还有相互协同或制约作用。

第二十章　遗传学
Chapter 20　Genetics

第一节　什么是遗传学
Section 1　What is Genetics

★ 遗传　heredity
★ 遗传学　genetics
★ 基因　gene

Genetics is the study of the mechanisms of heredity by which traits or characteristics are passed from generation to generation.
遗传学是性状怎样从一代传向下一代遗传机制的研究。

第二节　孟德尔定律
Section 2　Mendel's Laws

孟德尔定律　Mendel's law
自由组合定律　law of independent assortment
分离定律　law of segregation
★ 子一代　first filial generation(F1)
★ 正反交　reciprocal cross
★ 显性性状　dominant character
★ 隐性性状　recessive character
★ 颗粒式遗传　particulate inheritance
混合式遗传　blending inheritance
★ 遗传因子　hereditary determinant
★ 配子　gamete
合子　zygote
★ 等位基因　allele
★ 基因型　genotype
★ 表型　phenotype
★ 纯合体　homozygote [ˌhɔməˈzaigəut]
★ 杂合体　heterozygote [ˌhetərəuˈzaigəut]
★ 测交　test cross
★ 亲组合　parental combination
★ 重组合　recombination
★ 棋盘法　punnet square
★ 分支法　branching process
★ 孟德尔性状　Mendelian character

1 Dominant character is the trait that expresses itself in all the offspring.
显性性状是在后代表现出来的性状。

2 Recessive character is the trait that fails to express in all the offspring.
隐性性状是在后代不会表现出来的性状。

第三节　遗传的染色体学说
Section 3　Chromosome Theory of Inheritance

★ 初级缢痕　primary constriction
★ 次级缢痕　secondary constriction
★ 随体　satellite [ˈsætəlait]
★ 绳珠模型　beads on-a-string model
★ 有丝分裂　mitosis
核区　nuclear area
★ 纺锤体　spindle body
胞质分割　cytokinesis [ˌsaitəukaiˈni:sis]
★ 浓缩期　diakinesis [ˌdaiəkaiˈni:sis]
★ 同源染色体　homologous chromosome
双价体　bivalent [ˈbaiˌveilənt]
★ 姐妹染色单体　sister chromatid
★ 交叉端化　terminalization
★ 性原细胞　gonium [ˈgəunjəm]
★ 精原细胞　spermatogonium [ˌspə:mətəˈgəuniəm]
★ 卵原细胞　oogonium [ˌəuəˈgəuniəm]
初级精母细胞　primary spermatocyte [ˈpraiməri][spəˈmætəsait]
初级卵母细胞　primary oocyte [ˈpraiməri][ˈəuəsait]
精子形成　spermatogenesis [ˌspə:mətəuˈdʒenisis]
★ 小孢子母细胞　microsporocyte
★ 大孢子母细胞　macrosporocyte
反足细胞　antipodal [ænˈtipədl]

★ 极核 endosperm nuclei ['endəuspə:m]['nju:kliai]

★ 双重受精 double fertilization

★ 交配型 mating type

★ 染色体学说 chromosome theory of inheritance

1 Mitosis is regularly associated with nuclear division of vegetative or somatic cells.
有丝分裂通常与营养细胞或体细胞核分裂方式有关。

2 Meiosis occurs in conjunction with formation of reproductive cells in sexually reproducing species.
减数分裂发生在有性生殖物种形成生殖细胞的过程。

3 A chromatid is a longitudinal half chromosome, sharing a common centromere with its sister chromatid.
染色单体是与其姊妹染色单体共有一个着丝点的一半染色体。

第四节 基因的作用及其与环境的关系

Section 4 Effect of Gene and its Relations with the Environment

★ 表型效应 phenotypic effect

★ 多效现象 pleiotropism

★ 表现度 expressivity

成骨不全 osteogenesis [ˌɔstiə'dʒenisis] imperfecta

★ 外显率 penetrance ['penitrəns]

并显性 codominance [ˌkəu'dɔminəns]

★ 镰形细胞贫血症 sickle cell anemia [ə'ni:miə]

★ 表型模写 phenocopy ['fi:nəkɔpi]

★ 显性致死 dominant lethal

★ 配子致死 gametic lethal

★ 合子致死 zygotic lethal

★ 亚致死现象 partial lethality

★ 遗传背景 genetic background

凝集素 lectin

★ 新生儿溶血症 erythroblastosis [iˌriθrəublæs'təusis] fetalis

★ 互补基因 complementary genes

不相容现象 incompatibility

★ 强化基因 intensifier [in'tensifaiə]

★ 限制基因 restriction gene

★ 抑制基因 inhibitor

上位效应 epistasis [i'pistəsis]

血型 blood group

1 Phenocopy is an environmentally induced phenotype that resembles the phenotype produced by a mutation.
表型模写是指环境导致的一种表型变化和某种由突变产生的表型相似。

2 Expressivity is the degree to which a particular genotype is expressed in the phenotype.
表现度是指特定的基因型表达成表型的程度。

3 Penetrance is the proportion of individuals with a specific genotype who manifest that genotype at the phenotype level.
外显率是指在一个群体中带有特定基因型的个体显示出相应表型的比例。

4 Multiple allelism is the existence of several known alleles of a gene.
复等位现象是指一个基因存在几种已知等位基因的现象。

5 Epistasis is a situation in which the differential phenotypic expression of genotypes at one locus depends on the genotype at another locus.
上位性是指在一个座位上基因型表达为不同的表型取决于另一个座位上的基因的情况。

第五节 性别决定与伴性遗传

Section 5 Sex Determination and Sex-linked Inheritance

★ 性染色体 sex-chromosome

★ 常染色体 autosome ['ɔ:təsəum]

★ 异配性别 heterogametic sex

★ 同配性别 homogametic sex

★ 伴性遗传 sex-linked inheritance

★ 绞花遗传 criss-cross inheritance

★ X-连锁遗传 X-linked inheritance

A 型血友病 hemophilia [ˌhi:mə'filiə] A

X-连锁隐性遗传 X-linked recessive inheritance

★ 携带者 carrier

色盲 color blindness

★ 限性遗传 holandric [həu'lændrik] inheritance

1 Sex-linked genes are genes located exclusively on the X chromosome.
性连锁基因特指只位于 X 染色体上的基因。

2 Holandric genes are genes occurring only on the Y chromosome can produce their effects only in males.
限雄遗传基因是指 Y 染色体上的特有基因，性状只在雄性体现。

3 Sex-limited inheritance is mechanism whereby a given trait is limited to one sex.
限性遗传是指某给定性状只在一个性别个体中表现的机制。

4 Sex-influenced inheritance is mechanism in which dominance of a given allele depends on the sex of the bearer.
从性遗传是指某显性基因是否表现性状取决于个体性别的机制。

第六节 染色体和连锁群
Section 6 Chromosome and Linkage Group

交叉型假设 chiasmatype [kai'æzmə,taip] hypothesis
★ 交叉 chiasma [kai'æzmə]
对应片段 homologous segment
交换 crossing over
★ 基因重组 recombination
★ 连锁群 linkage group
★ 三点侧交 three-point test cross
图距 map distance
★ 重组子 recombinant
三杂合体 trihybrid [trai'haibrid]
★ 干涉 interference
★ 并发率 coincidence [kəu'insidəns]
★ 连锁图 linkage map
遗传学图 genetic map
重组率 recombination fraction
★ 交换率 crossing over percentage
交互过程 reciprocal process
★ 基因转变 gene conversion
★ 原养型 prototroph ['prəutətrɔf]
★ 营养缺陷型 auxotroph ['ɔ:ksətrɔf]
★ 标记基因 gene marker

1 Linkage group are all the genes carried on a given chromosome
连锁群是指某给定染色体上的所有基因构成一个连锁群。

2 Interference is a phenomenon that once a crossover occurs, the probability of another crossover in an adjacent region is reduced.
干扰是指一次交换发生后，染色体上临近区域发生另一次交换的概率降低的现象。

第七节 细菌和噬菌体的重组和连锁
Section 7 Recombination and Linkage of Bacteria and Phage

基因带 genophore ['dʒi:nəfɔ:]
★ 抗性 resistant
遗传信息 genetic information
★ 高频重组 high frequency of recombination
★ 影印培养 replica plating
★ 中断杂交技术 interrupted mating technique
相反的重组子 reciprocal recombinant
★ 性导 sexduction/F-duction
雄性菌株 male strain
过滤因子 filterable agent
★ 特异性转导 specialized transduction ['speʃəlaizd][træns'dʌkʃən]
特异转导者 transducer
超数感染 superinfection [,sju:pərin'fekʃən]

1 Fertility factor (F factor) is a bacterial episome whose presence confers donor ability (maleness).
可育因子(又称 F 因子)是细菌的一种附加因子，它的存在使宿主具有供体的能力(雄性)。

2 Sexduction is sexual transmission of donor E. coli chromosomal genes on the fertility factor.
性导是供体大肠杆菌的染色体基因在可育因子上的性传导。

3 Interrupted mating is aA technique used to map bacterial genes by determining the sequence in which donor genes enter recipient cells.
中断杂交是通过确定供体基因进入受体基因的时间来对细菌基因组绘图的技术。

4 Transduction is the movement of genes from a bacterial donor to a bacterial recipient with the use of a phage as the vector.
转导是以噬菌体为载体，基因从供体菌转移到受体菌的过程。

第八节 数量性状遗传

Section 8 Quantitative Character Inheritance

★ 质量性状 qualitative character
★ 数量性状 quantitative character
★ 多基因假说 multiple factor hypothesis
方差 variance [ˈvɛəriəns]
标准误 standard error
★ 广义遗传率 heritability in the broad sense
★ 狭义遗传率 heritability in the narrow sense
★ 株系 line
★ 纯系 pure line
★ 近亲繁殖或近交 inbreeding
纯合性 homozygosity [ˌhəuməuzaiˈgɔsəti]
衰退现象 depression
杂合性 heterozygosity
★ 自交 selfing
★ 近交系数 coefficient of inbreeding
★ 杂种优势 heterosis [ˌhetəˈrəusis]
★ 显性假说 dominance hypothesis
偏态分布 skewed distribution
正态分布 normal distribution
★ 超显性假说 overdominance hypothesis
等位基因的相互作用 intra-allelic interaction
多基因 polygenes

1 Quantitative character is the existence of a range of phenotypes for a specific character, differing by degree rather than by distinct qualitative differences.
数量性状是显示某一特征的连续的表型性状，不同于差别分明的质量性状。

2 Multiple-factor hypothesis is a hypothesis to explain quantitative variation by assuming the interaction of a large number of genes (polygenes), each with a small additive effect on the character.
多因子学说是一种解释数量变异的学说，认为数量变异是多个基因相互作用，而每个基因带有很小的相加效应。

3 Heritability in the broad sense is the proportion of total phenotypic variance at the population level that is contributed by genetic variance.
广义遗传率是在群体中由于基因变异引起表型变异的比例。

4 Heritability in the narrow sense is the proportion of phenotypic variance that can be attributed to additive genetic variance.
狭义遗传率是加性基因方差在表型方差中所占的比例。

5 Inbreeding coefficient is the probability of homozygosity that results because the zygote obtains copies of the same ancestral gene.
近交系数是同一个祖先的一个基因拷贝在某后代中得到纯合的概率。

第九节 遗传物质的改变（一）染色体畸形

Section 9 Change of Genetic Matter (1) Chromosomal Abnormality

★ 染色单体的断裂 breakage
★ 染色粒 chromomere
多线染色体 polytene chromosome
★ 缺失环 deletion loop
★ 断裂融合桥的循环 breakage-fusion-bridge cycle
★ 拟显性现象 pseudodominance
★ 猫叫综合征 cri-du-chat syndrome
★ 染色体组 chromosome complement
★ 不等交换 unequal crossover
★ 相互易位 reciprocal translocation
邻近分离 adjacent segregation
交换分离 alternate segregation
★ 假连锁现象 pseudolinkage
★ 平衡易位携带者 balanced translocation carrier
★ 染色体病 chromosomal disease
★ 永久杂种 permanent hybrid
★ 平衡致死品系 balanced lethal system
游离端 telomere [ˈteləmiə]
多倍体 polyploid [ˈpɔliˌplɔid]
★ 同源多倍体 autopolyploid [ˈɔːtəu ˈpɔliˌplɔid]

同源三倍体　autotriploid [ˌɔ:təu'triplɔid]
★异源多倍体　allopolyploid ['ælə'pɔliˌplɔid]
染色体组分析　genome analysis
★非整倍体　aneuploid ['ænjuplɔɪd]
★先天愚型　Down syndrome
发生率　incidence
★嵌合体　mosaic
表型性别　phenotypic sex
★整倍体　euploid ['ju:plɔid]

1 Euploidy is the loss or addition of entire sets of chromosomes.
整倍体以染色体组为单位增减。

2 Aneuploidy is the loss or addition of single chromosome.
非整倍体有单条染色体的增减。

3 Polyploids are euploids with three or more genomes.
多倍体是指具有三个及以上染色体组的整倍体。

4 Autotetraploid is a tetraploid developed by the colchicines treatment.
同源四倍体是指二倍体经秋水仙素处理得到的四倍体。

5 Allopolyploid is polyploids developed from hybrids between different species.
异源多倍体是指不同种杂交形成的多倍体。

6 Translocations is a phenomenon involving the shift of a part of one chromosome to another, nonhomologous chromosome.
易位是指一条染色体与另一条非同源染色体之间发生片断互换的现象。

第十节　遗传物质的改变(二)基因突变
Section 10　Change of Genetic Matter (2) Gene Mutation

形态突变　morphological mutation
★生化突变　biochemical mutation
可见突变　visible mutation
★半致死　semilethal [ˌsemi'li:θəl]
低活性　subvital [sʌb'vaitl]
回复品系　revertant stock
★生物素　biotin

1 Point mutation refers to mutation that can be mapped to one specific locus.
点突变指可归于特定位点的突点。

2 Spontaneous mutation is a mutation occurring in the absence of mutagens, usually due to errors in the normal functioning of cellular enzymes.
自发突变指由于细胞自身酶功能障碍引起的,而非诱导剂引发的突变。

3 Back mutation is the production of a wild-type gene come from a mutant gene.
回复突变是指使突变基因变成野生型基因。

4 Lethal gene is a gene whose expression results in the death of the individual expressing it.
致死基因是指表达的结果导致个体死亡的基因。

第十一节　遗传的分子基础
Section 11　Molecular Basis of Genetics

★多核苷酸　polynucleotide [ˌpɔli'nju:kljəˌtaid]
异源双链　heteroduplexes [ˌhetərəu'dju:plekses]
热变性　thermal denaturation
色差染色体　harlequin ['hɑ:likwin] chromosome
核酸外切酶活性　exonuclease activity
单链结合蛋白　single-strand binding protein
运载蛋白质　transport protein
促性腺物质　gonadotrope [gəu'nædətrəup]
全酶　holoenzyme [ˌhɔləu'enzaim]
起动区　promoter
终止区　terminator
★编码顺序　coding sequence
核仁的组织中心　nucleolar organizer
多核糖体或多体　polyribosome or polysome
三叶草结构　clover-leaf structure
★摆动假说　wobble hypothesis
起始蛋白质　initiation protein
拖尾序列　trailer ['treilə] sequence

转译装置 translation apparatus
生化先成论 biochemical predestination [pri:desti'neiʃ(ə)n]
★ 基因间区域 intergenic region
通读框 open reading frame
★ 隔裂基因 split gene
★ 重叠基因 overlapping gene
★ 高度重复序列 highly repetitive sequence
异染色体区域 heterochromatic ['hetərəukrəu'mætik] region
串联重复 tandem repeats
★ 念珠学说 bead theory
顺反试验 cis-trans test
★ 突变位点 mutational site
★ 复感染 double infection
★ 超感染 superinfection
两点杂交 two-point crosses
★ 重组作图法 recombinational mapping
广谱质粒 promiscuous [prə'miskjuəs] plasmid
★ 遗传工程 genetic engineering
★ 基因操作 gene manipulation
修饰酶 modification enzyme
★ 同裂酶 isoschizomers [ˌaisəuˌski'zəuməz]
噬菌斑杂交法 plaque hybridization
菌落杂交 colony hybridization
珠蛋白 globin ['gləubin]
蛋白基因簇 globin gene cluster
★ 原位杂交 in site hybridization
★ 限制性片段长度多态 restriction fragment length polymorphism(RFLP)
★ 地中海贫血 thalassemia

1 Structural gene is a gene encoding the amino acid sequence of a protein.
结构基因是编码蛋白质的氨基酸序列的基因。

2 Phenylketonuria (PKU) is a human metabolic disease caused by a mutation in a gene encoding a phenylalanine-processing enzyme, which leads to mental retardation if not treated; inherited as an autosomal recessive phenotype.
苯丙酮尿病是由于编码苯丙氨酸羟化酶基因突变所引起的人类代谢病，如果不治疗的话会导致精神障碍，为常染色体隐性遗传。

3 Hybridization in situ means finding the location of a gene by adding specific radioactive probes for the gene and detecting the location of the radioactivity on the chromosome after hybridization.
原位分子杂交指用放射性标记的特定探针寻找基因的位置，与染色体杂交后检测染色体上放射性位点。

第十二节 突变和重组机理
Section 12 Mechanism of Mutation and Recombination

★ 错义突变 missense mutation
★ 无义突变 nonsense mutation
★ 血红蛋白病 hemoglobinopathy
断裂愈合模型 breaking joining model
模写选择模型 copy-choice model
半染色单体转变 half-chromatid conversion
减数后分裂 post-meiotic segregation
断裂点 breakage point
转座遗传因子 transposable [træns'pəuzəbl] genetic element
控制因子 controlling element
解离因子 dissociation(Ds)
★ 颠倒重复序列 inverted repeat (IR) sequences
抗性转移功能基因 resistance-transfer functional gene
靶部位 target site
前病毒 provirus
末端重复顺序 long terminal repeats(LTR)
★ 致癌基因 retroviral oncogene [ˌretrəu'vaiərəl]['ɔnkəˌdʒi:n]

1 Frame-shift mutation is the insertion or deletion of a nucleotide pair or pairs, causing a disruption of the translational reading frame.
移码突变是由于碱基对的缺失或插入导致了翻译读框的改变。

2 Transition is a type of nucleotide-pair substitution in which a purine replaces another purine or a pyrimidine replaces another pyrimidine.
转换是碱基替换的一种类型，即嘌呤与嘌呤之间或嘧啶与嘧啶之间的转换。

3 Transversion is a type of nucleotide-pair substitution in which a purine replaces a pyrimidine or vice versa.
颠换指在碱基替换中嘌呤替换嘧啶,嘧啶替换嘌呤。

4 Missense mutation is a mutation alters a codon so that it encodes a different amino acid.
错义突变是使某个密码子改变以至于编码另一种不同的氨基酸的一种突变。

5 Nonsense mutation is a mutation alters a gene so as to produce a nonsense codon.
无义突变是基因发生突变,使编码区中的某个密码子变为无义密码子。

6 Insertion sequence(IS) is a mobile piece of bacterial DNA (several hundred nucleotide pairs in length) that is capable of inactivating a gene into which it inserts.
插入序列是在细菌DNA中的一段可移动的DNA片段(含几百个核苷酸),它插入基因后使基因失活。

7 Transposon is a mobile piece of DNA that is flanked by terminal repeat sequences and typically bears genes encoding transposition functions.
转座子是一种可移动的DNA片段,两侧有重复序列,典型的转座子携带有编码转座功能的基因。

8 Excision repair is the repair of a DNA lesion by removal of the faulty DNA segment and its replacement with a wild-type segment.
切除修复是切除错误DNA片段,以野生型DNA片段进行替换的修复过程。

第十三节 细胞质和遗传
Section 13 Cytoplasm and Inheritance

前质体 proplastid
毒素蛋白质 killer toxin protein
卡巴粒 kappa particle
★ 雄性不育 male sterility

第十四节 遗传与个体发育
Section 14 Inheritance and Development

分裂球 blastomere ['blæstəumiə]
胚盘 blastoderm ['blæstəudə:m]
极细胞质 pole plasm
★ 异固缩现象 heteropycnosis [ˌhetərəu 'piknəsis]
★ 异染色质化 heterochromatinization [ˌhetərəˌkrəumətiniˈzeiʃən]
血红蛋白 hemoglobin [ˌhi:məu'gləubin]
早期胚胎 first trimester [trai'mestə]
★ 同工酶 isoenzyme or isozyme
★ 降解物激活蛋白 catabolite activator protein
★ 增强子 enhancer
降钙素 calcitonin [ˌkælsi'təunin]
信息储存所 informational store house
★ 信使 messenger
表达部位 expression site
抗体产生细胞 antibody-producing cell
上位性的雄性决定因子 epistatic male-determing factors
自发装配过程 self-assembly process
浓缩蛋白质 condensing protein
孢子形成 sporogenesis

1 Differentiation is the changes in cell shape and physiology associated with the production of the final cell types of a particular organ or tissue.
分化是在发育中细胞的形态和生理功能发生了改变,最终产生了不同的组织和器官过程。

2 Heterochromatin is densely staining condensed chromosomal regions, believed to be for the most part genetically inert.
异染色质是指能被浓染的凝聚状态的染色体区域,通常认为在极大程度上无遗传活性。

3 Totipotency is the ability of a cell to proceed through all the stages of development and thus produce a normal adult.
全能性是细胞具有通过各发育阶段产生一个正常成体的能力。

4 Operator is a DNA region at one end of an operon that acts as the binding site for repressor protein.

操纵基因是操纵子一端的 DNA 片段，可作为阻遏蛋白结合位点。

5 Promoter is a regulator region a short distance from the 5 end of a gene that acts as the binding site for RNA polymerase.

启动子是基因 5' 端的转录调控区，是 RNA 聚合酶的结合位点。

6 Regulatory genes are genes that have roles in turning on or off the transcription of structural genes.

调节基因的作用是使结构基因的转录开始或终止。

7 Enhancer is a cis-regulatory sequence that can elevate levels of transcription from an adjacent promoter. Many tissue-specific enhancers can determine spatial patterns of gene expression in higher eukaryotes.

增强子是邻近启动子，能够提高转录水平的顺式调节因子。在高等真核生物中，许多组织特异性增强子能决定基因表达的立体模式。

第十五节　遗传和进化
Section 15　Inheritance and Evolution

★ 群体遗传学　population genetics
★ 分子遗传学　molecular genetics
相互分歧　divergence [daɪ'vɜːdʒəns, dɪ-]
纤维蛋白肽　fibrinopeptide ['faibrinəˌpeptaid]
单倍基因组　haploid genome
温度稳定性　thermal stability
编码制度　coding system
复合染色体　compound chromosome
★ 用进废退学说　theory of the inheritance of acquired character
痕迹奇观　rudimentary [ruːdi'mentəri] orans
★ 自然选择学说　theory of natural selection
催产素　oxytocin [ˌɔksi'təusin]
★ 遗传平衡定律　law of genetic equilibrium [iːkwi'libriəm]
★ 融合遗传概念　concept of blending inheritance
★ 适合度　fitness
适应值　adaptive value
★ 中性基因　neutral gene
★ 随机遗传漂变　random genetic drift
发病率　incidence
遗传多肽　genetic polymorphism [ˌpɔli'mɔːfizəm]
★ 隔离屏障　isolation barrier
★ 隔离机制　isolation mechanism
★ 异源多倍体或双二倍体　amphidiploid [ˌæmfi'diplɔid]

1 Denaturation is the separation of the two strands of a DNA double helix without breaking the major bonds of its chains.

变性是在不打破链的主要化学键情况下，DNA 双螺旋解链的过程。

2 Amphidiploid is a polyploid formed from the union of two separate chromosome sets and their subsequent doubling.

双二倍体是两套异源二倍体形成的多倍体。

第二十一章　生物制品
Chapter 21　Biologics

第一节　生物制品概述
Section 1　Biologics Overview

★ 生物制品　biologics [baiə'lɔdʒiks]
★ 药品生产质量管理规范　GMP (good manufacturing practice)
★ 细菌类疫苗　bacterial vaccines [bæk'tiəriəl][væk'si:n]
★ 病毒类疫苗　viral vaccines ['vaiərəl][væk'si:n]
★ 抗毒素　antitoxin [ˌænti'tɔksin]
★ 血液制品　blood products
诊断制品　diagnostic products
活性制剂　active agent
★ 基因治疗　gene therapy

Biologic: A preparation, such as a drug, a vaccine, or an antitoxin, that is synthesized from living organisms or their products and used as a diagnostic, preventive, or therapeutic agent.

生物制品：以生物体或其产品为原料制备，用于人类疾病的诊断、预防或治疗的药品，如药物、疫苗、抗毒素等。

第二节　生物制品的制备
Section 2　Preparation of Biologics

★ 超声破碎　ultrasonication [ˌʌltrəsɔni'keiʃən]
溶菌酶　lysozyme ['laisəzaim]
葡聚糖酶　dextranase [dek'stræneis]
蛋白酶　protease ['prəutieis]
甘露聚糖酶　mannanase [mæ'nəneis]
糖苷酶　glycosidase [glai'kəusideis]
肽键内切酶　endopeptidase [ˌendəu'peptideis]
超离心　ultracentrifugation
分子排阻层析　molecular-exclusion chromatography
★ 透析　dialysis [dai'ælisis]
疏水层析　Hydrophobic chromatography
糖复合物　glycoconjugate
复合碳水化合物　complex carbohydrate
水解法　hydrolysis [hai'drɔlisis]
吸附分离法　adsorptive separation

Biologics can be composed of sugars, proteins, or nucleic acids or complex combinations of these substances, or may be living entities such as cells and tissues. Biologics are isolated from a variety of natural sources — human, animal, or microorganism — and may be produced by biotechnology methods and other technologies.

生物制品包含糖类、蛋白质、核酸及这几类物质的复合物，也可以是生物体，例如细胞和组织。生物制品来源于丰富的天然资源——人体、动物或微生物——也可以通过生物技术和其他技术手段生产。

第三节　生物制品的质量管理、鉴定与标准化
Section 3　Quality Management, Certification and Standardization of Biologics

★ 真空度　vacuum degree
区带电泳　zone electrophoresis (ZE)
聚丙烯酰胺凝胶电泳　polyacrylamide gel electrophoresis(PAGE)
★ 热原　pyrogen ['paiərəudʒin]
★ 过敏性试验　sensitivity test
牛血清蛋白　bovine serum albumin (BSA)
血型物质　blood group substance
★ 效力　potency
★ 国际生物标准品　international biological standard
★ 国际生物参考品　international biological reference preparation
国际生物参考试剂　international biological reference reagent
★ 国际单位　international unit (IU)

1 Biopharmaceutical manufacturing is carried out under the philosophy and approach of good manufacturing practice (GMP). GMP is an aspect of quality assurance that ensures that the biological drug is consistently produced and controlled to a standard appropriate for its intended use.
生物制药遵循药品生产质量管理规范(GMP) 的理念和方法。GMP 能确保生物药品规范化生产，达到可供使用的质量标准。

2 Drug potency:The amount of drug required to produce a given percentage of its maximal effect.
药物效力:药物针对某一特定效应的剂量。

3 When a committee considers the possibility of establishing a new standard it has to bear in mind a number of important requirements. In the first place the substance must be such that a stable standard preparation can be made, i. e. , a preparation which does not lose its potency under proper conditions of storage. Secondly, laboratory methods must be available which will enable different laboratories to obtain comparable results when assaying other samples of the substance against the international standards preparation.
当一个委员会考虑建立新标准时，应该牢记几条重要原则：首先标准品必须可以制备得到，性质稳定，在适宜的存储条件下效力稳定。其次，对照国际标准品测定其他样品时，必须采用实验室测定手段，并使不同实验室都能得到相近的测定结果。

第四节 生物制品的包装、保存与运输

Section 4 Packaging, Preservation and Transportation of Biologics

★ 冷冻干燥 freeze-drying
★ 冷链 cold chain
★ 渗透剂 penetrant ['penitrənt]
★ 冷冻保护剂 cryoprotectant [kraiɔprə'tektənt]

1 The packaging must be strong enough that it enables easy transportation of biologics through all means i. e. air, water or land. There are some products which require certain minimum temperature level even when they are transported in order to protect them from getting spoiled and unsuitable for use. So, even this factor must be considered while carrying the packaging process for these products.
包装物必须牢固可靠，使生物制品无论从水、陆、空都便于运输。一些产品运输途中须维持较低的环境温度，以免被污染或失去效力。因此，这些产品的包装过程必须考虑到这一因素。

2 The cold chain refers to the transportation of temperature sensitive products along a supply chain through thermal and refrigerated packaging methods and the logistical planning to protect the integrity of these shipments.
冷链：通过保温和冷藏的包装方法和物流规划，经由供应链运输具有温度敏感性的产品，以保护运输品完好无损。

3 Freeze-drying, or lyophilization, is a well established technology used in the preservation of numerous pharmaceutical and biological products.
冷冻干燥是一种应用于多种药品和生物制品保存的成熟技术。

4 Some molecules protect freeze-dried material. Known as cryoprotectants, these molecules are typically polyhydroxy compounds such assugars (mono-, di-, and polysaccharides), polyalcohols, and their derivatives.
一些分子能够保护冻干后的物质，这些分子叫作冷冻保护剂，通常是多羟基化合物，如糖类(单糖、二糖和多糖)、聚醇类及其衍生物。

第五节 生物反应器及其检测和控制系统

Section 5 Bioreactor and Its Detection and Control System

★ 中试规模 pilot scale
★ 生产规模 industrial scale
★ 搅拌式生物反应器 stirred-tank bioreactor
★ 流加培养 fed-batch culture

灌流培养 perfusion culture [pə'fju:ʃən]['kʌltʃə]
中空纤维式反应器 hollow-fiber bioreactor
膜生物反应器 membrane bioreactor

1 A bioreactor is a container which is used to hold organisms for the purpose of harnessing their natural biochemical processes.
装有生物体并旨在利用其自身生物化学反应的容器称为生物反应器。

2 Most industrial bioreactor manufacturers use vessels, sensors and a control system networked together.
大多数工业化生物反应器都包含三部分：容器、传感器和控制系统，三者相互连通。

第六节 生物制品生产中的安全防护技术
Section 6 Security Protection Technology in Production of Biologics

★ 微生物感染 microbial infection
★ 气溶胶 aerosol ['eərə,sɔ:l]
★ 废弃物处理 waste disposal
★ 污水处理 sewage disposal ['su:idʒ][dis'pəuzəl]
★ 悬浮固体 suspended solid (SS)

1 World-wide, there are many documented cases of laboratory-acquired infections resulting from the production and inhalation of infectious aerosols. If adequate precautions are not taken, aerosol production can occur when using laboratory equipment, even when the equipment is used properly and under normal conditions.
全球范围内，有多起由于生产及吸入致病性的气溶胶而导致的实验室获得性感染记录在案。一旦未采取充足的预防措施，即便在正常条件下规范地使用实验室仪器也会产生气溶胶。

2 Materials contaminated with hazardous biological agents must be collected in the appropriate containers and sterilized or disinfected before disposal.
被有害生物制剂污染的材料必须收集于合适的容器中，灭菌处理后再加以处置。

第七节 疫苗概论
Section 7 Vaccine Introduction

疫苗 vaccine ['væksi:n]
免疫原性 immunogenicity
反应原性 reactivity
★ 佐剂 adjuvant ['ædʒuvənt]
★ 稳定剂 stabilizer ['steibə,laizə]
遗传重组疫苗 genetic recombinant vaccine
合成肽疫苗 synthetic peptide vaccine
表位疫苗 epitope vaccine ['epitəup][væk'si:n]
抗独特型抗体疫苗 anti-idiotype vaccine
微胶囊疫苗 microencapsulated vaccine
可控缓释疫苗 controlled release vaccine
遗传学稳定性 genetic stability
原始种子批 original seed lot
★ 原疫苗 original vaccine
★ 主种子批 master seed lot
★ 工作种子批 working seed lot
★ 免疫 immune
白喉类毒素 diphtheria toxoid [dif'θiəri:ə]['tɔksɔid]
抗原 antigen ['æntidʒən]
破伤风类毒素 tetanus toxoid ['tetənəs]['tɔksɔid]
百日咳菌苗 pertussis [pə'tʌsis]
口服脊髓灰质炎疫苗 oral poliomyelitis vaccine (OPV)
脊髓灰质炎灭活疫苗 inactivated poliomyelitis vaccine (IPV)
麻疹 measles ['mi:zəlz]
流行性腮腺炎 mumps [mʌmps]
霍乱毒素 cholera toxin(CT) ['kɔlərə]['tɔksin]
不耐热肠毒素 heat-labile enterotoxin (LT)
卡介苗 bacillus calmette-guerin
免疫刺激复合物 immunostimulating complexes
★ 细胞因子 cytokine ['saitə,kain]
抗原呈递细胞 antigen presenting cell (APC)
淋巴细胞 lymphocyte ['limfə,sait]
树突状细胞 dendritic cell (DC) [den'dritik sel]
粒细胞 granulocyte ['grænjuləsait]

1 Vaccine: A preparation of a weakened or killed pathogen, such as a bacterium or virus, or of a portion of the pathogen's structure that upon administration stimulates antibody production or cellular immunity against the pathogen but is incapable of causing severe infection.
疫苗：由毒性减弱或灭活的病原体，如细菌或病毒，或病原体结构的一部分制备的生物制品，可以诱导机体产生针对特定致病原的抗体或细胞免疫，但不会导致严重感染。

2 Vaccines are made with a variety of ingredients including antigens, stabilizers, adjuvants, and preservatives; they may also contain residual by-products from the production process.
疫苗成分包括抗原、稳定剂、佐剂、防腐剂，也包含生产过程中的副产物。

3 The immune system has two components: the innate system and the adaptive system. The innate immune system, also known as non-specific immune system and first line of defense, comprises the cells and mechanisms that defend the host from infection by other organisms in a non-specific manner. The adaptive immune system, also known as the specific immune system, is composed of highly specialized, systemic cells and processes that eliminate or prevent pathogenic growth.
免疫系统分为两部分：先天性免疫系统和适应性免疫系统。先天性免疫系统也称为非特异性免疫系统，是机体的第一道防线，由细胞组成，其机制为非特异性地防御其他生物体攻击宿主。适应性免疫系统也称为特异性免疫系统，它包括高度专一的体细胞，能消除或预防病原菌的滋生。

第八节 基因工程疫苗
Section 8 Gene Engineered Vaccines

★ 基因工程亚单位疫苗 gene engineered subunit vaccine
★ 载体疫苗 vectored vaccine
痘苗病毒 vaccinia virus [væk'siniə][ˈvaiərəs]
腺病毒 adenovirus [ˌædinəuˈvaiərəs]
★ 核酸疫苗 nucleic acid vaccine
★ 基因缺失活疫苗 gene deleted live vaccine
★ 蛋白工程疫苗 protein engineered vaccine

DNA vaccine: A type of noninfectious vaccine that directly injects deoxyribonucleic plasmids that express antigens of interest, resulting in foreign protein expression within the cells of the vaccine; however, the vaccine itself is unable to replicate.
DNA 疫苗：一种直接注入脱氧核糖核酸质粒，表达目标抗原的非感染性疫苗，能在疫苗细胞内表达外源蛋白，疫苗本身不可复制。

第九节 细菌类疫苗
Section 9 Bacterial Vaccines

霍乱疫苗 cholera vaccine [ˈkɔlərə][ˈvæksiːn]
伤寒疫苗 typhoid vaccine [ˈtai ˌfɔid][vækˈsiːn]
水肿 edema [iˈdiːmə]
★ 致死的 lethal [ˈliːθəl]
★ 保护性抗原 protective antigen (PA)
致死毒素 lethal toxin
水肿毒素 edema toxin
结核疫苗 tuberculosis vaccine [tuˌbəːkjəˈləusis][vækˈsiːn]
鼠疫疫苗 plague vaccine [pleig][vækˈsiːn]

1 Bacterial vaccines contain killed or attenuated bacteria that activate the immune system. Antibodies are built against that particular bacteria, and prevents bacterial infection later.
细菌类疫苗指能激活免疫系统的灭活或减毒细菌。机体产生抵抗相应细菌的抗体，避免后续细菌感染。

2 Toxins are poisons produced by living organisms. Toxins are classified as either exotoxins or endotoxins. Exotoxins are a diverse group of soluble proteins released into the surrounding tissue by living bacterial cells. Exotoxins have specific reaction sites in the host. Endotoxins are polysaccharide and phospholipid substances found in the cell walls of bacteria that are freed when the cells die and break up. Toxoids are protein toxins that have been heated or chemically treated to deprive them of their toxicity but not of the ability to induce the formation of antibodies.

毒素由活的生物体产生。毒素分为外毒素和内毒素。外毒素是一组可溶的蛋白质，由细菌活细胞释放到外界组织中。外毒素在宿主中有专一的作用位点。内毒素是细菌细胞壁上多糖和磷脂物质，细胞死亡后，随菌体崩解而释放出来。经加热或化学处理后去除毒性而仍保留其免疫原性的蛋白质毒素称为类毒素。

第十节 病毒类疫苗
Section 10 Viral Vaccines

★ 乙型肝炎病毒 hepatitis B virus (HBV)
甲型肝炎病毒 hepatitis A virus (HAV)
丙型肝炎病毒 hepatitis C virus (HCV)
★ 艾滋病 acquired immune deficiency syndrome (AIDS)
人体免疫缺陷病毒 human immunodeficiency virus (HIV)
原病毒 provirus [prəu'vaiərəs]
流行性乙型脑炎 epidemic encephalitis type B
狂犬病 rabies ['reibi:z]
★ 流感 influenza ['influ:'enzə]
抗原漂移 antigen drift
抗原转变 antigen shift
流行性出血热 epidemic hemorrhagic fever (EHF)
轮状病毒 rotavirus [ˌrəutə'vaiərəs]
乳头瘤病毒 papillomavirus [ˌpæpi'ləumə,vaiərəs]
单纯疱疹病毒 herpes simplex virus (HSV)
登革病毒 dengue virus
登革热 dengue fever
巨细胞病毒 cytomegalovirus [ˌsaitə,megələu'vaiərəs]
严重急性呼吸系统综合征 severe acute respiratory syndrome (SARS)
冠状病毒 coronavirus [ˌkɔrənə'vairəs]

Prevention of viral diseases can be achieved by the use of viral vaccines that induce active immunity or by administration of preformed antibody that provides passive immunity.

有两种方法预防病毒引起的疾病：使用病毒类疫苗，这类疫苗能诱导免疫活性；或调节预先产生的抗体产生被动免疫。

第十一节 血液及血液制品概述
Section 11 Overview of Blood and Blood Products

★ 血量 blood volume
循环血液 circulating blood
储备血液 reservoir blood
★ 血浆 blood plasma
粒细胞 granulocyte ['grænjuləsait]
单核细胞 monocyte ['mɔnəu,sait]
淋巴细胞 lymphocyte ['l imfə,sa it]
巨噬细胞 macrophage ['mækrə,fe idʒ]
自然杀伤细胞 natural killer cell
★ 血小板 platelet ['pletlit]
★ 黏度 viscosity [viskɔsiti:]
★ 渗透压 osmotic pressure
人白细胞抗原 human leukocyte antigen (HLA)
★ 凝血因子 clotting factors
★ 血型 blood type
★ 输血 blood transfusion

Blood plasma is the straw-colored/pale-yellow liquid component of blood that normally holds the blood cells in whole blood in suspension. It makes up about 55% of total blood volume. It is the intravascular fluid part of extracellular fluid. It is mostly water (93% by volume) and contains dissolved proteins, glucose, clotting factors, mineral ions, hormones and carbon dioxide (plasma being the main medium for excretory product transportation). Plasma also serves as the protein reserve of the human body. It plays a vital role in intravascular osmotic effect that keeps electrolyte in balance form and protects the body from infection and other blood disorders. Blood serum is blood plasma without fibrinogen or the other clotting factors (i. e., whole blood minus both the cells and the clotting factors).

血浆是血液中淡黄色的液体组分，使血液中的血细胞维持在悬浮状态。血浆约占血液容积的55%，是细胞外、血管内的流体。血浆中93%体积都是水分，其余为溶解蛋白质、葡萄糖、凝血因子、矿物离子、激素和二氧化碳(血浆是代谢产物运输的主要载体)。血浆也是人体蛋白质的储备站，血浆在维持血管内渗透压的稳定上发挥了重要作用，可保持电解质处于稳定状态，保护机体免受感染及其他血液疾病。血清是除去了纤维蛋白原或其他凝血因子的血浆（即去除血细胞和凝血因子的血液）。

第十二节 血液制品及其生产技术

Section 12 Blood Products and Their Production Technologies

人血白蛋白 human serum albumin
血浆蛋白成分 plasma protein fraction (PPF)
免疫球蛋白 immunoglobulin [i'mju:nəu'glɔbjulin]
静脉注射 intravenous injection [,intrə' vi:nəs][in' dʒekʃən]
★ 纤维蛋白原 fibrinogen [fai'brinədʒən]
抗凝血酶 antithrombin [,ænti'θrɔmbin]
纤维结合蛋白 fibronectin
抗胰蛋白酶 antitrypsin

1 Blood proteins, also termed serum proteins or plasma proteins, are proteins found in blood plasma. They serve many different functions, including (1)circulatory transport molecules for lipids hormones, vitamins and metals. (2)enzymes, complement components, protease inhibitors, and kinin precursors. (3)regulation of acellular activity and functioning in the immune system.

血液蛋白，也叫血清蛋白或血浆蛋白，是血浆中的蛋白质。其功能不尽相同，包括：(1) 针对脂类激素、维生素和金属离子的循环运输分子。(2) 酶、补体成分、蛋白酶抑制剂和激肽前体。(3) 脱细胞活性的调节及免疫系统的运作。

2 60% of plasma proteins are made up of the protein albumin, which are major contributors to osmotic pressure of plasma which assists in the transport of lipids and steroid hormones.

血浆蛋白中 60% 是白蛋白，这是维持血浆渗透压的主要因素，渗透压帮助运输脂类和类固醇激素。

第十三节 人血液代用品

Section 13 Human Blood Substitutes

★ 携氧剂 oxygen carrier
★ 全氟碳化合物 perfluorocarbon (PFC)
化学修饰血红蛋白 chemical modified hemoglobin
交联血红蛋白 crosslinked hemoglobin
多聚血红蛋白 polyhemoglobin
共轭血红蛋白 conjugated hemoglobin
★ 人工红细胞 artificial red blood cell (ARBC)
脂质体包裹血红蛋白 liposome encapsulated hemoglobin (LEH)
基因重组人血红蛋白 recombinant hemoglobin

A blood substitute (also called artificial blood or blood surrogates) is a substance used to mimic and fulfill some functions of biological blood, usually in the oxygen-carrying sense. It aims to provide an alternative to blood transfusion, which is transferring blood or blood-based products from one person into another. The main categories of such oxygen-carrying blood substitutes are hemoglobin-based oxygen carriers (HBOC) and perfluorocarbon-based oxygen carriers (PFBOC).

通常从携氧意义上说，血液替代品（又叫人造血液或血液代用品）能模拟生物血液并实现其某些功能。其目的在于提供输血的替代品，可以从一个人向另一个人转移血液或血液产品。携氧血液替代品主要分为基于血红蛋白的携氧剂和基于全氟碳化合物的携氧剂。

第十四节 生物技术药物概论

Section 14 Introduction to Biotechnology Drugs

★ 生物制药 biopharmaceutics ['baiəu,fa:mə'sju:tiks]
★ 药效学 pharmacodynamics [fa:mə,kəudai'næmiks]
★ 药理学 pharmacology [,fa:mə'kɔlədʒi:]
★ 药物代谢动力学 pharmacokinetics [,fa:məkəu,kai'netiks]

1 Biopharmaceutics is the field of study concerning biopharmaceuticals and medical drugs produced using biotechnology. They include proteins (including antibodies), nucleic acids (DNA, RNA or antisense oligonucleotides) and living microorganisms like virus and bacteria where the virulence of viruses and bacteria is reduced by the process of attenuation, they can be used for therapeutic or in vivo diagnostic purposes.

生物制药是研究应用生物技术生产生物药物和医疗药品的领域。生产的药物包括蛋白质(包括抗体)、核酸(DNA、RNA 或反义寡核苷酸)以及活微生物如病毒和细菌(毒力减活株)。生物药物制品可用于治疗和体内诊断。

2 Clinical trials involves three steps: Phase I trials, usually in healthy volunteers, determine safety and dosing; Phase II trials are used to get an initial reading of efficacy and further explore safety in small numbers of sick patients; Phase III trials, a large, pivotal trials to determine safety and efficacy in sufficiently large numbers of patients.

临床试验分为三期:一期临床试验,是在健康人身上进行新药研究的起始期,确定安全性及给药量;二期临床试验,是在少量病人身上考察新药疗效,并进一步探索安全性;三期临床试验,是大型的关键试验,在大量病人身上研究新药的安全性和疗效。

第十五节　细胞因子类药物
Section 15　Cytokine Drugs

★ 集落刺激因子　colony stimulating factor (CSF)

粒细胞集落刺激因子　granulocyte colony stimulation factor (G-CSF)

巨噬细胞集落刺激因子　macrophage colony stimulating factor (M-CSF)

多潜能集落刺激因子　multipotential-colony stimulating factor

促红细胞生成素　erythropoietin(EPO) [i,riθrə'pɔiətin]

干细胞因子　stem cell factor (SCF)

白血病抑制因子　leukemia inhibitory factor (LIF)

血小板生成素　thrombopoietin (TPO)

★ 肿瘤坏死因子　tumor necrosis factor (TNF)

★ 趋化因子　chemokine ['kiːməu,ka in]

胰岛素样生长因子　insulin-like growth factor (IGF)

表皮生长因子　epidermal growth factor (EGF)

血小板衍生生长因子　platelet-derived growth factor (PDGF)

成纤维细胞生长因子　fibroblast growth factor (FGF)

神经生长因子　nerve growth factor (NGF)

骨形态形成蛋白　bone morphogenetic protein (BMP) [bəun][,mɔːfədʒi'netik]['prəutiːn]

淋巴细胞毒素　lymphotoxin [,limfəu'tɔksin]

Cytokines are small cell-signaling protein molecules that are secreted by numerous cells and are a category of signaling molecules used extensively in intercellular communication. Cytokines can be classified as proteins, peptides, or glycoproteins. The cytokines includes the interleukins, lymphokines and cell signal molecules, such as tumor necrosis factor and the interferons, which trigger inflammation and respond to infections.

细胞因子是传递细胞信号的蛋白质小分子,由大量细胞分泌得到,是一类广泛用于细胞间交流的信号分子。细胞因子被划分为蛋白质、肽类和糖蛋白类。细胞因子包括白细胞介素、淋巴因子和细胞信号分子(如引发炎症并应对感染的肿瘤坏死因子和干扰素)。

第十六节　重组激素类药物
Section 16　Recombinant Hormone Drugs

★ 耐受　resistance [ri'zistəns]

★ 糖尿病　diabetes mellitus [,da iə'biːt is]['melitəs]

生长激素释放激素　growth hormone-releasing hormone (GHRH)

生长激素释放抑制素　growth hormone release inhibiting hormone (GHRIH)

卵泡刺激素　follicle-stimulating hormone (FSH)

黄体生成激素 luteinizing hormone (LH) ['lu:ti:ə,na iz iŋ]['hɔ:,məun]

黄体 corpus luteum ['kɔ:pəs]['lu:ti:əm]

A hormone is a chemical released by a cell or a gland in one part of the body that sends out messages that affect cells in other parts of the organism. Hormones are essential for every activity of life, including the processes of digestion, metabolism, growth, reproduction, and mood control. Vertebrate hormones fall into three chemical classes: peptide hormones, lipid and phospholipid-derived hormones.

激素是由机体内的细胞或腺体分泌的化合物,能够释放信息,影响机体内其他部分的细胞。激素参与各种生命活动,作用重大,包括消化、代谢、生长、繁殖和情绪控制。脊椎动物的激素分为三类:肽类激素、脂质和磷脂派生激素。

第十七节 重组溶血栓药物

Section 17 Recombinant Thrombolytic Drugs

溶解血栓的 thrombolytic [,θrɔmbəu'litik]

链激酶 streptokinase(SK)

尿激酶 urokinase [,juərəu'kaineis]

水蛭素 hirudin [hi'ru:dən]

★ 纤溶酶原 plasminogen [plæz'minədʒin]

★ 纤溶酶 plasmin ['plæzmin]

Thrombolytic drugs dissolve blood clots by activating plasminogen, which forms a cleaved product called plasmin. Plasmin is a proteolytic enzyme that is capable of breaking cross-links between fibrin molecules, which provide the structural integrity of blood clots. Because of these actions, thrombolytic drugs are also called "plasminogen activators" and "fibrinolytic drugs". There are three major classes of fibrinolytic drugs: tissue plasminogen activator (tPA), streptokinase (SK), and urokinase (UK).

溶血栓药物通过激活纤溶酶原,形成一种称为纤溶酶的裂解产物来溶解血块。纤溶酶是一种能打破纤维蛋白分子之间联系的蛋白水解酶,从而破坏血栓的结构完整性。由于这些作用,溶血栓药物也称为"纤溶酶原激活剂"和"溶纤维蛋白药物",分为三大类:组织纤溶酶原激活剂、链激酶和尿激酶。

第十八节 重组可溶性受体和黏附分子药物

Section 18 Recombinant Soluble Receptor and Adhesion Molecule Drugs

★ 膜结合型受体 membrane bound receptor

★ 可溶性受体 soluble receptor

细胞黏附分子 cell adhesion molecules (CAM)

成纤维细胞 fibroblast ['faibrəublɑ:st]

整合素 integrin

选择素 selectin

内皮细胞 endothelium [,endəu'θi:liəm]

多形核中性粒细胞 polymorphonuclear neutrophils (PMN)

钙黏附素 cadherin

上皮型钙黏附素 epithelial cadherin

A receptor is a molecule most often found on the surface of a cell, which receives chemical signals originating externally from the cell. Through binding to a receptor, these signals direct a cell to do something — for example to divide or die, or to allow certain molecules to enter or exit. Receptors are protein molecules, embedded in either the plasma membrane (cell surface receptors) or the cytoplasm or nucleus (nuclear receptors) of a cell, to which one or more specific kinds of signaling molecules may attach. A molecule which binds (attaches) to a receptor is called a ligand.

受体分子多数位于细胞表面,接受从细胞外部传来的化学信号。这些信号结合在受体上,指导细胞发挥生物功能,如分裂或死亡,或允许某些分子出入。受体分子是蛋白质,锚定于质膜上(细胞表面受体)或细胞质或细胞核上(核受体),可与一种或多种专一的信号分子结合。与受体结合的分子叫作配体。

第十九节 基因治疗与核酸药物

Section 19 Gene Therapy and Nucleic Acid Drugs

★ 反义技术 antisense technology

致癌基因 oncogene ['ɔnkə,dʒi:n]

单纯疱疹病毒胸苷激酶 herpes simplex thymidine kinase
病毒介导的 virus-directed
主要组织相容性复合体 major histocompatibility complex (MHC)
多药耐药基因 multi-drugresistance gene (MDR)
多肽核酸 peptide nucleic acid (PNA)

Gene therapy is the use of DNA as a pharmaceutical agent to treat disease. It derives its name from the idea that DNA can be used to supplement or alter genes within an individual's cells as a therapy to treat disease. The most common form of gene therapy involves using DNA that encodes a functional, therapeutic gene in order to replace a mutated gene. Other forms involve directly correcting a mutation, or using DNA that encodes a therapeutic protein drug (rather than a natural human gene) to provide treatment.

基因治疗利用DNA作为治疗疾病的药剂。研究者认为DNA可以在个体细胞内补充或替换基因,以此作为治疗疾病的疗法,这便是基因治疗一词的由来。基因治疗的最常见形式即利用编码功能性治疗基因的DNA来替换突变基因。其他形式包括直接修正突变,或利用编码治疗性蛋白药物的DNA(而不是天然的人类基因)进行治疗。

第二十节 抗体药物
Section 20 Antibody Drugs

★ 基因工程抗体 genetic engineering antibody
互补决定区 complementarily determining region (CDR)
框架区 framework region (FR)
基础免疫 basic immunization
免疫麻痹 immune paralysis
破伤风抗毒素 tetanus antitoxin
肉毒中毒 botulism [ˈbɔ tʃ əˌl izəm]
肉毒抗毒素 botulinum antitoxin
气性坏疽抗毒素 gas-gangrene antitoxin
示踪或导向作用 tracer or targeted effect
★ 拮抗作用 competitive inhibition
★ 细胞毒性 cytotoxicity
★ 细胞溶解 cytolysis [saiˈtɔlisis]
植物抗体 plantibody
人源化 humanization
双特异性抗体 bispecific Ab
免疫黏附素 immunoadhesin
催化抗体 catalytic Ab
噬菌体抗体库 phage antibody library
全套抗体库 repertoire antibody library
组合抗体库 combinatorial antibody library
核糖体展示技术 ribosome display technique

1 Monoclonal antibodies (mAb or moAb) are monospecific antibodies that are the same because they are made by identical immune cells that are all clones of a unique parent cell.
单克隆抗体是相同的单特异性抗体,因为它们是由来自同一母细胞的克隆,即相同的免疫细胞制备得到的。

2 An antitoxin is an antibody with the ability to neutralize a specific toxin. Antitoxins are produced by certain animals, plants, and bacteria. An antitoxin provides passive immunity.
抗毒素是能中和特定毒素的抗体,由某些动物、植物和细菌生产。抗毒素提供被动免疫功能。

第二十二章　植物病原微生物
Chapter 22　Plant Pathogenic Microbes

第一节　绪论
Section 1　Introduction

病害三角　disease triangle
爱尔兰饥馑　Irish famine
孟加拉饥荒　Bengal famine
症状　symptom ['simptəm]
侵填体　tylose ['tiləus]
胼胝质　callose ['kæləus]
★ 病症　sign
★ 变色　discoloration
★ 黄化　yellowing
花叶　mosaic
斑驳　mottle
★ 脉明　vein cleaning
★ 坏死　necrosis
★ 叶斑　spot
环斑　ring spot
叶烧　leaf scad
★ 猝倒　damping off
梢枯　die back
★ 溃疡　canker
★ 腐烂　rot
★ 萎蔫　wilt
★ 畸形　malformation [ˌmælfɔ:'meiʃən]
增长　hyperplasia [ˌhaipə(:)'pleiʒiə]
增大　hypertrophy [hai'pə:trəfi]
减生　hypoplasia [ˌhaipəu'pleizjə]
★ 矮缩　dwarf [dwɔ:f]
矮化　stunt
花变叶　phyllody ['filədi]
并发症　complex disease
拮抗现象　antagonism [æn'tægənizəm]
协生现象　synergism ['sinədʒizm]
★ 隐症现象　masking of symptom
寄生性　parasitism
★ 寄主　host
★ 活体营养型　biotrophe [ˌbaiəu'trɔf]
★ 腐生物　saprogen ['sæprədʒin]
★ 致病性　pathogenicity
★ 侵染力　aggressiveness [ə'gresivnis]
★ 侵染过程　infection process
侵入期　penetration period
病害循环　disease cycle
潜育期　incubation period
发病期　symptom period
越冬　over wintering
越夏　over summering
★ 初次侵染　primary infection
★ 再次侵染　reinfection
植物保卫素　phytoalexin [ˌfaitəuə'leksin]
避病性　avoidance
抗病性　resistance
耐病性　tolerance

第二节　植物病原真菌
Section 2　Plant Pathogenic Fungi

★ 植物病原真菌　plant pathogenic fungi
★ 鞭毛菌亚门　Mastigomycotina
无隔多核菌丝体　coenocytic [ˌsi:nə'sitik] mycelium
整体产果式　holocarpic [ˌhɔlə'ka:pik]
分体产果式　eucarpic [ju:'ka:pik]
薄壁游动孢子囊　zoosporangium [ˌzəuəspə'rændʒiəm]
休眠孢子囊　resting sporangium
休眠孢子　resting spore
★ 层出现象　proliferation [prəu'liksfə'ræʃən]
★ 两游现象　diplanetism [daiplə'netizəm]
休止孢　cystospore [ˌsistə'spɔ:]
★ 孢囊梗　sporangiophore
葡萄菌丝　stolon ['stəulɔn]
囊轴　columella
★ 子实层　hymenium [hai'mi:niəm]
★ 子囊果　ascocarp

★ 闭囊壳 cleistothecium [ˌklaistə'θi:siəm]
★ 子囊壳 perithecium [ˌperi'θiʃiəm]
★ 子囊盘 apothecium [ˌæpə'θisiəm]
★ 子囊座 ascostroma
侧丝 paraphysia [ˌpærə'fiziə]
顶侧丝 apicalparaphysis ['æpikəlpə'ræfisis]
拟侧丝 paraphysoid ['pæræfsɔid]
缘丝 periphysis [pə'rifəsis]
拟缘丝 periphysoid
★ 产囊体 ascogonium
产囊丝 ascogenoushypha
产囊丝钩 crozier ['krəuʒə]
附属丝 appendage
★ 子囊腔 locule [lɔkju:l]
★ 假囊壳 pseudoperithecium [sju:dəu ˌperi'θiʃiəm]
双囊壁 bitunicate [bai'tju:nikit]
担子菌亚门 Basidiomycotina
裸果型 gynocarpic
半被果型 hemi-angiocarpic
被果型 angiocarpic
冬孢菌纲 Teliomycetes
层菌纲 Hymenomycetes
腹菌纲 Gasteromycetes
★ 单主寄生 autoecism [ɔ:'ti:sizm]
★ 转主寄生 heteroecism [ˌhetə'ri:sizəm]
★ 多型现象 polymorphism [ˌpɔli'mɔ:fizəm]
★ 锈孢子 aeciospore ['i:siəspɔ:(r)]
★ 锈孢子器 aecium ['i:ʃiəm]
★ 夏孢子 urediospore [juə'ri:diəspɔ:(r)]
★ 冬孢子 teliospore ['ti:liəˌspɔ:]
★ 担孢子 basidiospore [bə'sidiəuspɔ:]
半知菌亚门 Deuteromycotina
半知菌 fungi-imperfecti
★ 分生孢子盘 acervulus
★ 分生孢子器 pycnidium [pik'nidiəm]
节孢子 arthrospore ['ɑ:θrəspɔ:]
★ 产孢细胞 conidiogenous
疫病 blight
霜霉病 downy mildew
白锈病 white rust
叶斑病 leaf spot
炭疽病 anthracnose
白粉病 powdery mildew ['paudəri]['mildju:]
煤烟病 sooty mould
霉病 mould
萎蔫病 wilt
干腐枝枯病 dryrot and blight
干腐 dry rot
枝枯 blight or die

There are four phyla within the fungi, divided from each other on the basis of differences in their mechanisms of sexual reproduction. The four phyla are the Mastigomycotina and Zygomycotina (these two phyla are termed the lower fungi) and the Ascomycotina and Basidiomycotina (termed the Higher fungi). A fifth group exists which contains fungi where sexual reproduction is not known but where asexual reproduction is seen. These fungi are placed in the phylum Deuteromycotina.

根据有性生殖的机制不同真菌可分为 4 个亚门。这 4 个亚门分别是鞭毛菌亚门、接合菌亚门(这两类称为低等真菌)、子囊菌亚门、担子菌亚门(这两类称为高等真菌)。第五个类群是只知其无性繁殖,而不知其有性繁殖,这类真菌被放在半知菌亚门。

第三节 植物病原原核生物
Section 3 Plant Pathogenic Prokaryotes

极鞭 polar flagella
周鞭 periphery [pə'rifəri] flagella
侵染途径 infection path
侵染源 source of infection
传播 dissemination
★ 薄壁菌门 Phylum Gracilicutes
★ 厚壁菌门 Phylum Firmicutes
★ 软壁菌门 Phylum Tenericutes
★ 疵壁菌门 Phylum Mendosicutes
基因种 genospecies [ˌdʒi:nə'spi:ʃi:z]
亚种 subspecies (subsp.)
★ 致病变种 pathovar (pv.)
致瘤质粒 tumor inducing plasmid(Ti)
致发根质粒 rhizogen inducing plasmidor(Ri)
★ 菌系 strain
生化型 biotype
生化变种 biovar

第四节 植物病毒
Section 4 Plant Virus

★ 植物病毒 plant virus
侵染性活液 contagium virum fluidum
病毒粒体 virus particle
二十面体 icosahedral [ˌaikəusəˈhedrəl] particle
联体病毒 gemini virus
植物弹状病毒 plant rhabdovirus
囊膜 envelope
基因传递 gene transmission
复制中间型 replicative intermediate(RI)
双义 RNA ambisense RNA
多聚蛋白 poly protein
通读蛋白 read-though
细胞质内含体 cytoplasmic inclusion
核围内含体 perinuclear inclusion
★ 复制增殖 multiplication
★ 介体传播 vector transmission
★ 获毒(取食)期 acquisition period
★ 循回期 incubation period
★ 接毒(取食)期 inoculation period
★ 增殖型 propagative [ˈprɔpəgeitiv]
★ 半持久性 semipersistent
★ 持久性 persistent
★ 机械传播 mechanical transmission
国际病毒分类委员会 International Committee on Taxonomy of Viruses(ICTV)
单链 single strand (ss)
分离株 isolate
确定种 definitive species
暂定种 tentative species
反向电泳 return-poly acryl amide gel electrophoresis (R-PAGE)
★ 鉴别寄主 indicator
★ 稀释限点 dilution end point(DEP)
★ 钝化温度 thermal inactivation point(TIP)
★ 体外存活期 longevity in vitro(LIV)

Virus infected plants can display a wide range of different symptoms. The whole plant can be stunted (dwarf-like) and infected leaves can show abnormal colorations like mosaics and yellowing. Very often infected leaves have a rolling or curly appearance and also the fruits can display various symptoms dependent on the virus. The yield of crop plants can be severely reduced by a virus infection. However, the age of the plants, the viral strain and many other factors are affecting both yield and symptoms of infected plants.
病毒侵染植物可以表现各种症状，整个植株矮化，被侵染的叶片表现变色如花叶和黄花，还经常表现皱缩和卷曲，被侵染的果实也依侵染的病毒不同而表现不同的症状。被病毒侵染后，作物严重减产。但是植物的龄期、病毒株系和许多其他因素均影响被侵染植物的产量和症状。

第五节 侵染过程
Section 5 Infection Process

★ 非侵染性病害 Noninfectious diseases
★ 接触期 touching period
趋触性 contact tropism [ˈtrəupizəm]
★ 死体营养型 necrotrophic [ˈnekrəu ˈtrɔfik]
★ 活体营养型 biotrophic [ˌbaiəuˈtrɔfik]
★ 病害循环 disease cycle
★ 侵染循环 infection cycle
★ 土壤寄居菌 soil invader
★ 土壤习居菌 soil inhabitant

1 Disease cycle is not synonymous with the life cycle of either the host or the parasite but rather, like the concept of pathodeme, is the result of both host and parasite life cycles functioning simultaneously.
病害循环与寄主或寄生物的生活史不是同义词，类似于病原组群的概念，病害循环同时包括寄主和寄生物生活史。

2 Diseases can be categorized into two types of cycles: monocyclic—initiated by primary inoculum only and polycyclic—additional infection cycle(s) initiated by secondary inoculum.
病害循环分为两种循环类型：单循环 - 只有初次侵染和多循环 - 由二次侵染引起的多次侵染。

第六节　植物病原物的寄生性和致病性
Section 6　Parasitism and Pathogenicity of Plant Pathogen

★ 共生关系　symbiosis
★ 共栖关系　commensalism [kə'mensəlizəm]
亲和性　compatibility
不亲和性　incompatibility
★ 致病因素　pathogenicity factor
侵入钉　penetration peg
寄主非选择性毒素　host-non-selective toxin
寄主非专化性毒素　host-non-specific toxin
★ 寄主专化性毒素　host specific toxin
生长调节物质　growth regulators
★ 植物生长素　auxin

第七节　植物的抗病性
Section 7　Plant Resistance

★ 主效基因抗病性　major gene resistance
★ 微效基因抗病性　minor gene resistance
小种专化抗病性　race-specific resistance
小种非专化抗病性　race-non-specific resistance
★ 基因对基因学说　gene-for-gene theory
★ 被动抗病性　passive resistance
★ 主动抗病性　active resistance
避病　disease escaping
耐病　disease tolerance
★ 诱导抗病性　induced resistance
蜡质层　wax layer
角质层　cuticle
木质化作用　lignification
★ 木栓化　suberization
木栓质　suberin ['sju:bərin]
★ 过敏性坏死反应　necrotic hypersensitive reaction
★ 植物保卫素　phytoalexin [ˌfaitəuə'leksin]
★ 获得抗病性　acquired resistance
★ 诱发接种　inducing inoculation
★ 挑战接种　challenge inoculation
★ 局部诱发抗病性　local induced resistance
★ 系统诱发抗病性　systemic induced resistance
系统获得抗病性　systemic acquired resistance
★ 病程相关蛋白　pathogenesis related protein

1 Horizontal resistance means in a given cultivar, there are similar levels of resistance to each of the races of a given pathogen.
水平抗性是一定的品种对一定的病原表现相似水平的抗性。

2 The hypersensitive response (HR) is a localized death of host cells at the site of infection. It is the result of a specific recognition of a pathogen attack by the host. The HR is considered to be a a type of 'programmed cell death'. It can be a very effective form of resistance against obligate parasites.
过敏性坏死反应是侵染部位的寄主细胞局部死亡。是寄主与病原间特异性识别的结果。过敏性坏死反应是一种程序细胞死亡类型，是一种抵抗专性寄生非常有效的形式。

第八节　植物病害的流行与预测
Section 8　Prevalence and Prediction of Plant Disease

植物病害流行学　botanical epidemiology [bə'tænik(ə)l][ˌepiˌdi:mi'ɔlədʒi]
★ 单循环病害　monocyclic ['mɔnəusaiklik] disease
★ 多循环病害　polycyclic disease
季节流行曲线　disease progress curve
★ 逻辑斯蒂增长期　logistic phase
表观侵染速率　apparent infection rate
中心式流行　focus epidemic
病害梯度　disease gradient
侵染梯度　infection gradient
★ 预测　prediction
prognosis [prɔg'nəusis]
★ 预报　forecasting ['fɔ:kɑ:stiŋ]
★ 临界期　critical period
负预测　negative prognosis
★ 损失估计　disease loss assessment
经济损害水平　economic injury level
★ 经济阈值　economic threshold ['θreʃhəuld]

第九节 植物病害的诊断和防治

Section 9 Diagnosis and Control of Plant Disease

★ 科赫氏法则 Koch's Rule
回避 avoidance
杜绝 exclusion
铲除 eradication [iˌrædiˈkeiʃən]
抵抗 resistance
治疗 therapy
★ 植物检疫 plant quarantine [ˈkwɔrənti:n]
★ 人为传播 man-associated dispersal
有害生物风险分析 pest risk analysis
检疫性有害生物 quarantine pest
★ 检疫法规 quarantine regulations
环境管理 management of the physical environment
栽培防治 culture control
★ 种子健康检验 seed health testing
病圃 disease nursery
重寄生 hyper parasitism
★ 交互保护 cross-protection

Rapid and accurate diagnosis of disease is necessary before proper control measures can be suggested. It is the first step in the study of any disease. Diagnosis is largely based on characteristic symptoms expressed by the diseased plant. Identification of the pathogens is also essential to diagnosis.

在采取适当的防治措施前进行快速而准确的诊断是十分必要的。这是研究任何病害的第一步。诊断主要依据感病植物表现的症状特点。病原物的鉴定对于诊断也很重要。

第二十三章　药学概论
Chapter 23　Introduction to Pharmacy

第一节　绪论
Section 1　Introduction

★ 药　drug/medicine
★ 药学　pharmacy [ˈfɑːməsi]
药厂　pharmaceutical [ˌfɑːməˈsjuːtikəl] factory
★ 药房　dispensary [disˈpensəri]/pharmacy [ˈfɑːməsi]
★ 药典　pharmacopoeia [ˌfɑːməkəˈpiːə]
药科大学　pharmaceutical university
★ 药剂师　pharmacist [ˈfɑːməsist]
★ 药品生产和质量管理规范　Good Manufacturing Practice(GMP)
处方　Prescription

The first drugs known to man were obtained from natural sources. They were either from plants like the poppy (opium) and foxglove (digitalis), or they were simple inorganic substances like Epsom salts and arsenic. Later, the development of organic chemistry led to the introduction of synthetic drugs, and at present some 100,000 compounds have been tested for pharmacological activity.

人类最初所认识的药物是从自然来源获得的，它们或来自诸如罂粟(鸦片)和毛地黄(洋地黄)一类植物，或者就是诸如泻盐和砷一类的简单无机物质。此后，由于有机化学的发展导致合成药物的引用，而目前约有 100,000 种化合物已进行过药理活性的实验。

第二节　中药与天然药物
Section 2　Traditional Chinese Medicines and Natural Drugs

★ 中药　traditional Chinese medicine
★ 西药　western medicine
★ 天然药物　natural drug
合成药物　synthetic drug
★ 中草药　Chinese herbal medicine
★ 药性　property of a drug
本草考证　herbological study
★ 中药学　The Chinese Material Medical
药引　guiding drug
★ 四气　four natures of a drug
★ 五味　five kinds of flavor
药渣　dreg of a decoction
清热泻火　heat-clearing and fire-purging
★ 生药学　pharmacognosy [ˌfɑːməˈkɔgnəsi]
植物志　herbal [ˈhəːbəl]
崩解剂　disintegrant [disˈintigrənt]
鉴定　identification
★ 植物化学　phytochemistry [ˌfaitəuˈkemistri]
次级代谢产物　secondary metabolite
盖伦制剂　Galenic preparation
泻药　purgative [ˈpəːgətiv]
香料　perfume [ˈpəːfjuːm]
质量控制　quality-control
★ 分类学　taxonomy [tækˈsɔnəmi]
分类　classification

1 One of the oldest of human activities is the study of plants and animals, particularly as sources of food. From the earliest times man had to distinguish between those plants which were poisonous and those which were not and there gradually developed a knowledge of naturally occurring drugs which was transmitted at one time orally, later in written form as papyri, baked clay tablets, parchments, manuscript herbals, printed herbals, pharmacopoeias and other works, and most recently by computerized information retrieval systems.

人类最古老的活动之一是对植物和动物的研究，特别是作为食物的来源。从远古时代起，人们不得不在一些有毒植物和无毒植物之间加以辨别。于是有关天然生长的药物的知识渐渐地增加起来了。这些知识在某个时期靠口头流传，继而记载于抄本、黏土焙制的书板、羊皮纸、本草书手抄本和印刷本、药典及其他著作；而新近则采用电子计算机信息检索系统。

2 The name pharmacognosy, derived from the Greek, pharmakon, a drug, and gignosco, to acquire a knowledge of, was not introduced until 1815. Whilst pharmacognosy is mainly concerned with naturally occurring substances having a medicinal action, it is not entirely limited to such substances.
"生药学"这一名称源出于希腊文 pharmakon,意为"药物",以及 gignosco,意为"获取某种知识",直至1815年才被引用。尽管生药学主要论及天然生长并有医疗作用的物质,但它并不完全限于此类物质。

3 Pharmacognosy is closely related to both botany and plant chemistry and its history entitles it to be regarded as the parent of both.
生药学同植物学及植物化学有着紧密的联系,而且根据其历史情况,它完全可被看成是两者的母体。

4 As a result of modern isolation and pharmacological testing procedures, new plant drugs usually find their way into medicine as purified substances rather than in the form of older Galenical preparations.
由于运用现代分离和药理实验方法的结果,新的植物药通常是设法制成提纯物应用于医疗,而不是采用较古老的盖伦制剂的形式。

第三节 药物化学
Section 3 Pharmaceutical Chemistry

★ 药物化学 pharmaceutical chemistry/medicinal chemistry
★ 药物 drug/medicament/medicine/pharmacon [ˈfɑːməkɔn] /remedy [ˈremidi]
★ 构效关系 structure-activity relationship
先导化合物 lead compound
★ 药物设计 drug design
★ 药物修饰 drug modification
药物基团 pharmacophore [ˈfɑːməkəfɔː]
★ 药理活性物质 pharmacologically active agent
药瘾 drug addiction
计算机辅助药物研究 computer-aided drug research
★ 化学治疗 chemotherapy
★ 分子药理学 molecular pharmacology
★ 阿托品 atropine [ˈætrəpiːn]
奎宁 quinine [kwiˈniːn]
★ 咖啡因 caffeine [ˈkæfiːn]
反应停 thalidomide [θəˈlidəˌmaid]
★ 致畸作用 teratogenesis [ˌterətəuˈdʒenisis]
★ 致癌作用 carcinogenesis [ˌkɑːsinəuˈdʒenisis]
酶的自杀性底物 enzyme suicide substrate
组合化学 combinatorial chemistry
高通量筛选 high throughput screening
化学信息学 chemoinformatics [keməuinfəˈmætiks]

1 Ultimately, the medicinal chemist may be able to depict the structure of a yet unsynthesized molecule, and by making certain apriori assumptions, predict both the type and degree of pharmacologic action that the compound will exhibit. There is still a long way to go to attain this goal. In the search for new drugs, the usual approach is to first have in mind some known active compound and then attempt various structural modifications in the hope of either altering or attenuating its known biologic activity. The only alternative to this is a random screening of all organic compounds, across a broad spectrum of possible actions — a monumental and wasteful technique. Molecular modification, too, could verge on the infinite were it not for guidelines which have been established by several generations of investigators.
最终,药物化学家也许能够描绘出一种尚未合成过的分子结构,并且运用演绎法做出某些假设,去预测该化合物将会显示的药理作用的类型和程度。要达到这个目标尚有一段遥远的路程。在寻求新药方面,通常的途径是,首先考虑到一些已知活性的化合物,然后尝试做各种结构上的改变,以期转变或减弱其已知的生物活性。对此唯一可用的办法是,在其可能作用的广阔范围内,对所有的有机化合物进行随机筛选,这是一项巨大而不经济的技术。分子结构的改变如果没有几代研究者所建立的准则,也将可能濒于永无休止的境地。

2 There are several sources from which we may derive the original structure activity reference point in our quest for new compounds. Originally, all prototypes came from nature. The adrenergic

phenethylamines all derive from ephedrine, a naturally occurring compound whose use in medicine goes back some 5000 years.

现有几种来源可以使我们在寻找新化合物方面去推导原始的构效关系基准点。最初，所有原型都是来自自然界。肾上腺素能药物苯乙胺类全都从麻黄碱衍生而来，后者是一种天然存在的化合物，在医药上的应用可以追溯到约 5000 年前。

第四节　药理学

Section 4　Pharmacology

★ 药理学　pharmacology [ˌfɑːməˈkɔlədʒi]
★ 治疗学　therapeutics [ˌθerəˈpjuːtiks]
临床药理学　clinical pharmacology
★ 生化药理学　biochemical pharmacology
★ 作用机制　mechanism of action
处置　disposition
★ 药效学　pharmacodynamics [fɑːməˌkəudaiˈnæmiks]
★ 药物不良反应　adverse drug reaction
药源性疾病　drug induced disease
★ 激动剂　agonist [ˈæɡənist]
拮抗剂　antagonist [ænˈtæɡənist]
★ 量效关系　dose-response relation
效价强度　potency [ˈpəutənsi]
中毒剂量　toxic dose
肾上腺素　epinephrine [ˌepiˈnefrin]
★ 药动学　pharmacokinetics [ˌfɑːməkəuˌkaiˈnetiks]
★ 生物利用度　bioavailability [ˌbaiəuəˌveiləˈbiliti]
★ 生物转化　biotransformation [ˌbaiəuˌtrænsfəˈmeiʃən]
★ 排泄　excretion
时量曲线　time-concentration curve
★ 达峰时　peak time
半衰期　half-life
★ 一级动力学　first-order kinetics
★ 稳态血药浓度　steady state plasma concentration
坪值　plateau
★ 肝脏首过效应　hepatic first-pass effect
★ 胎盘屏障　the placental barrier
★ 治疗指数　therapeutic index

1 Today, advances in pharmacology and in therapeutics have broadened our appreciation of (1) the mechanisms involved in disposition of drugs by the body, (2) the inherent ability of drugs to modify the physiologic integrity of the host, and (3) the interactions between drugs when present simultaneously in the body and the competition between them for each to influence or to be influenced by the organism. By being cognizant of these pharmacologic principles, by applying them fully, and by remaining vigilant to countless interactions between drug and ailing body, one can substantially reduce the side effects of numerous drugs.

今天，药理学和治疗学的进展加深了我们的以下认识：(1) 人体对药物性能的处置机制；(2) 药物对宿主生理完整性所具有的影响能力；(3) 当多种药物同时存在于体内时，药物之间的相互作用和相互竞争影响机体或被机体所影响。只要充分认识并运用这些药理学原理，并对药物与人体间的无数相互作用保持高度警惕，我们就能大大减少许多药物的副作用。

2 Pharmacodynamics may be defined as the study of the action and effects of drugs on organs and tissues at cellular and subcellular levels.

药效学是指从细胞和亚细胞水平研究药物对器官和组织的作用和效应。

3 Therefore, pharmacodynamics provides information on how drugs bring about their beneficial effects and how they cause their side effects.

因此，药效学应提供药物如何产生效用以及如何引起副作用的资料。

4 Pharmacokinetic principles, which deal with the fate of drugs, discuss the absorption, distribution, binding, biotransformation, and excretion of drugs and their metabolites in the body.

药代动力学原理涉及药物的命运，内容是讨论药物及其体内代谢产物的吸收、分布、结合、生物转化和排泄过程。

5 Since drugs are mostly absorbed from the upper part of the small intestine, the rate of gastric emptying plays a crucial role in drug absorption.
既然药物大多数从小肠上部吸收，所以胃排空速率在药物吸收中起决定性作用。

6 However, by far the most important reason an orally or a parenterally administered drug may not develop an adequate plasma concentration is the "first-pass effect", which may be defined as the loss of drug as it passes through the liver for the first time.
然而，口服或非肠道给药之所以不能获得足够的血浓，其最主要的原因是"首过效应"。首过效应系指药物首次通过肝脏的损失。

第五节 药物分析学
Section 5 Pharmaceutical Analysis

分析校准 analytical calibration
分析样品 analytic(al) sample
分析纯试剂 analytical reagent (AR)
检验 verify
★ 检测系统 detection system
检测器灵敏度 detector sensitivity
质量控制 quality control
标准试样 standard sample
★ 标定 standardization
标准物 standard substance
★ 药品稳定性 drug stability
★ 中国药典 *Chinese Pharmacopoeia* (ChP) [ˌfɑːməkə'piːə]
★ 美国药典 *The United States Pharmacopoeia* (USP)
★ 含量测定 determination
指纹图谱 finger print
★ 容量分析 volumetric analysis
重量分析法 gravimetry [grə'vimitri]

1 A study of the stability of pharmaceutical products and of stability testing techniques is essential for three main reasons. Firstly it is important from the point of view of the safety of the patient. The present trend in the pharmaceutical industry is towards production of highly specific, chemically complex, potent drugs. It is important, therefore, that the patient receives a uniform dose of drug throughout the whole of the shelf-life of the product. In addition, although a drug may have been shown to be safe for use, this is not necessarily true of the decomposition product(s). Secondly consideration must be given to the relevant legal requirements concerned with the identity, strength, purity and quality of the drug, and finally such a study is important to prevent the economic repercussions of marketing an unstable product.
研究药物制品的稳定性和稳定性试验技术是必要的，这有三个主要理由。首先，从病人安全的角度来看，这种研究是很必要的。制药工业的当前趋势，是倾向于生产高度专一性、化学结构复杂及强效的药物。因此，在制品贮藏的全部过程中，病人能够接受一种剂量均一的药物也是很重要的。此外，尽管一种药物在应用方面已经证明是安全，但其分解产物却未必如此。其次，必须针对有关药物的均一性、含量(效价)、纯度及质量的相应法定要求加以考虑。最后，为了防止出售不稳定制品而受到的经济损害，这一研究也是很重要的。

2 Perhaps the most significant addition to the law is the concept of "current good manufacturing practice"(GMP). This is a requirement that drugs, and the methods used in, or the facilities or controls used in their manufacture, processing, packing or holding conform with those practice which will assure that such drugs meet the requirements of the Act as to safety, and have the identity, strength, quality, and purity characteristics which they purport or are represented to possess. If they do not, they are adulterated.
"现行药品生产质量管理规范"(GMP) 的概念，也许是对法律的最有意义的增补。这是要使药物及在其制造、加工、包装和保管过程中所用的方法、设备或控制符合规定的操作程序，借以保证这些药物能满足"法令" 关于安全性的要求，并具备药物本身表明的或是据称含有的均一性、效价、质量和纯度等特性。如果药物达不到这些要求,那它们就是伪劣品。

第六节 药剂学

Section 6 Pharmaceutics

★ 制剂 preparation
药剂 medicament [meˈdikəmənt]
★ 药剂学 pharmaceutics [ˌfɑːməˈsjuːtiks]
★ 胶囊 capsule
★ 给药方式 administration of drug
丸剂 pill
★ 片剂 tablet
挥发性液体 volatile liquid
★ 结晶混悬液 crystalline suspension
气雾剂 aerosol [ˈɛərəsɔl]
★ 长效制剂 long-acting preparation
★ 脂质体 liposome [ˈlipəusəum]
★ 临床药学 clinical pharmacy
汤剂 decoction
物理药学 physical pharmacy
★ 生物药剂学 biopharmaceutics
工业药剂学 industrial pharmacy
注射剂 injection
漂浮片 floating tablet
复方乙酰水杨酸片 compound acetyl salicylate tablets
★ 缓释制剂 sustained-release preparations
结肠靶向释药剂型 colon-targeted release dosage forms
★ 经皮给药系统 transdermal drug delivery system
★ 靶向给药系统 targeting drug delivery system
灭菌法 methods of sterilization

1 Drugs are administered as a solid in the form of capsules, tablets, and pills (e. g. , clonidine), as a volatile liquid (e. g. , halothane, enflurane), as a solution (e. g. , chlorpromazine), as an aerosol (e. g. , beclomethasone), as a gas (e. g. , oxygen, nitrous oxide), and as a crystalline suspension (e. g. , insulin). The chosen route of administration is determined by the desired onset of action and duration of action of drugs, by the nature of the drugs, by special circumstances, and by the bioavailability of drugs.

药物可以胶囊、片剂和丸剂等固体形式(如可乐宁)、挥发性液体(如氟烷,安氟烷)、溶液(如氯丙嗪)、气雾剂(如比诺米塞松)、气体(如氧,氧化亚氮)和结晶混悬液(如胰岛素)的形式给予。给药途径的选择取决于所期望的药物显效时间和作用持续时间,取决于药物的性质、特殊的环境和药物的生物利用度。

2 Most modern drugs are available in tablets, capsules or injections which are supplied by the pharmaceutical industry. The process of dispensing has, therefore, become very much simplified, and whereas the pharmacist was formerly required to carry out manipulative processes in the preparation of medicines, he is less frequently called upon to do so nowadays.

近代药物大多以片剂,胶囊剂或注射剂的形式由制药工业来供应。因此,调剂过程变得简单多了,于是药师先前在药房制备药剂需要用于手工进行的操作过程,现今已经很少要求这样做了。

3 Medicaments or solutions or emulsions thereof suitable for injection, bearing titles of the form — Injection.

药剂或溶液或乳剂适于注射用者,可冠以"注射液" 剂型名称。

4 Most compressed tablets consist of the active ingredient and a diluent (filler), binder, disintegrating agent, and lubricant.

多数压制片剂是由有效成分和稀释剂(填充剂)、黏合剂、崩解剂及润滑剂组成。

5 Tablets are prepared by three general methods, wet granulation, dry granulation or slugging, and direct compression.

片剂制备的通法有三种:湿颗粒法,干颗粒法或重压法以及直接压片法。

第七节 生物制药
Section 7 Biotechnological Pharmaceuticals

★ 重组人生长激素 recombinant human growth hormone
生长激素释放激素 growth hormone-releasing hormone
★ 神经生长因子 nerve growth factor
细胞因子 cytokine [ˌsaitəukain]
淋巴细胞活化因子 lymphocyte [ˈlimfəsait] activating factor
★ 集落刺激因子 colony stimulating factor
★ 组织纤溶酶原激活剂 tissue plasminogen [plæzˈminədʒin] activator
★ 基因工程疫苗 gene engineered vaccine
★ 基因表达库免疫 expression library immunization
免疫原性复合物 immunogenic complex

Monoclonal antibodies can be produced by cells that represent the fusion products of an antibody-producing spleen cell and a cancerous (and thus immortal) myeloma cell. The fusion products, known as hybrid myeloma (hybridoma) cells, can be indefinitely cultured in a nutrient suspension that includes blood serum components. Scale-up production batches consist of approximately 35 L of solution.
单克隆抗体可通过将产生抗体的脾细胞与癌性骨髓瘤细胞融合而生产。该融合物,称为杂交骨髓瘤(杂交瘤)细胞,可在含血清成分的营养悬液中无限培养。分批放大生产体积约为 35 升溶液。

第二十四章　免疫学
Chapter 24　Immunology

第一节　绪论
Section 1　Introduction

表位　epitope [ˈepitəup]
特化　specialized [ˈspeʃəlaizd]
多变性　variability [ˌvɛəriəˈbiliti]
记忆性　memory [ˈmeməri]
★ 皮质　cortex [ˈkɔːteks]
★ 胸腺　thymus [ˈθaiməs]
★ 滤泡　follicle [ˈfɔlikl]
黑素细胞　melanocyte [ˈmelənəuˌsait]
★ 白细胞　leucocyte [ˈljuːkəˌsait]
★ 粒细胞　granulocyte [ˈgrænjuləsait]
★ 吞噬细胞　phagocytic cell [ˌfægəu ˈsitik sel]
★ 中性粒细胞　neutrophil [ˈnjuːtrəfil]
★ 嗜碱性粒细胞　basophil [ˈbeisəfil]
★ 嗜酸性粒细胞　eosinophil [ˌiːəuˈsinəfil]
调理素　opsonin [ˈɔpsənin]
树突状细胞　dendritic cell，DC [denˈdritik][sel]
免疫病理学　immunopatholgy [ˌimjunəupəˈθɔlədʒi]
趋化性介质　chemotactic mediator [ˌkeməuˈtæktik][ˈmiːdieitə]
★ 趋化因子　chemokine [ˈkiːməuˌkain]
★ 治疗性疫苗　therapeutic vaccine [θerəˈpjuːtik][ˈvæksiːn]
★ 超敏反应　hypersensitivity [ˌhaipəsensiˈtiviti]
★ 胚胎干细胞　embryonic stem cell，ES [ˌembriˈɔnik][stem][sel]

1 Infectious microbes and larger organisms such as worms are present in our environment. They range from being helpful (e. g. *E. coli*) to being major pathogens which can be fatal (e. g. HIV).
在我们的生活环境中存在有传染性的微生物和较大的微生物如蠕虫。这些微生物的范畴包括从有益的微生物(如大肠杆菌)到能致命的主要病原体(如免疫缺陷病毒 HIV)。

2 Microbes gain entrance into the body actively (penetration of skin), or passively (ingestion of food and inhalation). They have to pass across physical barriers such as the skin or epithelial cells which line the mucosal surfaces of the respiratory, gastrointestinal and genitourinary tracts.
微生物能主动侵入人体(透过皮肤)，或被动进入(通过食物的摄取或吸入)。它们必须越过生理屏障，如皮肤和位于呼吸道、胃肠道、泌尿生殖道表面的上皮细胞。

第二节　抗原
Section 2　Antigen

变应原 allergen [ˈælədʒən]
★ 耐受原　tolerogen /t [əlerəudʒin]
★ 异物性　foreignness [ˈfɔrinis]
★ 半抗原　hapten [ˈhæpten]
构象表位　conformational epitope [ˌkɔnfɔːˈmeiʃənəl ˈepitəup]
★ 多价性　polyvalency [ˌpɔliˈveilənci]
★ 独特位　idiotope [ˌidiəˈtɔp]

1 Protection against pathogenic microorganisms requires the generation of effective immune mechanisms. Thus, vaccines must be capable of targeting the immune system appropriately i. e. cellular and / or humoral mechanisms.
对抗致病微生物的保护作用依赖于形成有效的免疫机制。这样，疫苗必须准确地靶向免疫系统，包括细胞免疫和 / 或体液免疫机制。

2 Most vaccines consist of either attenuated organisms, killed organisms, inactivated toxins, or subcellular fragments and more recently genes for antigens in viral "vectors", and DNA itself.
大多数疫苗由减毒的生物、死的生物、灭活的毒素或者亚细胞片段组成，最近还有用病毒"载体"抗原的基因和病毒本身 DNA 作为疫苗的。

3 Invading organisms have antigens which are recognized by the immune system. Antigens are defined as substances which induce an immune response. They include proteins, carbohydrates, lipids and nucleic acid.
入侵机体的微生物具有能够被免疫系统识别的抗原。抗原被定义为能够诱导免疫应答的物质，它们包括蛋白质、碳水化合物、脂质和核酸。

第三节 免疫球蛋白
Section 3 Immunoglobulin

★ 恒定区 constant region, C ['kɔnstənt 'ri:dʒən]
★ 可变区 variable region, V ['vɛəriəbl 'ri:dʒən]
★ 分泌片 secretory component [si'kri:təri kəm'pəunənt]
★ 同种型 isotype ['aisəutaip]
同种异型 allotype ['ælətaip]
骨架区 framework region ['freimwə:k 'ri:dʒən]
亲和力成熟 affinity maturation [ə'finiti ˌmætju'reiʃən]
★ 多克隆抗体 polyclonal antibody [ˌpɔli'kləunəl 'æntiˌbɔdi]
类别转换 isotype switch ['aisəutaip switʃ]
★ 亲和力 affinity [ə'viditi]
嵌合抗体 chimeric antibody [kai'merik 'æntiˌbɔdi]
★ 噬菌体展示技术 phage display technology [feidʒ] [di'splei] [tek'nɔlədʒi]

1 Affinity is the tightness of binding of an antibody binding site to an antigenic determinant: the tighter the binding, the less likely the antibody is to dissociate from antigen. Different antibodies to an antigenic determinant vary considerably in their affinity for that determinant. Antibodies produced by a memory response have higher affinity than those in a primary response.
亲和力是指抗体结合部位与抗原决定簇之间的结合的紧密程度；结合越紧，抗体与抗原越不易分离。不同的抗体对一个抗原决定簇的亲和力是不同的。通过记忆应答产生的抗体，比初次应答中产生的抗体有更高的亲和力。

2 Antibodies, often termed "immunoglobulins", are glycoproteins that bind antigens with high specificity and affinity. In humans there are five chemically and physically distinct classes of antibodies (IgG, IgA, IgM, IgD, IgE).
抗体(常被称为"免疫球蛋白")是以高特异性、高亲和力与抗原结合的糖蛋白。人类有 5 类化学上和物理上不同类别的抗体(IgG, IgA, IgM, IgD, IgE)。

3 The valence of an antibody is the number of antigenic determinants with which it can react. Having multiple binding sites for an antigen dramatically increases its binding (avidity) to antigens on particles such as bacteria or virus. For example, two binding sites on IgG are 100 times more effective at neutralizing virus than two unlinked binding sites.
抗体效价是指能够与抗体起反应的抗原决定簇的数量。抗体对一种抗原有多重结合位点，则显著增强其与(如细菌或病毒)粒子上抗原的结合。例如，同时有两个结合位点的 IgG，其中和病毒的效力是单结合位点抗体的 100 倍。

第四节 补体系统
Section 4 Complement system

★ 补体受体 complement receptor, CR ['kɔmplimənt ri'septə]
凝集素途径 lectin pathway, LP ['lektin 'pɑ:θwei]
★ 纤维蛋白胶凝素 ficolin ['fi:kəulin]
急性期反应 acute-phase response [ə'kju:t- feiz ris'pɔns]
★ 旁路途径 alternative pathway, AP [ɔ:l'tə:nətiv 'pɑ:θwei]
攻膜复合体 membrane attack complex, MAC ['membrein ə'tæk 'kɔmpleks]
★ 补体依赖的细胞毒作用 complement depend-

ent cytotoxicity, CDC ['kɔmplimənt di'pendənt ˌsaitə'tɔksiti] /

免疫黏附 immune adherence [i'mju:n əd'hiərəns]

膜辅蛋白 membrane cofactor protein, MCP ['membrein kəu'fæktə 'prəuti:n]

★ 群集素 clusterin ['klʌstərin]

白细胞黏附缺陷 leukocyte adhesion deficiency, LAD

1 IgG or IgM antibodies can, on binding to antigen, activate the classical pathway of complement leading to lysis of the cell on which the antigen is located, and / or to attraction of immune cells(chemotaxis) which phagocytose the antigen expressing cells.

在与抗原结合时,IgG 和 IgM 抗体能激活补体经典途径,导致抗原侵染的细胞溶解和/或通过(趋化性)吸引免疫细胞吞噬具有抗原的细胞。

2 The complement system consists of a large number of interdependent proteins, which on sequential activation may mediate protection against microbial infection and contribute to the inflammatory response. Synthesized by hepatocytes and monocytes, these proteins help (complement) antibody responses and have a wide spectrum of activities including a pivotal role in innate defense mechanisms. Complement may be activated by either the classical pathway (through antibody) or the alternative pathway (innate).

补体系统是由很多种互相依赖的蛋白质组成,它们在按顺序激活时启动保护机制对抗微生物感染并引起炎症反应。由肝细胞和单核细胞合成的这些蛋白质能够辅助(补充)抗体反应,并有广泛的活性,在固有免疫中起关键性作用。补体系统也可以通过经典抗体途径激活,或通过固有免疫系统中的旁路途径活化。

第五节 细胞因子
Section 5 Cytokine

★ 单核因子 monokine ['mɔnəukain]

★ 淋巴因子 lymphokine ['limfəukain]

★ 生长因子 growth factor [grəuθ 'fæktə]

★ 转化生长因子 tumor transforming growth factor, TGF

胶原诱导的关节炎 collagen-induced arthritis, CIA ['kɔləˌdʒən in'dju:s ɑ:'θraitis]

禽流感病毒 influenza virus strains ['influ'enzə 'vaiərəs streinz]

系统性炎症反应综合征 systemicinflammatory response syndrome

1 Cytokines are small molecules that signal between cells, inducing growth differentiation, chemotaxis, activation, enhanced cytotoxicity and/or regulation of immunity.

细胞因子是具有细胞间传递信号功能的小分子,包括生长分化、趋化、激活、增强细胞毒性和/或免疫调节。

2 The same cytokine can be made by different cell types and may have different effects on different cell populations (i. e. activate macrophages to kill intracellular microbes and B cells to undergo antibody class switching).

相同的细胞因子可以由不同类型的细胞生成,对不同的细胞类群有不同的生理效应,如:激活巨噬细胞杀死细胞内的微生物,激活 B 细胞形成不同的抗体。

第六节 白细胞分化抗原
Section 6 Leukocyte differentiation antigen

★ 细胞表面标志 cell surface marker [sel 'sə:fis 'mɑ:kə]

白细胞分化抗原 leukocyte differentiation antigen

细胞黏附分子 cell adhesion molecules, CAM [sel əd'hi:ʒən 'mɔlikju:lz]

辅助受体 co-receptor [ri'septə]

★ 共刺激信号 co-stimulatory signal ['stimjuleitə 'signl]

淋巴细胞归巢受体 lymphocyte homing receptor，LHR [ˈlimfəsait ˈhəumiŋ riˈseptə]

★ 血管地址素　vascular addressin [ˈvæskjulə]

非霍奇金淋巴瘤　non-Hodgkin's lymphoma，NHL [nʌŋ ˈhɔdʒkin limˈfəumə]

第七节　主要组织相容性复合体

Section 7　Major histocompatibility complex

★ 次要组织相容性抗原　minor histocompatibility antigen [ˈmainə ˌhistəukəmˌpætiˈbiliti ˈæntidʒən]

近交系小鼠　inbred mice [ˈinˈbred]

人类白细胞抗原　human leukocyte antigen，HLA [ˈhjuːmən ˈljuːkəsait ˈæntidʒən] /

★ 主要组织相容性抗原　major histocompatibility antigen，MHA [ˈmeidʒə ˌhistəukəmˌpætiˈbiliti ˈæntidʒən]

★ 组织相容性　histocompatibility [ˌhistəukəmˌpætiˈbiliti]

单倍型遗传　genetic haplotype [dʒiˈnetik ˈhæpləutaip]

补体依赖性细胞毒试验　complement dependent cytotoxicity test，CDC [ˈkɔmplimənt diˈpendənt]

纯合分型细胞　homozygote typing cell，HTC [ˌhɔməˈzaigəut ˈtaipiŋ sel] /

连锁不平衡　linkage disequilibrium [ˈliŋkidʒ ˌdisikwiˈlibriəm]

预致敏淋巴细胞试验　primed lymphocyte test，PLT [praim ˈlimfəsait test]

共用基序　common motif [ˈkɔmən məuˈtiːf]

关联　association [əˌsəusiˈeiʃən]

相对危险性　relative risk，RR [ˈrelətiv risk]

Two classes of polymorphic MHC genes code for human leukocyte antigens (HLA) which act as peptide receptors and are critical to antigen presentation. Class Ⅰ genes encode two polypeptide chains, a polymorphic heavy chain and β2- microglobulin, expressed on the surfaces of all nucleated cells. The heavy chain has a binding groove for peptides to be recognized by T cells. Class Ⅱ genes encode molecules composed of two dissimilar polymorphic polypeptide chains, both of which contribute to the peptide binding groove.

人类白细胞抗原(HLA)的两类多态性 MHC 基因编码，能作为多肽受体，对抗原提呈是至关重要的。Ⅰ类基因在有核细胞表面表达，编码两条链，一条多态性重链和一条 β2 微球蛋白。重链有一个用于 T 细胞识别的肽“结合槽”。Ⅱ 类基因编码两条不同的多肽链，两条链均用于构筑肽结合槽。

第八节　固有免疫

Section 8　Innate immunity

★ 固有免疫系统　innate immunity system [ˈineit iˈmjuːniti ˈsistəm]

皮肤黏膜屏障　epidermis and mucosal surface [ˌepiˈdəːmis][mjuːˈkəusə ˈsəːfis]

★ 防御素　defensin [diˈfensin]

急性期蛋白　acute-phase protein [əˈkjuːt feiz ˈprəutiːn]

抗菌肽　antimicrobial peptides [ˌæntimaiˈkrəubiəl ˈpeptaɪdz]

血清淀粉样蛋白　serum amyloid protein，SAP [ˈsiərəm ˈæmiˈlɔid ˈprəutiːn]

呼吸爆发　respiratory burst [risˈpaiərətəri bəːst]

尘细胞　dust cell [dʌst sel]

单核细胞　monocyte，Mo [ˈmɔnəuˌsait]

组织细胞　histiocytes [ˌhistiəˈsitis]

大颗粒淋巴细胞　large granule lymphocytes [lɑːdʒ ˈgrænjuːl ˌlimfəuˈsitik]

多糖识别区　carbohydrate-recognition domain，CRD

固有样淋巴细胞　innate-like lymphocytes，ILLs [ˈineit laik ˈlimfəsait]

肥大细胞　mast cell [mɑːst sel]

★ 模式识别受体　pattern recognition receptors，PRRs [ˈpætən ˌrekəgˈniʃən riˈseptə] s/

危险模式理论　danger signal model [ˈdeindʒə ˈsignl ˈmɔdl]

甘露糖结合凝集素　mannose-binding lectin，MBL [ˈmænəus ˈbaindiŋ ˈlektin]

损伤相关的分子模式　damage associated molecular patterns，DAMPs [ˈdæmidʒ əˈsəuʃieitid məuˈlekjulə ˈpætən]

甘露糖受体 mannose receptor，MR ['mænəus ri'septə]
清道夫受体 scavenger receptor ['skævindʒə ri'septə]
白细胞介素受体相关激酶 interleukin receptor-associated kinase，TRAK [ˌintə'ljukin ri'septə ə'səuʃieitid 'kineis]
死亡结构域 death domain，DD [deθ dəu'mein]

1 Inherent in other cells of the immune system: Of every hue of other cells, including eosinophils, platelets and red blood cells, plays a role in immune defense. Eosinophils are granular leukocytes. They attack and kill the parasite through the releases of toxins. Platelet activation, releases of activated complement medium, resulting in leukocyte attraction. Red cells can be combined and remove small immune complexes.
免疫系统其他细胞的内在特点：其他细胞的共性特点，包括嗜酸性粒细胞、血小板和红细胞，是都在免疫防御中起作用。嗜酸性粒细胞是颗粒白细胞。它们通过释放毒素攻击和杀伤寄生虫。血小板活化后，释放活化补体的介质，导致吸引白细胞。红细胞可结合并去除小的免疫复合物。

2 Response termination and repair: Once the removal or control of invasive damage (such as microbial), proinflammatory cytokine inhibitor, can inhibit inflammation, and activation of tissue repair mechanisms. Macrophages are also produced in the repair process which is very important in collagen and growth factor.
应答的终止和修复：一旦除去或控制侵入的危害(如微生物)、促炎细胞因子的抑制物，即可抑制炎症，并激活组织修复机制。巨噬细胞也产生在修复过程中，它对胶原蛋白和生长因子是很重要的。

第九节 黏膜免疫系统
Section 9 Mucosal immune system

肠上皮细胞 epithelial cell，EC [ˌepi'θi:ljəl sel]
★ 转吞作用 transcytosis
肠道相关淋巴组织 gut-associated lymphoid tissues. GALTs [gʌt ə'səuʃieitid 'limfɔid 'tiʃju:lə(r)]
★ 滤泡相关上皮 follicle-associated epithelium ['fɔlikl ə'səuʃieitid ˌepi'θi:ljəm]
上皮层 epithelial layer [ˌepi'θi:ljəl 'leiə]
★ 外周淋巴结地址素 peripheral lymph node addressin，PNAd [pə'rifərəl limf nəud ə'dresiŋ]
★ 经口耐受 oral tolerance ['ɔ:rəl 'tɔlərəns]

1 Nature killed (NK) cells are found throughout the tissues of the body but mainly in the circulation. They contain cytotoxic substances which are important for protection against viruses and some tumors.
自然杀伤(NK)细胞遍及全身组织，但主要在循环系统中。它们含有细胞毒性物质，而它们对保护机体不受病毒和某些肿瘤的侵害是很重要的。

2 Changes in the surface molecules of cells as the result of virus infection allow NK cell to bind to infected cells.
病毒感染引起细胞表面分子的改变，导致 NK 细胞对感染细胞的结合。

第十节 T淋巴细胞
Section 10 T lymphocyte

★ 有丝分裂源 mitogen ['mitədʒən]
★ 效应性记忆 T 细胞 effector memory T cell，T_{EM} [i'fektə 'meməri][sel]
★ 辅助性 T 细胞 helper T lymphocyte，Th ['helpə T 'limfəsait]
★ 细胞毒性 T 细胞 cytoxic T lymphocyte，CTL 或 cytoxic T cell，Tc
★ 肿瘤浸润淋巴细胞 tumor infiltrating lymphocyte，TIL ['tju:mə ˌinfil'treitiŋ 'limfəsait]
★ 颗粒酶 granzyme ['grænzaim]

Cell mediated immunity is due to the direct action of T cell, which distinguishes it from immunity mediated by antibodies (humoral immunity). These terms evolved from the finding that immunity to certain antigens could be transferred to other animals by either cells, if they were of the same inbred

strain, or antibodies.
细胞介导的免疫是由于T细胞的直接作用，这把它与抗体介导的免疫(体液免疫)区分开来。这些术语是由发现对某些抗原的免疫性能通过细胞或抗体传递给其他动物而推论的。

第十一节 B淋巴细胞
Section 11 B lymphocyte

类别或同种型转换 class or isotype switching [klɑ:s ɔ: 'aisəutaip 'switʃiŋ]
★ 替代剪切 alternative splicing [ɔ:l'tə:nətiv 'splaisiŋ]
T细胞受体库 TCR repertoire [ðɔɔ 'repətwɑ:]
红细胞生成素 erythropoietin [i,riθrə'pɔiətin]
淋巴样干细胞 lymphoid stem cell ['limfɔid stem sel]
★ 髓样干细胞 myeloid stem cell ['maiə,lɔid stem sel]
爆式红系前体形成单位 burst forming unit-erythroid precursor
巨核细胞集落形成单位 colony forming unit-megacaryocyte, CFU-Meg
双阳性细胞 double positive cell, DP ['dʌbl 'pɔzətiv sel]
双阴性细胞 double negative cell, DN ['dʌbl 'negətiv sel]
成熟初始B细胞 mature native B [mə'tjuə 'neitiv B]
早期祖B细胞 early pro-B cell ['ə:li B sel]
法氏囊 bursa of Fabricius ['bə:sə əv fə'briʃəs]
假性轻链 surrogate light chain ['sʌrəgit lait tʃein]
★ 天然抗体 natural antibody ['nætʃərəl 'ænti,bɔdi]
边缘带 marginal zone, MZ ['mɑ:dʒinəl zəun]

The BCR complex consists of the antigen receptor, Ig, in association with two other polypeptides, Igα and Igβ. Igα and Igβ are signaling molecules for the BCR and are also required for assembly and expression of Ig.
B细胞受体(BCR)复合物由抗原受体(Ig)和与之相关联的另外两条多肽Igα和Igβ(CD79a和CD79b)组成。Igα和Igβ是BCR的信号分子，也是Ig装配与表达所必需的。

第十二节 抗原提呈细胞
Section 12 Antigen presenting cell

★ 抗原提呈细胞 antigen presenting cell, APC ['æntidʒən pri'zentiv sel]
专职性APC professional APC [prə'feʃənl]
浆细胞样DC plasmacytoid DC, pDC [,plæzməsai'tɔid] /
经典DC conventional DC, cDC [kən'venʃənl DC]
淋巴样DC lymphoid DC ['limfɔidDC]
髓样DC myeloid DC ['maiə,lɔid DC]
MHC Ⅰ类途径 MHC class Ⅰ presentation pathway [,prezen'teiʃən 'pɑ:θwei]
MHC Ⅱ类途径 MHC class Ⅱ presentation pathway [,prezen'teiʃən 'pɑ:θwei]
★ 内化 internalization [in,tɜ:nəlai'zeiʃən]
受体介导的内吞作用 receptor-mediated endocytosis [ri'septə midieitid ,endəusai'təusis]
MHC Ⅲ类途径 MHC class Ⅲ presentation pathway [,prezen'teiʃən 'pɑ:θwei]
恒定链 invariant chain, Ii [in'vɛəriənt tʃein]
恒定链多肽 class Ⅱ-associated invariant chain peptide, CLIP [klɑ:s Ⅱ in'vɛəriənt tʃein 'peptaɪd]
★ 内体 endosome ['endə,səum]
低分子量多肽或巨大多功能蛋白酶 low molecular weight polypeptide or larger multifunctional protease, LMP [ləu məu'lekjulə weit ,pɔli'peptaid ɔ: 'prəutieis]
抗原加工相关转运体或抗原肽转运体 transporter associated with antigen processing or transporter of antigenic peptides, TAP

Some pathogens replicate in cellular vesicles of macrophages, others are endocytosed from the environment into endocytic vesicles (exogenous pathway). In both cases peptides from proteins associated with these microbes are primarily presented on MHC class Ⅱ molecules to $CD4^+$ helper T cell.
一些病原体在巨噬细胞的细胞囊泡中复制，另一些病原体从环境进入内吞细胞时被吞噬(外源途径)。在这两种情况下，来源于微生物相关的蛋白质的多肽都最初呈现到Ⅱ型MHC分子表面并提呈给 $CD4^+$ 辅助T细胞。

第十三节 特异性T淋巴细胞的应答
Section 13 Immune response of specific T cells

★ 记忆T细胞 memory T cell ['meməri T sel]
静息T细胞 resting T cell ['restiŋ T sel]
抗原识别 antigen recognition ['æntidʒən ˌrekəg'niʃən]
★ 免疫突触 immunological synapse [si'næps]
★ 黏附分子 adhesion molecules [əd'hi:ʒən]
★ 自身MHC限制性 self MHC restriction [self] [ris'trikʃən]
交叉提呈 cross-presentation [krɔs ˌprezen'teiʃən]
★ 信号转导 signal transduction ['signl træns'dʌkʃən]
蛋白酪氨酸激酶 protein tyrosine kinase, PTK ['prəuti:n 'tirəsi:n 'kineis]
★ 分子簇化 molecular clustering [məu'lekjulə 'klʌstəriŋ]
★ 适配蛋白 adapter protein [ə'dæptə(r) 'prəuti:n]
丝裂原活化的蛋白 mitogen-activated protein, MAP ['mitəˌdʒən 'æktiveitid 'prəuti:n]
甘油二酯 diacylglycerol, DAG [daiæsil'glisərɔl]
磷脂酶 phospholipase, PLC [ˌfɔsfəu'laipeis]
磷脂酰肌醇二磷酸 phosphatidylinositol 4,5-bisphosphate, PIP_2 [ˌfɔsfə'taidilsəri:n]
三磷酸肌醇 inositol 1,4,5-trisphosphate, IP3 [i'nəusitəul]
细胞外活化激酶 extracellularreceptor-activated kinase, ERK
蛋白激酶 protein kinase
核因子 nuclear factor, NF ['nju:kliə 'fæktə]
活化T细胞核因子 nuclear factor of activated T cells. NFAT ['nju:kliə 'fæktə ɔv 'æktiveitid T sel]
免疫酪氨酸的转换基序 immunoreceptor tyrosine-based switch motif, ITSM ['tirəsi:n beist swift]
多肽—MHC四聚体 peptide-MHC tetramers ['peptaɪd - MHC te'træmitə]
中心性记忆T细胞 central memory T cell [sen'trəl 'meməri T sel]
活化诱导的细胞死亡 activation induced cell death, AICD [ˌækti'veiʃən in'dju:s sel deθ]
自身稳定 homeostasis [ˌhəumiəu'steisis]

1 The T cell repertoire selection: In the thymus, the expression of MHC and its weaker binding to TCR T cells undergo positive selection. The expression of MHC and its solid combination of TCR T cells are eliminated (negative selection). In addition, T cells identifying the MHC and their peptides are removed (negative selection), while leaving the identification of modified their MHC molecules on T cells survive, and become mature in peripheral lymphoid tissue has the function of T cells.
T细胞库的选择：在胸腺中，那些表达与自身MHC较弱结合的TCR的T细胞经历了阳性选择。而那些表达与自身MHC牢固结合的TCR的T细胞被消除(阴性选择)。另外识别自身MHC以及自身肽的T细胞也被除去(阴性选择)，而留下那些识别被修饰的自身MHC分子的T细胞得以存活、成熟并变成外周淋巴组织中有T细胞功能的部分。

2 Accessory molecules: T cells via the T cell antigen receptor start recognition processing of antigens.
辅助分子：T细胞通过T细胞抗原受体启动抗原的识别过程。

第十四节 特殊的B细胞的免疫应答
Section 14 Immune response of specific B cells

TNF受体相关因子 TNF receptor-association, TRAF [TNF ri'septə ə,səusi'eiʃən]
★ 脂筏 lipid rafts ['lipid ræfts]
明区 light zone [lait zəun]
Ig类别转换 Ig class switch [klɑ:s switʃ]
★ 浆细胞 plasma cell, PC ['plæzmə sel]
亲和力成熟 affinity maturation [ə'finiti ,mætju'reiʃən]
★ 受体编辑 receptor editing [ri'septə 'editiŋ]
侧链理论 side chain theory [said tʃein 'θiəri]
记忆性B细胞 memory B cell ['meməri B sel]
★ 回忆应答 anamnestic response [,ænæm'nestik ris'pɔns]

The initial response and memory response: Initial contact with antigen, is to produce IgM antibody immune response, usually in the 3-4d and later IgG immune response. This response is self-limited, and when no stimulation of B cells by antigen is stopped. When antigens enter again, they produce more differentiation into effector-stronger antigen-specific memory B cells, leading to more rapid response, and often causes IgG antibody.

初次应答和记忆应答:初次接触抗原时,导致产生IgM抗体的初次免疫应答。通常在3～4d内接着发生IgG免疫应答。这种应答是自限的,并且当没有抗原刺激B细胞时即停止。当抗原再次进入时,产生更多已分化成效应更强的抗原特异性记忆B细胞,导致更迅速的应答,且通常引起IgG抗体产生。

第十五节 免疫系统调节
Section 15 The regulation of immune system

独特型—抗独特型网络 idiotype and anti-idiotype network ['idiətaip ænd ,ænti 'idiətaip 'netwə:k]
免疫调节 immunoregulation [,imjunəu,regju'leiʃən]
内环境稳定 homeostasis [,həumiəu'steisis]
抗体反馈性抑制 antibody feedback inhibition ['ænti,bɔdi 'fi:dbæk ,inhi'biʃən]
被动细胞死亡 passive cell death ['pæsiv sel deθ]
★ 凋亡 apoptosis ['æpəp'təusis]
反式自杀 suicide in trans ['sjuisaid in træns]
★ 顺式自杀 suicide in cis ['sjuisaid]
免疫偏离 immune deviation [i'mju:n ,di:vi'eiʃən]
★ 天然存在的调节性T细胞 naturally existing regulatory T cell, nTreg ['nætʃərəli]
★ 诱导性调节性T细胞 induced regulatory T cells, iTreg ['regjulətəri sel]

1 The immune system must be subject to strict regulation, by heterologous biological threat starting response, and then finely adjusted, when out of threat it closed. In addition, It also required for the regulation of the immune system components, and for one hundred years they do not responses to their own.

免疫系统必须受严格的调节,受异种生物威胁时启动应答,然后进行精细的调整,在除去威胁时关闭。另外,也需要调节免疫系统的组分,一百年它们也不对自身应答。

2 In the innate immune system, phagocytic cells can only in its damage or death to be identified; natural killer cells usually inhibited receptor inhibition, lest the killer cell.

在固有免疫系统中,吞噬细胞只在其自身细胞受损伤或即将死亡时才能识别它们;自然杀伤细胞通常受抑制受体抑制,减少自然杀伤细胞。

3 Antigen triggers and drives the immune response, its strength under genetic control. Removal of antigen leads to responses to quell. Helper T cells involved in the regulation of immune responses and adjustment of other cellular functions, IgG antibody can inhibit subsequent antibody production. Lymphocyte and antibody against their own tolerance in development at the same time begins,

peripheral mainly through the lack of a common stimulus nuclei via activation induced cell death triggered. The neuroendocrine system in the regulation of the immune response is very important.
抗原引发并驱动免疫应答，其强度受遗传控制。取出抗原导致应答平息。辅助T细胞参与调节免疫应答并具有调整其他细胞的功能，IgG抗体能抑制其后的抗体产生。淋巴细胞和抗体对自身的耐受在个体发育的同时就同步开始（中枢耐受），而外周系统中缺乏共同刺激的细胞核进而通过活化引发细胞死亡而激发（免除这些细胞对机体自身组织的破坏）。神经内分泌系统在调节免疫应答过程中具有重要作用。

第十六节 免疫耐受
Section 16 Immunotolerance

★ 免疫耐受 immunological tolerance or immunotolerance [ˈtɔlərəns ɔː ˌimjunəˈtɔlərən]
★ 外周耐受 peripheral tolerance [pəˈrifərəl ˈtɔlərəns]
★ 中枢耐受 central tolerance [senˈtrɑːl ˈtɔlərəns]
★ 免疫麻痹 immunologic paralysis [pəˈrælisis]
★ 免疫缺陷 immunodeficiency [ˌimjunəudiˈfiʃənsi]
★ 免疫抑制 immunosuppression [ˌimjunəusəˈpreʃən]
★ 禁忌克隆 forbidden clone [fəˈbidn kləun]
低带耐受 low zone tolerance [ləu zəun ˈtɔlərəns]
高带耐受 high zone tolerance [hai zəun ˈtɔlərəns]
克隆禁忌 clonal forbidden [ˈkləunəl fəˈbidn]
克隆流产 clonal abortion [ˈkləunəl əˈbɔːʃən]
克隆清除 clonal deletion
克隆失能 clonal anergy [ˈkləunəl ˈænədʒi]
克隆忽略 clonal ignorance [ˈkləunəl ˈignərəns]
免疫赦免区 immune privileged site [iˈmjuːn ˈprivilidʒd sait]

1 Central tolerance: Central tolerance is immature T cells and B cells in the mature period, in primary lymphoid organs, tissue obtained on autoantigens tolerance process.
中枢耐受：中枢耐受是未成熟的T细胞和在成熟期间的B细胞，在初级淋巴器官和组织内获得对自身抗原的耐受性的过程。

2 Peripheral tolerance: Since not all of the autoreactive lymphocytes can through the central tolerance mechanisms be cleared (mainly is because in the primary lymphoid organ in the absence of most of their antigen), so in the peripheral tissue.
外周耐受：由于并非所有的自身反应淋巴细胞都能通过中枢耐受机制被清除（主要原因是初级淋巴器官缺乏大多数的自身抗原），因此（成熟的T、B细胞）在外周组织中也形成耐受。

3 Peripheral T cell because of the lack of antigen presenting cells of co-stimulatory molecules are second signal (T cell activation required), and cause no response. Due to the lack of T cell costimulatory signal results, peripheral B cells can become no response, and cannot develop into plasma cells.
因为缺乏抗原呈递细胞提供的共同刺激分子（T细胞活化所必需的），外周T细胞成为第二信号，因此不能产生应答反应。由于缺乏来自T细胞的共同刺激信号的结果，外周B细胞不形成（免疫）应答反应，也不能发育成浆细胞。

4 The maturation of the immune system: Tolerance before birth or in the early neonatal are relatively easy to obtain, which may be related to T, B and / or antigen-presenting cells has not been mature. T cells than B cells are more easily tolerated.
免疫系统的成熟：耐受在出生前或在新生儿早期比较容易获得，这可能与T、B和/或抗原呈递细胞尚未成熟有关。T细胞比B细胞更容易耐受。

第十七节 超敏反应
Section 17 Hypersensitivity

★ 超敏反应 hypersensitivity [ˌhaipəsensiˈtiviti]
★ 变态反应 allergy [ˈælədʒi]
★ 反应素 reagin [riˈeidʒin]
过敏症 anaphylaxis [ˌænəfiˈlæksis]
特应症 atopy [ˈætəpi]
白三烯 leukotriene, LT [ˌljuːkəˈvaiərəs]
系统性自身免疫性疾病 systematic autoimmunity disease [sisˈtemik ˌɔːtəuiˈmjuːniti diˈziːz]
重症肌无力 myasthenia gravis, MG [ˈmaiæsˈθiːniə]
多发性硬 化 multiple sclerosis, MS [ˈmʌltipl skliəˈrəusis]
风湿热 rheumatic fever [ru(ː)ˈmætik ˈfiːvə]
★ 表位扩展 epitope spreading [ˈepitəup ˈprediŋ]
内皮细胞 vascular endothelial cell, VEC [ˈvæskjulə sel]
★ 包容性 inclusivity [ˈinkluˈsiviti]
西罗莫司 sirolimus [siˈrəuliməs]
★ 胞内抗体 intrabody [intrəˈbɔdi]
预防性育苗 prophylactic vaccine [ˌprɔfiˈlæktik ˈvæksiːn]
肿瘤疫苗 tumor vaccine [ˈtjuːmə ˈvæksiːn]
胸腺肽 thymic peptide [ˈθaimik ˈpeptaid]
转移因子 transfer factor [trænsˈfəː ˈfæktə]
卡介苗 Bacille Calmette-Guerin, BCG [bəˈsilai]
配方 formulation [ˌfɔːmjuˈleiʃən]

1 Damage of host tissue can occur as an overreaction by the immune system to a variety of both inert antigens and infectious organisms. This hypersensitivity occurs only after antigen sensitization of the host and is therefore the effect of the adaptive immune system. Overreaction to microbial antigens can occur in the natural immune system although these reactions are not currently classified as hypersensitivity reaction.

免疫系统对各种无活性的抗原和感染性生物过度反应时，可发生对宿主组织的损伤。这种超敏感应仅发生在宿主被抗原致敏以后，因此是适应性免疫系统的效应。在自然免疫系统中能发生对微生物抗原的过度反应，虽然目前没有把这些反应归类为超敏反应。

2 Hypersensitivity reactions have long been classified into four types, with an additional type recently added. Types Ⅰ、Ⅱ、Ⅲ and Ⅴ depend on antibodies, alone or with complement, and because they are evident within hours, are termed immediate hypersensitivities. Type Ⅳ is mediated by T cells and the cytokines they produce when activated. As this response requires at least a day to develop, it is termed delayed hypersensitivity.

早就将超敏反应分为 4 种类型，最近又加上 1 种。Ⅰ、Ⅱ、Ⅲ 及 Ⅴ 型完全依赖于抗体，或同时依靠补体，由于它们在几个小时内表现出来而被称为速发型超敏反应。Ⅳ 型超敏反应是由 T 细胞及其活化时产生的细胞因子介导的。由于这种反应至少需要一天才显现，因而被称为迟发型超敏反应。

第二十五章 生态学
Chapter 25 Ecology

第一节 绪论
Section 1 Introduction

★ 生态学 ecology
★ 环境生物学 environmental biology
★ 光周期现象 photoperiodism [ˌfəutəuˈpiəriədizm]
★ 生态系(统) ecosystem [ˈi:kəˌsistəm]
种群，繁殖群 breeding flock
形态学 morphology
生理学 physiology
遗传学 genetics
分类学 systematics [ˌsistiˈmætiks]
气象学 meteorology [ˌmi:tjəˈrɔlədʒi]
地质学 geology
古生物学 paleontology [ˌpæliɔnˈtɔlədʒi]
现代人，智人 homosapiens [ˌhəuməuˈseipinz]
海洋学 oceanography [ˌəuʃiəˈnɔgrəfi]
★ 无脊椎动物 invertebrate [inˈvə:tibrit]
★ 低气压 barometric [ˌbærəuˈmetrik] depression
★ 食物链，营养链 food chain
★ 环境阻力 environmental resistance
★ 生态位，小生境 niche
自然保护区 nature reserve
自然史 natural history
生物区系，生物相 biota
★ 进化论 evolutionism
★ 自然选择 natural selection
人口统计 vital statistics
★ 演替 succession [səkˈseʃən]
★ 个体生态学 autoecology [ˌɔ:təuˌi:ˈkɔlədʒi]
★ 群体生态学 synecology [ˌsiniˈkɔlədʒi]
★ 生物群系，生物带 biome [ˈbaiəum]

1 Ecology, the study of how organisms relate to one another and to their environments, is a complex and fascinating area of biology.
生态学研究生物如何与其他生物个体或环境相联系，是一个复杂而令人着迷的生物学领域。

2 All the organisms that live together in a place are called a community.
聚集生活在同一个地方的所有生物称为一个群落。

第二节 个体生态
Section 2 Individual Ecology

气候因子 climatic factor
密度制约因子 density dependent factor
非密度制约因子 density independent factor
捕食 predation
天敌 natural enemy
流行病 epidemic
★ 光周期现象 photoperiodism [ˌfəutəuˈpiəriədizm]
日照长度 length of day
春化作用 vernalization [ˌvə:nəliˈzeiʃən]
★ 春化阶段 stage of vernalization
幼虫/幼体 larva [ˈlɑ:və]
成虫期/成虫 imago [iˈmeigəu]
腐殖土 humus soil
大气圈 ambience [ˈæmbiəns]
北极熊 polar bear
企鹅 penguin [ˈpeŋgwin]
秃鹫 vulture [ˈvʌltʃə]
岩石圈 geosphere [ˈdʒi:əuˌsfiə]
水圈 hydrosphere [ˈhaidrəsfiə]
寒带 frigid zone
针叶林 coniferous forest
热带雨林 tropical rain forest
抑制因子 inhibitor
抑制剂 katastaltic [kətəˈstɔltik]
绝对零度 absolute zero
广温性生物 eurytherm [ˈjuəriθə:m]
恒温生物 homoiotherm [həuˈmɔiəθə:m]
变温动物 poikilothermal [pɔiˌkiləuˈθə:ml] animal
★ 休眠期，静止期 repose period
★ 光合作用，光能合成 photosynthesis

苔藓,地衣 moss
光化圈,臭氧层 chemosphere ['keməsfiə]
★生物钟 biochronometer
指示生物 biological indicator
海底生物,底栖生物 benthic organism
最小法则 law of the minimum
忍受法则 law of tolerance
★限制因素,限制因子 limiting factor
★中间寄主 intermediate host
消化系统 digestive system
虫寄养性,类寄生 parasitoidism [ˌpærə'saitɔidzəm]
重寄生性,重寄生的状态 hyperparasitism [haipə'pærəˌsaitizəm]
★寄生链 parasitic chain
★中性现象 neutralism ['nju:trəlizəm]
★啮齿动物 rodent ['rəudənt]
★昼夜节律 circadian rhythms [sə:'keidiən]['riðəmz]
★自运周期 freerunning cycle
★时间调节器 timesetter [taim'setə]
★时间依存机制 time-dependent mechanism
日照中性生物 day-neutral organism
长日照生物 long-day organism
短日照生物 short-day organism
长夜植物 long-night plant
短夜植物 short-night plant
★黄体激素 luteinizing [ˌlu:ti:inai'zeiiŋ] hormone
★滞育现象 diapause ['daiə'pɔ:z]
化蛹 pupate ['pju:peit]
羽化 eclosion [i'kləuʒən]
月球日 lunar day
★物候学 phenology [fi'nɔlədʒi]
★季节性 seasonality [si:zenæliti]
★雨季 rainy season

第三节 种群生态
Section 3 Population Ecology

★(动物的)种群 population
★年龄结构 age structure
出生率 birth rate
天敌 natural enemy
出芽 budding
体外产卵,卵生 oviparity [ˌəuvi'pæriti]
★趋光反应 phototactic reaction
★分裂 cleavage
★腔肠动物 coelenterate [si'lentəreit]
交尾 coition
虫害 insect pest
栖息地 habitat
★自然调节 natural regulation
霜冻 frost
消灭,(生物等的)灭绝 extinction
★野生动物保护区 a wild life sanctuary ['sæŋktjuəri]
野生动植物 wildlife
偷猎者 poacher ['pəutʃə]
杂合性,异型接合性 heterozygosity [ˌhetərəuzai'gɔsəti]
★人工繁殖 artificial propagation
杀虫剂,农药,杀虫活动 insecticide
耐药性 drug resistance
接触除草剂 contact herbicides
内吸除草剂 systemic herbicides
土壤杀菌剂 soil sterilant
★恒温生物 homoiotherm [həu'mɔiəθə:m]
综合治理 comprehensive treatment
★对联种群 metapopulation ['metəˌpɔpju'leiʃən]

1 The growth rate of a population is a sensitive function of its age structure; populations with many young individuals grow rapidly as these individuals enter reproductive age.
一个种群的生长率是其年龄结构的敏感指标;年轻个体多的种群由于它们进入生殖期而生长迅速。

2 Populations grow if the birthrate exceeds the death rate until they reach the carrying capacity of their environment.
如果出生率超过了死亡率,种群数量就会增大,直到达到它们所处环境的承载能力为止。

第四节 群落生态
Section 4 Community Ecology

★季相、周期变化 aspection
★沼泽地 marshland

苔原,冻原,寒漠,冻土带 tundra [ˈtʌndrə]
★ 食草动物 herbivore [ˈhəːbivɔː]
杂食动物 omnivorous [ɔmˈnivərəs] animal
干草原 steppe [step]
穴居,穴居生活 troglodytism [ˌtrɔgləubaiˈɔtism]
★ 海洋水生物 halobios [ˌhæləuˈbaiɔs]
潮汐 tide
★ 浮游生物 plankton [ˈplæŋkt(ə)n]
★ 甲壳类的(动物) crustacean [krʌsˈteiʃjən]
★ 次级演替 secondary succession
地衣 lichen [ˈlaikən]
草本植物(尤指牧草) herbage [ˈhəːbidʒ]
裸岩 bare rock
阔叶树 broadleaved tree
灌木(丛) shrub
★ 演替顶极 climax [ˈklaimæks]
生态型,生态变种 ecotype [ˈiːkəutaip]
★ 生态位分离 niche separation
生态物宽度 niche width
★ 取食生态位 feeding niches
★ 生态位压缩 niche compression
★ 生态释放 ecological release

1 Plants have evolved many mechanisms to defend themselves from herbivores.
植物进化出了许多可以抵御食草动物的机制。

2 This relationship led American ecologist F. E. Clements, at about the turn of the century, to propose the concept of a final climax community.
这个关系使得美国生态学家 F. E. Clements 在世纪之交提出了最终的演替顶级的概念。

3 This kind of succession, which occurs in areas where an existing community has been disturbed, is called secondary succession
这种发生于现存群落被打乱的区域的演替被称为次级演替。

第五节 生态系统
Section 5 Ecosystem

★ 分解体,分解者 decomposer
洋流 ocean current
地表径流 overland runoff
★ 太阳辐射 solar radiation
根系 descending axis
降雨量 rainfall precipitation
★ 化学自养细菌 chemosynthetic [ˈkeməuˈsinˈθetik] bacteria
肉食动物,食虫植物 carnivore [ˈkɑːnivɔː]
★ 物质循环 circulation of materials
能量流动 energy flow
★ 食物网 food web
金字塔现象 pyramid
★ 生物量 biomass
★ 叶面蒸发,蒸腾(作用) transpiration
★ 初级生产者 primary producer
底栖无脊椎动物 benthonic [benˈθɔnik] invertebrate
浮游生物 plankton [ˈplæŋkt(ə)n]
能流 energy stream
地下水 ground water
自流井 artesian well [ɑːˈtiːzjən][wel]
★ 软体动物 mollusk [ˈmɔləsk]
★ 原生动物 protozoan [ˌprəutəuˈzəuən]

Organisms can be either producers or consumers in terms of energy flow through an ecosystem.
从生态系统中能量流动的情况来看,生物要么是生产者,要么是消费者。

第六节 人口、环境、资源
Section 6 Population Environment and Resource

★ 生物圈,生命层,生物大气层 biosphere [ˈbaiəsfiə]
★ 计划生育 family planning
★ 沙(漠)化 desertification [ˌdezəːtifiˈkeiʃən]
★ 温室效应 greenhouse effect
臭氧 ozone [ˈəuzəun]
含酸雨 acid rain
海平面 sea level
冷却剂,冷却介质 cooling agent
★ 免疫功能 immunologic function
光化学烟雾 photochemical [ˌfəutəuˈkemikəl] smog
洗涤剂,清洁剂 abluent [ˈæbluənt]
★ 赤潮 red current

热污染 heat pollution
★ 合成纤维 synthon ['sinθɔn]
★ 天然纤维 natural fabric
再用，回收利用 reutilization [riˌjuːtilaiˈzeiʃən]
地壳 earth crust
太阳能 solar power
★ 地热 terrestrial heat
生物能 bioenergy
水土流失 soil and water loss
针叶林 coniferous forest
常绿阔叶树 broadleaf evergreen
风化 efflorescence [ˌeflɔːˈresəns]
侵蚀强度 erosion intensity [iˈrəuʒən][inˈtensiti]
侵蚀率，侵蚀指数 erosion index
牧童式经济体系 cowboy economy
太空式经济体系 spaceman economy
★ 环境污染 environmental pollution

Environmental Pollution is an international journal that addresses issues relevant to the nature distribution and ecological effects of all types and forms of chemical pollutants in air, soil and water.
《环境污染》是一个国际性刊物，它出版关于自然的分布状态、各种类型的生态效应以及空气、土壤和水中的各种化学污染的形式。

第二十六章 动物生理学
Chapter 26 Animal Physiology

第一节 绪论
Section 1 Introduction

生理学 physiology [ˈfɪzɪˈɑlədʒɪ]
内环境 internal environment
同化作用 anabolism [əˈnæbəlɪzəm]
异化作用 catabolism [kəˈtæbəlɪzəm]
神经调节 nervous regulation
反射弧 reflex arc
体液调节 humoral regulation
内分泌腺 endocrine gland
旁分泌 paracrine [ˌpærəˈkrain]
自身调节 autoregulation [ˈɔːtəʊˌregjʊˈleɪʃən]
正反馈 positive feedback
组织液 interstitial fluid
神经分泌 Neurosecretion [ˌnjʊərəʊsɪˈkriːʃən]
负反馈 negative feedback
兴奋性 excitability [ɪkˌsaɪtəˈbɪləti]

Physiology is a branch of biological sciences. The discipline of physiology explores and explains the vital functions of living organisms, including the functions of the component organs, cells, and molecules of the body.

生理学是生物科学的一个分支,也是探讨和解释生命体的关键生理功能的学科,包括身体器官的、细胞的和分子水平的功能研究。

第二节 细胞的基本功能
Section 2 The Basic Function of Cells

失活 inactivation [inˌæktiˈveiʃən]
钠一钾泵 sodium-potassium pump
化学门控通道 chemically gated channel
跨膜信号转导 transmembrane signal transduction
电压门控通道 voltage-gated channel
促离子型受体 ionotropic receptor
环一磷酸腺苷 cycle adenosine monophosphate, cAMP
第二信使 second messenger
三磷酸肌醇 inositol triphosphate, IP3
G-蛋白耦联受体 G-protein coupled receptor
酪氨酸激酶受体 tyrosine kinase receptor
兴奋性 excitability [ɪkˌsaɪtəˈbɪləti]
阈值 threshold [ˈθrɛʃhold]
动作电位 action potential
超极化 hyperpolarization [ˈhaɪpəˌpolərɪˈzeʃən]
复极化 repolarization [ˈriˌpolərɪˈzeʃən]
锋电位 spike potential
膜片钳 patch clamp
局部兴奋 local excitation
时间性总和 temporal summation
跳跃式传导 saltatory conduction
终板电位 end-plate potential
胆碱酯酶 cholinesterase [koləˈnɛstəˌrez]
滑行理论 sliding theory
等长收缩 isometric contraction
后负荷 afterload
长度一张力曲线 length-tension curve
磷脂酶 C phospholipase C [ˌfɑsfoˈlɪpˌes]
二酰甘油 diacylglycerol, DG [daiæsilˈglisərɔl]
促代谢型受体 metabotropic receptor
一氧化氮 nitric oxide, NO
兴奋 excitation [ˌɛksaɪˈteʃən]
去极化(除极化) depolarization [ˈdiˌpolərɪˈzeʃən]
阈电位 threshold (membrane) potential
电紧张性扩布 electrotonic propagation
空间性总和 spatial summation
量子式释放 quantal release
乙酰胆碱 acetylcholine, ACh [ˌæsətɪlˈkolin]
等张收缩 isotonic contraction
单收缩 single twitch
前负荷 preload [ˈpriːˈləud]
初长度 initial length
(肌肉)收缩能力 contractility (of muscle)
张力一速度(关系)曲线 force velocity (relation)

curve | 钙调蛋白 calmodulin，CaM [kæl'madʒəlɪn]

1 The membrane that surrounds the cell is a remarkable structure. It is made up of lipids and proteins, allowing some substances to pass through but not others, and hence is semipermeable. However, its permeability can also be varied because it contains numerous regulated ion channels and other transport proteins that can change the amounts and variety of substances moving across it. It is generally referred to as the plasma membrane.

包围细胞的膜是一个非常重要的结构，它由脂质和蛋白质组成，允许一些物质通过，而其他物质则不能通过，因此称为半透膜。但是由于膜内有许多离子通道，还有能改变物质形状和数量的转运蛋白，因此膜的渗透性是变化的。它通常被称为等离子体膜。

2 G proteins commonly have two families. One is the larger heterotrimetric G proteins, which couple cell surface receptors to catalytic units that catalyze the intracellular formation of second messengers or couple the receptors directly to ion channels. Another family of G proteins, the small G proteins, are involved in many celluar functions.

G 蛋白通常分为两类，一个是大的异质三聚体 G 蛋白，能与细胞表面受体耦合，来催化细胞内产生第二信使，或者直接与受体结合形成离子通道。另一类是小 G 蛋白，与许多细胞功能相关。

第三节 神经与肌肉组织的一般生理学
Section 3 The General Physiology of Nervous and Muscle Tissue

兴奋组织 excitable tissue
阈强度 threshold intensity
极性法则 law of polarity
绝对不应期 absolute refractory period
相对不应期 relative refractory period
超常期 supranormal period
阈下总和 subliminal summation
生物电 Bioelectricity [ˌbaɪoɪlɛk'trɪsɪti]
损伤电位 injury potential
局部电流学说 local circuit theory
跳跃传导 saltatory conduction
神经肌肉接点 neuromuscular junction
突触前终端 presynaptic terminal
突触间隙 synaptic cleft
兴奋—分泌耦联 excitation-secretion coupling
肌节 sarcomere ['sɑrkəˌmɪr]
肌管系统 sarcotubular system
兴奋—收缩耦联 excitation-contraction coupling
不完全强直收缩 incomplete tetanus
完全强直收缩 complete tetanus
临界融合频率 critical fusion frequency
"全或无"现象 "all or none" phenomenon

Once threshold intensity is reached, a full-fledged action potential is produced. Further increases in the intensity of a stimulus produce no increment or other change in the action potental as long as the other experimental conditions remain constant. The action potential fails to occurs if the stimulus is subthreshold in magnitude, and it occurs with a constant amplitude and form regardless of the strength of the stimulus at or above threshold intensity. The action potential is therefore "all or none" in character and is said to obey the all-or-none law.

一旦刺激达到阈强度，就会产生一个完全的动作电位。只要其他实验条件不变，再增强电刺激的强度，动作电位并不能增大或者有其他反应。阈下刺激无动作电位，阈刺激和阈上刺激都出现恒幅、相同的动作电位。动作电位具有全或者无的特性，叫做"全或无"定律。

第四节 血液
Section 4 Blood

黏滞性 viscosity [vɪ'skɑsɪti]
胶体渗透压 colloid osmotic pressure
造血微环境 hemopoietic microenvironment
血细胞比容 hematocrit [hɪ'mætəˌkrɪt]
晶体渗透压 crystal osmotic pressure
渗透脆性 osmotic fragility
悬浮稳定性 suspension stability
红细胞沉降率 erythrocyte sedimentation rate, ESR
集落刺激因子 colony stimulating factor, CSF

内因子 intrinsic factor
促红细胞生成素 erythropoietin，EPO
血小板黏附 thrombocyte adhesion
血液凝固 blood coagulation
凝血因子 blood clotting factor
外源性凝血 extrinsic coagulation
纤维蛋白溶解系统 fibrinolytic system
爆式促进因子 burst promoting activator，BPA
肝素 heparin [ˈhɛpərɪn]
生理性止血 physiological hemostasis
血小板聚集 thrombocyte aggregation
前列环素 prostacyclin，PGI2 [ˌprɑstəˈsaɪklɪn]
血清 serum [ˈsɪrəm]
内源性凝血 intrinsic coagulation
凝血酶 thrombin [ˈθrɑmbɪn]
纤溶酶原 plasminogen [plæzˈmɪnədʒən]
组织因子途径抑制物 tissue factor pathway inhibitor，TFPI
纤溶酶(血浆素) plasmin [ˈplæzmɪn]
(红细胞)凝集 agglutination [əˌglutiˈneʃən]
凝集素 agglutinin [əˈglutinɪn]
交叉配血试验 cross-match test
血型 blood group
凝集原 agglutinogen [ˌæglʊˈtɪnədʒən]
人体白细胞抗原 human leukocyte antigen，HLA
成分输血 transfusion of blood component

1 The red，oxygen-carrying pigment in the red blood cells of vertebrates is hemoglobin，a protein with a molecular weight of 64450. The average normal hemoglobin content of blood is 16 g/dl in men and 14 g/dl in women，all of it in red cells. Hemoglobin is a globular molecule made up of four subunits. They are two α and two β polypeptide chains.
脊椎动物红细胞中红色、携带氧气的色素物质就是血红蛋白，它是分子量为 64450 的一个蛋白。男性血液中正常血红蛋白的平均含量是 16 g/dl，而在女性中的平均含量是 14 g/dl，所有的血红蛋白都在红细胞中。血红蛋白是由四个亚基球状分子组成的，即两个 α 多肽链和两个 β 多肽链。

2 The A and B antigens are inherited as mendelian dominants，and individuals are divided into four major blood types on this basis. Type A individuals have the A antigen，type B have the B，type AB have both，and type O have neither. The A and B antigens are actually complex oligosaccharides that differ in their terminal sugar.
A 和 B 抗原继承了孟德尔优势群落，在此基础上将人分为四个主要血型。A 血型的人有 A 抗原、B 血型的人有 B 抗原、AB 血型的人两种抗原都有，而 O 血型的人两种抗原均没有。A 和 B 抗原实际是在其终端含有不同糖的复杂低聚糖。

第五节 血液循环
Section 5 Blood Circulation

心动周期 cardiac cycle
心室舒张期 ventricular diastole
快速射血期 rapid ejection period
心音 cardiac sound
每搏输出量(搏出量) stroke volume output
心房收缩期 atrial systole
等容收缩期 isovolumic systole
减慢充盈期 reduced filling period
心音图 phonocardiogram [ɪˌlɛktroˈkɑrdɪogræm]
射血分数 ejection fraction
心指数 cardiac index
心脏做功(量) myocardial work
心力储备 cardiac reserve
有效不应期 effective refractory period
代偿间歇 compensatory pause
起搏点 pacemaker [ˈpesˌmekɚ]
异位心律 ectopic rhythm
超速驱动压抑 overdrive suppression
心电图 electrocardiogram，ECG [ɪˌlɛktroˈkɑrdɪogræm]
阻力血管 resistant vessel
容量血管 capacitance vessel
血流量 blood flow
血流速度 velocity of blood flow
顺应性 compliance [kəmˈplaɪəns]
收缩压 systolic pressure
脉搏压(脉压) pulse pressure
中心静脉压 central venous pressure
微循环 microcirculation
直捷通路 thoroughfare channel
迂回通路 circuitous channel
重吸收 reabsorption [ˌriːəbˈsɔːpʃən]

缩血管神经纤维 vasoconstrictor fiber
心血管中枢 cardiovascular center
压力感受器 baroreceptor [ˌbærərɪ'sɛptɚ]
心肺感受器 cardiopulmonary receptor
化学感受器 chemoreceptor [ˌkimorɪ'sɛptɚ]
脑内渗透压感受器 brain osmoreceptor
心室功能曲线 ventricular function curve
自动节律性(自律性) autorhythmicity [ɔːtəhɪð'mɪsɪtɪ]
平台(期) plateau [plæ'to]
期前收缩 premature systole
起搏离子流 pacemaker current
窦性心律 sinus rhythm
抢先占领 capture, preoccupation
房一室延搁 atrioventricular delay
弹性储器血管 windkessel vessel
毛细血管前阻力血管 precapillary resistant vessel
短路血管 shunt vessel
血压 blood pressure
外周阻力 peripheral resistance
舒张压 diastolic pressure
动脉脉搏 arterial pulse
静脉回心血量 venous return
毛细血管前括约肌 precapillary sphincter
动一静脉短路 arteriovenous shunt
有效滤过压 effective filtration pressure, EFP
舒血管神经纤维 vasodilator fiber
压力感受性反射 baroreceptor reflex
缓冲神经 buffer nerve
容量感受器 volume receptor
防御反应 defence reaction
肾素一血管紧张素系统 renin-angiotensin system
循环系统平均充盈压 mean circulatory filling pressure
血管紧张肽转换酶 angiotensin-converting enzyme, ACE
肾上腺素 epinephrine, E [ˌɛpə'nɛfrɪn]
去甲肾上腺素 norepinephrine, NE, or noradrenalin, NA
血管升压素 vasopressin, VP [ˌvezo'prɛsɪn]
抗利尿激素 antidiuretic hormone, ADH
血管活性肠肽 vasoactive intestinal polypeptide, VIP
内皮细胞舒张因子 endothelium-derived relaxing factor
激肽释放酶一激肽系统 kallikrein-kinin system
血管舒张素 kallidin [kælɪdɪn]
前列腺素 prostaglandin, PG
冠脉循环 coronary circulation
血一脑屏障 blood-brain barrier, BBB
缓激肽 bradykinin [ˌbrædɪ'kaɪnɪn]
心房钠尿肽 atrial natriuretic peptide, ANP
组胺 histamine ['hɪstəˌmin]
腺苷 adenosine [ə'dɛnəˌsin]
血-脑脊液屏障 blood-cerebrospinal fluid barrier
毛细血管 capillaries ['kæpɪlərɪz]

1 The circulatory system is the transport system that supplies O_2 and substances absorbed from the gastrointestinal tract to the tissue, returns CO_2 to the lungs and other products of metabolism to the kidneys, functions in the regulation of body temperature, and distributes hormones and other agents that regulate cell function.

循环系统是为组织提供氧气和从胃肠道吸收物质的,同时将二氧化碳返回到肺部,以及其他代谢产物运回肾脏的运输系统,还具有调节体温、分布激素和细胞功能的作用。

2 The tight junctions between capillary endothelial cells in the brain and between the epithelial cells in the choroids plexus effectively prevent proteins from entering the brain in adults and slow the penetration of smaller molecules. An example is the slow penetration of urea. This uniquely limited exchange of substances into the brain is referred to as the blood-brain barrier.

大脑内的毛细血管内皮细胞之间的紧密连接,以及脉络膜丛中的上皮细胞之间的紧密连接,都能有效地防止蛋白进入大脑并减慢小分子的渗透作用,例如尿素的缓慢渗透。这种物质独特的有限制的物质交换称为血脑屏障。

第六节 呼吸
Section 6 Respiration

腹式呼吸 abdominal breathing
平静呼吸 eupn(o)ea, or quiet breathing

呼吸困难　dyspnea
气胸　pneumothorax [ˌnumoˈθɔˌræks]
比顺应性　specific compliance
气道阻力　airway resistance
肺通气　pulmonary ventilation
胸式呼吸　thoracic breathing
肺内压　intrapulmonary pressure
胸膜腔内压　intrapleural pressure
肺表面活性物质　pulmonary surfactant
潮流气量　tidal volume，TV
贮备吸气量(吸气储备量)　inspiratory reserve volume，IRV
补呼气量(呼气储备量)　expiratory reserve volume，ERV
功能残气量(有效余气量)　functional residual capacity，FRC
残气体积(余气量)　residual volume，RV
肺活量　vital capacity，VC
时间肺活量　timed vital capacity
肺总气量　total lung capacity，TLC
解剖无效腔　anatomical dead space
肺泡通气量　alveolar ventilation
气体扩散速率　diffusion rate of gas
氧合　oxygenation [ˌɔksidʒiˈneiʃən]
血氧含量　oxygen content of blood
氧解离曲线　oxygen dissociation curve
肺牵张反射　pulmonary stretch reflex
深吸气量　inspiratory capacity，IC
用力肺活量　forced vital capacity，FVC
用力呼气量　forced expiratory volume，FEV
每分通气量　minute ventilation volume
肺泡无效腔　alveolar dead space
气体交换　gas exchange
通气血流比值　ventilation/perfusion ratio
氧容量　oxygen capacity
氧饱和度　oxygen saturation
呼吸调整中枢　pneumotaxic center
吸气切断机制　inspiratory off-switch mechanism

1 Three important conditions affect the oxygen-hemoglobindissociation curve: the pH, the temperature, and the concentration of 2,3-diphosphoglycerate.
三个重要条件影响了氧—血红蛋白解离曲线:pH 值、温度和 2,3—二磷酸甘油酸的浓度。

2 Hiccup is a spasmodic contraction of the diaphragm and other inspiratory muscles that produces an inspiration during which the glottis suddenly closes. The glottic closure is responsible for the characteristic sensation and sound. Hiccups occur in the fetus in utero as well as throughout extrauterine life. Their function is unknown.
呃逆是指膈膜和呼吸肌间歇性的收缩，声门突然关闭造成空气突然被吸入。声门关闭具有特别的感觉和声音。不仅子宫内的胎儿，人从出生后的整个一生都会发生呃逆，它们的功能是未知的。

第七节　胃肠功能

Section 7　Gastrointestinal Functions

消化　digestion [daɪdʒɛstʃən]
基本电节律　basic electrical rhythm，BER
黏膜下丛　submucosal plexus
胃肠激素　gastrointestinal hormone
胆囊收缩素　cholecystokinin，CCK [ˌkolɪˌsɪstəˈkaɪnɪn]
肠抑胃肽　gastric inhibitory polypeptide
唾液　saliva [səˈlaɪvə]
胃液　gastric juice
慢波　slow wave
肠肌丛;肌间神经丛　myenteric plexus
胃泌素　gastrin [ˈgæstrɪn]
胰泌素　secretin [sɪˈkriːtɪn]
营养作用　trophic action
蠕动　peristalsis [ˌpɛrɪˈstælsɪs]
胃蛋白酶原　pepsinogen [pɛpˈsɪnədʒən]
黏液—碳酸氢盐屏障　mucus-bicarbonate barrier
假饲　sham-feeding
促胃泌素释放肽　gastrin-releasing peptide，GRP
肠胃反射　entero-gastric reflex
胃排空　gastric emptying
胰液　pancreatic juice
脂肪酶　lipase [ˈlɪpeɪz]
胰蛋白酶　trypsin [ˈtrɪpsɪn]
糜蛋白酶　chymotrypsin [ˌkaɪməˈtrɪpsɪn]
胆汁　bile，or gall
胃蛋白酶　pepsin [ˈpɛpsɪn]
生长激素抑制素　somatostatin [soˌˌmætəˈstætɪn]
蛙皮素　bombesin [bɔmbəsin]

肠抑胃素 enterogastrone [ˌentərəuˈgæstrəun]
容受性舒张 receptive relaxation
迷走—迷走反射 vago-vagal reflex
胰淀粉酶 pancreatic amylase
胰蛋白酶原 trypsinogen
糜蛋白酶原 chymotrypsinogen [kaiməuˈtripˈsinədʒən]
肠激酶 enterokinase [ˌɛntərəˈkaɪnes]
胆盐的肠肝循环 enterohepatic circulation of bile salt
排便 defection [dɪˈfɛkʃən]
分节运动 segmentation contraction

1 90% to 95% of the bile salts are absorbed from the small intestine. Some are absorbed by nonionic diffusion, but most are absorbed from the termina ileum by an extremely efficient Na^+ -bile salt cotransport system powered by basolateral Na^+ -K^+ ATPase. One of the Na^+ -bile salt cotransporters involved in this secondary active transport system has been cloned, and there is evidence that at least one more exists.

90% ～ 95% 的胆盐是在小肠位置被吸收的。一些是通过非离子扩散吸收的，但大多数的胆盐是在末端回肠以基底侧的 Na^+ —K^+ 三磷酸酶腺苷为驱动力，通过非常高效的钠 — 胆盐协同运输系统被吸收的。第二主动运输系统的一个 Na^+ — 胆盐协同体已经被克隆，至少有一个证据表明它的存在。

2 The absorptive surface of the smallintestine is increased about 600-fold by the valvula conniventes, villi, and microvilli. It has been estimated that the inner surface area of a mucosal cylinder the size of the small intestine would be about 3300 cm^2, that the valvula increase the surface area to 10000 cm^2, that the villi increase it to 100000 cm^2, and that the microvilli increase it to 2 million cm^2.

环形皱褶、绒毛和微绒毛使小肠表面的吸收面积增加了 600 倍。估计小肠的黏膜柱状内表面面积约为 3300 平方厘米，环形皱褶使表面积增加到 1 万平方厘米，绒毛使它增加到 10 万平方厘米，而微绒毛使它增加到 200 万平方厘米。

第八节 能量代谢和体温

Section 8 Energy Metabolism and Body Temperature

能量代谢 energy metabolism
食物的热价 thermal equivalent of food
氧债 oxygen debt
呼吸商 respiratory quotient, RQ
氧热价 thermal equivalent of oxygen
食物特殊动力效应 specific dynamic effect of food
体温 body temperature
生物节律 biorhythm [ˈbaɪorɪðəm]
战栗产热 shivering thermogenesis
寒冷性肌紧张 thermal muscle tone
辐射散热 thermal radiation
对流散热 thermal convection
不感蒸发(不显汗) insensible perspiration
体温调节 regulation of body temperature
热敏神经元 warm-sensitive neuron
基础代谢率 basal metabolism rate, BMR
昼夜节律 circadian rhythm
生物钟 biologic clock
产热 heat production, or thermogenesis
散热 thermolysis, or body heat loss
传导散热 thermal conduction
蒸发(散热) evaporation [ɪˌvæpəˈreʃən]
发汗 perspiration, or sweating
调定点 set point
冷敏神经元 cold-sensitive neuron
致热源 pyrogen [ˈpaɪrədʒən]
视前区—下丘脑前部 preoptic-anterior hypothalamus area

1 The O_2 is consumed and CO_2 is produced when the foods - carbohydrates, fats and proteins are oxidized in the cells. This ratio in the steady state of the volume of CO_2 produced to the volume of O_2 consumed per unit of time is called the respiratory quotient. The RQ of carbohydrate is 1.00, and that of fat is about 0.70.

当食物 —— 碳水化合物、脂肪和蛋白质 —— 在细胞内被氧化时，氧气被消耗的同时产生了二氧化碳。在平稳状态时，单位时间内呼出二氧化碳的量和消耗氧气的量的比值叫做呼吸商。碳水化合物的呼吸商是 1.00，而脂肪的约为 0.70。

2 Curling up"in a ball" is a common reaction to cold in animals and has a counterpart in the position which some people assume on climbing into a cold bed. Curling up decreases the body surface exposed to the environment. Shivering is an involuntary response of the skeletal muscles, but also causes a semiconscious general increase in motor activity. Examples include foot stamping and dancing up and down on a cold day. Increased catecholamine secretion is an important endocrine response to cold. Thyroid-stimulating hormone secretion is increased by cold and decreased by heat in laboratory animals, but of the change in TSH secretion produced by cold in adult humans is small, it is of questionable significance.

蜷缩成"一个球"是动物对冷的常见反应，一些人认为爬进冰冷的床也有一样的反应，蜷缩能减少身体与环境的接触面积。颤抖是一个骨骼肌的无意识反应，但是冷还能导致半意识肌动活动的普遍增加，比如在冷天里跺脚、上下蹦。儿茶酚胺分泌的增加是对冷的重要内分泌反应。在实验动物身上，冷的时候促甲状腺激素的分泌增加，热时减少；但是因为冷而产生促甲状腺激素分泌的变化很小时，则是有问题的标志。

第九节　尿的生成和排出

Section 9　Formation and Excretion of Urine

尿生成　urine formation
球旁器(近球小体)　juxtaglomerular apparatus
肾血流量　renal blood flow
肾小球滤过率　glomerular filtration rate, GFR
滤过平衡　filtration equilibibrium
分泌　secretion [sɪ'kriʃən]
尿的浓缩　urine concentration
渗透性利尿　osmotic diuresis
定比重吸收　constant fraction reabsorption
内移(内化)　internalization [ɪn'tənəlɪ'zeʃən]
醛固酮　aldosterone [æl'dɑstə,ron]
菊粉(菊糖)　inulin [ɪnjəlɪn]
尿排出　urine excretion
致密斑　macula densa
管—球反馈　tubuloglomerular feedback
滤过分数　filtration fraction
肾血浆流量　renal plasma flow
肾糖阈　renal glucose threshold
尿的稀释　urine dilution
球—管平衡　glomerulotubular balance
回漏　back leak
水利尿　water diuresis
排尿　micturition [,mɪktʃə'rɪʃən]

1 Glomerular filtration rate (GFR) can be measured in intact experimental animals and humans by measuring the excretion and plasma level of a substance that is freely filtered through the glomeruli and neither secreted nor reabsorbed by the tubules. The GFR in an average-sized normal man is approximately 125 mL/min.

肾小球滤过率可以通过在实验动物和人身上测量一种物质的分泌和血浆水平来获得，这种物质可以自由通过肾小球，但肾小管不能分泌也不能再吸附。一般来说正常人肾小球滤过率是约 125 毫升 / 分。

2 Adrenal mineralocorticoids such as aldosterone increase tubular reabsorption of Na^+ in association with secretion of K^+ and H^+ and also Na^+ reabsorption with Cl^-. When these hormones latent period of 10-30 minutes occurs before their effects on Na^+ reabsorption become manifest, because of the time required for the steroids to alter protein synthesis via their action on DNA. Mineralocorticoids may also have more rapid membrane-mediated effects, but these are not apparent in terms of Na^+ excretion in the whole animal. The mineralocorticoids act primarily in the collecting ducts to increase the number of active epithelial sodium channels in this part of the nephrons.

肾上腺盐皮质激素(如醛固酮) 能增加 Na^+ 的肾小管重吸收，相关联的是 K^+ 和 H^+ 的分泌以及伴随 Na^+ 的 Cl^- 重吸收。这些激素对 Na^+ 吸纳有 10 ～ 30 分钟的潜伏期。Na^+ 重新吸纳变得明显之前，因为类固醇通过作用于 DNA 改变蛋白质的合成是需要时间的。盐皮质激素也具有可能更快的膜介导作用，但这些在整个动物体内的 Na^+ 排泄都不是明显的。盐皮质激素在收集管中的主要作用是增加在肾单元这部分的活跃上皮细胞钠离子通道的数量。

第十节 感觉器官的功能

Section 10 Functions of the Sense Organs

感觉器官 sense organ
适宜刺激 adequate stimulus
发生器电位 generator potential
编码(作用) encoding [ɪn'koudɪŋ]
视觉 vision ['vɪʒən]
视敏度(视力) visual acuity
近点 near point
眼球会聚 convergence of the eyeball
视杆细胞 rod [rɑd]
晚光觉系统 nightlight perceptive system
视紫红质 rhodopsin [ro'dɑpsɪn]
视网醛 retinene ['rɛtn,in]
暗适应 dark adaptation
视野 visual field
气传导 air conduction
螺旋器(柯蒂器) Corti's organ
换能作用 transducer function
感受器电位 receptor potential
简化眼 reduced eye
(眼的)调节 accommodation (of the eye)
瞳孔近反射 pupillary near reflex
瞳孔对光反射 pupillary light reflex
视杆系统 rod system
视锥细胞 cone [kəun]
昼光觉系统 daylight perceptive system
视蛋白 opsin ['ɑpsən]
传递蛋白 transducin [trænzdju:'sɪn]
明适应 light adaptation
听觉 audition [ɔ'dɪʃən]
骨传导 bone conduction
行波理论 travelling wave theory
(耳蜗)微音器电位 (cochlear) microphonic potential
前庭自主神经性反应 vestibular autonervous reaction
平衡感觉 equilibrium sensation
嗅觉 olfactory sensation
皮肤感觉 cutaneous sensation
温度觉 temperature sensation
眼球震颤 nystagmus [nɪs'tægməs]
味觉 gustatory sensation
触一压觉 touch and pressure sensation
痛觉 pain [peɪn]

1 The photosensitive compounds in the rods and cones of the eyes of humans and most other mammals are made up of a protein called an opsin, and $retinene_1$, the aldehyde of vitamin A1. The term $retinene_1$ is used to distinguish this compound from $retinene_2$, which is found in the eyes of some animal species. Since the retinenes are aldehydes, they are also called retinals. The A vitamins themselves are alcohols and are therefore called retinols.
光敏化合物主要存在于人和大多数哺乳动物眼睛的视杆细胞和视锥细胞中,由一个叫视蛋白的蛋白质和视黄素1(维生素A1的醛化合物)组成的。视黄素1区别于视黄素2在于它存在于某些动物的眼睛内。由于视黄素都是醛类,也叫视黄醛,A类维生素自身就是乙醇类,因此也叫视黄醇。

2 The ear converts sound waves in the external environment into action potentials in the auditory nerves. The waves are transformed by the eardrum and auditory ossicles into movements of the footplate of the stapes. These movements set up waves in the fluid of the inner ear. The action of waves on organ of Corti generates action potentials in the nerve fibers.
耳朵将外部环境中的声波转换成听觉神经的动作电位。声波由鼓膜和听小骨变成内耳的镫骨的足板运动,这些运动促使内耳的液体形成波动,波动在柯蒂器官的运动中产生了神经纤维中的动作电位。

第十一节 神经系统的功能

Section 11 Functions of the Nervous System

神经冲动 nerve impulse
神经营养因子 neurotrophin, NT [,nuro'traufən]
顺向轴浆运输 anterograde axoplasmic transport
逆向轴浆运输 retrograde axoplasmic transport
兴奋性突触后电位 excitatory postsynaptic potential, EPSP
抑制性突触后电位 inhibitory postsynaptic potential, IPSP

突触后抑制　postsynaptic inhibition
突触前易化　presynaptic facilitation
回返性抑制　recurrent inhibition
强直后增强　posttetanic potentiation
敏感化　sensitization [ˌsɛnsɪtɪ'zeʃən]
长时程抑制　long-term depression, LTD
突触前抑制　presynaptic inhibition
传入侧支性抑制　afferent collateral inhibition
可塑性　plasticity [plæ'stɪsəti]
习惯化　habituation [həˌbɪtʃu'eʃən]
长时程增强　long-term potentiation, LTP
突触延搁　synaptic delay
非突触性化学传递　non-synaptic chemical transmission
电突触传递　electrical synaptic transmission
局部神经元回路　local neuronal circuit, LNC
交互性突触　reciprocal synapse
神经调质　neuromodulator [ˌnjuərəu'mɔdjuleitə]
递质共存　coexistence of transmitters
同源脱敏　homologous desensitization
胆碱能纤维　cholinergic fiber
毒蕈碱式作用　muscarine-like action
烟碱样作用　nicotine-like action
肌肉型烟碱受体　muscle-type nicotinic receptor
儿茶酚胺　catecholamine [ˌkætɪ'koləˌmin]
肾上腺素能纤维　adrenergic fiber
局部回路神经元　local circuit neuron, LCN
串联性突触　serial synapse
神经递质　neurotransmitter ['nʊrotrænzmɪtɚ]
调制作用　modulation [ˌmɑdʒu'leʃən]
突触前受体　presynaptic receptor
异源脱敏　heterologous desensitization
胆碱能受体　cholinergic receptor
毒蕈碱受体　muscarinic receptor
烟碱受体　nicotinic receptor
神经元型烟碱受体　neuronal-type nicotinic receptor
多巴胺　dopamine, DA ['dopəˌmin]
肾上腺素能受体　adrenergic receptor
5-羟色胺　5-hydroxytryptamine, 5-HT [haiˌdrɔksi'triptəmiːn]
谷氨酸　glutamic acid, or glutamate
辐散状联系　divergence connection
后放　after discharge
多突触反射　polysynaptic reflex
非特异投射系统　non-specific projection system
-氨基丁酸　-aminobutyric acid, GABA
条件反射　conditioned reflex
聚合状联系　convergence connection
单突触反射　monosynaptic reflex
特异投射系统　specific projection system
感觉柱　sensory column
网状结构上行激动系统　ascending reticular activating system
本体感觉　peoprioception [prəupriə'sepʃən]
皮层诱发电　evoked cortical potential
牵涉痛　referred pain
脑电图　electroencephalogram, EEG [ɪˌlɛktroɛn'sɛfələˌgræm]
自发脑电活动　spontaneous electric activity
波阻断　alpha blocking
脑电觉醒　electroencephalogram arousal
快波睡眠　fast wave sleep, FWS
快速眼动　rapid eye movements, REM
最后公路　final common path
腱反射　tendon reflex
肌梭　muscle spindle
随意运动　voluntary movement
锥体系　pyramidal system
脊髓休克　spinal shock
屈肌反射　flexor reflex
去大脑僵直　decerebrate rigidity
翻正反射　righting reflex
静止性震颤　static tremor
意向震颤　intention tremor
交感神经　sympathetic nerve
行为觉醒　behavioral arousal
慢波睡眠　slow wave sleep, SWS
异相睡眠　paradoxical sleep, PS
上行抑制系统　ascending inhibitory system
牵张反射　stretch reflex
肌紧张　muscle tonus
梭内肌纤维　intrafusal fiber
运动柱　motor column
锥体外系　extrapyramidal system
姿势反射　postural reflex
对侧伸肌反射　crossed extensor reflex
状态反射　attitudinal reflex
震颤性麻痹　paralysis agitans
舞蹈病　chorea [kə'rɪə]
小脑性共济失调　cerebellar ataxia
副交感神经　parasympathetic nerve

下丘脑调节肽 hypothalamus regulatory peptide，HRP
边缘系统 limbic system
情绪反应 emotional reaction
饱中枢 satiety center
假怒 sham rage
动机 motivation [ˌmotəˈveʃən]
操作条件反射 operate conditioned reflex
长时记忆 long term memory
失写症 agraphia [əˈgræfɪə]
失读症 alexia [əˈlɛksɪə]
失用症 apraxia [eˈpræksɪə]
本能行为 instinctual behavior
摄食中枢 feeding center
防御反应区 defense zone
强化 reinforcement [ˌriɪnˈfɔrsmənt]
短时性记忆 short term memory
运动性失语症 motor aphasia
感觉性失语 sensory aphasia
一侧优势 laterality cerebral dominance
裂脑 split brain

1 Axoplasmic transport occurs along microtubules. Fast transport responsible for organelles that covered with plasmalemma, such as the granule, vesicle and mitochondria, occurs at about 400 mm/d, and slow anterograde transport for dissoluble substances occurs at 0.5-10 mm/d. Fast anterograde transport is mediated by kinesin, a form of molecular motors resemble the myosin that is an ATPase.
轴浆沿微管传递，快速运输负责具有质膜细胞器的运输，如颗粒体、囊泡和线粒体，传递速度大约为400mm/d，慢速顺向运输的是可溶性物质，速度为0.5～10mm/d。快速顺向运输的介质是驱动蛋白，类似于一个三磷酸腺苷酶肌球蛋白形式的分子马达。

2 The excitatorypostsynaptic potential (EPSP) is produced by depolarization of the postsynaptic cell membrane immediately under the presynaptic ending. The excitatory transmitter opens Na^+ or Ca^+ ion channels in the postsynaptic membrane, producing an inward current. The EPSP due to activity in one synaptic knob is small, but the depolarizations produced by each of the active knobs summate.
兴奋性突触后电位（EPSP）是由突触前膜兴奋刚结束时，立即由细胞突触后膜的去极化而产生的。兴奋性物质使突触后膜的Na^+或者Ca^+离子通道开放，产生了内向电流。虽然EPSP的每个突触小体的电位很小，但所有突触小体的总和就产生了去极化。

3 When a skeletal muscle with an intact nerve supply is stretched, it contracts. This response is called the stretch reflex. The stimulus that initiates the reflex is stretch of the muscle, and the response is contraction of the muscle being stretched. The sense organ is the muscle spindle. The impulses originating in the spindle are conducted in the CNS by fast sensory fibers that pass directly to the motor neurons which supply the same muscle. The neurotransmitter at the central synapse is glutamate. Stretch reflexes are the best known and studied monosynaptic reflexes in the body.
当具有完好神经分布的骨骼肌被拉伸时，它会产生收缩，这种反应叫做牵张反射。引起反射的刺激是肌肉的拉伸，响应是被拉伸肌肉的收缩，感觉器官是肌梭。原产于肌梭的冲动在中枢神经系统中通过快速感官纤维进行传递，直接传递给分布在同一肌肉的运动神经元。中央突触的神经递质是谷氨酸。牵张反射在人体中是众所周知的、被研究的单突触反射。

第十二节 内分泌学
Section 12 Endocrinology

内分泌 endocrine [ˈɛndəkraɪn]
自分泌 autocrine [ɔːtəukrain]
允许作用 permissive action
增量调节(上调) up regulation
远距分泌 telecrine [ˈtelɪkrain]
神经激素 neurohormone [ˌnʊroˈhɔrmon]
亲和力 affinity [əˈfɪnɪti]
减量调节(下调) down regulation
促甲状腺激素释放激素 thyrotropin-releasing hormone，TRH
促性腺激素释放激素 gonadotropin-releasing hormone，GnRH
生长素释放抑制激素 Growth hormone releasing-inhibiting hormone，GHRIH

生长激素抑制素 somatostatin [soˌmætəˈstætn]
生长素释放激素 growth hormone releasing hormone, GHRH
促肾上腺皮质激素释放激素 corticotropin releasing hormone, CRH
催乳素释放抑制因子 prolactin releasing-inhibiting factor, PIF
促黑(素细胞)激素释放抑制因子 melanophore-stimulating hormone releasing-inhibiting factor, MIF
促黑(素细胞)激素释放因子 melanophore-stimulating hormone releasing factor, MRF
促甲状腺激素 thyroid stimulating hormone, TSH
促卵泡激素 follicle stimulating hormone, FSH
促黑(素细胞)激素 melanophore stimulating hormone, MSH
催乳素释放因子 prolactin releasing factor, PRF
促肾上腺皮质激素 adrenocorticotropin, ACTH [əˈdriːnəʊˌkɔːtɪkəʊˈtrɒpɪn]
胰岛素样生长因子 insulin-like growth factor, IGF
黄体生成素 luteinizing hormone, LH
催乳素 prolactin, PRL [proˈlæktɪn]
生长调节素 somatomedin, SM [soˌmætəˈmidn]
甲状腺素(四碘甲腺原氨酸) thyroxin, or tetraiodothyronine, T4
盐皮质激素 mineralocorticoid [ˌmɪnərəloˈkɔrtɪˌkɔɪd]
雌二醇 estradiol, E2 [ˌɛstrəˈdaɪˌɔl]
应急反应 emergency reaction
胰高血糖素 glucagon [ˈglukəˌgɑn]
褪黑素 melatonin, MT [ˌmɛləˈtonɪn]
催产素 oxytocin [ˌɑksɪˈtosɪn]

1 Glucagons secretion is increased by hypoglycemia and decreased by a rise in plasma glucose. The B cells contain GABA, and evidence suggests that coincident with the increased insulin secretion produced by hyperglycemia, GABA is released and acts on the A cells to inhibit glucagons secretin by activating $GABA_A$ receptors. Secretion is also increased by stimulation of the sympathetic nerves to the pancreas, and this sympathetic effect is mediated via β-adrenergic receptors and cAMP.
血糖过低会增加胰高血糖素的分泌，血糖上升会减少胰高血糖素的分泌。B细胞含有氨基丁酸(GABA)，血糖升高会导致胰岛素升高，从而降低血糖长度，GABA的释放会活化A细胞通过活化$GABA_A$受体来抑制胰高血糖素分泌素。交感神经刺激胰腺也能使胰高血糖素的分泌上升，交感神经作用的介质是β-肾上腺素受体和环磷酸腺苷(cAMP)。

2 The principal hormones secreted by the thyroid are thyroxine (T4) and triiodothyronine (T3). T3 is also formed in the peripheral tissues by deiodination of T4. The normal total plasma T4 level in adults is approximately 8 ug/dL (103nmol/L), and the plasma T3 level is approximately 0.15 ug/dL (2.3nmol/L).
甲状腺分泌的主要激素是甲状腺素(T4)和三碘甲状腺原氨酸(T3)，T3还能在外围组织中由T4脱碘形成。成人正常总血浆T4含量大约为8 ug/dL (103nmol/L)，正常总血浆T3含量大约为0.15 ug/dL (2.3nmol/L)。

第十三节 生殖功能
Section 13 Reproduction Function

睾酮 testosterone, T [tɛsˈtɑstəˌron]
抑制素 inhibin [ɪnˈhɪbɪn]
排卵 ovulation [ˌɒvjuˈleɪʃn]
月经周期 menstrual cycle
孕激素 progestogen [proˈdʒɛstədʒən]
妊娠 pregnancy [ˈprɛgnənsi]
着床 implantation [ˌɪmplænˈteʃən]
生精形成 spermatogenesis [spɚˌmætəˈdʒɛnɪsɪs]
雄激素 androgen, or androgenic hormone
生卵形成 oogenesis [ˌoəˈdʒɛnɪsɪs]
月经 menstruation [ˌmɛnstruˈeʃən]
雌激素 estrogen [ˈɛstrədʒən]
孕酮 progesterone, P [proˈdʒɛstəˌron]
分娩 parturition [ˌpartjuˈrɪʃən]
卵子 ovum [ˈovəm]
卵巢周期 ovarian cycle
β-雌二酮 β-estradiol [ˌɛstrəˈdaɪˌɔl]

雌酮 estrone [ˈɛsˌtron]
雌三醇 estriol [ˈɛstraɪˌɔl]
子宫 uterus [ˈjutərəs]

The naturally occurring estrogens are 17β-estradiol, estrone, and estriol. They are C_{18} steroids which do not have an angular methyl group attached to the position 10 or a Δ^4-3-keto configuration in the A ring. They are secreted primarily by the granulose cells of the ovarian follicles, the corpus luteum, and the placenta. Theca cells have many luteinizing hormone (LH) receptor and LH acts via cAMP to increase conversion of cholesterol to androstenone. Some of the androstenone is converted to estradiol, which enters the circulation. The theca interna cells also supply androstenedione to the granulose cells. The granulose cells make estradiol when provided with androgens, and it appears that the estradiol they form in primates is secreted into the follicular fluid.

自然产生的雌激素包括 17β- 雌二醇、雌酮和雌三醇，它们是不具有角甲基附加到位置 10 或 A 环中的 Δ^4-3- 酮结构的 C^{18} 类固醇。雌激素主要是由卵泡、黄体和胎盘的卵巢颗粒细胞分泌的。卵泡内膜细胞具有许多黄体生成素受体，黄体生成素通过环一磷酸腺苷作用增加了胆固醇向雄甾烯酮转化，一些雄甾烯酮转化成雌二醇进入循环。卵泡内膜细胞供应雄甾烯酮给颗粒细胞。当有雄激素时颗粒细胞产生雌二醇，就像灵长类产生的雌二醇分泌到卵泡液中了一样。

第二十七章　药理学

Chapter 27　Pharmacology

第一节　总论

Section 1　Pandect

★ 药理学　pharmacology [ˌfɑrmə'kɑlədʒi]
★ 药物效应动力学(药效学)　drug effect dynamics
★ 药物代谢动力学(药动学)　pharmacokinetic
作用　action ['ækʃən]
作用机制　mechanism of action
适应症　indication [ˌindi'keiʃən]
不良反应　adverse reaction
禁忌症　contraindication
毒理学　toxicology [ˌtɔksi'kɔlədʒi]
临床药理学　clinical pharmacology
★ 毒性测试　toxicity test
化学治疗学　chemotherapy
生化药理学　biochemical pharmacology
分子药理学　molecular pharmacology
量子药理学　quantum pharmacology
神经药理学　neuropharmacology ['njuərəuˌfɑ:mə'kɔlədʒi]
免疫药理　immunopharmacology ['ɪmjʊnəʊprɪˌsɪpɪ'teɪt]
遗传药理学　pharmacogenetics [ˌfɑ:məkəudʒi'netiks]
临床试验　clinical trials
安慰剂　placebo [plə'si:bəu]
★ 药物作用　drug action
★ 药物效应　drug effect
兴奋　stimulation or excitation
兴奋药　stimulators or excitants
抑制药　depressants or inhibitors
局部作用　local action
导泻作用　purgative action
全身作用　general action
吸收作用　absorptive action
★ 系统作用　systemic action
治疗作用　therapeutic action
对因治疗　etiological treatment
对症治疗　symptomatic treatment
副作用或副反应　side effect/side reaction
毒性反应　toxic reaction
急性毒性　acute toxicity
慢性毒性　chronic toxicity
变态反应　allergy reaction
继发性反应　secondary reaction
后遗效应　residual reaction
致畸作用　teratogenesis
沙利度胺　thalidomide [θə'lidəˌmaid]
致癌　carcinogenesis [kɑ:sɪnəʊ'dʒenɪsɪs]
致突变　mutagenesis [ˌmjutə'dʒenisis]
受体拮抗剂　receptor antagonist
★ 配体　ligand ['ligənd]
受点　receptor site
内源性配体　endogenous ligand
离子带受体　ionotropic receptors
核受体　nuclear receptors
受体调节　receptor regulation
向下调节(衰减性调节)　down regulation
向上调节　up regulation
同种调节　homospecific regulation
异种调节　heterospecific regulation
占领学说　occupation theory
亲和力　affinity [ə'finiti]
内在活性　intrinsic activity
效能　efficacy ['efikəsi]
备用受体　spare receptor
静息受体　silent receptor
速率学说　rate theory
变构学说或二态学说　allosteric theory or two state theory
静息态　resting state
活化态　active state
激动剂或完全激动剂　agonist or full agonist
部分激动剂　partial agonist
两重性　dualism ['dju:əlizəm]
★ 拮抗剂　antagonist [əæn'tægənist]
竞争性拮抗剂　competitive antagonist
★ 药物作用机制　mechanism of drug action
非特异性药物作用机制　non-specific drug
络合剂　chelating agent
★ 特异性药物　specific drug

量效关系 dose-response relationship
最小有效量或阈剂值 minimum effective dose or threshold dose
最大效应 maximum effect
最小中毒量 minimum toxic dose
量反应 graded response
半数有效量 median effective dose, ED_{50}
半数致死量 median lethal dose, LD_{50}
治疗指数 therapeutic index, TI
安全指数 safety index
安全界限 safety margin
安全范围 margin of safety
药物处置 drug disposition
药物转运 transportation of drug
脂溶扩散 lipid diffusion
药物的脂/水分配系数 lipid/aqueous partition coefficient
★水溶扩散 aqueous diffusion
电压依赖性通道 voltage dependent channel, VDC
化学依赖性通道 chemical dependent channel
★肌内注射 intramuscular injection, IM
首关效应 first-pass effect
重分布 redistribution [ˈriːdistriˈbjuːʃən]
脑脊液 cerebrospinal fluid, CSF
胎盘屏障 placental barrier
灭活 inactivation [inˌæktiˈveiʃən]
活化 activation [ˌæktiˈveiʃən]
非微粒体酶系 non-microsomal enzyme system
脑核性黄疸 kernicterus [kəˈniktərəs]
解离型药物 dissociative drug
★潜伏期 latent period
★持续期 persistent period
★残留期 residual period
★药峰浓度 peak concentration, max
给药间隔时间 dosing interval time
积蓄中毒 cumulative intoxication
生物利用度 fraction of bioavailability
试验制剂 test formulation
消除半衰期 elimination half-life
稳态血药水平或坪值 steady state level or plateau
清除率 clearance CL [ˈkliərəns]
房室模型 compartment model
稳态血浆浓度或坪浓度 steady state plasma concentration, Css or plateau concentration
负荷量 loading dose
维持量 maintenance dose
个体差异 individual variation
高敏性 hypersensitivity
耐受性 tolerance [ˈtɔlərəns]
过敏反应 hypersensitivity reaction
特异质 idiosyncrasy [ˌidiəˈsiŋkrəsi]
原发性血色素沉着病 hemochromatosis [ˌhiːməˌkrəuməˈtəusis]
种属差异 species variation [ˌsɜːkəluːˈnɑːdɪən]
交叉耐受性 cross tolerance
★药物依赖性 drug dependence
躯体依赖性 physical dependence
★成瘾性 addiction [əˈdikʃən]
生理依赖 physiological dependence
欣快感 euphoria [juːˈfɔːriə]
戒断综合征 abstinence syndrome
精神依赖性 psychic dependence
习惯性 habituation [ˌhæbitjuˈeiʃən]
心理依赖性 psychological dependence
药物相互作用 drug reaction
相加作用 additive effect, summation
增强作用 potentiation [pəˌtenʃiˈeiʃən]
增敏作用 sensitization [ˌsensitaiˈzeiʃən]
★拮抗作用 antagonism [ænˈtægənizəm]
药理性拮抗 pharmacological antagonism
生理性拮抗 physiological antagonism
生化性拮抗 biochemical antagonism
化学性拮抗 chemical antagonism

1 In pharmacology, a receptor is, more generically, any molecule that is a target for a drug.
在药理学里，受体通常是一个药物靶分子。

2 A particular priority should be integrating drug metabolism and pharmacokinetic (DMPK) studies into research on traditional medicines.
一个特别的优先事项是把药物代谢和药代动力学(DMPK)研究整合到传统药物的研究中。

3 To discuss The mechanism of drug resistance of nucleotides in the treatment of chronic hepatitis B.
探讨应用核苷类药物治疗慢性乙型肝炎过程中的耐药机制。

4 Weight was found to significantly influence both clearance (CL) and volume of distribution (VD).

体重明显影响药物的清除率和分布容积。

5 For the elderly, aminoglycosides should be administrated in a way of lower dose or longer interval between dosing.
老年人应减少对氨基糖苷类抗生素的服用剂量或延长服药间隔时间。

6 Allergic bronchopulmonary aspergillosis results from a hypersensitivity reaction to the fungus in asthmatic patients.
过敏性支气管肺曲菌病是由真菌哮喘患者的过敏反应导致的。

第二节　外周神经系统药理
Section 2　Peripheral nervous system pharmacology

★传出神经系统　efferent nervous system
★自主神经系统　autonomic nervous system
★运动神经系统　somatic motor nervous system
★交感神经　sympathetic nervous system
肠神经系统　enteric nervous system
神经递质　neuro-transmitter
乙酰胆碱　acetylcholine, Ach [ə,si:təl'kəulin]
去甲肾上腺素　noradrenaline, NA [,nɔ:rə'drenəlin]
胆碱能神经　cholinergic nerve
去甲肾上腺素能神经　noradrenergic nerve
肾上腺素能神经　adrenergic never
多巴胺能神经　dopaminergic never
胞裂外排　exocytosis [,eksəusai'təusis]
毒蕈碱型胆碱受体　muscarinic Ach receptor
烟碱型胆碱受体　nicotinic Ach receptor
肾上腺素受体　adrenoceptor [ə'drenəkrəum]
多巴胺受体　dopamine receptor
阻断药　blocking drug, blocker
密胆碱　hemicholine [,hemɪ'kɔ:deɪt]
α-甲基酪氨酸　α-methyltyrosine
胆碱受体激动药　cholinoceptor agonists
拟胆碱药　cholinomimetic drug
氨甲酰胆碱　carbamylcholine
毒蕈碱　muscarine
毛果芸香碱　Pilocarpine [,pailəu'kɑ:pin]
★烟碱　nicotine ['nikəti:n, -tin]
新斯的明　neostigmine [,ni:əu'stigmi:n]
重症肌无力　myasthenia gravis
毒扁豆碱　physostigmine [,faisəu'stigmi:n, -min]
有机磷脂类　organophosphates [,ɔ:gənəu'fɔsfə-rəs, ɔ:,gænəu]
胆碱酯酶复活剂　cholinesterase reactivators
碘解磷定　pralidoxime iodide
氯磷定　pralidoxime iodide, PAM-cl
胆碱受体阻断药　cholinoceptor-blocking drugs
M胆碱受体阻断药　muscarinic receptor-blocking drugs
N_1胆碱受体阻断药　N_1-cholinoceptor blocking drugs
N2胆碱受体阻断药　N2-cholinoceptor blocking drugs
阿托品　atropine, [æt 'ætrəpi:n]
东莨菪碱　scopolamine [skəu'pɔləmin]
山莨菪碱　anisodamine
樟柳碱　anisodine
后马托品　homatropine ['hæmbitən]
托吡卡胺　tropicamide
丙胺太林　propantheline
双环胺　dicyclomine
★神经节阻断药　ganglion blocking drugs
骨骼松弛药　skeletal muscular relaxants
★神经肌肉阻断药　neuromuscular blocking drugs
去极化型　polarizing drugs
非去极化型　nondepolarizing drugs
琥珀胆碱　succinylcholine [,sʌksɪnɪl'kəuli:n]
竞争型肌松药　competitive muscular relaxants
筒箭毒碱　d-tubocurarine
肾上腺素受体激动药　adrenoceptor-activating drugs
拟交感药　sympathomimetic drugs
★儿茶酚类　catecholamines, CA
α受体激动药　α-adrenoceptor agonists
α、β受体激动药　α、β-adrenoceptor agonists
β受体激动药　β-adrenoceptor agonists
★去甲肾上腺素　noradrenaline, NA
Norepinephrine, NE
间羟胺　metaraminol [,metə'ræmɪnɔl]
去氧肾上腺素　phenylephrine [,fenə'lefrin]
可乐定　clonidine ['klɔnɪdi:n]

★ 肾上腺素　adrenaline, AD [ə'dreːnəlaɪz]
心血管系统　Cardiovascular System
★ 多巴胺　dopamine, DA ['dəupəˌmiːn]
★ 麻黄碱　ephedrine [e'fedrin]
异丙肾上腺素　Isoprenaline ['aisəu'prenəli(:)n]
多巴酚丁胺　dobutamine
沙丁胺醇　salbutamol
特布他林　terbutaline
肾上腺素受体阻断药　Adrenoceptor blocking drugs
α 受体阻断药　α-receptor antagonists or α-receptor blocking drugs
肾上腺素的翻转　adrenoceptor reversal
酚妥拉明　Phentolamine [fen'tɔləmiːn,-mɪn]
妥拉唑林　tolazoline
酚苄明　phenoxybenzamine [ˌfiːnɔksiˌmeθilˌpeni'silin]
哌唑嗪　prazosin
β 受体阻断药　β-receptor antagonists or β-receptor blocking drugs
普萘洛尔　Propranolol [prəu'prænəlɔl]
噻吗洛尔 timolol
纳多洛尔　nadolol
吲哚诺尔　pindolol ['pɪndəlɔl]
阿替洛尔　atenolol
美托洛尔　metoprolol [mɪ'trældʒɪə]
醋丁洛尔　acebutolol [ə'siːdiə]
拉贝洛尔　labetalol
★ 局部麻醉药　local anaesthetics
★ 表面麻醉　topical anaesthesia
浸润麻醉　infiltration anaesthesia
传导麻醉或阻滞麻醉　conduction anaesthesia or block anaesthesia
蛛网膜下腔麻醉(脊麻、腰麻)　subarachnoid anaesthesia, spinal anaesthesia
硬脊膜为爱麻醉　epidural anaesthesia
普鲁卡因　procaine ['prəukein, prəu'kein]
丁卡因　tetracaine ['tetrəˌkein]
利多卡因　lidocaine ['lɪdəukeɪn]
布比卡因　bupivacain

1 This leads to subtle changes in the autonomic nervous system, which controls the contraction and relaxation of these vessels.
这会导致控制这些血管收缩和舒张的自主神经系统发生微妙变化。

2 The sympathetic nervous system — a web of nerves and neurons running through the spinal cord and spread to virtually every organ in the body — is responsible for arousal and excitement.
相对应的交感神经系统是一张贯穿脊椎的由神经和神经元组成的网,它会将信息传递到身体的每个器官,是形成觉醒与兴奋的原因。

3 Their approximate duration of action is 20 to 35 minutes, or 33% to 50% shorter than that of long-acting neuromuscular blocking drugs.
其作用时间为 20 ~ 35 分钟,比长效神经肌肉阻滞剂的作用时间短 33% ~ 50%。

4 The antagonism of the neuromuscular blockade produced by nondepolarizing neuromuscular blocking drugs is achieved through the intravenous administration of anticholinesterases.
静脉应用胆碱酯酶抑制剂可实现非去极化肌松药的拮抗。

5 The snag with most local anaesthetics is their lack of specificity.
大多数局部麻醉剂的问题是,它们的作用没有针对性。

第三节　中枢神经系统药理
Section 3　The central nervous system pharmacology

★ 神经递质　neurotransmitter [ˌnjuərətræns'mitə]
★ 神经调质　neuromodulator [ənjuərəu'məutə]
兴奋性突触后电位　excitatory postsynaptic potential, EPSP
★ 全身麻醉药　general anaesthetics
吸入性麻醉药　inhalational anaesthetics
最小肺泡浓度　minimum alveolar concentration, MAC
恩氟烷　enflurane [en'fluəreɪn]
异氟烷　isoflurane
地氟烷　desflurane
七氟烷　sevoflurane
氧化亚氮　nitrous oxide
★ 静脉麻醉药　intravenous anaesthetics
硫喷妥钠　thiopental sodium
氯胺酮　ketamine ['ketəmiːn]

依托咪酯 etomidate
丙泊酚 propofol
羟丁酸钠 sodiumγ-hydroxybutyrate
复合麻醉 combined anaesthetics
麻醉前给药 preanaesthetic medication
基础麻醉 basal anaesthesia
诱导麻醉 induction anaesthesia
低温麻醉 hypothermal anaesthesia
★ 控制性降压 controlled hypotension
神经安定镇痛术 neuroleptanalgesia
★ 镇静催眠药 sedative-hypnotics
★ 镇静药 sedatives
★ 催眠药 hypnotics
苯二氮卓类 Benzodiazepines,BZ
快波睡眠 fast wave sleep,FWS
★ 巴比妥类 barbiturates
★ 抗癫痫 antiepileptic
水合氯醛 chloral hydrate
佐匹克隆 zopiclone
扎来普隆 zaleplon
★ 癫痫 epilepsy [ˈepilepsi]
苯妥英钠 phenytoin sodium
★ 苯巴比妥 Phenobarbital
扑米酮 primidone
卡马西平 carbamazepine [ˌkɑːbəˈmæzəpiːn]
丙戊酸钠 sodium valproate
乙琥胺 ethosuximide [ˌeθəʊˈsʌksɪmaɪd]
加巴喷丁 gabapentin
拉莫三嗪 lamotrigine
托吡酯 topiramate
氯硝西泮 clonazepam
硫酸镁 magnesium sulfate
精神失常 psychopathy
抗精神病药 antipsychotic drugs
抗狂躁和抑郁症药 antimanic and antidepressive drugs
抗焦虑药 antianxiety drugs
精神分裂症 schizophrenia
★ 狂躁抑郁症 manic depression
氯丙嗪 chlorpromazine [klɔːˈprəuməzin]
氯普噻吨 chlorprothixene
氟哌啶醇 haloperidol
氟哌利多 droperidol
氯氮平 clozapine
舒必利 sulpiride
泰必利 tiapride
五氟利多 penfluridol
利培酮 risperidone
麦普替林 maprotiline
米安舍林 mianserin
舍曲林 sertraline
氟西汀 fluoxetine
曲唑酮 trazodone
碳酸锂 lithium carbonate
丁螺环酮 buspirone
★ 帕金森病 Parkinson disease
左旋多巴 levodopa,L-dopa
卡比多巴 carbidopa
苄丝肼 benserazide
金刚烷胺 amantadine
溴隐亭 bromocriptine
吡贝地尔 piribedil
苯海索 trihexyphenidyl
阿尔茨海默症 Alzheimer's disease,AD
多奈哌齐 donepezil
石杉碱甲 huperzine
加兰他敏 galantamine
卡巴拉汀 rivastigmine
占诺美林 xanomeline
★ 中枢兴奋药 central stimulants
★ 咖啡因 caffeine
哌甲酯 methylphenidate
匹莫林 pemoline
阿屈非尼 adrafinil
吡拉西坦 piracetam
甲氯芬酯 meclofenoxate
胞磷胆碱 citicoline
尼可刹米 nikethamide
二甲弗林 dimefline
洛贝林 lobeline
贝美格 bemegride
★ 镇痛药 analgesics
阿片 opium
★ 吗啡 morphine
★ 可待因 codeine
丁丙诺啡 buprenorphine
丙烯吗啡 nalorphine
哌替啶 pethidine
芬太尼 fentanyl
美沙酮 methadone
曲马朵 tramadol
喷他佐辛 pentazocine [penˈtæzəsiːn]

罗通定　rotundine
奈福泮　nefopam
高乌甲素　lappaconitine
纳洛酮　naloxone
纳曲酮　naltrexone
★解热镇痛抗炎药　antipyretic analgesic and anti-inflammatory drugs
★非甾体抗炎药　non-steroidal anti-inflammatory drugs,NSAIDs
阿司匹林　aspirin [ˈæspərin]
瑞夷综合征　Reye's syndrome
二氟尼柳　diflunisal
双水杨酯　salsalate
对乙酰氨基酚　acetaminophen
保泰松　phenylbutazone
羟布宗　oxyphenbutazone
吲哚美辛　indomethacin
舒林酸　sulindac
布洛芬　ibuprofen
萘普生　naproxen
双氯芬酸　diclofenac
吡罗昔康　piroxicam
塞来昔布　celecoxib
尼美舒利　nimesulide
★痛风　gout
秋水仙碱　colchicines
丙磺舒　probenecid
别嘌醇　allopurinol [ˌæləˈpjuːrinɔl]
苯溴马隆　benzbromarone

1 Dopamine is a neurotransmitter critical to the experience of pleasure and desire, and sends a signal to the brain: Pay attention, this is important.
多巴胺是一种神经递质，对于人感受快乐和欲望至关重要，而且会向大脑发送这样一个信号：注意了，这很重要。

2 To investigate the effects of panax notoginseng saponins (PNS) on fast excitatory postsynaptic potential in stellate ganglion (SG) of rats and its mechanism.
探究三七总皂甙对大鼠星状神经节快兴奋性突触后电位的影响及其机制。

3 This paper focuses on the potential role of second messenger systems as general anaesthetic targets.
本文的重点是第二信使系统作为全身麻醉目标的潜在作用。

4 To evaluate the effect of intravenous propofol on the minimum alveolar concentration (MAC) of isoflurane.
评价静脉异丙酚对异氟醚肺泡最低有效浓度的影响。

5 Caffeine is a stimulant, and many people find that it improves their alertness — that's caffeine's affect on the central nervous system.
咖啡因是一种兴奋剂，许多人认为它会提高他们的警觉性，那是咖啡因对中枢神经系统的作用。

第四节　内脏系统药理
Section 4　Digestive system pharmacology

★心律失常　arrhythmia [əˈriθmiə]
膜反应性　membrane responsiveness
有效不应期　effective refractory period,ERP
★抗心律失常药　antiarrhythmic drugs
奎尼丁　quinidine
普鲁卡因胺　procainamide
丙吡胺　disopyramide
安博律定　aprindine
利多卡因　lidocaine
妥卡尼　tocainide
苯妥英钠　phenytoin sodium
普罗帕酮　propafenone
恩卡尼　encainide
胺碘酮　amiodarone
溴苄胺　bretylium
维拉帕米　verapamil
地尔硫卓　diltiazem
★心功能不全　cardiac insufficiency
充血性心力衰竭　congestive heart failure,CHF
★代偿期　compensatory stage
非代偿期　Non compensatory stage
强心苷　cardiac glycosides
★负性频率作用　negative chronotropic effect
氨力农　amrinone
米力农　milrinone
多巴酚丁胺　dobutamine

★心绞痛 angina pectoris
★抗心绞痛药 antianginal drugs
三酰甘油 triacylglylevol
★硝酸甘油 glceryl trinitrate
硝酸异山梨酯 isosorbide dinitrate
硝酸戊四醇酯 pentaerythritoltetranitrate
单硝酸异山梨酯 isosorbide mononitrate
硝苯地平 nifedipine
★动脉粥样硬化 atherosclerosis
考来烯胺 cholestyramine
氯贝丁酯 clofibrate
吉非贝齐 gemfibrozil
烟酸 nicotinic acid
烟酸肌醇酯 inositol nicotinate
洛伐他汀 lovastatin
★高血压 hypertension
★原发性高血压 primary hypertension, essential hypertension
★继发性高血压 secondary hypertension
★症状性高血压 symptomatic hypertension
高血压危象 hypertension crisis
可乐定 clonidine [ˈklɔnɪdiːn]
莫索尼定 moxonidine
α-甲基多巴 α-methyldopa
利舍平 reserpine
胍乙啶 guanethidine
哌唑嗪 prazosin
特拉唑嗪 terazosin
肼屈嗪 hydralazine
米诺地尔 minoxidil
二氮嗪 diazoxide
硝普钠 sodium nitroprusside
酮舍林 ketanserin
硝苯地平 nifedipine
氨氯地平 amlodipine
非洛地平 felodipine
氢氯噻嗪 hydrochlorothiazide
卡托普利 captopril
依那普利 enalapril
赖诺普利 lisinopril
苯那普利 benazepril
洛沙坦 losartan
★利尿药 diuretics
袢利尿药 loop diuretics
呋塞米 furosemide
托拉塞米 torasemide
吲达帕胺 indapamide
螺内酯 spironolactone
阿米洛利 amiloride
氨苯蝶啶 triamterene
双氯非那胺 diclofenamide
脱水药 osmotic diuretics
甘露醇 mannitol
山梨醇 sorbitol
高渗葡萄糖 glucose
★贫血 anemia [əˈniːmiə]
抗贫血药 antianemic drugs
硫酸亚铁 ferrous sulfate
枸橼酸铁铵 ferric ammonium citrate
右旋糖酐铁 iron dextran
叶酸 folic acid
维生素 B_{12} vitamin B_{12}
促凝血药 coagulants
抗凝血药 anticoagulants
维生素 K vitamin K
抗纤维蛋白溶解药 antifibrinolytics
氨甲苯酸 P-aminomethyl benzoic acid, PAMBA
氨甲环酸 tranexamic acid, AMCHA
肝素 heparin
低分子量肝素 low molecular weight heparins, LMWHs
替地肝素 fragmin
依诺肝素 enoxaparin lovenox
藻酸双酯钠 polysaccharide sulfate, PSS
香豆素类抗凝剂 Coumarin anticoagulants
双香豆素 bishydroxycoumarin, dicoumarol
醋销香豆素 acenocoumarol
新双香豆素 ethyl biscoumacetate
华法林 warfarin [ˈwɔːfərin]
枸橼酸钠 sodium citrate
纤维蛋白溶解药 fibrinolytic drugs
溶栓药 thrombolytic drugs
★链激酶 streptokinase
★尿激酶 urokinase / [ˌjuərəuˈkaineis]
组织纤维蛋白溶酶原激活剂 tissue-type plasminogen activator
水蛭素 hirudin [hiˈruːdən, ˈhirudin]
前列环素 prostacyclin [ˌprɔstəˈsaɪklɪn]
双嘧达莫 dipyridamole
噻氯匹定 ticlopidine
维生素 B_4 vitamin B_4
肌苷 carnine

地菲林葡萄糖苷 Cleistanthin-B
重组人红细胞生成素 recombinant human erythropoietin
重组人粒细胞集落刺激因子 recombinant human granulocyte colony-stimulation factor, rh-G-CSF
重组人粒细胞/巨噬细胞集落刺激因子 recombinant human granulocyte-macro-phage colony-stimulation factor, rh-GM-CSF
重组人血小板生成素 recombinant human thrombopoietin rHu-TPO
右旋糖酐 dextran
助消化药 digestants
消化性溃疡 peptic ulcer
抗酸药 antiacids
丙谷胺 proglumide
奥美拉唑 omeprazole
米索前列素 misoprostol
恩前列醇 enprostil
硫糖铝 sucralfate
铝碳酸镁 hydrotalcite
枸橼酸铋钾 bismuth potassium citrate
胶体果胶铋 colloidal bismuth pectin
★ 泻药 laxatives, cathartics
酚酞 phenolphthalein
蒽醌类 anthraquinones
蓖麻油 castor oil
液体石蜡 liquid paraffin
★ 止泻药 antidiarrheal drugs
地芬诺脂 diphenoxylate
药用碳 medicinal active charcoal
钾氧氯普安 metoclopramide
多潘立酮 domperidone
西沙比利 cisapride
昂丹司琼 ondansetron
鹅去氧胆酸 chenodeoxycholic acid
熊去氧胆酸 ursodeoxycholic acid
苯丙醇 phenylpropanol
乳果糖 lactulose
★ 平喘药 antiasthmatic drug
克伦特罗 clenbuterol
氨茶碱 aminophylline
胆茶碱 choline theophyllinate
二羟丙茶碱 diprophylline
色甘酸钠 sodium cromoglycate
奈多罗米钠 nedocromil sodium
酮替芬 ketotifen
氮卓斯汀 Azelastine
二丙酸倍氯米松 beclomethasone dipropionate
布地缩松 budesonide
★ 祛痰药 expectorants
愈甘醚 glyceryl guaicolate
溴己胺 bromhexine
★ 镇咳药 antitussives
★ 可待因 codeine [ˈkəudiːin]
右美沙芬 dextromethorphan
喷托维林 pentoxyverine
苯佐那脂 benzonatate [bənˈzəunəteit]
苯丙哌啉 benproperine
组胺 histamine [ˈhistəmiːn, -min]
组胺受体阻断药 histamine receptor antagonist
西咪替丁 cimetidine [saɪˈmetɪdiːn]
雷尼替丁 ranitidine
法莫替丁 famotidine
尼扎替丁 nizatidine
罗沙替丁 roxatidine
子宫平滑肌兴奋药 oxytocics
缩宫素 oxytocin [ˌɔksiˈtəusin, -ˈtɔsin]
★ 催产素 pitocin
垂体后叶素 pituitrin [piˈtjuːitrin]
★ 抗利尿激素 antidiuretic hormone
麦角生物碱 ergot alkaloid
硫酸沙丁醇 salbutamol sulfate
利托君 ritodrine

1 Anemia can lead to a rapid or irregular heartbeat — an arrhythmia. Your heart must pump more blood to compensate for the lack of oxygen in the blood when you're anemic.
贫血会导致心律失常。由于贫血时血液的供氧能力下降，因此作为代偿，心脏必须在单位时间内泵出更多的血。

2 Potassium channels which are targets of antiarrhythmic drugs (class III) play an important role in cardiomyocyte membrane.
钾离子通道是心肌细胞膜上具有重要功能的离子通道，是 Ⅲ 类抗心律失常药物的作用靶点。
It is explanation that the therapy group have significant clinical effects in angina pectoris and AsMI after PCI compare with control group.

说明经过PCI后，治疗组与对照组相比，在心绞痛和无症状心肌缺血的治疗上具有显著的临床效果。

3 Conclusion: Histamine receptor antagonist can prevent and ameliorate airway remodeling and acid-base imbalance in asthma of guinea pig.

结论：组胺在哮喘气道重塑中起介导作用，组胺受体拮抗剂对防治哮喘气道重塑及酸碱平衡紊乱有一定作用。

第五节 药物对内分泌系统和其他代谢的影响

Section 5 Endocrine system and other metabolic effects of the drug

★肾上腺皮质激素 adrenocortical hormones
★皮质激素 corticosteroids
盐皮质激素 mineralocorticoids
糖皮质激素 glucocorticoids
超射 overshoot
褪黑激素 melatonin [ˌmelə'təunin]
促皮质素 corticotrophin [ˌkɔ:tikəu'trəufin]
★性激素 sex hormone
避孕药 contraceptives
★雌激素 estrogens
雌激素反应元件 hormone response elements
性激素结合球蛋白 sex hormone binding globulin, SHBG
乳腺癌 breast cancer
★孕激素类 progestins
黄体酮 progesterone [prəu'dʒestərəun]
先兆性流产 Aura of abortion
习惯性流产 habitual abortion
他莫西芬 tamoxifen [tə'mɔksɪfen]
雷洛昔芬 raloxifene
氯米芬 clomifene
★雄激素 androgens
睾酮 testosterone [tes'tɔstərəun]
促红细胞生成素 erythropoietin [iˌriθrə'pɔiətin]
★心力衰竭 cardiac failure
米非司酮 mifepristone
★甲状腺激素 thyroid hormones
甲状腺球蛋白 thyroglobulin
★甲状腺素 thyroxine [θai'rɔksi:n]
碘甲腺原氨酸 triiodothyronine, T_3
★促甲状腺激素 thyrotropin, TSH
促甲状腺释放激素 thyrotropin releasing hormone, TRH
呆小症 cretinism ['kretɪnɪzəm]
丙硫氧嘧啶 propylthiouracil
甲巯咪唑 thiamazole
卡比马唑 carbimazole
复方碘溶液 Lugol's solution
★胰岛素依赖型 insulin-dependent diabetes mellitus IDDM
非胰岛素依赖性 noninsulin-dependent diabetes mellitus, NIDDM
★胰岛素 insulin ['insjulin]
★胰岛素原 proinsulin [ˌprəu'insjulin]
单峰胰岛素 single-peak insulin
单组分胰岛素 single-component insulin
磺酰脲类 sulfonylureas
甲苯磺丁脲 tolbutamide
氯磺丙脲 chlorpropamide
格列苯脲 glibenclamide
格列波脲 glibonuride
格列喹酮 gliquidone
格列美脲 glimepiride
瑞格列奈 repaglinide
苯乙双胍 phenformin
二甲双胍 metformin
★动脉粥样化 atheroma
阿卡波糖 acarbose
伏格列波糖 voglibose
曲格列酮 troglitazone
罗格列酮 rosiglitazone
吡格列酮 pioglitazone
骨质疏松症 osteoporosis [ˌɔstiəupɔ:'rəusis]
绝经后 postmenopausal
★雌激素类 estrogens
双磷酸盐类 bisphosphonates
依替磷酸钠 etidronate disodium
帕米磷酸 pamidronate
阿伦磷酸 alendronate
★降钙素 calcitonin [ˌkælsi'təunin]
畸形性骨炎 Paget's disease
★高血钙症 hypercalcemia [ˌhaipəˌkæl'si:miə]
双氢速甾醇 dihydrotachysterol
甲状旁腺素 parathyroid hormone, PTH
★体重指数 body mass index, BMI
安非拉酮 amfepramone

马吲哚　mazindol
西布曲明　sibutramine
芬氟拉明　fenfluramine
有芬氟拉明　dexfenfluramine
氟西汀　fluoxetine
奥利斯特　orlistat

1 Whether endogenous or iatrogenic, excesses of adrenocortical hormones have long been known to be associated with potassium deficiency.
不论内源性或医源性，肾上腺皮质激素过多均可引起缺钾，这一点早已被人们所熟知。

2 That may be because, over the long term, it leads to the ongoing release of stress hormones, such as glucocorticoids.
这也许是因为，长期的压力过大会造成身体持续释放糖皮质激素（肾上腺皮质素）等压力荷尔蒙。

3 This makes it hard to measure how any individual compound might affect the endocrine system.
这就很难衡量任何一种化合物到底会如何影响内分泌系统。

4 Hypercalcemia is an abnormally high level of calcium in the blood.
高钙血症是一种在血液中钙含量较高的现象。

5 Bisphosphonates are used in the treatment of hypercalcemia of malignancy, skeletal complications associated with metastatic bone disease, Paget's disease, and osteoporosis.
二磷酸盐是用于治疗恶性高钙血症，以及与转移性骨骼疾病、佩吉特氏病和骨质疏松症相关的骨并发症。

第六节　抗病原微生物药物药理
Section 6　The disease resistance microbial medicine pharmacology

★ 抗微生物药物　antimicrobial agents
化疗指数　chemotherapeutic index
安全指数　safety index
抗菌谱　antibacterial spectrum
★ 抗菌剂　bacteriostatic
杀菌剂　bactericide [bæk'tɪərɪsaɪd]
★ 最低杀菌浓度　minimal bactericidal concentration, MBC
抗生素后效应　postantibiotic effect, PAE
青霉素结合蛋白　penicillin binding proteins, PBPs
配接　conjugation [ˌkɔndʒu'geiʃən]
转移-转导　transduction [træns'dʌkʃən]
★ 主动外排系统　active efflux system
★ 交叉耐药性　cross resistence
喹诺酮类　quinolones
诺氟沙星　norfloxacin
依诺沙星　enoxacin
培氟沙星　pefloxacin
氧氟沙星　ofloxacin
左氧氟沙星　levofloxacin
环丙沙星　ciprofloxacin
洛美沙星　lomefloxacin
氟罗沙星　fleroxacin
司帕沙星　sparfloxacin
磺胺类　sulfonamides
磺胺嘧啶　sulfadiazine, SD
磺胺甲恶唑　sulfamethoxazole, sinomin, SMZ
磺胺多锌　sulfadoxine
柳氮磺吡啶　sulfasalazine
磺胺醋酰　sulfacetamide
磺胺嘧啶银　silver sulfadiazine
甲氧苄啶　trimethoprim, TMP
硝基呋喃类　nitrofurans
呋喃妥因　nitrofurantoin
呋喃唑酮　furazolidone
β内酰胺类抗生素　β-lactam antibiotics
青霉素 V　penicillin V
氨苄西林　ampicloxacillin
阿莫西林　amoxicillin
匹氨西林　pivampicillin
羧苄西林　carbenicillin
羧苄西林茚满酯　carbenicillin indanyl
磺苄西林　sulbenicillin
替卡西林　ticarcillin
呋苄西林　furbenicillin
美洛西林　mezlocillin
阿帕西林　apalcillin
阿洛西林　azlocillin
头孢菌素类　cephalosporins
亚胺培南　imipenem
西司他丁　cilastatin

美洛培南 meropenem
氨曲南 aztreonam
克拉维酸 clavulanic acid
舒巴坦 sulbactam
他唑巴坦 tazobactam
★大环内酯类 macrolides
★红霉素 erythromycin [iˌriθrəu'maisin]
克拉霉素 clarithromycin
罗红霉素 roxithromycin
阿奇霉素 azithromycin
乙酰螺旋霉素 acetylspiramycin
麦迪霉素 medecamycin
麦白霉素 meleumycin
米欧卡霉素 miocamycin
交沙霉素 josamycin
克林霉素 clindymycin
★万古霉素 vancomycin
去甲万古霉素 demethylvancomycin
替考拉宁 teicoplanin
利奈唑烷 linezolid
氨基糖苷类 aminoglycosides
★链霉素 streptomycin
★庆大霉素 gentamycin
妥布霉素 tobramycin
小诺米星 micronomicin
阿米卡星 amikacin
奈替米星 netilmicin
★卡那霉素 kanamycin
异帕米星 isepamicin
★新霉素 neomycin
大观霉素，壮观霉素 spectinomycin
多粘菌素 B polymyxin B
★粘菌素，多粘菌素 E colistin，polymycin E
★四环素 tetracycline
土霉素 oxytetracycline
多四环素 doxycycline
米诺环素 minocycline
★氯霉素 chloramphenicol [ˌklɔːræm'fenikɔl, -kəul]
甲砜霉素 thiamphenicol
真菌感染 fungal infections
两性霉素 B amphotericin B
氟胞嘧啶 flucytosine
酮康唑 ketoconazole
伊曲康唑 itraconazole
氟康唑 fluconazole
特比奈芬 terbinafine
灰黄霉素 griseofulvin
克霉唑 clotrimazole
咪康唑 miconazole
阿莫罗芬 amorolfine
制霉菌素 nystatin
逆转录病毒 retroviruses
阿昔洛韦 aciclovir
伐昔洛韦 valaciclovir
泛昔洛韦 famciclovir
更昔洛韦 ganciclovir
磷甲酸钠 foscarnet sodium
阿糖腺苷 vidarabine，Ara-A
齐多夫定 zidovudine
拉米夫定 lamivudine
阿波卡韦 abacavir
扎西他滨 zalcitabine
地丹诺辛 didanosine
对氨水杨酸 para-aminosalicylic acid，PAS
乙硫异烟胺 ethionamide
奈韦拉平 nevirapine
地拉韦定 delavirdine
利巴韦林 ribavirin
金刚烷胺 amantadine
金刚乙胺 rimantadine
干扰素 interferon
★乙肝疫苗 hepatitis B vaccine
★结核病 tuberculosis，TB
★抗结核病药 antituberculosis agents
异烟肼 isoniazid
利福平 rifampicin
乙胺丁醇 ethambutol
吡嗪酰胺 pyrazinamide
利福喷汀 rifapentine
利福定 rifandin
麻风病 lepriasis [lep'raiəsis]
抗麻风病药 antileprotic drugs
氨苯砜 dapsone，DDS ['dæpsəun]
苯丙砜 solasulfone
醋氨苯砜 acedapsone
氯法齐明 clofazimine

1 For these patients, our treatment options effectively go back to the era that predates the advent of antibiotics.

对于这类患者，我们的治疗选择实际上回到了抗生素出现以前的时期。

2 In contrast, antimicrobial agents kill susceptible organisms with profound and lasting consequences.
相反，抗菌制剂会杀死具有敏感性的生物，造成深远和持久的后果。

3 They have a wide antibacterial spectrum, and some even have effects on fungi, virus and cancer cells.
抗菌肽不仅有广谱抗细菌能力，而且对真菌、病毒及癌细胞也有作用。

4 The mechanism based on which persimmons cure of bacteria caused pyogenic infection is discussed using the disc diffusion method, minimal inhibitory concentration (MIC), drug resistance.
通过抗菌谱、最低抑菌浓度、耐药性实验等手段，对柿子抑制化脓性感染菌的机制进行了初步研究。

第七节　抗寄生虫病药的药理
Section 7　Parasitic disease resistance of the medicine pharmacology

抗原虫药　antiprotozoal drugs
抗蠕虫药　antihelminthic drugs
抗疟药　antimalarial drugs
氯喹　chloroquine [ˌklɔ(:)rəˈkwi:n]
奎宁　quinine [kwɪˈni:n; (US) ˈkwaɪnaɪn]
咯萘啶　pyronaridine
甲氟喹　mefloquine
★青蒿素　artemisinin
本芴醇　benflumetol
伯氨喹　primaquine
乙胺嘧啶　pyrimethamine
硝喹　nitroquine
★阿米巴病　amebiasis [ˌæmiˈbaiəsis]
抗阿米巴病药　amebicides
★甲硝唑　metronidazole
替硝唑　tinidazole
哌硝噻唑　piperanitrozole
二氯尼特　diloxanide
喹碘仿　chiniofon [kiˈniəfən]
氯碘羟喹啉　clioquinol
双碘喹啉　diiodohydroxyquinoline
依米丁　emetine
★滴虫病　trichomoniasis
乙酰胂胺　acetarsol [ˌæseˈtɑ:sɔl]
碘古霉素　trichomycin [ˌtrikəˈmaisin]
血吸虫病　bilharziasis [ˌbilhɑ:ˈzaiəsis]
吡喹酮　praziquantel
★丝虫病　filariasis [ˌfiləˈraiəsis]
乙胺嗪　diethylcarbamazine
伊维菌素　ivermectin
呋喃嘧酮　furapyrimidone
甲苯达唑　mebendazole
苯并咪唑　benzimidazole
阿苯达唑　albendazole
噻苯达唑　thiabendazole
左旋咪唑　levamisole
噻嘧啶　pyrantel
哌嗪　piperazine [piˈperəzi:n]
氯硝柳胺　niclosamide
恩波维铵　pyrvinium embonate

1 This is the use of a single drug (or monotherapy) of chloroquine, sulfadoxine-pyrimethamine, or another antimalarial medicine to fight malaria.
这种疗法使用氯喹、磺胺多辛-乙胺嘧啶或另一种抗疟药物等单一药物治疗疟疾（亦称单一疗法）。

2 Albendazole composite exhibits a synergistic effect of both albendazole and pyrantel pamoate.
复方阿苯达唑具有阿苯达唑和噻嘧啶两药的协同作用。

第八节　抗恶性肿瘤药和影响免疫功能药
Section 8　Resistance to malignant tumor medicine and affect the immune function medicine

★周期素　cyclins
宿主基因-原癌基因　proto-oncogenes
固有威胁　inbuilt deterrent
氮芥　chlormethine
苯丁酸氮芥　chlorambucil [klɔˈræmbjusil]
★环磷酰胺　cyclophosphamide, CTX
异环磷酰胺　ifosfamide
白消安　busulfan
卡莫司汀　carmustine [ˈkɑ:mjustain]
洛莫司汀　lomustine

达卡巴嗪 dacarbazine
替莫唑胺 temozolomide
甲氨蝶呤 methotrexate [ˌmeθəˈtrekseɪt]
氟尿嘧啶 fluorouracil [ˌfluːərəuˈjurəsil]
替加氟 tegafur
巯嘌呤 mercaptopurine [məˌkæptəuˈpjuəriːn, -rɪn]
喷司他丁 pentostatin
羟基脲 hydroxyurea
阿糖胞苷 cytarabine
安西他滨 ancitabine
★丝裂霉素 mitomycin [ˌmaitəˈmaisin]
博莱霉素类 bleomycins
平阳霉素 pingyangmycin
★放线菌素 D dactinomycin D
★柔红霉素 daunorubicin
多柔比星 doxorubicin [ˌdɔksəuˈruːbɪsɪn]
表柔比星 epirubicin
普卡霉素 plicamycin
顺铂 cisplatin
卡铂 carboplatin
奥沙利铂 oxaliplatin
长春碱类 vinca Alkaloids
紫杉醇 paclitaxel
高三尖杉酯碱 homoharringtonine
喜树碱 camptothecin
羟喜树碱 hydroxyl camptothecin
依托泊苷 etoposide
替尼泊苷 teniposide
氨鲁米特 aminoglutethimide
氟他胺 flutamide
他莫昔芬 tamoxifen
亮丙瑞林 leuprorelin
戈舍瑞林 goserelin
奥曲肽 octreotide
依西美坦 exemestane
维A酸，维甲酸 tretinoin
★免疫系统 immune system
★感应阶段 inductive stage
增殖分化阶段 proliferative and differentiation stage
★效应阶段 effective stage
环孢素 cyclosporine [ˌsaikləuˈsp ɔːriːn]
他克莫司 tacrolimus
硫唑嘌呤 azathioprine [ˌæzəˈθaɪəpriːn]
西罗莫林 sirolimus
霉酚酸酯 mycophenolate mofetil
★卡介苗 bacillus Calmette-Guerin vaccine
短小棒状杆菌菌苗 corynebacterium parvum vaccine
★胸腺素 thymosin [ˈθaiməsin]
白介素-2 interleukin-2

1 However, it is unlikely that the rheumatoid arthritis patients but not those with osteoarthritis were protected from malignancy by NSAID use.
然而，对于类风湿性关节炎而又不是骨关节炎患者来说，通过应用非类固醇类抗炎药预防恶性肿瘤的可能性很小。

2 From the cancer's point of view that is a very good choice, as one of the other jobs of the mitochondria is to kill a cell if it goes bad — a process known as apoptosis.
对于肿瘤而言，这再好不过，因为线粒体另外还有一个功能，那就是当细胞衰老或受损时杀死细胞，即人们所熟知的细胞凋亡过程。

3 Certain inherited diseases of the skin and the lungs are also caused by telomerase defects.
某些皮肤和肺部的遗传性疾病也是由端粒酶缺陷引起的。

4 Carmustine and busulfan can increase risk for scarring and inflammation of the lungs.
卡莫司汀和白消安可增加肺部瘢痕与炎症的风险。

第二十八章 药剂学
Chapter 28 Pharmaceutics

第一节 绪论
Section 1 Introduction

★ 药剂学 pharmaceutics [ˌfɑːməˈsjuːtiks]
药物剂型 dosage form
★ 药物制剂 pharmaceutical preparation
★ 制剂学 pharmaceutical engineering
生命质量 quality of life，QOL
★ 工业药剂学 industrial pharmaceutics
★ 物理药剂学 physical pharmaceutics
药用高分子材料学 polymers in pharmaceutics
★ 生物药剂学 biopharmaceutics
药物动力学 pharmacokinetics
★ 临床药剂学 clinical pharmaceutics
★ 药物的传递系统 drug delivery system，DDS
自调式释药系统 self-adjusted system
经皮传递系统 transdermal drug delivery system，TDDS
★ 首过效应 first-pass effect
★ 药典 pharmacopoeia [ˌfɑːməkəˈpiːə]
★ 处方药 prescription drug or ethical drug
非处方药 non Prescription drug
可在柜台上买到的药物 over the counter，OTC
★ 药品生产质量管理规范 good manufacturing-practice，GMP
★ 药物非临床研究质量管理规范 good laboratory practice，GLP
★ 药物临床试验管理规范 good clinical practice，GCP

1 Drug absorption is the process of drug movement from the administration site to the systemic circulation. Drug absorption is determined by physicochemical properties of drugs，their formulations，and routes of administration.
药物吸收是指药物由用药部位进入体循环的转运过程。药物的吸收是由药物的理化性质、药剂的配方及给药途径所决定的。

2 GMP refers to the Good Manufacturing Practice Regulations promulgated by the US Food and Drug Administration under the authority of the Federal Food，Drug，and Cosmetic Act.
GMP 指药物生产质量管理规范，它是由美国食品药品管理局颁布的，该管理局遵守联邦食品、药品和化妆品法案。

3 Pharmacogenomics plays an important role in clinical pharmaceutics and new drug research and development.
药物基因组学在临床药物治疗学和新药开发中具有重要意义。

4 Short-term and long-term safety evaluation and prediction are important，as is the evaluation of the pharmacokinetics of the adjuvant alone.
短期和长期安全性评价和预测非常重要，同样重要的还有佐剂本身的药物代谢动力学评价。

第二节 液体制剂
Section 2 Liquid preparations

★ 均相液体制剂 homogeneous liquid preparation
非均相液体制剂 nonhomogeneous liquid preparation
★ 混悬剂 suspensions
★ 极性溶剂 polar solvent
★ 半极性溶剂 semi polar solvent
非极性溶剂 non-polar solvent
★ 增溶剂 solubilizer [ˈsɔljubɪlaɪzə(r)]
★ 助溶剂 hydrotropy agent
潜溶剂 cosolvent [kəuˈsɔlvənt]
防腐剂 preservative [priˈzəːvətiv]
着色剂 colorants
溶液剂 solutions

★ 芳香水剂 aromatic waters
糖浆剂 syrups ['sirəpi]
★ 酯剂 spirits
涂剂 paint [peint]
★ 酊剂 tincture ['tiŋktʃə]
高分子溶液剂 polymer solution
★ 分散法 dispersion method
凝聚法 coagulation [kəu,ægju 'leiʃən]
★ 助悬剂 suspending agents
乳剂 emulsions
相比 phase volume ratio
油中乳化剂法 emulsifier in oil method
水中乳化剂法 emulsifier in water method
新生皂法 nascent soap method
★ 两相交替加入法 alternate addition method
搽剂 liniments
涂抹剂 paints
洗剂 lotions
滴鼻剂 nasal drops
滴耳剂 ear drops
含漱剂 gargle
滴牙剂 drop dentifrices
★ 合剂 mixture

1 Colorants are often added to tablet formulations for esthetic value. Most colorants are photosensitive and they fade when exposed to light.
为了产品的美观，着色剂经常被添加到片剂中。大多数着色剂具有感光性，当被暴露在光线下，颜色会褪色。

2 Oleine oil ZR is prepared through optimization of formulations based on spindle oil, emulsifier EM, Turker red oil and solubilizer as raw materials.
和毛油 ZR 是以锭子油、乳化剂 EM、太古油、增溶剂等为主要原料，通过优化配方制成的。

第三节 灭菌剂与无菌制剂

Section 3 Sterilizing agent and sterile products

灭菌法 the technique of sterilization
无菌操作法 aseptic technique
★ 注射剂 injection [in'dʒekʃən]
★ 皮内注射 kntradermal route
★ 皮下注射 subcutaneous route
★ 肌内注射 intramuscular route
★ 静脉注射 intravenous route
★ 脊椎腔注射 vertebra caval route
★ 动脉内注射 intra-arterial route
等渗溶液 isoosmotic solution
等张溶液 isotonic solution
★ 热源 pyrogen ['paiərəudʒin]
滴眼剂 eye drop
洗眼剂 collyr
植入给药系统 implantable drug delivery systems
海绵剂 spongia
★ 内毒素 endotoxin [,endəu'tɔksin]
输液 infusion solution

1 Parenteral articles are prepared scrupulously by methods designed to ensure that they meet pharmacopeial requirements for sterility, pyrogen, particulate matter, other contaminants, and where appropriate, contain inhibitors of the growth of microorganisms.
制备注射剂的方法应能保证其符合无菌、无热源、无不溶性微粒和其他污染物的药典要求。

2 A suitable substance or mixture of substances to prevent the growth of microorganisms must be added to preparations intended for injection that are packaged in multiple-dose containers, regardless of the method of sterilization employed.
多剂量容器包装的注射用制剂必须添加防止微生物生长的适宜物质，不论其灭菌方法。

3 In severs chronic cases, treatment is started with daily 1 to 2 ampoules of Bilocid continued with one ampoule every other day, injected slowly by the intravenous route.
治疗严重的慢性疾病时，治疗从每日 1～2 支利胆素注射剂开始，继续治疗时每隔一日 1 安瓶缓慢静脉注射。

第四节 固体制剂
Section 4 Solid dosage forms

散剂 powders
粉碎 crushing [ˈkrʌʃiŋ]
筛分法 sieving method
对流混合 convective mixing
剪切混合 shear mixing
扩散混合 diffusive mixing
★ 临界相对湿度 critical relative humidity，CRH
★ 颗粒剂 granules
★ 片剂 tablets
普通片 compressed tablets
包衣片 coated tablets
糖衣片 sugar coated tablets
薄膜衣片 film coated tablets
肠溶衣片 enteric Coated tablets
★ 泡腾片 effervescent tablets
咀嚼片 chewable tablets
★ 分散片 dispersible tablets
★ 缓释片 sustained release tablets
多层片 multilayer tablets
★ 舌下片 sublingual tablets
★ 口含片 toroches
★ 颊额片 buccal tablets
★ 皮下注射用片 hypodermic tablets
★ 溶液片 solution tablets
★ 阴道片 vaginal tablets
★ 稀释剂 diluents
糖粉 sugar [ˈʃugə]
可压性淀粉 pregelatinized starch
微型纤维素 microcrystalline cellulose，MCC
润湿剂 moistening agent
★ 蒸馏水 distilled water
乙醇 ethanol [ˈeθənɔːl, -nəul]
黏合剂 adhesives
★ 崩解剂 disintegrants
润滑剂 lubricants
助流剂 glidants
★ 抗黏剂 antiadherent
湿法制粒 wet granulation
胶囊剂 capsules
★ 硬胶囊剂 hard capsules
软胶囊剂 soft capsules
肠溶胶囊剂 enteric capsules
★ 滴丸剂 drop pills
★ 膜剂 films

1 The Solid drug products (e. g. , tablets) disintegrate and disaggregate，but absorption can occur only after drugs enter solution.
固体药物制品(如片剂)需经崩解和解聚，但其吸收则要等进入溶液后才能发生。

2 Capsules are solid dosages forms in which the drug is enclosed within either a hard or fort shell. The shells are usually formed from gelatin；however，they also may be made from starch or other suitable substances.
胶囊是一种固体剂型，其将药物填充装于硬质或软质的胶囊壳中。胶囊壳通常是由凝胶制成的；然而，胶囊壳同样也能由淀粉或其他适当的物质制成。

3 Although tests were done for many times (try to add various disintegrants)，but without any best result.
虽然曾多次试验(试加各种崩解剂)，但效果不是很好。

第五节 半固体制剂
Section 5 Semisolid dosage forms

软膏剂 Ointments
★ 基质 Bases [ˈbeisiz]
★ 眼膏剂 Eye ointments
★ 凝胶剂 Gel [dʒel]
★ 栓剂 Suppository [səˈpɔzitəri]
冷压法 Cold compression method
热熔法 Fusion method

1 Microbicides are a gel，cream or suppository that releases a drug.
杀菌药是一种能够释放药物的凝胶、乳膏或者栓剂。

2 The unguent enjoys a good quality in adhesion，spread stretchability and portability. It's convenient

and easy to carry with.

软膏剂有着良好的附着性、涂展性,且携带、使用方便。

第六节 表面活性剂

Section 6 surfactant

★表面活性剂 surfactant [sə:'fæktənt]

离子表面活性剂 Ionic surfactant

非离子表面活性剂 non-ionic surfactant

★临界胶束浓度 Criticalmicelle concentration, CMC

★亲水亲油平衡值 Hydrophile-lipophile, HLB

1 Nonsurfactant additives known as "builders" have long been used to improve detergent action of surfactants.

非表面活性剂的添加剂称作"助洗剂",长期以来用以改进表面活性剂的去污功能。

2 The combination of water and nonionic surfactant increases the amount of pigment that is extracted and solubilised.

水与非离子表面活性剂的结合增加了萃取及增溶的颜料量。

3 According to the knowledge of surfactants, the emulsifiers were selected and mixed by the rule of Hydrophile-Lipophile-Balance(HLB) value.

根据表面活性剂知识,采用 HLB 值法筛选和复配乳化剂。

4 Each surfactant manifests a characteristic concentration — critical micelle concentration (CMC) — at which micelles of characteristic size will form.

每一种表面活性剂有一个特有的浓度 —— 临界胶团浓度(CMC)—— 形成特定尺寸胶团的浓度。

第七节 粉体学

Section 7 Micromeritics

★粉体学 micromeritics [,maɪkrəumɪ'rɪtɪks]

几何学粒子径 geometric diameter

筛分径 sieving diameter

有效径 effect diameter

★比表面积等价径 equivalent specific surface diameter

粒度分布 particle size distribution

频率分布 frequency size distribution

★累积分布 cumulative size distribution

数基准 count basis

质量基准 mass basis

面积基准 surface basis

体积基准 volume basis

长度基准 length basis

显微镜法 microscopic method

★库尔特技术法 coulter counter method

★沉降法 sedimentation method

比表面积法 specific surface area method

筛选法 sieving method

球形度 degree of sphericility

圆形度 degree of circularity

★粒子的比表面积 specific surface area

气体吸附法 gas adsorption method

真密度 true density

颗粒密度 granule density

松密度 bulk density

振实密度 tap density

空隙率 porosity [pɔ:'rɔsiti]

★休止角 angle of repose

★流出速度 flow velocity

压缩度 compressibility

吸湿性 moisture absorption

润湿性 wetting ['wetiŋ]

粘附性 adhesion [əd'hi:ʒən]

★凝聚性 cohesion [kəu'hi:ʒən]

★压缩性 compressibility [kəm,presi'biliti]

★成形性 compactibility [kəm,pækti'biliti]

压缩功 compressive work

★弹性功 elastic work

1 The result shows that breathing intensity, particle diameter and particle density are the three key factors influencing the deposition of inhalational particles.

研究结果表明,呼吸强度、颗粒粒径以及颗粒密度是影响可吸入颗粒物沉积的关键因素。

2 Gas Adsorption is one of the most optimal theory models to express specific surface area and pore structure of porous materials.
气体吸附法是测量多孔材料表面积和孔结构的常用方法。

3 Such a powder with good flowability and high tap density, after heat treatment below 1250℃, is suitable for thermal spray in the deposition of nanostructured coatings.
低于 1250℃ 热处理后的粉末流动性好，振实密度高，适于喷涂制备纳米结构涂层。

4 Compared with other measure techniques, the sedimentation method has been applied abroad because of its simple principle and easy operation.
与其他的测定技术相比，沉降法以其原理简单、易于操作而被广泛采用。

第八节　药物制剂的设计
Section 8　Design of the drug preparation

有效性　effectiveness [i'fektivnis]
★ 可控性　controllability
★ 稳定性　stability [stə'biliti]
顺应性　compliance [kəm'plaɪəns]
★ 处方前工作　preformulation
分配系数　partition coefficient
多晶形　polymorphism [ˌpɔli'mɔːfizəm]
★ 相对湿度　relative humidity, RH
单纯形优化法　simplex method
效应面优化法　response surface methodology
★ 效应面　response surface
析因设计　factorial design
★ 星点设计　central composite design, CCD

1 Drug design is the approach of finding drugs by design, based on their biological targets. Typically a drug target is a key molecule involved in a particular metabolic or signaling pathway that is specific to a disease condition or pathology, or to the infectivity or survival or a microbial pathogen.
药物设计是一种发现新药的方法，这种设计是基于生物学的靶点。具有代表性的药物靶点是一个参与特殊的新陈代谢或是信号通路的关键分子，这种分子对某种病情或病理，或对传染性和抵御微生物致病菌有特效。

2 The results show that the partition coefficient increases with the increase of the temperature and decrease with the pressure.
结果表明，分配系数随着系统压力的降低和系统温度的升高而增大。

3 Its good hygroscopicity makes it have fine substantivity, thus it can reduce the production costs in printing and dyeing mill.
其良好的吸湿性使其具有很好的着色效果，可为染厂降低生产成本。

第九节　缓释、控释制剂
Section 9　Slow release, controlled release formulation

★ 控释制剂　controlled-release preparations
★ 生物利用度　bioavailability
★ 阻滞剂　retardants
脉冲释药　pulsed/pulsatile release
定时钟　time clock
时控-控释系统　time controlled explosive system
★ 口服定位释药系统　oral site-specific drug delivery system
★ 胃定位释药系统　oral stomach-retained drug delivery system
★ 靶向制剂　targeting drug system, TDS
被动靶向制剂　passive targeting preparation
主动靶向制剂　active targeting preparation
★ 药物载体　drug carrier
反向靶向　inverse targeting
前体药物　prodrug
长循环脂质体　long-circulating liposome

1 Hyaluronic acid (HA) is a very good drug carrier due to its excellent biocompatibility and biodegradability.
透明质酸(HA)具有良好的生物相容性和生物降解性，是优良的药物载体。

2 And the system also can be prepared into corresponding slow-release and controlled-release solid preparations by selecting an appropriate skeleton material and a coating technique.
也可选用适宜的骨架材料和包衣技术制备相应的缓释、控释固体制剂。

3 Every proton pump inhibitor so far developed is administered in a prodrug form.
迄今为止每个质子泵抑制剂都被研究开发为一个药的管理形式。

第十节 经皮给药制剂
Section 10 Transdermal drug delivery systems

★经皮给药系统 transdermal drug delivery systems, TDDS
棘层 prickle cell layer
粘胶分散型 adhesive dispersion-type
骨架扩散型 matrix-diffusion type
微贮库型 microreservoir-type
经皮吸收促进剂 penetration enhancers
★物理学方法 physical approach
★化学方法 chemical approach
★生物学途径 biochemical approach
★压敏胶 pressure sensitive adhesive, PSA
黏附力 adhesive strength
快粘力 tacking strength

1 Base on the requirement of hydrophilic pressure-sensitive adhesives (PSA) in transdermal drug delivery systems (TDDs), a hydrophilic polyurethane PSA were prepared.
针对目前经皮给药制剂发展急需性能优良的亲水性新型压敏胶基质，制备了亲水性聚氨酯压敏胶。

2 Transdermal drug delivery systems (TDDS) have broad market prospects. However, transdermal delivery is limited by the inability of the large majority of drugs to cross skin.
经皮给药制剂是一种有着广阔市场前景的给药途径，但皮肤使多数药物难以渗透吸收。

3 Acrylic pressure sensitive adhesive was selected as A component, polyurethane adhesive was selected as B component, two-component acrylic adhesive was prepared by mixing A and B component.
丙烯酸酯压敏胶粘剂作为A组分，聚氨酯胶粘剂作为B组分，把A组分和B组分混合，制备双组分丙烯酸酯胶粘剂。

第二十九章　制药工程

Chapter 29　Pharmaceutical Engineering

第一节　绪论

Section 1　Introduction

★ 制药的　pharmaceutical [ˌfɑːməˈsjuːtikəl]
★ 药理学的　pharmacological [ˌfɑːməkəˈlɔdʒikəl]
★ 剂量　dosage [ˈdəusidʒ]
合成的　synthetic [sinˈθetic]
★ 前体　precursor [pri(ː)ˈkəːsə]
★ 临床的　clinical [ˈklinikəl]
★ 处方　prescription [priˈskripʃən/]

1 A drug is any natural or synthetic substance that alters the physiological state of a living organism.
药物是能改变生物机体生理状态的天然或合成物质。

2 Perhaps one of the most exciting fields of modern enzymology is the application of enzyme inhibitors as drugs in human and veterinary medicine.
也许现代酶学最令人激动的领域之一就是酶抑制剂在人药和兽药中的应用。

第二节　先导化合物的发现和筛选

Section 2　Discovery and screening of leading compound

★ 先导化合物　leading compound [ˈliːdiŋ][ˈkɔmpaund]
★ 高通量筛选　high throughput screening [hai][ˈθruːput][ˈskriːniŋ]
虚拟筛选　virtual screening [ˈvəːtjuəl][ˈskriːniŋ]
★ 组合化学　combinatorial chemistry [ˌkɔmbinəˈtɔːriəl][ˈkemistri]
组合生物合成　combinatorial biosynthesis [ˌkɔmbinəˈtɔːriəl][ˌbaiəuˈsinθisis]
★ 结构生物学　structural biology [ˈstrʌktʃərəl][baiˈɔlədʒi]
表面展示技术　surface display technology [ˈsəːfis][diˈsplei][tekˈnɔlədʒi/]

The lead compound is not intended to be used as a clinical agent. It is the starting point from which a clinically useful compound can be developed.
先导化合物不能被用作临床药物，它是被开发为临床有用化合物的起点。

第三节　天然产物的分离与纯化

Section 3　Separation and purification of natural compound

★ 细胞破碎　cell disruption [sel][disˈrʌpʃən]
★ 离心　centrifugation [senˌtrifjuˈgeiʃən]
★ 沉淀　precipitation [priˌsipiˈteiʃən]
★ 过滤　filtration [filˈtreiʃən]
★ 层析　chromatography [ˌkrəuməˈtɔgrəfi]
★ 透析　dialysis [daiˈælisis]
★ 结晶　crystallization [ˈkristəlaiˈzeiʃən]
反胶束提取纯化　reversed micelles purification [riˈvəːst][miˈselz][ˌpjuərifiˈkeiʃən]

1 Filtration is defined as the process in which particles are separated from a liquid by passing the liquid through a permeable material. The porous filter medium is the permeable material that separates particles from the liquid passing through it and is known as a filter.
过滤可定义为使液体通过可渗透材料从而除去其中微小颗粒的过程。这种可渗透材料是一种多孔物质，称为过滤介质，液体通过它后固体颗粒会被截留下来。

2 Understanding crystallization processes is important for the rational formulation, process development, and stability of pharmaceutical products.
理解结晶过程对合理的制药产品设计、工艺开发以及制药产品的稳定性都很重要。

第四节 微生物制药
Section 4 Microorganism pharmaceutics

★ 微生物转化 microbial transformation [mai'krəubiəl][ˌtrænsfə'meiʃən]
★ 手性药物 chiral drug ['tʃirəl][drʌg]
立体异构体 stereoisomer [ˌstiəriəu'aisəmə]
对映异构体 enantiomorph [i'næntiəumɔːf]
不对称合成 asymmetric synthesis [æsi'metrik]['sinθisis]
不对称放大 asymmetric amplification [æsi'metrik][ˌæmplifi'keiʃən]
外消旋 racemic [rə'siːmik]
立体选择性反应 stereoselective reaction [ˌstiəriəusi'lektiv][ri(ː)'ækʃən]
立体专一性反应 stereospecific reaction [ˌstiəriəuspi'sifik][ri(ː)'ækʃən]
拆分 resolution [ˌrezə'ljuːʃən]
外消旋体转换 racemic switch [rə'siːmik][switʃ]

1 In general, fungi are capable of high-level expression of various proteins, many of which secrete into their extracellular media. The extracellular production of a biopharmaceutical would be distinctly advantageous in terms of subsequent downstream processing.
通常情况下,真菌能够高效表达多种蛋白,许多蛋白都会分泌到胞外培养基中。生物药物的胞外生产在接下来的下游过程中显然具有极大的优势。

2 Many microorganisms represent attractive potential production systems for therapeutic proteins. They can usually be cultured in large quantities, inexpensively, and in a short time, by standard methods of fermentation.
许多微生物拥有能够产生药用蛋白的生产体系。它们通常可以用常规的发酵方法大量、廉价、短时间地培养出来。

第五节 基因工程制药
Section 5 Genetic engineering pharmaceutics

显微注射 microinjection [ˌmaikrəuin'dʒekʃən]
★ 定点突变 site-directed mutagenesis [sait][di'rektid][ˌmjutə'dʒenisis]
集落刺激因子 colony stimulating factors ['kɔləni]['stimjuleitiŋ]['fæktəz]
★ 人生长激素 human growth hormone ['hjuːmən][grəuθ]['hɔːməun/]

The bulk of biopharmaceuticals currently on the market are produced by genetic engineering using various recombinant expression systems.
目前市场上的大量生物技术药物都是通过基因工程运用各种重组表达系统制造的。

第六节 细胞工程制药
Section 6 Cell engineering pharmaceutics

灌注技术 perfusion technology [pə(ː)'fjuːʒən][tek'nɔlədʒi]
★ 贴壁依赖性细胞 anchorage-dependent cell ['æŋkəridʒ][di'pendənt][sel]
★ 悬浮细胞 suspension cell [səs'penʃən][sel]
★ 细胞克隆培养 cell clonal culture [sel]['kləunəl]['kʌltʃə]
★ 无血清培养基 serum-free medium ['siərəm][friː]['miːdjəm]
★ 基因显微注射 gene microinjection [dʒiːn][ˌmaikrəuin'dʒekʃən]
★ 反转录病毒感染 retrovirus-mediated gene transfer [ˌretrəu'vaiərəs]['miːdjətid][dʒiːn][træns'fəː]
★ 胚胎干细胞 embryo stem cell ['embriəu][stem][sel]
精子载体导入 sperm-mediated gene transfer [spəːm]['miːdjətid][dʒiːn][træns'fəː]
★ 基因打靶 gene targeting [dʒiːn]['tɑːgitiŋ]

酵母人工染色体 yeast artificial chromosome
[ji:st][ˌɑ:ti'fiʃəl]['krəuməsəum]

乳腺生物反应器 mammary gland bioreactor
['mæməri][glænd][ˌbaiəuri(:)'æktə]

细胞核移植技术 nuclear transplantation
['nju:kliə][trænsplɑ:n'teiʃən]

促红细胞生成素 erythropoietin
[iˌriθrə'pɔiətin/]

When compared to*E. coli*, animal cells display a very complex nutritional requirement, grow more slowly and are far more susceptible to physical damage. In industrial terms, this translates into increased production costs.

与大肠杆菌相比,动物细胞有着相当复杂的营养需求,生长更为缓慢,并且更容易受到物理伤害。在工业生产中,这些意味着生产成本的增加。

第七节 化学制药工程
Section 7 Chemical pharmaceutical engineering

★ 相似物 analogue ['ænəlɔg]

★ 全合成 total synthesis ['təutl]['sinθisis]

★ 半合成 semisynthesis [ˌsemi'sinθisis]

★ 衍生物 derivative [di'rivətiv]

★ 逆向合成分析 retrosynthetic analysis
['retrəusin'θetic][ə'nælisis]

★ 立体化学 stereochemistry
[ˌstiəriə'kemistri/]

Semisynthesis is usually used when the precursor molecule is too structurally complex, too costly or too inefficient to be produced by total synthesis.

当前体分子结构太复杂,使用全合成成本太高或者效率太低时,通常使用半合成。

第八节 中药制药工程
Section 8 Chinese medicine pharmaceutical engineering

补气药 invigorating vital energy drugs

补血药 hematonic drugs
[ˌhi:mə'tɔnik][drʌgz]

补阴药 invigorating yin drugs
[in'vigəreitiŋ][jin][drʌgz]

补阳药 invigorating yang drugs
[in'vigəreitiŋ][jɑ:ŋ][drʌgz]

理气药 regulating vital energy drugs
['regjuleitiŋ]['vaitl]['enədʒi][drʌgz]

活血药 activate blood circulation drugs
['æktiveit][blʌd][ˌsə:kju'leiʃən][drʌgz]

止血药 hemostatic drugs
[ˌhi:mə'stætɪk][drʌgz]

安神药 tranquilizers ['træŋkwilaizəz]

平肝药 claming the liver drugs
[klæmiŋ][ði:]['livə][drʌgz]

明目药 improving eyesight drugs
[im'pru:viŋ]['aisait][drʌgz]

化痰止咳药 eliminating phlegm and relieving cough drugs

消导药 promoting digestion and relieving dyspepsia drugs
[ænd][ri'li:viŋ][dis'pepsiə][drʌgz]

制酸药 relieving hyperacidity drugs
[ri'li:viŋ][ˌhaipə(:)rə'siditi][drʌgz]

收涩药 inducing astringency and arresting discharge drugs

催吐药 emetic drugs [i'metik][drʌgz]

清热药 clearing heat drugs
['kliəriŋ][hi:t][drʌgz]

温里药 warming up the interior drugs
['wɔ:miŋ][ʌp][ði:/in'tiəriə][drʌgz]

祛湿药 eliminating dampness drugs
[i'limineit][dæmpnis][drʌgz]

软坚散结药 softening and resolving the hardmasses

驱虫药 anthelmintic drugs
[ˌænθel'mintik][drʌgz]

开窍药 resuscitation-promoting drugs
[riˌsʌsi'teiʃən][prə'məutiŋ][drʌgz]

止消渴药 diabetes-arresting drugs
[ˌdaiə'bi:ti:z][ə'restiŋ][drʌgz]

第九节 药品生产质量管理工程

Section 9 Management engineering of pharmaceutical production quality

★ 良好作业规范 good manufacturing practice (GMP) [gud][ˌmænju'fæktʃəriŋ]['præktis]

传播 promulgate ['prɔməlgeit]

★ 预防性的 proactive [ˌprəu'æktiv]

★ 污染 contamination [kənˌtæmi'neiʃən]

★ 卫生 sanitation [sæni'teiʃən]

处方药 prescription drug [pri'skripʃən][drʌg]

★ 无菌度 sterility [ste'riliti]

★ 致热原 pyrogen ['paiərəudʒin]

表 1 GMP 中的生产区域

Table 1 Production area in GMP

英文	中文	英文	中文
first changing room	一更室	corridor	走廊
hand disinfection room	手消毒室	processing room	加工车间
cleanroom	洁净室	filling room	灌装车间
gowning room	更衣室	immediate package room	内包装车间
airlock room	气闸室	emergency door	安全门

第十节 制药工程设备

Section 10 Equipment of pharmaceutical engineering

★ 丹剂 dan forms [dæn][fɔːmz/]

★ 颗粒剂 granule ['grænjuːl]

★ 口服液 oral liquid ['ɔːrəl]['likwid]

★ 胶囊剂 capsule ['kæpsjuːl]

★ 注射剂 injection [in'dʒekʃən]

★ 黏合剂 binder ['baində]

★ 润滑剂 lubricant ['luːbrikənt]

★ 甜味剂 sweetening agent ['swiːtniŋ]['eidʒənt]

★ 崩解剂 disintegrating agent [dis'intigreitiŋ]['eidʒənt]

泡腾片 effervescent tablet [ˌefə'vesnt]['tæblit/]

1 Most compressed tablets consist of the active ingredient and a diluent, binder, disintegrating agent, and lubricant. Pigments, flavors and sweetening agents may also be present.

大多数压制片由活性成分、稀释剂、黏合剂、崩解剂和润滑剂组成。其中也可能含有色素、芳香剂和甜味剂成分。

2 The two main types of capsules are hard-shell capsules, which are normally used for dry, powdered ingredients, and soft-shell capsules, primarily used for oils and for active ingredients that are dissolved or suspended in oil.

胶囊的两种主要类型为硬胶囊剂和软胶囊剂，硬胶囊剂通常适用于干燥的粉末状成分，而软胶囊剂主要适于油状物及能够溶解或悬浮于油状物中的活性成分。

第三十章　药物合成反应
Chapter 30　Drug Synthesis Reactions

第一节　卤化反应
Section 1　Halogenation Reactions

碳-卤键　halogen-carbon bond
保护基团　protective group
阻断剂　blocker
反应机理　reaction mechanism [ri'ækʃən][ˈmekənizəm]
氟化　fluorination [ˌflu:əri'neiʃən]
氯化　chlorination [ˌklɔ:ri'neiʃən]
溴化　bromination [ˈbrəumi'neiʃən]
碘化　iodination [ˌaiəudi'neiʃən]
反式加成　trans addition
受限旋转　restricted rotation
立体化学　stereochemistry [ˌstiəriə'kemistri]
诱导偶极　induced-dipole [ˈdaipəul]
环溴翁离子　cyclic bromonium ion [ˈbrəuməniəm]
立体位阻　steric hindrance [ˈhindrəns]
马尔科夫尼可夫规则　markovnikov rule
乙烯型碳正离子　ethylene-typecarbocations
二卤代烷　dihalogenoalkane
吸热的　endothermic [ˌendəu'θə:mik]

1 Halogenation: A chemical process or reaction in which a halogen element is introduced into a substance, generally by the use of the element itself.
卤化：将卤素原子引入化合物的化学过程或反应，一般使用卤素单质作为反应物。

2 Halogenation reactions can be subdivided in several ways, for example, according to the type of halogen (fluorine, chlorine, bromine, or iodine), type of material to be halogenated (paraffin, olefin, aromatic, hydrogen, and so on), and operating conditions and methods of catalyzing or initiating the reaction.
卤化反应可以根据多种方法进行细分：例如，根据卤素原子的种类（氟、氯、溴、碘）分类，根据被卤化的材料分类（石蜡、烯烃、芳烃、氢气等），根据操作条件、催化和引发反应的方法等进行分类。

3 Free Radical Halogenation occurs when either bromine or chlorine is added to an alkane in the presence of light. Free radical halogenation with fluorine or iodine is rarely used because flourine is so reactive that it is not very selective, and the addition of Iodine is so endothermic to be effective as a synthesis of alkyl halides.
在紫外光照射下，将溴或氯气加入到烷烃中可以引发自由基卤化反应。在自由基卤化反应中，氟或碘很少用于自由基卤化反应。因为氟太活泼而导致选择性差；而碘化反应是强吸热反应，不是一个制备碘代烷的有效方法。

第二节　烷化反应
Section 2　Alkylation

★ 碳负离子　carbanion [ˈkɑ:bənaiən]
★ 卡宾　carbene [ˈkɑ:bi:n]
★ 烷基化试剂　alkylating agent
金属有机化合物　organometallic compound
耦合反应　coupling
有机卤化物　organichalide
醇盐离子　alkoxide ion [æl'kɔksaid]
乙醇钠　sodiumethoxide [i:' θɔksaid]
卤代醇　halohydrin
环氧化合物　epoxide [e'pɔksaid]
对甲苯磺酸盐　tosylate [təuzi'leit]
酚醚　phenol ether
硫酸二甲酯　dimethyl sulfate
氨基化反应　amination reaction
环六亚甲基四胺　methenamine [mə'θi:nəmi:n]
还原氨基化反应　reductive alkylation reaction
季碳原子　tertiary carbon
★ 傅列德尔-克拉夫茨　friedel-Crafts reaction
乙炔钠　sodium acetylide [ˌæsiti'laid]
★ 格利雅试剂　Grignard reagent

活性亚甲基化合物 active methylene compound
烯胺 enamine [i'næmin]
相转移催化剂 phase transfer catalyst
二烷基铜锂 lithium dialkylcuprate
芳基化反应 arylation [æri'leiʃən]

1 Alkylation is the transfer of an alkyl group from one molecule to another.
烷化反应是将一个烷基从一个化合物转移到另一个化合物的过程。

2 Friedel - Crafts alkylation involves the alkylation of an aromatic ring with an alkyl halide using a strong Lewis acid catalyst.
傅列德尔-克拉夫茨反应是指在强路易斯酸催化下,使用卤代烃对一个芳环进行烷基化的反应。

3 The Williamson ether synthesis is an organic reaction, forming an ether from an organohalide and an alcohol.
威廉姆逊醚合成是将有机卤化物和醇反应生成醚的有机反应。

4 Reductive alkylation is an effective method to synthesize primary and higher amines from aldehydes and ketones and hence is commercially used for the synthesis of a variety of fine and specialty chemicals. The amine (or ammonia) is condensed with a carbonyl group to yield the intermediate imine (a Schiff's base) which can then be catalytically hydrogenated to the desired amine. The main challenge in these types of reaction is the selectivity to the desired amine while minimizing the reduction of the carbonyl group.
还原烷基化是从醛和酮合成初级和高级胺的一种有效方法,因此在商业上广泛用于合成多种精细与专用化学品。在此反应中,胺(或铵)与羰基缩合生成中间产物亚胺(希夫碱),然后亚胺被催化氢化生成目标胺。这类反应的主要挑战是如何高选择性生成所需的胺同时最大限度减少羰基的还原。

第三节 酰基化反应
Section 3 Acylation Reaction

★ 酰化作用 acylation [,æsi 'leiʃən]
★ 酰化剂 acylating agent
★ 酰卤 acylhalide
琥珀酸 succinic acid
二环己基碳二亚胺 dicyclohexylcarbodiimide (DCC)
乙烯酮 ketene ['ki:ti:n]
甲酰化 formylate ['fɔ:mileit]
乙酰化 acetylation [ə,seti'leiʃən/]

1 In chemistry, acylation is the process of adding an acyl group to a compound. The compound providing the acyl group is called the acylating agent.
在化学中,酰化反应是指在一个化合物上引入酰基的过程。提供酰基的化合物称为酰基化试剂。

2 Acyl halides and anhydrides of carboxylic acids are also commonly used acylating agents to acylate amines to form amides or acylate alcohols to form esters. The amines and alcohols are nucleophiles; the mechanism is nucleophilic addition-elimination.
羧酸的酰卤和酸酐衍生物常常被用作从胺制备酰胺或者从醇制备酯的酰基化试剂。胺和醇是亲电试剂,反应机理是亲电加成-消除反应机理。

第四节 缩合反应
Section 4 Condensation Reaction

★ 分子间缩合 intermolecularcondensation [,kɔnden'seiʃən]
★ 分子内缩合反应 intramolecular condensation
狄克曼缩合反应 Dieckmann condensation
羟醛缩合反应 aldol condensation
克莱森缩合 Claisen condensation
环加成 cycloaddition [,saikləuə'diʃən]
亲二烯体 dienophile [dai'enəfail]
α,β-不饱和羰基化合物 enone
β-二酮酯 β-ketoester
安息香缩合 Benzoin condensation
雷福尔马茨基反应 Reformatsky reaction
普林斯反应 Prins reaction

1 A condensation reaction is a chemical reaction in which two molecules or moieties (functional groups) combine to form one single molecule, together with the loss of a small molecule.
两个分子或两个基团相互作用后结合成一个大分子，并伴有失去小分子的化学反应称为缩合反应。

2 An aldol condensation is an organic reaction in which an enol or an enolate ion reacts with a carbonyl compound to form a β-hydroxyaldehyde or β-hydroxyketone, followed by a dehydration to give a conjugated enone.
羟醛缩合反应是指烯醇或烯醇离子与一个羰基化合物反应生成β-羟基醛或β-羟基酮，然后再脱水生成共轭α，β-不饱和醛酮的有机化学反应。

3 The Claisen condensation is a carbon - carbon bond forming reaction that occurs between two esters or one ester and another carbonyl compound in the presence of a strong base, resulting in a β-keto ester or a β-diketone.
克莱森缩合反应是指在强碱催化下，两个酯分子或一个酯和羰基化合物形成C-C共价键，生成β-酮酸酯或者β-二酮的反应。

4 The Prins reaction is an organic reaction consisting of an electrophilic addition of an aldehyde or ketone to an alkene or alkyne followed by capture of a nucleophile.
普林斯反应是将醛或酮与烯烃或炔烃发生亲电加成，然后和亲核试剂结合的反应。

5 The Reformatsky reaction is an organic reaction which condenses aldehydes (or ketones), with α-halo esters, using a metallic zinc to form β-hydroxy-esters. It was discovered by Sergey Nikolaevich Reformatsky.
雷福尔马茨基反应是醛或酮与α-卤代酸酯在金属锌的作用下发生缩合得到β-羟基酸酯的反应。它由俄国化学家雷福尔马茨基首次发现。

6 The Wittig reaction or Wittig Olefination is a chemical reaction of an aldehyde or ketone with a triphenyl phosphonium ylide to give an alkene and triphenylphosphine oxide.
威悌反应或威悌烯化作用是指一个醛或酮和三苯基膦叶立德试剂反应生成烯烃和三苯基氧膦的有机反应。

第五节 重排反应
Section 5 Rearrangement Reaction

★ 烯丙基重排 allylic rearrangement [ˈæl əˌlik] [riːəˈreindʒmənt]
★ 周环反应 pericyclic reactions [ˌperiˈsaiklik]
烯烃复分解反应 olefin metathesis [meˈtæθəsis]
频那醇重排 Pinacol rearrangement
沃尔夫重排反应 Wolff rearrangement
库尔提斯重排反应 Curtius rearrangement
落森重排 Lossen rearrangement
斯蒂文森重排 Stevens rearrangement
克莱森重排 Claisen rearrangement
科普重排 Cope rearrangement
椅式构象 chairconformation [ˈkənfəːˈmeiʃən]

1 A rearrangement reaction is a broad class of organic reactions where the carbon skeleton of a molecule is rearranged to give a structural isomer of the original molecule.
重排反应主要指分子的碳骨架发生重排生成原化合物的一个构造异构体的一类范围很广的有机反应。

2 The Beckmann rearrangement is an acid-catalyzed rearrangement of an oxime to an amide.
贝克曼重排反应是在酸催化下从肟到酰胺的重排反应。

3 A pericyclic reaction is a type of reaction with multiple carbon-carbon bond making and breaking wherein the transition state of the molecule has a cyclic geometry, and the reaction progresses in a concerted fashion.
周环反应是指通过多个C-C键的断裂和形成过程，并且在反应过程形成环状过渡态的协同化学反应。

第六节　氧化反应

Section 6　Oxidation Reaction

★ 化学氧化　chemical oxidation [ˌɔksiˈdeiʃən]
★ 电解氧化　electrolytic oxidation [iˌlektrəuˈlitik]
★ 催化氧化　catalytic oxidation [ˌkætəˈlitik]
二氧化锡　Tin dioxide
亲电烯反应　ene reaction
活化硫盐　sulfonium salt [sʌlˈfəuniəm]

1 Oxidation: chemistry, the loss of electrons, gain of oxygen, or loss of hydrogen by an atom, ion, or molecule during a chemical reaction. At a simple level, oxidation may be regarded as the reaction of a substance with oxygen. For example, rusting, respiration, and combustion (burning) are all oxidation reactions. When carbon is burned in air it is oxidized to carbon dioxide.

氧化：一个原子、离子或分子在化学反应中失去一个电子，得到一个氧原子，或失去一个氢原子的化学反应为氧化反应。简单地说，氧化反应是指一个化合物和氧气的反应。例如生锈，呼吸和燃烧过程都是氧化反应。碳在空气中燃烧，就被氧化成二氧化碳。

2 Chemical oxidation is being used at hundreds of sites across the country. It destroys pollution underground without having to dig it up or pump it out for transport to a treatment system. This saves time and money. Often chemical oxidation is used to clean up pollution that other methods can't reach, like pollution deep within the groundwater.

化学氧化在全国范围内在很多方面都有应用。它可以消除地下的污染而不需要把污染物挖掘出来或者抽出来转移到特定的处理系统内。这节省了时间和金钱。化学氧化经常可以清除使用其他方法无法清除的污染，例如很深的地下水。

3 Biological oxidation is the process in which substances are oxidized in living organism.

生物氧化是指在生物体内发生的化合物氧化过程。

第七节　还原反应

Section 7　Reduction Reaction

多相催化氢化　heterogeneous hydrogenation [ˌhetərəˈdʒiːniːəs]
转移氢化　transfer hydrogenation
均相催化氢化　homogeneous hydrogenation [ˌhɔməˈdʒiːnjəs]
电解加氢　electrolytic hydrogenation
受氢体　hydrogen acceptor
毒化剂　poison [ˈpɔizən]
克莱门森还原　Clemmensen reduction
还原胺化　borch reductive amination
罗森蒙德还原反应　Rosenmund reduction
施陶丁格反应　Staudinger reaction
脱苄反应　debenzylation [diːˌbenziˈleiʃən]

1 Hydrogenation is a chemical reaction between molecular hydrogen and another compound or element, usually in the presence of a catalyst. The process is commonly employed to reduce or saturate organic compounds.

氢化反应是指在催化剂作用下，氢气分子和其他化合物之间的化学反应。这个过程一般用来还原或饱和有机化合物。

2 Transfer hydrogenation is the addition of hydrogen to a molecule from a source other than gaseous H_2.

转移氢化反应是指给一个化合物分子加一个氢分子，但氢源不是氢气的反应。

3 Clemmensen reduction is a chemical reaction described as a reduction of ketones (or aldehydes) to alkanes using zinc amalgam and hydrochloric acid.

克莱门森还原反应是在指在锌汞齐和盐酸催化下，酮或醛被还原成烷烃的化学反应过程。

第八节 合成设计原理

Section 8 Principle of Synthesis Design

★ 靶分子 target molecule
★ 合成子 synthon [ˈsinθən]
等价试剂 equivalent reagent [iˈkwivələnt]
合成砌块 building block
烃化合成子 alkylating synthon
自由基合成子 radical synthon
周环反应合成子 pericyclic synthon
极性反转 umpolung
等电性反应 isohypsic reaction
★ 逆向切断 antithetical disconnection [ˌæntiˈθetikəl/ˌdiskəˈnekʃən]
逆向链接 antithetical connection
逆向重排 antithetical rearrangement
逆向官能团转换 antithetical functional group interconversion
逆向官能团加成 antithetical functional group addition
逆向官能团除去 antithetical functional group remove
变换靶分子 alternative target molecule
合成树 synthesis tree
会聚式合成 convergent synthesis [kənˈvəːdʒənt]
直线式合成 linear synthesis
多重建架反应 multiple construction
自动连贯式过程 self-consistent sequence
原子经济性 atom economy
逆合成分析 retro-synthetic analysis
多样性而定合成 diversity-oriented synthesis
反合成分析 antithesis [ænˈtiθisis]
策略性键 strategic bond
对映选择性合成 enantioselective synthesis
立体电子效应 stereoelectronic effect

第三十一章　天然药物化学

Chapter 31　Natural Medicinal Chemistry

第一节　绪论

Section 1　Introduction

利血平　reserpine [ri'sə:pin]

雌性信息素　female pheromone ['fi:meil]['ferəməun]

沙海葵毒素　palytoxin [ˌpæli'tɔksin]

神经毒素　neurotoxin [ˌnjuərəu'tɔksin]

桂皮酸途径　cinnamic acid pathway [si'næmik]

醋酸-丙二酸途径　acetic acid-malonic acid pathway [ə'si:tik][mæləu'nik]

氨基酸途径　amino acid pathway ['æminəu]

茴香脑　anethole ['ænəθəul]

丁香酚　eugenol ['ju:dʒinɔl]

苯丙素类　phenylpropanoid ['fenil'prəupeinɔid]

香豆素类　coumarin ['ku:mərin]

木质素类　lignin ['lignin]

液-液萃取法　liquid-liquid extraction [ik'strækʃən]

反流分布法　reflux distribution method

液滴逆流色谱法　droplet countercurrent chromatography

高速逆流色谱　high-speed countercurrent chromatography

气-液分配色谱　gas-liquid partition chromatography

液-液分配色谱　liquid-liquid partition chromatography

第二节　糖和苷

Section 2　Sugars and Glycosides

苷元　aglycone ['əglaikɔn]

呋喃型糖　furanose ['fjuərəˌnəus]

吡喃型糖　pyranose ['paiərənəus]

差向异构体　epimer ['epimə]

糖醛酸　uronic acid ['jurɔn]

戊醛糖　aldopentose [ˌældə'pentəus]

己醛糖　aldohexose [ˌældə'heksəus]

己酮糖　ketohexose [ˌki:təu'heksəus]

氨基糖　amino sugar

同多糖　homopolysaccharide [ˌhəuməuˌpɔli'sækəraid]

杂多糖　heteropolysaccharide [ˌhetərəˌpɔli'sækəraid]

硫酸软骨素　chondroitin sulfate [kən'drəuitin]['sʌlfeit]

酚苷　phenolic glycoside [fi'nɔlik]['glaikəˌsaid]

氰苷　cyanogenic glycosides [ˌsaiənəu'dʒenik]

乙酰解反应　acetolysis reaction [ˌæsi'tɔlisis]

转化糖酶　invertase [in'və:teis]

麦芽糖酶　maltase ['mɔ:lteis]

季铵盐沉淀法　quaternary ammonium salt precipitation method

纤维素柱色谱　cellulose column chromatography

凝胶柱色谱　gel column chromatography

制备区域电泳　regional preparation electrophoresis

1 Although heparin is used principally in medicine for anticoagulation, its true physiological role in the body remains unclear, because blood anti-coagulation is achieved mostly by heparan sulfate proteoglycans derived from endothelial cells.

虽然肝素主要用于抗凝药，但其在体内真正的生理作用尚不清楚，因为血液中抗凝血能力大多来自内皮细胞硫酸乙酰肝素蛋白多糖的作用。

2 Cassava was the main crop rich in starch, but with high content of the toxic cyanogenic glucosides linamarin and lotaustralin in all tissues, and cytochrome p450 generegulated the biosynthesis of cyanogenic glucoside.

木薯是含有丰富淀粉的主要作物，但是亚麻苦苷和百脉根甙等氰化物是其组织中的剧毒物质，细胞色素 p450 基因调控氰化物生物合成。

3 The intestinal mucous from giant pandas contained significant sucrase, lactase and maltase activities.
大熊猫小肠黏膜中存在显著量的蔗糖酶、乳糖酶和麦芽糖酶活力。

第三节 苯丙素类
Section 3 Phenylpropanoid

羟基桂皮酸 hydroxycinnamic acid [hai'drɔksi'næmik]
咖啡酸 caffeic acid [kə'fi:ik]
绿原酸 chlorogenic acid
奎宁酸 quinate ['kwaineit]
莱蓟素 cynarin ['sinərin]
松柏苷 coniferin ['kɔniferin]
七叶内酯 esculetin [,eskju'li:tən]
当归内酯 angelica lactone [æn'dʒelikə]['læktəun]
白芷内酯 isopsoralen
亮菌甲素 armillarisin [ɑ:mi'lærisin]
海棠果内酯 callophylloide [kælə'filəlaid]
逆没食子酸 ellagic acid [i'lædʒik]
东莨菪内酯 scopoletin
黄曲霉素 aflatoxin [,æflə'tɔksin]
光敏作用 photosensitization ['fəutəu,sensitai'zeiʃən]
木脂素 lignan ['lignæn]
新木脂素 neolignan [nju'lignæn]
倍半木脂素 sesquilignan ['ses'kwi'lignæn]
去甲木脂素 norlignan [nɔ:'lignæn]
黄酮木脂素 flavonolignan ['fleivənɔ'lignæn]
二芳基丁烷类 dibenzylbutanes [dai'benzil'bju:tein]
二芳基丁内酯类 dibenzyl butyrolactones [dai'benzil bju:tərəu'læktəunz]
芳基萘类 arylnaphthalenes ['æril'næfθəli:nz]
四氢呋喃类 tetrahydrofurans [,tetrə,haidrə'fjuərənz]
苯并呋喃类 benzofurans [,benzəu'fjurənz]
联苯类 biphenylenes [bai'fenəli:nz]

1 The lignans are a group of chemical compounds found in plants. Lignans are one of the major classes of phytoestrogens, which are estrogen-like chemicals and also act as antioxidants. Other classes of phytoestrogens are the isoflavones and coumestans.
木脂素是一组在植物中发现的化合物。木脂素是植物雌激素的主要类别，是雌激素类似的化合物并具有抗氧化剂作用。其他类植物雌激素是异黄酮和香豆雌酚。

2 There are 24 different lignan precursors that have been identified so far. They are found in different concentrations in different plants. Their concentrations are highest in flax and sesame seeds as well as several types of bran.
到目前为止，已确定的不同的木脂素前体有 24 个。这些前体在不同植物中浓度不同。它们的浓度最高是在亚麻和芝麻以及几种类型的麸皮中。

3 Caffeic acid Ge has antiproliferative and proapoptotic effect on U14 tumor cells in vivo and in vitro.
咖啡酸锗在体内、外均能有效抑制小鼠 U14 瘤细胞的增殖，并诱导肿瘤细胞凋亡。

4 With the observations above, cynarin of Echinacea and effective fraction of Schisandra can be used in the future as potential agents for curing auto- immune diseases.
据上述观察，松果和五味子有效成分莱蓟素未来可能用作治疗自身免疫性疾病的潜在药剂。

第四节 醌类化合物
Section 4 Quinones

苯醌类 benzoquinones [,benzəu'kwinəun]
密花醌 rapanone [rə'pænən]
信筒子醌 embelin ['imblin]
紫草素 shikonin ['ʃaikɔnin]
竹红菌素 hypocrellin [haipək'rəlin]
维生素 K vitamin K ['vaitəmin]
菲醌 phenanthrenequinone [fə'nænθri:nkwi'nəun]
丹参醌类 transhinones
蒽醌类 anthraquinones [,ænθrəkwi'nəun]

茜草素 alizarin [ə'lizərin]
柯桠素 chrysarobin [ˌkrisə'rəubin]
大黄素 emodin ['emədin]
泻下作用 purgative effect ['pə:gətiv]
胡桃醌 juglone [dʒu:'gləun]
蓝雪醌 plumbagin [plʌmbid'ʒn]
金丝桃素 hypericin [haipə'raisin]

1 Benzoquinone and naphthaquinone derivatives undergo Feigl reaction to turn the colorless ethanol solution of formaldehyde and 1,2-dinitrobenzene to produce purple products.
苯醌和萘醌衍生物发生的费格尔反应可以将无色的甲醛和1,2-二硝基苯的乙醇溶液转化为紫色。

2 Sennoside with different concentration hadn't direct effects on the contraction of small intestinal smooth muscle, but could increase the reactivity of smooth muscle cells to Ach. RhoA was activated and translocated from cytosol to plasma membrane.
不同浓度的番泻甙对小肠环行平滑肌没有直接收缩作用,但可以增加乙酰胆碱对平滑肌细胞的反应性,RhoA蛋白被激活,并从胞浆转运到质膜。

第五节 黄酮类化合物
Section 5 Flavonoids

黄酮醇类 flavonols ['fleivənɔl]
鱼藤酮类 rotenoids ['rəutnəun]
查尔酮类 chalcone ['kælˌkəun]
橙酮类 aurones [ɔ:rəun]
花色素类 anthocyanidin [ˌænθəsai'ænidin]
葛根黄素 puerarin ['pjuərərin]
儿茶素 catechin ['kætitʃin]
营实苷 multiflorin [mʌltif'lɔrin]
香叶木苷 diosmin [dai'ɔzmin]
羟乙基卢丁 hydroxyethylrudin [ˌhai'drɔksieθi'lrudin]
棉花皮苷 gossypin ['gɔsipin]
染料木素 genistein [dʒi'nistin]
桑色素 morin ['mɔ:rin]
山奈酚 kaempferol ['ki:mpfərɔl]
异甘草素 isoliquiritigenin [aisəlikwɪə'ritaidʒnɪn]
罂粟碱 papaverine [pə'pævərin]
色原酮 chromone ['krəuməun]
2-苯基色原酮 phenylchromone ['fenil'krəuməun]

Even if flavonoid content is given, scientists claim that results can vary from lab to lab.
即使给定类黄酮的含量,科学家们声称,由于实验室的不同其结果也有所差异。

第六节 萜类和挥发油
Section 6 Terpenes and Volatile Oil

异戊二烯法则 isoprene rule ['aisəupri:n][ru:l]
焦磷酸异戊烯酯 isopentene pyrophosphate [aisəu'penti:n][ˌpairəu'fɔsfeit]
胡萝卜素 carotene ['kærəti:n]
橡胶 caoutchouc ['kautʃuk]
艾里木酚酮 eremophilone [ˌerimǝu'filəun]
土青木香酮 aristolone [əˌristə'ləun]
扁柏酚 hinokitol [hi'nəukitəul]
薄荷烷 menthane ['menθein]
香叶醇 geraniol [dʒi'reiniɔl]
橙花醇 nerol ['nirəul]
香茅醇 citronellol [ˌsitrə'nelɔl]
芳樟醇 linalool [li'næləˌəul]
柠檬醛 citral ['sitrəl]
香叶醛 geranial [dʒi'reiniəl]
紫罗兰酮 ionone ['aiənəun]
斑蝥素 cantharidin [kæn'θærɪdɪn]
梓醇 catalpol [kə'tælpɔl]
龙胆碱 gentianine ['dʒeʃənin]
当药苷 sweroside [swə:rɔˌsaid]
金合欢烯 farnesene ['fɑ:nisi:n]
青蒿素 artemisinin [ˌɑ:ti:mi:sinin]
棉酚 gossypol ['gɔsiˌpɔl]
山道年 santonin ['sæntənin]
愈创木醇 guaiol ['gwaiɔl]
紫杉烷 taxanes ['tæksein]
松香烷 abietane ['æbiətein]
乌头烷 aconane [ə'kɔnein]
赤霉烷 gibberellane [ˌdʒibə'relein]
银杏内酯 ginkgolide ['giŋkgɔlid]

紫杉醇 taxol ['tæksl]
甜菊苷 stevioside ['sti:viə,said]
呋喃海绵素 furanospongin ['fjuəræn'spʌndʒin]
精油 essential oils [i'senʃəl][ɔil]
百里香草酚 thymol ['θaiməul]
茴香醚 anisole ['æni,səul]
丁香酚 eugenol ['ju:dʒinɔl]
芥子油 mustard oil ['mʌstəd]
大蒜油 garlic oil ['gɑ:lik]
杏仁油 almond oil ['ɑ:mənd]

1 As the use of essential oils has declined in evidence-based medicine, one must consult older textbooks for much information on their use. Modern works are less inclined to generalize; rather than refer to "essential oils" as a class at all, they prefer to discuss specific compounds, such as methyl salicylate, rather than"oil of wintergreen".

由于精油的使用在循证医学方面下降,需要参照旧教科书取得更多的应用指导,而现代著作倾向于讨论特定化合物,较少概括像"精油"这样的类,比如讨论水杨酸甲酯,而不是"冬青油"。

2 Taxol has been studied thoroughly in clinical trials, in which a group of people taking the drug have side effects documented. In this way, it is possible to see what side effects occur, how often they occur, and how they compare to people not taking this drug.

紫杉醇在临床上已被深入研究,在研究中一组人服用这种药物有副作

用记录。通过这种方式,可以看到会有哪些副作用发生以及副作用出现的频度如何,与不服用此药的人比较有哪些差别。

第七节 三萜及其苷类

Section 7 Triterpenoids and Their Glycosides

三萜皂苷 triterpenoid saponins [trai'tə:pinɔid]['sæpənin]
酯皂苷 ester saponins ['estə]['sæpənin]
次皂苷 prosapogenins [prə,seipə'dʒenin]
羊毛甾醇 lanosterol [læ'nɔstərɔl]
鲨烯环化酶 squalene cyclase ['skweili:n]['sikleis]
四环三萜 tetracyclic triterpenoids [,tetrə'saiklik][trai'tə:pinɔid]
达玛烷 dammarane ['dæmərein]
人参皂苷 ginsenoside ['dʒinsən'ɔ'said]
酸枣仁皂苷 jujuboside ['dʒu:dʒu(:)bə'said]
羊毛脂烷 lanostane ['lænəstein]
黄芪 astragalus [æ'strægələs]
环黄芪醇 cycloastragenol ['saikləuæ'stædʒi:nɔl]
黄芪醇 astragenol [æ'stædʒi:nɔl]
葫芦烷 cucurbitane [kju(:)'kəbitein]
肺结核 tuberculosis [tju,bə:kju'ləusis]
气管炎 tracheitis [,treiki'aitis]
楝烷 meliacane [mi:'liəkein]
大戟烷 euphane ['ju:fein]
四降三萜 tetranortriterpenoids ['tetrənɔ:trai'tə:pinɔid]
五环三萜 pentacyclic triterpenoid [,pentə'saiklik][trai'tə:pinɔid]
甘草 licorice ['likəris]
柴胡 bupleurum [bju:'pluərum]
乌苏烷 ursane ['ə:sein]
积雪草酸 asiatic acid [,eiʃi'ætik]

The therapeutic potential of ginseng has been studied extensively, and ginsenosides, the active components of ginseng, are shown to be involved in modulating multiple physiological activities.

人参的治疗潜力已被广泛地研究,人参皂苷是人参的有效成分,参与调节多种生理活动。

第八节 甾体及其苷类

Section 8 Steroidal and Glycosides

环戊烷多氢菲 cyclopentano-perhydrophenanthrene [saɪkləupen'tɑ pəhaɪdrəfə'nænθri:n]
强心苷类 cardiac glycosides ['kɑ:diæk]['glaikə,said]
孕甾烷 pregnane ['pregnein]
夹竹桃 oleander [,əuli'ændə]
甾体生物碱 squalamine [skwæ'læmain]
蟾酥毒类 bufotoxins [,bju:fə'tɔksin]
毛地黄毒苷元 digitoxigenin ['didʒi,tɔksi'dʒenin]

利尿 diuretic [ˌdaijuə'retik]
升压 boost [buːst]
止痛 odynolysis [ədɪ'nɔlisis]
镇咳 antitussive ['ænti'tʌsiv]
螺缩酮 spiroketal [spaɪə'rəukiːtæl]
强心甾 cardenolide ['kɑːdinəlid]
海葱甾 scillanolide [si'lənɔlid]
蟾酥甾 bufanolide [bjuː'fɑːnəlid]
加拿大麻苷 cymaroside [sɪmæ'rəusaid]
米替非林 mitiphyllin ['mitifilin]
D-鸡纳糖 D-quinovose ['kwinəvəus]
D-6-去氧阿洛糖 D-6-deoxyallose [diːɒksi'æləus]
D-毛地黄糖 D-digitalose [ˌdidʒi'tæləus]
L-齐墩果糖 L-oleandrose [əɔlnd'rɔz]
L-黄花夹竹桃糖 L-thevetose [ðəviː'təus]

Because cardenolides bind to the sodium pump, they affect multiple signaling pathways and thus have a number of marked effects on tumor cell behavior.
因为强心甾结合钠泵,它们会影响多种信号通路,从而对肿瘤细胞的行为有明显影响。

第九节 生物碱
Section 9 Alkaloids

胡椒碱 piperine ['pipərin]
苯丙胺类 amphetamine-type [æm'fetəˌmiːn]
那可丁 narcotine ['nɑːkətiːn]
阿马林 ajmaline [ədʒmə'lin]
阿替生 atisine ['ætisin]
新士的宁 neostrychnine [niːəust'rɪkniːn]
伪士的宁 pseudostrychnine [ˌsjuːdəu'strikniːn]
哈满尼 harmanine ['hɑːməniːn]
酚氧化偶联 oxidative coupling of phenols ['ɔksideitiv]['kʌpliŋ]['fiːnɔl]
配对反应 pairing reaction ['pɛəriŋ]
吡咯类 pyrrolidines [pi'rəuliˌdiːn]
托品烷类 tropanes [t'rəupeɪn]
吡咯里西丁类 pyrrolizidines [pirəli'zaɪdɪn]
哌啶类 piperidines [pi'peridiːn]
洛贝林 lobeline ['ləubiˌliːn]
石杉碱 huperzine [hjuː'peəzɪn]
一叶萩碱 securinine [si'kjuːriniːn]
喹诺里西丁类 quinolizidines [kwinəli'zaɪdɪn]
羽扇豆碱 lupinine ['luːpəˌniːn]
金雀花碱 cytisine ['sitisiːn]
苦参碱 matrine [meit'riːn]
石松碱 lycopodine [laikə'pəudɪn]
石榴碱 pelletierine [peli'tɪəriːn]
吖啶酮类 acridone ['ækrədəun]
厚朴碱 magnocurarine [mægˌnəukju'rɑːrin]
去甲乌药碱 higenamine [hidʒiː'neimin]
千金藤碱 stephanine [stefei'nain]
小檗碱 berberine ['bəːbəriːn]
长春碱 vinblastine [vin'blɑːstiːn]
金鸡宁 cinchonine ['sinkəniːn]
奎宁碱 quinine [kwi'niːn]
石斛碱 dendrobine ['dendrəubain]

1 Camptothecin (CPT) is a cytotoxic quinoline alkaloid which inhibits the DNA enzyme topoisomerase I (topo I). It was discovered in 1966 in systematic screening of natural products for anticancer drugs.
喜树碱是一种细胞毒性喹啉类生物碱,能抑制DNA拓扑异构酶。1966年它在从天然产物系统筛选抗癌药物中被发现。

2 Medicinal use of barberry dates back more than 2500 years. It has been used in Indian folk medicine to treat diarrhea, reduce fever, improve appetite, relieve upset stomach, and promote vigor as well as a sense of well being.
小檗药用历史可以追溯到2500年前。在印度民间它被用于治疗腹泻,减少发热,增进食欲,缓解胃部不适,促进活力以及舒适感。

第十节 海洋天然药物
Section 10 Marine Natural Medicines

西加毒素 ciguatoxin [ˌsiːgwə'tɔksin]
扇贝毒素 yessotoxin [jesə'tɔksin]
前列腺素 prostaglandin [ˌprɔstə'glændin]
珊瑚 coral ['kɔrəl]

海参皂苷 holothurin [ˌhɔləu'θjuərin] | 海绵毒素 spongistatin [s'pʌndʒisteitin]

1 Several projects run at the university of Nottingham are finding out how to create new, effective anti-cancer therapies inspired by natural products. These include fungal compounds which disrupt cancer-causing pathways, and marine natural products that appear to help the body's immune system reject tumours.

在英国诺丁汉大学进行的几个项目在寻找如何创造新的、有效的利用天然产物的抗癌方法。这些项目包括真菌化合物破坏致癌途径,发现海洋天然产物以帮助人体的免疫系统抑制肿瘤。

2 Many marine products contain tetrodotoxin, and they also have many important medical values; they will be more widely used in 21 century.

许多海洋产品中含有河豚毒素,它们也有许多重要的药用价值,将在 21 世纪获得更多应用。

第十一节 天然药物的研究开发

Section 11 Research and Development of Natural Medicines

亲缘关系 phylogenetic relationship [ˌfailəudʒi'netik][ri'leiʃənʃip]

药理学 pharmacology [ˌfɑːmə'kɔlədʒi]

毒理 toxicology [ˌtɔksi'kɔlədʒi]

新颖性 novelty ['nɔvəlti]

Terrestrial actinomycetes are noteworthy producers of a multitude of antibiotics, however, the marine representatives are much less studied in this regard.

土壤放线菌是多种抗生素值得注意的来源,但是来自海洋的代表在这方面的研究要少得多。

第三十二章　海洋生物制药学
Chapter 32　Marine Biopharmacy

第一节　绪论
Section 1　Introduction

★ 海洋生物制药　marine biopharmacy [mə'ri:n][ˌbaiəu'fɑ:məsi]
海洋药源生物　marine organism as pharmaceutical resource
棘皮动物　echinoderm [i'kainəˌdə:m]
腔肠动物　coelenterate [si'lentəreit]
软体动物　mollusk ['mɔləsk]

During the last 30 to 40 years, numerous novel compounds have been isolated from marine organisms and many of these substances have been demonstrated to possess interesting biological activities.
最近 30 ～ 40 年，从海洋生物中分离出来为数众多的新颖化合物，这些化合物中有很多已经被证明具有我们所感兴趣的生物活性。

第二节　海洋药用生物
Section 2　Marine Medicinal Organism

表 1　海洋药用生物代表品种

海洋水生植物类	绿藻类	*Ulva lactuca* Linnaeus 石莼
	褐藻类	*Laminaria japonica* Aresch 海带
		Sargassum fusiforme (Harv) Setchell 羊栖菜
	蓝藻类	*Calothrix parasitica* (Chanv) 苔垢菜
		Brachytrichia quoyi (C Ag) 海雹菜
	红藻类	*Porphyra haitanesis* T J Chang et B F Zheng 坛紫菜
		Gelidium amansii (Lamx) Lamx 石花菜
海洋动物类	水母类	*Physalia utriculus* La Martiniere 僧帽水母
		Rhopilema esculentum Kishinouye 海蜇
	珊瑚类	*Galaxea aspera* Quelch 粗糙盔形珊瑚
		Antipathes virgata Esper 多小枝黑角珊瑚
	星虫类	*Sipunculus nudus* Linnaeus 光裸星虫
	腹足类	*Haliotis diversicolor* Reeve 杂色鲍
		Notarchus leachii (Cirrosus Stimpson) 蓝斑背肛海兔
	双壳类	*Mytilus edulis* Linnaeus 贻贝
		Pinctada martensi (Dunker) 马氏珠母贝
	头足类	*Sepia esculenta* Hoyle 金乌贼
	甲壳类	*Penaeus orientalis* Kishinouye 东方对虾
		Panulirus stimpsoni Holthuis 中国龙虾
	肢口类	*Tachypleus tridentatus* (Leach) 中国鲎
	海参类	*Stichopus japonicus* Selenka 刺参
	海星类	*Craspidster hesperus* 镶边海星
		Asterias amurensis Lutken 多棘海盘车
	海胆类	*Anthocidaris crassispina* (A. Agassiz) 紫海胆
		Hemicentiotus pulcherrimus (A. Agassiz) 马粪海胆
	鱼类	*Fugu vermicularis* 虫纹东方豚
		Pegasus laternarius Cuvier 海蛾
	爬行类	*Chelonia mydas* (Linnaeus) 海龟
		Pelamis platurus (Linnaeus) 长吻海蛇
	哺乳类	*Physeter catodon* Linnaeus 抹香鲸
		Callorhinus ursinus Linnaeus 海狗

第三节 海洋生物活性物质
Section 3 Marine Bioactive Substance

★ 河豚毒素 tetrodotoxin [ˌtetrəu'tɔksin]
★ 海参毒素 holotoxin [ˌhɔləu'tɔksin]
海葵毒素 palytoxin [ˌpæli'tɔksin]
石房蛤毒素 saxitoxin [ˌsæksi'tɔksin]
西加毒素 ciguatoxin [ˌsigjuə'tɔksin]
赤潮毒素 brevetoxin [ˌbriːv'tɔksin]
膝沟毒素 gonyautoxin
刺尾鱼毒素 maitotoxin
海兔毒素 aplysiatoxin
扇贝毒素 pectenotoxin
冠柳珊瑚毒素 lophotoxin
芋螺毒素 conotoxin
章鱼毒素 cephalotoxin [ˌsefələ'tɔksin]
海兔素 dolastatin
膜海鞘素 didemnin
苔藓虫素 bryostatin

1 Cell growth inhibitory activity in vitro has been demonstrated by a group of cyclic and linear peptides and depsipeptides known collectively as dolastatins. These were isolated from the sea hare *dolabella auricularia*. At the time of publication dolastatin-10 was reported to be the most active antineoplastic substance known.

一些环肽、线性肽和大家都知道的海兔毒素类缩肽在体外具有抑制细胞生长活性已得到证明。海兔毒素是从截尾海兔中分离，海兔毒素 -10 被公布时曾被认为是最具抗肿瘤活性的物质。

2 Anticoagulant activity associated with polysaccharides from marine algae has been known for many years. The earliest report is from 1936 when an anticoagulant effect of a galactan sulphuric acid from*Iridaea laminarioides* was recorded.

海藻多糖有抗凝活性很多年前就为人所知，最早的是 1936 年有关红杉藻中磺化半乳糖酸的抗凝效能的报道。

第四节 海洋生物活性物质研究方法
Section 4 Research Method of Marine Bioactive Substance

有机溶剂提取 organic solvent extraction
水提取 water extraction
沉淀法 deposition [ˌdepə'ziʃən]
透析法 dialysis [dai'ælisis]
重结晶法 recrystallization
薄层层析法 thin layer chromatography
柱层析法 column chromatography
旋光光谱 optical rotatory dispersion
圆二色光谱 circular dichroism spectrum

The experimental work needed in a biopharmaceutical investigation has necessitated a thorough exploitation of many of the newer techniques of instrumental analysis, such as fluorimetry, chromatography and the use of radiotracers.

生物药剂学观测中的实验工作需要充分利用大量新的仪器分析技术，比如荧光法、色谱法和放射性示踪物的使用。

第五节 海洋生物制药的生物技术
Section 5 Biotechnology of Marine Biopharmacy

基因工程 gene engineering [dʒiːn][ˌendʒi'niəriŋ]
细胞工程 cell engineering [sel][ˌendʒi'niəriŋ]
发酵工程 fermentation engineering [ˌfəːmen'teiʃən][ˌendʒi'niəriŋ]
酶工程 enzyme ['enzaim] engineering
赤潮 red tide [red][taid]
蛋白质工程 protein engineering ['prəutiːn][ˌendʒi'niəriŋ]
生物反应器 bioreactor [ˌbaiəuri(ː)'æktə]
核酸探针技术 nucleic acid probe technology ['njuːkliik]['æsid][prəub][tek'nɔlədʒi]
★ 免疫促进剂 immunostimulator [i'mjuː nəu'stimjuleitə]

A relatively new method to investigate the diversity of marine microbes is the use of clone libraries, obtained from natural population by the polymerase chain reaction.
一个相对较新的研究海洋微生物多样性的方法是应用通过聚合酶链式反应从天然菌群中获得克隆文库。

第六节　海洋生物新药的筛选与评价
Section 6　Screening and Evaluation of New Marine Biopharmacy

★ 抗心脑血管疾病　anti-cerebral-cardiovascular-disease
['ænti]['seribrəl][ˌkɑːdiəu'væskjulə] [di'ziːz]
★ 抗艾滋病　anti-AIDS ['ænti][eidz]
★ 临床前研究　preclinical research
['priː'klinikəl][ri'səːtʃ]
老年性痴呆症　senile dementia
['siːnail][di'menʃiə]

1 A series of novel bi-indoles were isolated from the marine cyanobacterium, *Rivularia firma*. One of the major compounds was (+)-7′-methoxy-2,3,5,5′-tetrabromo-3,4′-bi-1H-indole, which is active in both the carrageenan-included rat paw oedema and kaolin rat-pae oedema tests and also in central nervous system tests.
从海洋蓝细菌 *Rivularia firma* 中能分离得到一系列新颖双吲哚结构的化合物，其中主要的一个化合物就是(+)-7′-甲氧基-2,3,5,5′-四溴苯-3,4′-双 1H 吲哚，它在角叉(菜)胶诱导的鼠爪水肿和高岭土鼠爪水肿试验中显示了相关活性，并且在中枢神经系统试验中也显示了活性。

2 A feature of many metabolites of marine organisms is the frequent occurrence of halogens, in particular bromine. Many of these compounds exhibit antimicrobial activity, but are either not sufficiently potent or too toxic, or both, to render them of clinical value.
海洋生物很多代谢物有这样一个特征，即卤素化特别是溴化现象很普遍。很多这样的化合物显示了抗菌活性，但是它们要么没有足够的效力，要么毒性太大，或两者兼而有之，抵消了它们在临床上的价值。

第七节　海洋生物新药的药学评价
Section 7　Pharmaceutical Evaluation of New Marine Biopharmacy

★ 强化试验　stress testing
★ 加速试验　accelerated testing
★ 长期试验　long-term testing
★ 药动学　pharmacokinetics
[ˌfɑːməkəuˌkai'netiks]
★ 消除　elimination [iˌlimi'neiʃən]
★ 处置　disposition [dispə'ziʃən]

1 Pharmacokinetics deals with the absorption, distribution, biotransformation and excretion of drugs.
药动学涉及药物的吸收、分布、生物转化以及排泄等方面。

2 The study of the biochemical and physiological effects of drugs and their mechanisms of action is termed as pharmacodynamics. As a border science, pharmacodynamics borrows freely from both the subject matter and the experimental techniques of physiology, biochemistry, microbiology, immunology, genetics and pathology.
对药物的生理生化作用及其活性机制的研究称为药效学。作为边缘学科，药效学从生理学、生物化学、微生物学、免疫学、遗传学和病理学等诸学科中吸取了大量的内容，不仅是科目实体，还包括实验技术。

第八节　海洋生物新药临床前安全性评价技术
Section 8　Preclinical Safety Evaluation of New Marine Biopharmacy

★ 特殊毒理学　special toxicology
['speʃəl][ˌtɔksi'kɔlədʒi]
★ 毒代动力学　toxicokinetics
[ˌtɔksikəu'dai'netiks]

药物依赖性　drug dependence

★ 积蓄毒性　accumulative toxicity [ə'kju:mjulətiv][tɔk'sisiti]

★ 致突变试验　mutagenicity test [ˌmju:tədʒi'nisiti]

刺激性试验　irritant test ['iritənt][test]

过敏性试验　anaphylaxis test [ˌænəfi'læksis]

光敏试验　photosensitive test [ˌfəutəu'sensitiv]

溶血性试验　hemolytic test [hi:'mɔlətik]

围产期试验　perinatal period test [ˌperi'neitl]['piəriəd][test]

As new drugs may be marketed after studies in only a few hundred or thousand patients, special vigilance is required in the first few years of use to determine low-frequency, but potentially serious adverse effects.

由于新药只是在数百或数千患者身上试用后便被推向市场，所以，在其使用的头几年应予以特别关注，从而确定其是否具有发生频率较低但却可能很严重的副作用。

第三十三章 药用高分子材料

Chapter 33 Medical Polymer Materials

第一节 绪论

Section 1 Introduction

★ 聚合作用 polymerization [ˌpɔlimərai'zeiʃ ən]
★ 单聚物 homopolymer [ˌhəumə'pɔlimə]
★ 共聚物 copolymer [kəu'pɔlimə]
聚电解质 polyelectrolyte [ˌpɔlii'lektrəuˌlait]
支化高分子聚合物 branched polymer
星型高分子 star polymer
梳状高分子 brush polymer
树枝状聚合物 dendrimer ['dendrimər]
数均分子量 number average molecular weight
重量平均分子量 weight average molecular weight
多分散性 polydispersity index [ˌpɔlidis'pə:səti]
持续长度 persistence length [pə'sistəns]
★ 交替共聚物 alternating copolymers
★ 接枝共聚物 graft copolymers
★ 嵌段共聚物 block copolymers
聚合物形貌 polymer morphology [mɔː'fɔlədʒi]
三元共聚体 terpolymers
★ 结晶度 crystallinity [ˌkristə'linəti]
热重分析法 thermogravimetry [ˌ θə:məugrə'vimitri]
乳液聚合 emulsion polymerization

1 A polymer is a large molecule (macromolecule) composed of repeating structural units. These sub-units are typically connected by covalent chemical bonds.
高分子化合物是指由重复结构单元组成的分子量很大的一类化合物。这些结构单元通常通过化学共价键相连。

2 Polymerization is the process of combining many small molecules known as monomers into a covalently bonded chain or network. During the polymerization process, some chemical groups may be lost from each monomer.
将多个小分子也就是单体以共价键的形式组成链状或网状的大分子化合物的过程，我们称之为聚合。在聚合过程中可能会伴随着每个单体上化学官能团的消失。

3 Block copolymers have two or more homopolymer subunits linked by covalent bonds. Polymers with two or three blocks of two distinct chemical species are called diblock copolymers and triblock copolymers, respectively. Polymers with three blocks, each of a different chemical species are termed triblock terpolymers.
嵌段共聚物包含两个或两个以上的以共价键相连的均聚物单元。由两段或三段两种化学性质不同的物种组成的嵌段链组成的聚合物，分别称为二嵌段共聚物和三嵌段共聚物。如果三嵌段共聚物的每个嵌段链是不同的化学物种组成，则称为三嵌段三元共聚物。

第二节 高分子材料的性能

Section 2 Properties of Polymers

★ 玻璃化 vitrification ['vitrifi'keiʃ ən]
★ 高弹态(橡胶态) rubbery state ['rʌbəri]
膨胀测量术 dilatometry [dilei'təumitri]
★ 黏性 viscosity [vi'skɔsəti]
差热扫描量热法 differential scanning calorimetry
热释电流 thermal stimulated current
光学性能 optical property
★ 光学各向异性 optically anisotropic
抗冲击性 impact resistance
★ 生物相容性 biocompatibility [ˌbaiəukəmˌpætə'biliti]
可控缓释 controlled Release

1 A polymer is a large molecule (macromolecule) composed of repeating structural units. These sub-units are typically connected by covalent chemical bonds. Although the term polymer is sometimes taken to refer to plastics, it actually encompasses a large class of compounds comprising both natural and synthetic materials with a wide variety of properties.
高聚物是指包含重复结构基本单元的大分子化合物。这些重复基本单元通常是通过共价键连接在一起的。虽然有时候聚合物是指塑料，但事实上聚合物种类繁多，包括多种具有不同性质的天然化合物和合成化合物。

2 The glass transition temperature (T_g) of a non-crystalline material is the critical temperature at which the material changes its behavior from being "glassy" to being "rubbery". "Glassy" in this context means hard and brittle (and therefore relatively easy to break), while "rubbery" means elastic and flexible.
非晶聚合物的玻璃化转变温度是指材料从玻璃态转变到高弹态时的临界温度。玻璃态是指坚硬和易碎的状态。高弹态是指有弹力的、柔韧性好的状态。

3 Polymer morphology generally describes the arrangement and microscale ordering of polymer chains in space.
高聚物的形貌一般是指聚合物链的堆积状态以及在微观上的空间排布方式。

4 Optical properties of polymers are important for a wide range of applications, ranging from packaging, where the aesthetics of an underlying product must be maintained, to glazing products in both the automotive and the construction industry.
高聚物的光学性能对这类材料在许多方面的应用都有重要作用，例如在包装业方面的应用，需要保持被包裹物包装的美观度，还用于汽车和建筑行业的玻璃等。

第三节　高分子材料在药物制剂中的应用原理

Section 3 Application Principles of Polymers in Pharmaceutical Preparation

聚合物膜　polymer membrane
药物控释膜　drug controlled release membrane
乙基纤维素　ethylcellulose [eθil'seljə'ləus]
溶胀度　swelling capacity
药物制剂　pharmaceutical preparation
微胶囊　microencapsulation
聚合物胶束　polymeric micelles
微凝胶　microgel
智能水凝胶　smart hydrogel
给药系统　drug delivery system
纳米粒子　nanoparticle

1 Polymers are becoming increasingly important in the field of drug delivery. The pharmaceutical applications of polymers range from their use as binders in tablets to viscosity and flow controlling agents in liquids, suspensions and emulsions. Polymers can be used as film coatings to disguise the unpleasant taste of a drug, to enhance drug stability and to modify drug release characteristics.
高聚物在药物传递领域发挥着越来越重要的作用。聚合物的药用应用范围从药片的包衣，到液体、分散液以及乳液的黏度和流速的控制剂。聚合物还可以作为膜包衣来消除药物的难闻气味，提高药物的稳定性，改善药物缓释性能。

2 Hydrogel is a network of polymer chains that are hydrophilic, sometimes found as a colloidal gel in which water is the dispersion medium. Hydrogels are highly absorbent natural or synthetic polymers. Hydrogels also possess a degree of flexibility very similar to natural tissue, due to their significant water content.
水凝胶是一种亲水性的网状聚合物，有时在水中以胶态凝胶的形式存在。天然或合成的水凝胶吸附能力很强。由于高水含量，同天然组织一样，水凝胶同样具有一定程度的柔韧性。

3 Drug delivery is the method or process of administering a pharmaceutical compound to achieve a therapeutic effect in humans or animals. Drug delivery technologies modify drug release profile, absorption, distribution and elimination for the benefit of improving product efficacy and safety, as

well as patient convenience and compliance.
药物传递是指通过控制药物从而达到对人或动物治疗效果的过程或方法。药物传递技术改善了药物缓释、吸收、分配和降解过程，从而提高了产品的效力和安全性，从而为病人提供方便，提高病人的配合度。

第四节 天然药物高分子材料及其衍生物
Section 4 Natural-based Polymers for Medical Applications

海藻酸 alginic acid [æl'dʒinik]
均多糖 homosaccharide [ˌhəumə'sækəraid]
杂多糖 heterosaccharide [ˌhetərəu'sækəraid]
羧甲基淀粉钠 sodiumcarboxymethylstarch
粉状纤维素 powered cellulose
胶态纤维素 colloidal cellulose [kə'lɔidəl]
生物黏附 bioadhesion
醋酸纤维素 cellulose acetate
纤维醋法酯 cellulose acetate phthalate
肠溶包衣水分散体 aqueous enteric coating dispersion
醋酸纤维素丁酸酯 cellulose acetate butyrate
白蛋白 albumin ['ʌlbjumin]

1 Polymers from natural sources are particularly useful as biomaterials and in regenerative medicine, given their similarity to the extracellular matrix and other polymers in the human body.
由于天然聚合物具有和细胞外基质以及人体中的其他聚合物类似的性质，天然高分子材料是非常有用的生物材料，在可再生医药中也非常重要。

2 Chitin may be employed for affinity purification of recombinant protein. A chitin binding domain is genetically fused to a protein of interest and then contacted to bead coated with chitin. The immobilized protein is purified and released from the beads by cleaving off the chitin binding domain.
甲壳素可能用于亲和纯化重组蛋白。首先几丁质的结合区域融合一个感兴趣的蛋白质分子，然后和一个涂上了甲壳素的固定床作用。目的蛋白被固载在固定床上，然后通过断开几丁质的结合区域，来纯化和释放目的蛋白。

3 Albumin is playing an increasing role as a drug carrier in the clinical setting. Principally, three drug delivery technologies can be distinguished: coupling of low-molecular weight drugs to exogenous or endogenous albumin, conjugation with bioactive proteins and encapsulation of drugs into albumin nanoparticles.
白蛋白作为药物载体在临床方面日益发挥重要作用。其主要应用在三个药物输送技术中：在外生或内源性白蛋白上耦合低分子量的药物；在白蛋白纳米粒子上耦合生物活性蛋白质；在白蛋白纳米粒子内封装药物。

第五节 药用高分子合成材料
Section 5 Synthetic Polymers for Medical Applications

聚丙烯酸 polyacrylic acid
交联聚丙烯酸钠 cross-linked sodium polyacrylate
卡波沫(羧基乙基共聚物) carbomer
甲基丙烯酸酯共聚物 polymethacrylate copolymer
丙烯酸树脂 acrylic acid resin
最低成膜温度 minimum film-forming temperature
聚甲基丙烯酸羟乙酯 poly (2-hydroxyethyl methacrylate)
聚乙烯醇 polyvinyl alcohol [ˌpɔli'vainil]
聚醋酸乙烯 polyvinyl acetate
交联聚乙烯吡咯烷酮 cross-linked polyvinylpyrrilidone
乙烯-醋酸乙烯共聚物 ethylene-vinylacetate copolymer
乙烯-乙烯醇共聚物 ethylene-vinylalcohol copolymer
聚乳酸 polyactide [ˌpɔli'læktaid]
配位-插入聚合 coordination-insertion polymerization
聚酸酐类 polyanhydride
聚山梨醇/吐温 polysorbate/ Tween

二甲基硅油 dimethicone [ˈdaiməˈθikəun]
硅橡胶 silicone rubber
氟苯布洛芬 Flurbiprofen
聚谷氨酸 polyglutamic acid
索奥佛林 Ofloxacin
氯苯布洛芬 chlorobenzene ibuprofen

1 Tissue engineering has long held promise for building new organs to replace damaged livers, blood vessels and other body parts.
组织工程学长期以来期望能通过构建新的器官来取代损坏的肝脏、血管以及其他人体器官。

2 Many opportunities exist for the application of synthetic biodegradable polymers in the biomedical area particularly in the fields of tissue engineering and controlled drug delivery.
人工合成的、生物可降解聚合物在生物医学领域有许多应用，特别是在组织工程和可控药物释放领域。

3 In tissue engineering, biodegradable polymers can be designed similar to tissues, providing a polymer scaffold that can withstand mechanical stresses, provide a suitable surface for cell attachment and growth, and degrade at a rate that allows the load to be transferred to the new tissue. In the field of controlled drug delivery, biodegradable polymers offer tremendous potential either as a drug delivery system alone or in conjunction to functioning as a medical device.
在组织工程方面，生物可降解聚合物可以设计成组织的相似结构，提供一种可以承受机械应力的聚合物支架，从而提供一个适合表面细胞黏附和生长的平台，并且通过控制其降解速度而允许负载转移到新组织。在可控药物释放方面，不管是单独作为药物输送系统还是结合其他功能作为医疗设备，生物可降解聚合物都具有巨大潜力。

第六节 高分子药物
Section 6 Polymeric Drug

肝素 heparin [ˈhepərin]
胃蛋白酶 pepsin
L-天门冬酰胺酶 L-asparaginase [ˈæspəˈrædʒineis]
激肽释放酶 kallikrein [ˌkæliˌkriːin]
弹性蛋白酶 elastase [iˈlæsteis]
促皮质素 adrenocorticotropic hormone [əˈdriːnəuˌkəːtikəuˈtrɔpik]
胸腺肽 thymus peptide
降钙素 calcitonin [ˌkælsiˈtəunin]
水蛭素 hirudin [hiˈruːdən]
绒膜促性激素 human chorionic gonadotrophin
人丙种球蛋白 γ-immunoglobulin [iˈmjuːnəuˈglɔbjulin]
胃膜素 gastric mucin
生物体前体药物 bioprecursor prodrug
载体前药 carrier prodrug
高分子前药 macromolecular prodrug
轭合物 conjugate [ˈkɔndʒəˌgeit]
大分子前药轭合物 polymer prodrug conjugate
寻靶器 homing device
间隔臂 spacer

1 A major challenge in cancer chemotherapy is the selective delivery of small molecule anti cancer agents to tumor cells. Water-soluble polymer-drug conjugates exhibit good water solubility, increased half-life, and potent anti tumor effects. By localizing the drug at the desired site of action, macromolecular therapeutics have improved efficacy and enhanced safety at lower doses.
癌症化疗的一个重大挑战是如何选择性将抗癌小分子运输到肿瘤细胞。水溶性聚合物-药物复合物具有良好的水溶性，更长的半衰期和强有力的抗肿瘤作用。通过将药物放置在特定作用部位、大分子药物在低剂量时已经表现出更好的治疗效果。

2 Prodrug is a new drug delivery system based on polymer compounds. Recent progresses in the studies of macromolecular prodrugs are summarized on types of main chain of macromolecular. The design and synthesis of macromolecular prodrugs, choice of polymer carrier, active drug and linker are introduced.
大分子前药是以大分子化合物为载体的一种新的药物传输体系。按大分子主链分类，介绍了近年来大分子前药的研究进展。对大分子前药的设计与合成、载体、活性药物和连接基的选择进行了论述。

3 Brufen, a nonsteroidal anti-inflammatory drug, was linked with 2-hydroxyethyl methacrylate (HEMA) through covalent linkage; the drug-linked HEMA(monomer) was homopolymerized or copolymerized to obtain polymeric drug.
非甾体消炎药布洛芬以共价键连接到含双键的甲基丙烯酸 2- 羟基乙酯上，制成含布洛芬药物的单体，进而通过自聚或共聚，合成了含布洛芬的高分子药物。

第七节 药品包装与贮运材料

Section 7 Drug Packaging and Transport Materials

塑料制品 plastic ['pla:stik]
聚酰胺/尼龙 polyamide [pɔli'æmaid]
聚碳酸酯 polycarbonate [pɔli'ka:bəneit]
塑料盒 carton ['ka:tən]
泡罩包装 blister ['blistə] packs
条形包装 strip packaging
显窃包装 tamper-evident packaging

1 In recent years, as a result of rapid development of drug that can be infused, demand for the packing material is on the increase and the quality requirement is becoming higher and higher. Packaging for pharmaceutical products can be either internal (integral) or external. External packaging surrounds the pharmaceutical product, such as a blister pack, while integral packaging is part of the actual drug delivery, for example, tablet coating.
近年来，随着输液药品的快速发展，对其包装材料的需求量越来越大，对药品包装材料质量的要求也越来越高。包装对制药产品可以是内部(整体)的或外部的。外包装就是包裹着制药产品，如泡罩包装，而整体包装是实际药物输送的一部分，如片剂包衣。

2 These two categories of packaging can be described as either active or passive. Active materials respond to external factors (e. g. pH triggers for drug release) while passive materials just have a pre-defined inherent functionality. One of the key challenges is to develop new materials that will meet industry's need for increasing functionality and decreasing production costs, while at the same time rising to the challenge of increasing environmental and health legislation for packaging and products.
这两类包装可以被描述为主动的或被动的。活跃材料能够对外部因素做出响应(如 pH 值变化引起药物释放)，而被动材料只是有一个预定义的固有功能。一个关键的挑战是如何开发能满足工业需要，具有更多功能(高附加值)，并能降低生产成本的新材料。更进一步的挑战是如何让包装和产品更加环保，符合卫生立法的要求。

3 The technology becoming available means that external packaging can be much more than a passive plastic container. Active polymers applied as thin films can be a cost-effective method for adding functionality to the packaging. Two examples are: a) Polymers that can be used as sensors for biological, chemical or environmental effects and thus help to predict shelf life or warn of product contamination or degradation; b) Polymers with selective permeability, for example, that let water in and carbon dioxide out to ensure the survival of the required "live" micro-organisms. Looking to the future, cheap and flexible carbon-based electronics will allow the functionality of electronic circuits and displays to be incorporated into packaging, whether this is for promotional purposes or security and tracking.
这项技术可应用性的实现意味着外部包装不仅仅是一个被动的塑料容器。将活性聚合物做成薄膜是用来增加包装功能的一个经济有效的方法。下面是两个例子：a) 聚合物可以用作生物、化学和环境传感器，从而有助于预测产品的保质期或提示产品的污染和降解；b) 具有选择性渗透能力的高分子材料，例如可以选择性允许水进入而让二氧化碳排出去，可以确保所需活性微生物的存活。将来不仅为了宣传，更为了产品的安全和跟踪，廉价的柔韧性好的、以碳为基础的电子器件的发展将促使电子电路和显示功能被引入包装中。

4 Packing chemicals and drugs into polymeric materials can allow the transport and delivery of

substances through severe environments to specific sites. Several functional systems for drug delivery have been explored at the Polymer Centre: hydrogels, dendrimers, hyper-branched polymers and vesicles (microscale polymer bags). Additional functionality can then be incorporated by using responsive polymers which can be triggered by a change in pH, pressure, temperature or light to release the active components that they carry.

将化学品和药物包裹在高分子材料中，可以将它们通过恶劣环境运输到特定位点。聚合物研究中心已经开发了几个用于药物输送的系统：水凝胶、树枝状分子、高支化聚合物和囊泡(微型聚合物包)。可以使用响应性聚合物来增加运输材料的额外功能，例如可以通过 pH 值、压力、温度、光等的变化来引发材料释放出所携带的药物。

第三十四章　动物解剖学
Chapter 34　Animal Anatomy

第一节　绪论
Section 1　Introduction

解剖学　anatomy
上　superior
下　inferior
前　anterior
腹侧　ventralis
背侧　posterior
内侧　medialis
外侧　lateralis
内　internus
外　externus
浅　superficies
深　profundus
腋前线　anterior axillary line
前正中线　anterior median line
腋中线　midaxillary line
锁骨中线　midclavicular line
胸骨旁线　parasternal line
腋后线　posterior axillary line
后正中线　posterior median line
肩胛线　scapular line
内脏学　splanchonlogy
胸骨线　sternal line
内脏　viscera [ˈvɪsərə]

1 The term anatomy, with invention of microscope for observing structure, has spread into encompassing those more than that can be seen with naked eyes.
术语解剖学，随着观察显微结构的显微镜的发明，已涵盖了更多肉眼不能看到的。

2 Gross anatomy itself can be studied regionally, where a definite region of the body can be classified into head, neck, back, thorax, abdomen, pelvis, perineum and the upper and lower limbs.
大体解剖本身可以分区域研究，身体的不同区域可分为：头、颈、背、胸、腹部、骨盆、会阴和上肢、下肢。

3 Organs of human body can be divided into 10 systems based on their functional variations. They are locomotor system, digestive system, respiratory system, urinary system, reproductive system, cardiovascular system, lymphatic system, sensory organs, nervous system and endocrine system.
人类身体的器官根据其功能的差异可以分为 10 个系统，分别是运动系统、消化系统、呼吸系统、泌尿系统、生殖系统、心血管系统、淋巴系统、感觉器官、神经系统和内分泌系统。

第二节　运动系统
Section 2　Locomotor System

长骨　os longum
短骨　os breve
扁骨　os planum
不规则骨　os compositum
软骨　cartilage
骨膜　periosteum
骨密质　substantia compacta
骨松质　substantia spongiosa
骨髓　medulla ossium
关节　articulatio
关节囊　capsula articularis
韧带　ligamentum
脊骨　columna vertebralis
颈椎　vertebra cervicales
胸椎　vertebra thoracalis
腰椎　vertebra lumbalis
骶骨　sacrum
尾骨　os coccygis
髋臼　acetabulum
椎间盘　discus intervertebralis
胸廓　thorax
肋骨　ossa costale
脑颅　neurocranium
面颅　viscerocranium

眶 orbit
额窦 frontal sinus
上颌窦 maxillary sinus
筛窦 ethmoidal sinus
锁骨 clavicular
肩胛骨 scapula
肱骨 humerus
桡骨 radius
尺骨 ulna
肩关节 articulation humeri
肘关节 articulation cubiti
腕骨 carpus
掌骨 metacarpal bone
指骨/趾骨 phalanges
髂骨 ilium
耻骨联合 pubic symphysis
股骨 femur
髋关节 articulatio coxae
膝关节 pelvis
骨骼肌 skeletal muscle
筋膜 fascia
滑液囊 bursa synovialis
腱鞘 vagina tendinea
半月板 meniscus
斜方肌 trapezius
背阔肌 latissimus dorsi
竖脊肌 erector spinae muscles
膈 diaphragm
腹直肌鞘 sheath of rectus abdominis
肱二头肌 biceps brachii

1 The skeletal system of an adult consists of approximately 206 bones arranged to form a strong, flexible body framework. The skeleton is divided into axial skeleton and appendicular skeleton portions.
成年人有大约 206 块骨的骨骼系统排列组成一个强壮、灵活的身体框架，分为中轴骨和四肢骨。

2 Bone tissue is composed of several types of bone cells embedded in the matrix of ground substance, inorganic salt(calcium and phosphorus), and collagenous fibers.
骨组织是由包埋在基本物质、无机盐(磷和钙)和胶原纤维基质中的几种类型细胞构成的。

3 More than 600 skeletal muscle make up the muscular system, and technically each one is an organ—it is composed of skeletal muscle tissue, connective tissue, and nervous tissue.
肌肉系统由 600 多骨骼肌组成，从技术上讲每一块肌肉都是一个器官 —— 它由骨骼肌肉组织、结缔组织和神经组织组成。

第三节 消化系统
Section 3 Digestive System

消化系统 digestive system
咽峡 isthmus of fauces
人中 philtrum
腭 palate
咽 pharynx
牙 dentes
舌 lingua
口腔 oral cavity
固有口腔 cavum oris proprium
口腔前庭 vestibule oris
口咽 oropharynx
咽喉 laryngopharynx
味蕾 taste bud
集合淋巴滤泡 aggregated lymphatic follicles
扁桃体 tonsil
食管 oesophagus
胃 stomach
贲门 cardia
幽门 pylorus
小肠 intestinum tenue
十二指肠 duodenum
空肠 jejunum
回肠 ileum
大肠 intestinum crassum
盲肠 caecum
阑尾 vermiform appendix
结肠 colon
升结肠 ascending colon
横结肠 transverse colon
降结肠 descending colon
乙状结肠 colon sigmoideum
直肠 rectum
肛门 anus
消化腺 alimentary gland

腮腺 parotid gland
下颌下腺 submandibular gland
舌下腺 sublingual gland
肝 hepar
肝门 porta hepatis
胆囊 gall bladder
胆囊管 ductus cysticus
胰 pancreas
腹膜 peritonaeum

The alimentary system (or digestive system) consists of the alimentary canal and the alimentary glands. The alimentary canal includes the mouth, pharynx, esophagus, stomach, small intestine (subdivided into the duodenum, jejunum and ileum) and large intestine (including the cecum, appendix, colon, rectum and anal canal). The alimentary glands are subdivided into the large alimentary glands and small alimentary glands. The large alimentary glands are located mainly outside the alimentary canal, forming a secretory duct connecting the alimentary canal. The small alimentary glands are usually located in the mucosa, submucosa of the alimentary canal. They contain lip glands, cheek glands, lingual glands and others.

消化道系统(或消化系统)包括消化道和消化腺。消化道包括口、咽、食道、胃、小肠(分为十二指肠、空肠、回肠)和大肠(包括盲肠、阑尾、结肠、直肠、肛门)。消化腺分为大消化腺和小消化腺。大消化腺主要位于消化道外,形成分泌导管与消化道连接。小消化腺通常位于黏膜、消化道的黏膜下层,包含唇腺、颊腺、舌腺等。

第四节 循环系统
Section 4 Circulating System

心血管系统 cardiovascular system
总血液循环 circulus sanguinis totus
动脉 artery
毛细血管 capillary
血窦 sinusoid
静脉 vein
静脉瓣 venous valve
心脏冠状沟 coronary sulcus
心底 basis cordis
心尖 cardiac apex
右心房 right atrium
右心室 ventriculus dexter
左心房 atrium sinistrum
左心室 ventriculus sinister
心肌膜 myocardium
冠状动脉 coronary artery
心包 pericardium
肺动脉 pulmonary artery
主动脉 aorta
腹主动脉 aorta abdominals
主动脉弓 arch aorta
胸主动脉 thoracic aorta
肺静脉 pulmonary vein
上腔静脉 vena cava superior
下腔静脉 vena cava inferior
主动脉窦 aortic sinus
主动脉瓣 aortic valve
主动脉前庭 aortic vestibule
动脉韧带 arterial ligament
升主动脉 ascending aorta
房室结 atrioventricular node
房室隔 atrioventricular septum
颈动脉窦 carotid sinus
肝总动脉 common hepatic artery
动脉圆锥 conus arteriosus
冠状窦 coronary sinus
房间隔 interatrial septum
室间隔 interventricular septum
左房室瓣 left atrioventricular valve
二尖瓣 mitral valve
节制索 moderator band
肺动脉干 pulmonary trunk
肺动脉瓣 pulmonary valve
右房室瓣 right atrioventricular valve
右纤维三角 right fibrous trigone
窦房结 sinuatrial node
三尖瓣 tricuspid valve
三尖瓣复合体 tricuspid valve complex
脐动脉 umbilical artery
椎动脉 vertebral artery
奇静脉 azygos vein
半奇静脉 hemiazygos vein

肝门静脉 hepatic portal vein
肝静脉 hepatic vein
肠系膜下静脉 inferior mesenteric vein
下腔静脉 inferior vena cava
髂内静脉 internal iliac vein
颈内静脉 internal jugular vein
椎内静脉丛 internal vertebral plexus

1 The right atrium receives systemic venous blood from the superior vena cava, which drains the upper portion of the body, and from the inferior vena cava, which drains the lower portion.
右心房接收全身静脉血液，身体的上部分来自上腔静脉，而身体的下部来自下腔静脉。

2 Blood from the right atrium passes through the right atrioventricular (AV) valve to fill the right ventricle. The cavity of the ventricle is divided into inflow tract and outflow tract by the supraventricular crest, which is situated between the pulmonary and the atrioventricular orifices.
来自右心房的血液通过右房室(AV)瓣来充满右心室。心室腔由室上嵴分为流入道及流出道，位于肺部和房室口之间。

3 After gas exchange has occurred within the capillaries of the lungs, oxygenated blood is transported to the left atrium through two right and two left pulmonary veins.
气体在肺毛细血管内发生交换后，氧合血通过两个右腔和两个左腔肺静脉运到左心房。

4 The left ventricle receives blood from the left atrium. These two chambers are separated by the left atrioventricular valve, which attaches to the mitral annulus at the left atrioventricular orifice.
左心室接收来自左心房的血液。左房室瓣把心室分隔两个内庭，属于在左房室口二尖瓣环。

第五节 淋巴系统
Section 5 Lymphatic System

淋巴系统 lymphatic system
毛细淋巴管 lymphatic capillary
淋巴管 lymphatic vessel
淋巴干 lymphatic trunk
淋巴导管 ductus lymphaticus
淋巴结 lymphonodus
腰淋巴结 lymph nodes
乳糜池 cisterna chili
脾 spleen
脾门 splenic hilum
脾切迹 splenic notch
胸导管 thoracic duct
胸腺 thymus
甲状腺淋巴结 thyroid lymph node
胸导管 thoracic duct
右淋巴导管 right lymphatic duct
脾切迹 splenic notch

Lymph from various parts of the body ultimately enters either the thoracic duct or the right lymphatic duct. The thoracic duct commences in the abdomen, usually as a dilated sac, the cisterna chili. The cisterna chili is about 5-8 mm long and lies in front of the body of the first lumbar vertebra, to the right of the abdominal aorta.
身体各个部位的淋巴液最终进入胸导管或右淋巴导管。胸导管开始于腹部，通常作为膨大囊，即乳糜池。乳糜池长约 5 ～ 8 毫米，而在第一腰椎的前面，到腹主动脉的右侧。

第六节 呼吸系统
Section 6 Respiratory System

外鼻 external nose
鼻腔 nasal cavity
鼻前庭 vestibularis nasi
固有鼻腔 cavum nasi proprium
嗅部 olfactory region
鼻旁窦 sinus paranasales
喉 larynx
喉腔 laryngeal cavity
喉肌 laryngeal muscle
喉结 laryngeal prominence
甲状软骨 thyroid cartilage
环状软骨 cricoid cartilage
会厌软骨 epiglottic cartilage
杓状软骨 arytaenoidea cartilago

喉腔 cavum laryngis
前庭襞 vestibular fold
前庭裂 rima vestibule
声襞 vocal fold
声带 vocal cord
声门裂 rima glottidis
声门 glottis
气管 trachea
支气管 bronchi
肺 pulmo
肺门 hilus pulmonis
肺根 root of lung
呼吸性细支气管 respiratory bronchiole
肺泡管 alveolar ducts
肺泡囊 alveolar sac
肺泡 pulmonary alveoli
肺泡膈 alveolar septa
胸膜 pleura
胸膜腔 pleural cavity
支气管动脉 bronchial artery
会厌 epiglottis
筛泡 ethmoidal bulb
筛骨迷路 ethmoidal labyrinth
左主支气管 left principal bronchus
左肺小舌 lingual of left lung
肺叶支气管 lobar bronchi
肺动脉 pulmonary artery
肺韧带 pulmonary ligament
肺小叶 pulmonary lobule
肺段 pulmonary segment
右主支气管 right principal bronchus
甲状软骨 thyroid cartilage

1 The respiratory system can be divided structurally into upper and lower divisions, and functionally into a conducting division and a respiratory division. The upper respiratory tract consists of the nose, pharynx and larynx, and the lower respiratory tract consists of trachea, bronchi and lung.
呼吸系统从结构上分为上呼吸道和下呼吸道,从功能上分为传导和呼吸。上呼吸道包括鼻、咽、喉,下呼吸道包括气管、支气管、肺。

2 The shape of left lung is different from that of the right one. The former is long and narrow, but the latter is short and wide. Each lung has one apex, one base, two surfaces and three borders.
左右肺的形状是不同的,左肺是窄而长,右肺宽而短。左右肺都有一个“尖”、一个“底”、两个“面”和三个“缘”。

第七节 泌尿系统
Section 7 Urinary System

肾 kidney
肾纤维膜 tunica fibrosa renis
肾脂肪囊 renal adipose capsule
肾筋膜 renal fascia
肾门 renal hilum
肾窦 sinus renalis
肾蒂 renal pedicle
肾柱 renal column
肾锥体 renal pyramid
肾乳头 renal papillae
肾小盏 calyces renales minores
肾大盏 calyx calyces majores
肾盂 renal pelvis
泌尿小管 uriniferous tubule
肾小体 renal corpuscle
肾皮质 renal cortex
肾髓质 renal medulla
肾小管 renal tubulus
输尿管 ureter
膀胱 vesica urinaria
膀胱三角 trigonum vesicae
尿道 urethra
膀胱 urinary bladder
盆腔 pelvic cavity
肾前筋膜 pretrnal fascia
膀胱前隙 prevesical space
肾素 rennin
尿道旁腺 skeins gland
上狭窄 superior stricture
肾上腺 suprarenal gland
膀胱襞 vesical plica
膀胱垂 vesical uvula

1 The urinary system consists of two kidneys,two ureters, the urinary bladder, and the urethra. The urinary system maintains the composition and property of the body fluid that establishes the internal

environment of the body cells.
泌尿系统由两个肾、两个输尿管、膀胱、尿道组成。泌尿系统维护体液的组成和特性,确保了身体细胞的内环境。

2 The urinary bladder is a musculomembranous sac which acts as a reservoir for the urine. Its shape, size, situation and the thickness of wall varies according to amount of fluid it contain. The normal volume of adult's urinary bladder is 350 to 500 ml, with the largest volume of 800 ml. The urinary bladder's volume of infant is about 1/10 of adult's, and the urinary bladder of female is smaller than that of male, while the urinary bladder's volume of the old people become for low tension.
膀胱是贮存尿液的肌性囊,其形状、大小、位置和壁的厚度均随尿液充盈程度而变化。成人膀胱的正常容量是350至500毫升,最大容量可达800毫升;婴儿的膀胱容量大约为成年人的1/10;女性膀胱的容量比男性的小;老年人由于肌紧张力降低,则容量增大。

第八节 神经系统
Section 8 Nerve System

神经元 neuron
突触 synapse
轴突 axon
树突 dendrite
单极神经元 unipolar neuron
双极神经元 bipolar neuron
假单极神经元 pseudounipolar neuron
多极神经元 multipolar neuron
反射弧 reflex arc
施万细胞 Schwann cell
感觉神经元 sensory neurone
运动神经元 motor neuron
中间神经元 internuncial neuron
星形胶质细胞 astrocyte
少突胶质细胞 oligodendrocyte
小胶质细胞 microglia
神经递质 neurotransmitter
中枢神经系统 central nervous system
脊髓圆锥 conus medullaris
终丝 filum terminale
前正中裂 anterior median fissure
后正中沟 posterior median sulcus
前角 anterior hom
灰质 Substantia grisea
白质 Substantia alba
脑干 brainstem
迷走神经三角 vagal triangle
延髓 medulla oblongata
脑桥 pons
中脑 mesencephalon
第四脑室 fourth ventricle
间脑 diencephalon
上丘脑 epithalamus
下丘脑 hypothalamus
松果体 pineal body
小脑 cerebellum
蚓部 vermis
小脑半球 hemisphaerium cerebelli
前叶 anterior lobe
后叶 posterior lobe
大脑 cerebrum
额叶 frontal lobe
枕叶 occipital lobe
岛叶 insula
顶叶 parietal lobe
颞叶 temporal lobe
基底核 basal nuclei
尾状核 caudate nucleus
豆状核 lentiform nucleus
屏状核 claustrum
胼胝体 corpus callosum
穹窿 fornix
侧脑室 lateral ventricle
大脑皮质 cerebral cortex
边缘系统 limbic system
周围神经 peripheral nervous
内脏神经 visceral nerve
杏仁体 amygdaloid body
听觉区 auditory area
大脑纵裂 cerebral longitudinal fissure
大脑横裂 cerebral transverse fissure
味觉区 gustatory area
海马 hippocampus
垂体 hypophysis

漏斗 infundibulum
边缘叶 limbic lobe
乳头体 mammilary body
运动性语言中枢 motor speech area
嗅觉区 olfactory area
视交叉 optic chiasma
投射纤维 projection fibers
端脑 telencephalon
平衡觉区 vestibular area
视觉区 visual area
视觉性语言中枢 visual speech area
书写中枢 writing area
脊髓 spinal cord
原小脑 archicerebellum
基底沟 basilar sulcus
脑干 brain stem
延髓脑桥沟 bulbopontine sulcus
小脑皮质 cerebellar cortex
小脑核 cerebellar nuclei
蜗神经核 cochlear nucleus
锥体交叉 decussation of pyramid
面神经丘 facial colliculus
绒球 flocculus
前脑 forebrain
下丘 inferior colliculus
正中沟 median sulcus
新小脑 neocerebellum
旧小脑 paleocerebellum
黑质 substantia nigra
界沟 sulcus limitans
上丘 superior colliculus
斜方体 trapezoid body
三叉丘系 trigeminal lemniscus
小脑蚓 vermis [ˈvɜːmis]
前庭区 vestibular area
前庭神经核 vestibular nucleus
前庭小脑 vestibulo cerebellum
血-脑屏障 blood-brain barrier(BBB)
交通支 communicating branches
白交通支 white communicating branches
听辐射 acoustic radiation
肾上腺能通路 adrenergic pathway
胺能通路 aminergic pathway
听觉传导通路 auditory pathway
丘脑中央辐射 central radiation of thalamus
胆碱能通路 cholinergic pathway
传导通路 conductive pathway
多巴胺能通路 dopaminergic pathway
平衡觉传导通路 equilibrium pathway
锥体束 pyramidal tract
视觉传导通路 visual pathway

1 The human nervous system is a control system that regulates and coordinates all function of the organ. Besides, it is also the site of all mental activity, including consciousness, memory, and thinking.
人的神经系统是一个控制系统，调节和协调所有的器官功能。此外，它是一切心理活动的网站，包括意识、记忆和思维。

2 The cavity between the pons, medulla and cerebellum is called the fourth ventricle, which is connected with the third ventricle by the mesencephalic aqueduct and extends into the central canal of the medulla and spinal cord downwards.
脑桥、延髓和小脑之间的空腔被称为第四脑室，通过中脑水管与第三脑室连接，延伸到延脑中央管，向下到脊髓。

第九节 感觉器
Section 9 Sensory Organs

外感受器 exteroceptor
内感受器 interoceptor [ˌintərəˈseptə]
感觉器 sensory organs
视器 visual organ
睫状体 ciliary body
结膜 conjunctiva
角膜 cornea
虹膜 iris
泪器 lacrimal apparatus
泪小管 lacrimal ductile
泪腺 lacrimal gland
泪湖 lacrimal lacus
泪乳头 lacrimal papilla
泪囊 lacrimal sac

晶状体 lens
黄斑 macula lutea
鼻泪管 nasolacrimal duct
眼睑 palpebrae
睑结膜 palpebral conjunctiva
眶骨膜 periorbita
瞳孔 pupil
视网膜 retina
房水 aqueous humor
玻璃体 vitreous body
听器 auditory apparatus
听小骨 auditory ossicles
咽鼓管 auditory tube
耳廓 auricle
耳垂 auricular lobule
骨迷路 bony labyrinth
蜗管 cochlear duct
外耳 external ear
砧骨 incus
内耳 internal ear
膜迷路 membranous labyrinth
中耳 middle ear
嗅器 olfactory organ
岬 promontory
球囊 saccule
膜半规管 semicircular ducts
螺旋器 spiral organ
前庭器 vestibular apparatus
前庭蜗器 vestibulocochlear organ
鼓膜 tympanic membrane
鼓室 tympanic cavity
耳蜗 cochlea
半规管 semicircular canal
表皮 epidermis
基底层 stratum basale
角化层 stratum corneum
真皮 dermis
皮下组织 hypodermis
汗腺 sweat gland
皮脂腺 sebaceous

1 The optic nerve is about 5 cm in length, uniting the optic chiasma and the eyeball. This nerve is slightly longer than the distance it travels; thus it permits free movement of the eyeball.
视神经大约5厘米长，与视交叉神经和眼球相连接。此神经稍大于它运动的距离，因此允许眼球自由转动。

2 The vestibule is connected anteriorly to the bony cochlea, posteriorly to the semicircular canals, and communicates with the posterior cranial fossa through the aqueduct of the vestibule. This bony canal, containing the endolymphatic duct, extends to the posterior surface of the petrous part of the temporal bone, where it opens posterolaterally to the internal acoustic meatus.
前庭与前面的骨性耳蜗相连，其后到半规管，并与颅后窝透过前厅渡槽进行通信。此骨性管含淋巴导管，延伸到其后的颞骨岩部位置，其后外侧通到内耳道。

第十节 生殖系统
Section 10 Reproductive System

睾丸 testis
附睾 epididymis
前列腺 prostate
阴囊 scrotum
前列腺部 prostatic part
膜部 membranacea part
海绵体部 cavernous part
输精管 ductus deferens
卵巢 ovary
输乳管 lactiferous duct
子宫 uterus
子宫体 uterus body
前庭球 bulb of vestibule
乳房 mamma, breast
子宫颈 neck of uterus
子宫口 orifice uterus
卵巢伞 ovarian fimbria
子宫附件 uterine appendages
输卵管 uterine tube

The ovaries are the paired primary sex organs of the female that produce gametes, or ova, and the sex hormones, estrogens and progesterone. The ovaries of a sexually mature female are solid, ovoid structures about 4 cm long, 2 cm wide, and 1 cm thick.

卵巢是女性成对主性器官，能产生配子，或者卵、性激素、雌激素和孕激素的。性成熟女性的卵巢是固体、卵形结构约 4 厘米长、2 厘米宽、1 厘米厚。

第十一节　内分泌系统
Section 10　Endocrine System

内分泌腺　endocrine gland
垂体　hypophysis
胰岛　pancreatic islets
肾上腺　suprarenal gland
胸腺　thymus
甲状腺　thyroid gland
内分泌系统　endocrine system
甲状腺素　thyroxine
降钙素　calcitonin
甲状旁腺素　parathyroid hormone
盐皮质激素　mineralocorticoid
糖皮质激素　glucocorticoid
肾上腺素　adrenaline
去甲肾上腺素　noradrenalin
垂体　hypophysis
生长激素　somatotropin
催乳激素　lactotropic hormone
卵泡刺激素　follicle-stimulating hormone
黄体生成素　luteinizing hormone
促甲状腺激素　thyroid stimulating hormone

The thyroid gland weight 20. 7 g in adult male and 25. 3 g in female. The H-shape gland is located in the anterior neck, just below the larynx. Its right and left lateral lobes, each about 5 cm long, are positioned on either lateral side of the inferior larynx and the superior trachea and connected anteriorly by a narrow portion called the isthmus of thyroid gland.

成年男性的甲状腺重量 20.7 克，而女性的为 25.3 克。H 形腺位于颈前，喉的下方。左右侧叶都约为 5 厘米长，贴于喉下方和气管上方的侧面，与称为甲状腺腺体的峡部相连。

第二篇　实验课程部分

第一章　无机与分析化学实验
Chapter 1　Inorganic & Analytical Chemistry Experiment

实验项目：水中钙、镁含量的测定(水的硬度的测定)

实验目的

1. 掌握用配位滴定法测定水的硬度的原理。
2. 掌握铬黑 T、钙指示剂的使用条件和终点变化。
3. 掌握水的硬度的测定方法。
4. 了解水的硬度的表示方法。

实验原理

天然水流过岩石和土壤时溶解了许多无机盐，含有大量 Ca^{2+}、Mg^{2+}、Fe^{2+} 和 Fe^{3+} 的水被称为硬水。其中，Ca^{2+} 和 Mg^{2+} 是硬水中最常见的离子，水的硬度可用水样中钙镁离子的总浓度表示，离子浓度高的水叫硬水，离子浓度低的水叫软水。

硬水常常给生活和工业生产带来不便。例如，Ca^{2+} 和 Mg^{2+} 离子与肥皂反应生成不溶于水的盐，其结果使肥皂失去清洁功效，同时，不溶的 Ca^{2+} 和 Mg^{2+} 的脂肪酸盐形成水垢后会沾在容器、织物或衣服上，显得更脏。当加热热水器、茶壶或锅炉里含有较大量的 $Ca(HCO_3)_2$ 及 $Mg(HCO_3)_2$ 的硬水时，会产生不溶于水的 $CaCO_3$ 及 $MgCO_3$，形成"锅垢"沉淀在管道和容器内壁，造成安全隐患。但硬水对人体没有害处，钙镁是人体必需的营养元素，饮用硬水可以提供人体每天所需的微量钙镁元素的一部分。

锅炉用水或制备蒸馏水的用水都需要测定其硬度，配位滴定法是最常用的测定方法。

用 EDTA 测定 Ca^{2+} 和 Mg^{2+} 时，通常在两个等分溶液中分别测定 Ca^{2+} 量及 Ca^{2+} 和 Mg^{2+} 的总量，Mg^{2+} 量则从两者所用 EDTA 量的差数求出。

在测定 Ca^{2+} 离子时，先用 NaOH 调节溶液到 pH＝12～13，使 Mg^{2+} 离子生成难溶的 $Mg(OH)_2$ 沉淀。加入钙指示剂与 Ca^{2+} 离子配位呈红色。滴定时，EDTA 先与游离的 Ca^{2+} 离子配位，然后夺取已和钙指示剂配位的 Ca^{2+} 离子，使溶液由红色变成蓝色为终点。从 EDTA 标准溶液用量可计算 Ca^{2+} 离子的含量。

测定 Ca^{2+} 和 Mg^{2+} 离子总量时，在 pH＝10 的 NH_3-NH_4Cl 缓冲溶液中，以铬黑 T 为指示剂，用 EDTA 滴定。因稳定性 $CaY^{2-} > MgY^{2-} > MgIn^{-} > CaIn^{-}$，铬黑 T 先与部分 Ca^{2+} 和 Mg^{2+} 离子配位形成 $CaIn^{-}$ ($MgIn^{-}$)(酒红色)。而当 EDTA 滴入时，EDTA 首先与游离的 Ca^{2+} 和 Mg^{2+} 离子配位，然后再夺取 $CaIn^{-}$ ($MgIn^{-}$) 中的 Ca^{2+} 和 Mg^{2+} 离子，使铬黑 T 游离，因此到达终点时，溶液由酒红色变为蓝色。从 EDTA 标准溶液的用量，即可计算出样品中的钙镁总量，然后换算成相应的硬度单位。反应式如下：

滴定前：$Ca^{2+}(Mg^{2+}) + HIn^{2-}$(蓝色)$= CaIn^{-}(MgIn^{-})$(酒红色)$+ H^{+}$

终点前：$Ca^{2+}(Mg^{2+}) + H_2Y^{2-} = CaY^{2-}(MgY^{2-})$(无色)$+ 2H^{+}$

终点时：$CaIn^{-}(MgIn^{-})$(酒红色)$+ H_2Y^{2-} = CaY^{2-}(MgY^{2-}) + HIn^{2-}$(蓝色)$+ H^{+}$

表示硬度的方法通常有以下两种形式：

1. 将测得的 Ca^{2+}、Mg^{2+} 折算成 $CaCO_3$ 的质量，以每升水中含有 $CaCO_3$ 的毫克数表示硬度。现在我国《生活饮用水卫生标准》GB5749-85 规定城乡生活饮用水总硬度(以碳酸钙计)不得超过 450 mg/L。

2. 将测得的 Ca^{2+}、Mg^{2+} 折算成 CaO 的质量(德国硬度)，以每升水中含 10 mg CaO 为 1 度(1°)表示水的硬度。

仪器和试剂

仪器：酸式滴定管，25 mL；锥形瓶，250 mL；移液管，25 mL；量筒，100 mL；洗瓶。

试剂：EDTA 标准溶液，0.01 mol/L；6mol/L NaOH；NH_3-NH_4Cl 缓冲溶液，pH≈10；钙指示剂；铬黑 T 指示剂；水样。

实验步骤

1. Ca^{2+} 离子的测定

用移液管准确吸取水样 50.00 mL，于 250 mL 锥形瓶中，加入 50 mL 去离子水，2 mL 6 mol/L NaOH(此时溶液的 pH=12～13)，8 滴钙指示剂。用 EDTA 标准溶液(0.01mol/L)滴定，溶液由酒红色转变为蓝色，即为终点。记下所用 EDTA 的体积 V_1。平行实验三次。做空白实验进行校正。

2. Ca^{2+}、Mg^{2+} 总量的测定

用移液管准确吸取水样 50.00 mL，于 250 mL 锥形瓶中，加入 50 mL 去离子水，5 mL NH_3-NH_4Cl 缓冲溶液，3 滴铬黑 T 指示剂，用 EDTA 标准溶液(0.01 mol/L)滴定，溶液由酒红色转变为纯蓝色，即为终点。记下所用 EDTA 的体积 V_2。平行实验三次。做空白实验进行校正。

按下式分别计算 Ca^{2+}、Mg^{2+} 含量。

$$Ca^{2+}\text{含量}=\frac{CV_1\times M(\text{Ca})}{50}\times 1000$$

$$CaCO_3\text{ 含量}=\frac{CV_2\times M(\text{CaCO}_3)}{50}\times 1000$$

$$\text{CaO 含量}(1^\circ)=\frac{CV_2\times M(\text{CaO})}{50\times 10}\times 1000$$

$$Mg^{2+}\text{含量}=\frac{C(V_2-V_1)\times M(\text{Mg})}{50}\times 1000$$

注意事项

1. 应注意水样采集的时间、方式、所用容器等。

2. 为了消除实验所用去离子水及指示剂可能带来的误差，要做空白实验，计算水样硬度时应扣除滴定空白时所消耗的 EDTA 的体积。

3. 水的硬度有两种：暂时硬度和永久硬度。如果水中含有 HCO_3^-，水硬度由碳酸氢钙或碳酸氢镁引起，这种水属于暂时性硬水。暂时性硬度可以通过加热水至沸腾除去 CO_2，产生碳酸盐沉淀而解除其硬度。反应如下：

$$Ca(HCO_3)_2\text{ 或 }Mg(HCO_3)_2(aq)\xrightarrow{\Delta}CaCO_3(MgCO_3)(s)+H_2O+CO_2(g)$$

永久硬度则是由水中含有钙、镁的硫酸盐或氯化物引起的，这种硬度不能通过使水沸腾的方式除去，系永久性硬水。总硬度就是暂时硬度和永久硬度之和。

思考题

1. 如果只有铬黑 T 指示剂，能否测定 Ca^{2+} 含量？

2. Ca^{2+}、Mg^{2+} 与 EDTA 的配合物，哪个稳定？为什么滴定 Mg^{2+} 时要控制 pH=10，而 Ca^{2+} 则需控制 pH=12～13？

3. 测定的水样中若含有少量 Fe^{3+}、Cu^{2+} 离子时，对终点会有什么影响？如何消除其影响？

4. 如果一个学生做了空白实验，但计算时忘记扣除空白，那么其计算出的硬度值是高了，还是低了？为什么？

5. EDTA 和其他类似的螯合剂常常用于洗涤产品中。简要说明含有 EDTA 的洗涤产品是怎样去除容器中的硬水肥皂浮垢的。肥皂粉中存在的 EDTA 能否阻止硬水中的肥皂浮垢生成?

Experimental projects: Determination of calcium and magnesium content in water (Determination of water hardness)

Purpose

1. To master the principle of determination of water hardness by complexometric titration.
2. To master the conditions of use and endpoint change of eriochrome black T and calcium indicator.
3. To master the method of determination of water hardness.
4. To understand the way of expressing water hardness.

Principle

Natural water can dissolve many inorganic salts when it flows through rocks and soil. Water containing significant amounts of calcium ions (Ca^{2+}), magnesium ions (Mg^{2+}), ferrous ions(Fe^{2+}), or ferric ions (Fe^{3+})is known as hard water. Ca^{2+} and Mg^{2+} ions are the most common sources of hardness in water. Water hardness is an expression for the sum of the calcium and magnesium cation concentration in a water sample. High levels of these ions produce what is called hard water, otherwise is called soft water.

Hardness in water causes problems at home as well as in industry. For example, Ca^{2+} and Mg^{2+} ions react with soap to form water-insoluble salts, as a result, the soap loses some of its cleaning power, and the insoluble Ca^{2+} and Mg^{2+} salts of fatty acids form a scum that sticks to sinks, bathtubs and fabrics to make them even more dirty. When hard water containing relatively large amounts of $Ca(HCO_3)_2$ and $Mg(HCO_3)_2$ is heated, insoluble $CaCO_3$ and $MgCO_3$ precipitates as "boiler scale" inside pipes and vessels such as hot water heaters, teakettles, and commercial boilers. This may results in a safety hazard. However, hard water is not a health hazard. Calcium and magnesium are essential nutrient elements. Drinking hard water contributes a small amount of calcium and magnesium toward the total human dietary needs of calcium and magnesium.

So, water hardness should be determined when water is heated in boilers or is used for preparation of distilled water. The complexometric titration is the common method of determination of water hardness.

Determination of Ca^{2+} and Mg^{2+} content by using EDTA as titrant, in the usual case, determination of Ca^{2+} and the total amount of Ca^{2+} and Mg^{2+} in two equal parts solution, respectively. Mg^{2+} ions content can be obtained from the difference between the total amount of Ca^{2+} and Mg^{2+} and the amount of the Ca^{2+}.

In the determination of Ca^{2+} ions, first, adjusted the solution to $pH=12\sim13$ with NaOH, in the meanwhile, Mg^{2+} ions is formed insoluble $Mg(OH)_2$ precipitation. Then, calcium indicator and Ca^{2+} ions can form a red complex when add the calcium indicator into above solution. Titration by with EDTA, EDTA and free Ca^{2+} ions form complexes at first, and then snatch the Ca^{2+} ions in $CaIn^-$ complex. Thus, the solution changes color from red to blue as the endpoint. At last, the content of Ca^{2+} ions can be calculated from the amount of EDTA standard solution.

Adjust the solution to $pH=10$ with NH_3-NH_4Cl buffer solution, determination of the total concentration of Ca^{2+} and Mg^{2+} ions with EDTA using Eriochrome Black T. As stability is $CaY^{2-} > MgY^{2-} > MgIn^- > CaIn^-$, Eriochrome Black T and part of the Ca^{2+} and Mg^{2+} ions form $CaIn^-$ ($MgIn^-$) complex (wine red). To use EDTA titration, EDTA and free Ca^{2+} and Mg^{2+} ions form complexes at first, then, snatch the Ca^{2+} and Mg^{2+} ions in $CaIn^-$ ($MgIn^-$)complex to release the Eriochrome Black T free, the solution changes color from wine red to blue as the endpoint. The total concentration of Ca^{2+}

and Mg^{2+} ions can be calculated from the amount of EDTA standard solution. At last, the total concentration of Ca^{2+} and Mg^{2+} ions can be translated into the corresponding hardness units. The reactions are shown as follows:

Before titration: $Ca^{2+}(Mg^{2+}) + HIn^{2-}$(blue) $= CaIn^{-}(MgIn^{-})$(wine red) $+ H^{+}$

Before endpoint: $Ca^{2+}(Mg^{2+}) + H_2Y^{2-} = CaY^{2-}(MgY^{2-})$(colorless) $+ 2H^{+}$

Endpoint: $CaIn^{-}(MgIn^{-})$(wine red) $+ H_2Y^{2-} = CaY^{2-}(MgY^{2-}) + HIn^{2-}$(blue) $+ H^{+}$

Two ways of expressing hardness in common:

1. Measured Ca^{2+} and Mg^{2+} converted into $CaCO_3$ quality, the hardness of water is expressed to the number of milligrams per liter of water contains $CaCO_3$. China's "drinking water health standards" GB5749-85 Provision, urban and rural drinking water hardness (calcium carbonate) should not exceed 450mg/L.

2. The hardness of water is expressed in units of 10mg per liter of an equivalent amount CaO is called 1 degree(1°)(Germany hardness).

Apparatus and Chemicals

Apparatus: Acidic buret, 25mL; Erlenmeyer flask, 250mL; Pipette, 25mL; Graduated cylinder, 100mL; Bottle.

Chemicals: standard EDTA solution, 0.01mol/L; NaOH, 6mol/L; NH_3-NH_4Cl buffer solution, pH≈10; Calcium indicator; Eriochrome Black T; Water sample.

Procedure

1. Determination of Ca^{2+} ions

Add a 50.00mL of the water sample to a erlenmeyer flask(250mL) with a graduated cylinder, add 50mL of deionized water, 2.0mL of NaOH(6mol/L) and 8 drops calcium indicator, titration with standard EDTA solution(0.01mol/L) until the solution color changes from red to blue as the endpoint, and note of the volume V_1 of the EDTA. Repeat the titration three times. Perform a blank determination and make any necessary correction.

2. Determination of the total concentration of Ca^{2+} and Mg^{2+} ions

Add a 50.00mL of the water sample to a erlenmeyer flask(250mL) with a graduated cylinder, add 50mL of deionized water, 5.0mL of the ammonia-ammonium chloride (NH_3-NH_4Cl) buffer solution and 3 drops Eriochrome Black T, titration with standard EDTA solution(0.01mol/L) until the solution color changes from red to pure blue as the endpoint, and note of the volume V_2 of the EDTA. Repeat the titration three times. Perform a blank determination and make any necessary correction.

The contents of Ca^{2+} and Mg^{2+} are calculated by following formulas:

$$\text{The content of } Ca^{2+} = \frac{CV_1 \times M(\text{Ca})}{50} \times 1000$$

$$\text{The content of } CaCO_3 = \frac{CV_2 \times M(CaCO_3)}{50} \times 1000$$

$$\text{The content of CaO}(°) = \frac{CV_2 \times M(\text{CaO})}{50 \times 10} \times 1000$$

$$\text{The content of } Mg^{2+} = \frac{C(V_2 - V_1) \times \text{M(Mg)}}{50} \times 1000$$

Notes

1. Notice the collection time, collection method and collection apparatus of water sample.

2. In order to correct for any error attributable to the deionized water and the indicator color transition, you will be analyzing a blank solution. The volume of EDTA used to titrate the blank will be sub-

tracted from all other titration volumes.

3. There are two types of water hardness: temporary and permanent hardness. Temporary hardness is due to the $Ca(HCO_3)_2$ and $Mg(HCO_3)_2$, HCO_3^- being present in the water. This type of hardness can be removed by boiling the water to expel the CO_2 and precipitate the $CaCO_3$ ($MgCO_3$) as indicated by the following equation:

$$Ca(HCO_3)_2 \text{ or } Mg(HCO_3)_2(aq) \xrightarrow{\Delta} CaCO_3(MgCO_3)(s) + H_2O + CO_2(g)$$

Permanent hardness is due to the presence of the ions Ca^{2+}, Mg^{2+} and SO_4^{2-} or Cl^-, this type of hardness cannot be eliminated by boiling. The water with this type of hardness is said to be permanently hard. Total hardness is the sum of permanent and temporary hardness.

Questions

1. If only Eriochrome Black T, whether the determination of Ca^{2+} content?

2. Which is stable between CaY^{2-} and MgY^{2-} complexes? Why titration of Mg^{2+} to control the pH=10, and Ca^{2+} is required to control the pH=12~13?

3. What impact to the endpoint if water samples containing a small amount of Fe^{3+} and Cu^{2+}? How to eliminate its impact?

4. If a student has a non-zero blank correction, but forgets to include that correction in his/her calculations, calculated hardness value is higher? Or is lower, why?

5. EDTA and other similar chelating agents are frequently used in cleaning products. Briefly explain how a cleaning product containing EDTA can aid in removing water soap scum from a sink. Would the presence of EDTA in a soap powder help prevent soap scum formation in hard water?

第二章　有机化学实验

Chapter 2　Organic Chemistry Experiment

实验项目：乙酸正丁酯的制备

实验目的

1. 掌握浓硫酸催化醇和有机酸制备酯的基本原理；
2. 掌握共沸蒸馏方法及其在有机合成中的应用；
3. 了解分水器的使用方法。

实验原理

乙酸正丁酯可通过浓硫酸催化乙酸和正丁醇一步反应制备。由于反应是可逆的，所以可通过分水器移除反应中生成的水，破坏反应平衡，使反应进行完全。

主要反应：

$$CH_3\overset{\overset{\displaystyle O}{\|}}{C}OH + CH_3CH_2CH_2CH_2OH \xrightarrow{\text{浓硫酸}} CH_3\overset{\overset{\displaystyle O}{\|}}{C}OCH_2CH_2CH_2CH_3 + H_2O$$

主要反应物、试剂和产物的物理常数

名称	分子量	密度 g/mL	比旋光度	熔点℃	沸点℃	闪点℃	溶解度(g/100 g)		
							H_2O	EtOH	Et_2O
正丁醇	74.12	0.8097^{20}	1.3993^{20}	−89.5	117.7	37	7.4	微溶	微溶
乙酸	137.02	1.2686^{25}	1.4374^{25}	−112.4	101.6	18	不溶	可溶	可溶
乙酸正丁酯	116.16	0.8813^{25}	1.3941^{20}	−77	126	126	0.43	微溶	微溶

实验装置

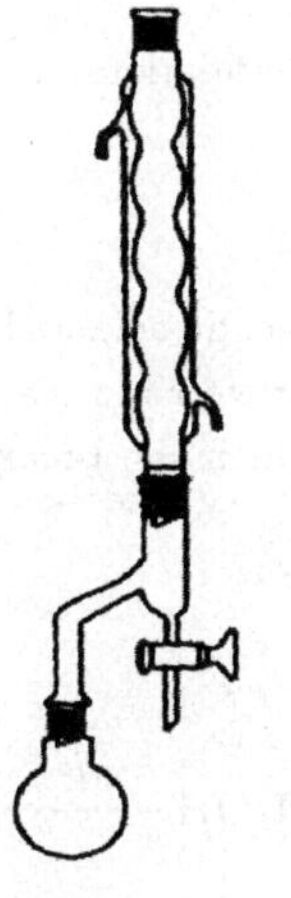

仪器和试剂

仪器：球形冷凝管；圆底烧瓶，250 mL；分水器，10 mL。

试剂：醋酸 7.3 mL；正丁醇 11.5 mL；浓硫酸少许。

反应流程

$$\underset{11.5\ \text{mL}}{n\text{-}C_4H_9OH} + \underset{7.3\ \text{mL}}{CH_3COOH} + \underset{3\sim4\ \text{滴}}{\text{浓硫酸}} \xrightarrow[10\sim15\ \text{分钟}]{\text{中度沸腾}} \xrightarrow{\text{回流，水分}} \xrightarrow{\text{冷却至室温}} \xrightarrow{10\ \text{mL 水洗除酸}}$$

$$\xrightarrow{10\ \text{mL}10\%Na_2CO_3\ \text{洗除酸}} \xrightarrow{10\ \text{mL 水洗除碱}} \xrightarrow{\text{无水硫酸镁干燥}} \xrightarrow[124\sim127℃]{\text{蒸馏}} CH_3COOCH_2CH_2CH_2CH_3$$

注意事项

1. 浓硫酸为反应的催化剂，硫酸用量对反应非常重要。反应中使用一个催化剂量的浓硫酸进行催化，同时摇动反应瓶，以避免反应物碳化；

2. 分水器中应保有适量的水，在 3～5 毫米以下，以保证正丁醇可以回流到反应器中。注意：液面达到支管口时，方可排水；

3. 反应开始后，中度沸腾 10～15 分钟，至冷凝管中无蒸汽溢出后，加热回流；

4. 判断反应是否完成的方法：(a)分水器中有没有水滴下沉；(b)比较水实际生成量与反应理论产量。

思考题

1. 酯化反应有哪些特点？本实验中如何提高产品收率？又如何加快反应速度？

2. 计算反应完全时应分出多少水？

3. 在提纯粗产品的过程中，用碳酸钠溶液洗涤主要除去哪些杂质？若改用氢氧化钠溶液是否可以？为什么？

Experimental project: Preparation of *n*-Butyl Acetate

Purpose

1. Master the principle of preparation of ester from alcohol and organic acid catalyzed by the amount of sulfuric acid;

2. Master the principle of azeotropic distillation and its use in organic synthesis;

3. Learn the use of water segregator.

Principle

n-Butyl acetate can be prepared from acetic acid and n-butyl alcohol by a single step reaction using a catalytic amount of sulfuric acid. This is a reversible reaction. The water generated from the reaction was removed through azeotropic distillation, which can break the reaction equilibrium and make the reaction proceed thoroughly.

Main reaction

$$CH_3\overset{\overset{\displaystyle O}{\|}}{C}OH + CH_3CH_2CH_2CH_2OH \xrightarrow{H^+} CH_3\overset{\overset{\displaystyle O}{\|}}{C}OCH_2CH_2CH_2CH_3 + H_2O$$

Physical constants of the main reactants, reagents and products

Name	Formula weight	Density g/mL	Refractive index	Mp. (℃)	Bp. (℃)	Fp. (℃)	Solubility(g/100 g)		
							H_2O	EtOH	Et_2O
1-Butanol	74.12	0.8097^{20}	1.3993^{20}	−89.5	117.7	37	7.4	misc	misc
Acetic acid	137.02	1.2686^{25}	1.4374^{25}	−112.4	101.6	18	no	vs	vs
n-Butyl acetate	116.16	0.8813^{25}	1.3941^{20}	−77	126	126	0.43	misc	misc

Experimental device

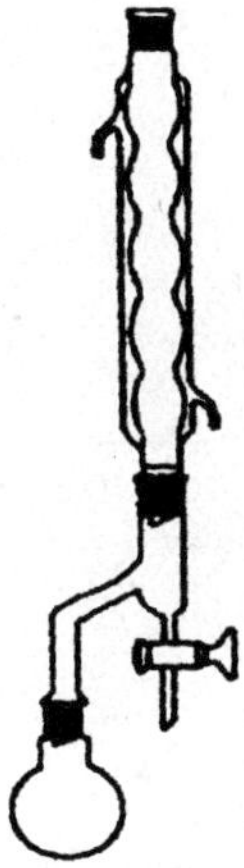

Apparatus and Chemicals

Apparatus: spherical condenser; round-bottomed flask, 250 mL; manifold, 10 mL.
Chemicals: acetic acid 7.3 mL; *n*-butanol 11.5 mL; little concentrated sulfuric acid.

Procedure

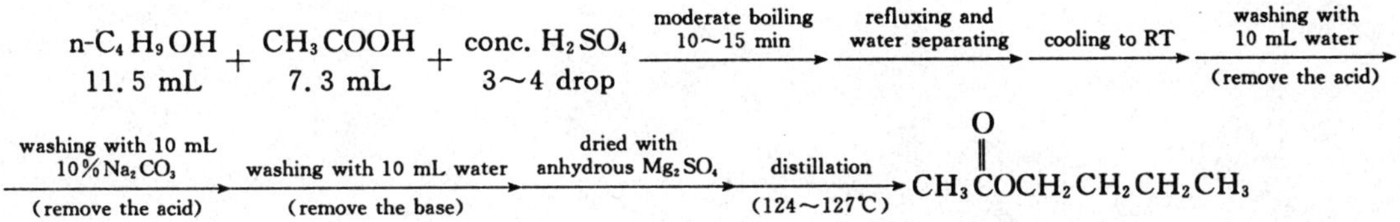

Notes

1. The concentrated sulfuric acid is a catalyst in this reaction, but the amount of the sulfuric acid is very important in the reaction. You must not use too much just a catalytic amount. You must shake the bottle when the concentrated sulfuric acid was added dropwise to avoid the carbonization;

2. The amount of water should be added to water segregator to make liquid level below the branch pipe 3~5 mm. So the n-butanol could return to the reactor. Notice: The water was not discharged until

the liquid level reached the branch pipe;

3. Be sure that no steam stream to condenser tube at the beginning of this reaction. After 10～15 min you could raise the reaction temperature to reflux;

4. There are two ways to judge whether the reaction has finished: (a) There are no more water drops sinking; (b) Compare the amount of the water from the separator with the theoretical amount of the water produced in the reaction.

Questions

1. What are the characteristics of the esterification reaction? How to increase product yield in this experiment? How to accelerate the reaction rate?

2. How much volume of water is the separation determined by calculating the reaction is complete?

3. In the process of purification of the crude product, washed with sodium carbonate solution is mainly to remove impurities? Can switch to sodium hydroxide solution? Why?

第三章　物理化学实验
Chapter 3　Physical Chemistry Experiment

实验项目：燃烧热的测定

实验目的

1. 了解氧弹量热计各主要部件的作用及拆装方法。
2. 测定有机物质燃烧热。
3. 学会应用雷诺图解法校正温度改变值。

实验原理

在燃烧反应中，碳氢化合物和氧气反应生成二氧化碳和水。测量化学反应热的仪器称为量热计，本实验采用氧弹式量热计测量反应燃烧热。氧弹充以氧气并密封保证完全燃烧和阻止生成物外漏。将氧弹放入盛有一定量水的内桶中，内桶整个系统与外层通过空气隔热。样品燃烧是通过电气装置引燃氧弹中的金属丝来完成的。

如果量热计可以看作是一个绝热系统的话，那么体系和周围环境之间没有热交换，即 $Q=0$。很明显，反应放出的热量一定是完全以热的形式传给了水和量热计的其他部分，引起体系温度升高。如果知道量热计的热容和燃烧前后温度的改变，我们就可以计算反应的燃烧热。

$$\frac{m}{M}\cdot Q_v + m_{丝}\cdot q_{丝} = -C\cdot \Delta T \tag{1}$$

式中：m：待测物的质量，g；M：待测物的摩尔质量，$g\cdot mol^{-1}$；Q_v：待测物的恒容燃烧热，$J\cdot mol^{-1}$；$m_{丝}$：燃烧的金属丝的质量，g；$q_{丝}$：燃烧的金属丝的燃烧热，$J\cdot g^{-1}$；C：量热计的热容，$J\cdot K^{-1}$；ΔT：温度的改变值，K。

通常用已知恒容摩尔燃烧热的苯甲酸标准物（$Q_{v,苯甲酸}=-3238.6\ kJ\cdot mol^{-1}$）来标定量热计的热容 C，其燃烧反应方程式如下：

$$C_6H_5COOH(s) + (15/2)\,O_2(g) == 7\,CO_2(g) + 3\,H_2O(l)$$

由于氧弹中的反应是在恒容条件下进行的，恒容燃烧热 Q_v 等于体系内能变化 ΔU，反应过程的焓变为：

$$\Delta H = \Delta U + \Delta(pV) \tag{2}$$

如果忽略固体和液体体积的 $\Delta(pV)$ 项，并将产生的气体按照理想气体处理，则上式变为：

$$Q_p = Q_v + \Delta n_g RT \tag{3}$$

式中　Δn_g——产物与反应物中气体物质的量之差，单位为 mol。

实验中，将氧弹置于内桶的水中，内桶中还装有温度探头和搅拌器。理想的量热计与外界完全隔离，反应前和反应后水的温度应保持不变。可实际上，由于热漏和搅拌会导致温度的改变。设量热计与周围环境的热交换的速率是不变的，那么温度随时间的变化速率是常数。图 1 为量热计反应前、反应期和反应后的温度-时间图。

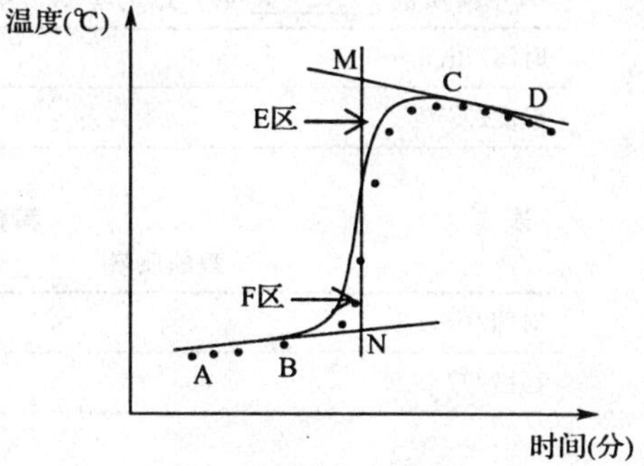

图 1　温度-时间图

量热计内桶的水温通常略低于室温，所以反应前期的温度-时间图上会出现一段斜率略正的线段 AB，反应开始后，由于反应热的放出导致水温明显升高，见线段 BC。反应结束后，水温通常高于室温，在温度-时间图上会出现一段斜率略负的线段CD。延长AB和CD线,然后作一条垂线使E区和F区的面积大致相等,两条延长线和垂线的交点M和N所对应的温度分别为反应前和反应后的温度值。

仪器与试剂

仪器:氧弹式量热计;压片机;氧气钢瓶;温度计;燃烧丝; 天平;剪刀;容量瓶,1000 mL。

试剂:萘(AR);苯甲酸(AR)。

实验步骤

1. 称取0.7 g左右的苯甲酸,用压片机压成片状,精确测定其质量并记录。将片状样品放在干净的燃烧杯中。

2. 将氧弹盖放在弹头支架上,剪取大约10 cm长的燃烧丝,将燃烧丝的两端穿过电极上的小孔分别紧绕在两根电极上。将装有样品的燃烧杯放在两电极间的O形架上。将燃烧丝的中间绕成环状,并将螺旋部分紧贴在样品的表面上。

3. 用手旋紧氧弹盖,用万用表检查两电极间的电阻值,电阻值应和10 cm长的燃烧丝的电阻值相近。

4. 连接氧气钢瓶和氧气表,并用高压铜线管与氧弹进气管连接,在减压阀关闭状态下(松开状态),打开氧气钢瓶的总开关,然后打开减压阀(拧紧),使氧气缓缓进入氧弹内。待减压表指示压力达到1.5 MPa时,关闭总阀门。放出剩余气体后,关闭减压阀。为将氧弹内氮气排出,拧开氧弹放气阀,重复充氧3次。

5. 再次用万用表检查两电极间的电阻值,如变化不大,将氧弹放入内桶底座上并接好引火导线。用容量瓶准确量取2000 ml自来水(温度低于室温1 ℃)注入内桶中,观察氧弹是否漏气。装好搅拌头,盖上量热计盖子,将温度测量探头通过盖上的小孔竖直地插入内桶水中。开动搅拌马达并将量热计与点火控制箱相连。

6. 为获得反应的初始温度,每隔30秒记录一次温度值,精确到0.001 ℃。当升温速率达到恒定时,继续记录5分钟温度值后点火,点火后每隔15秒读一次温度值,直到温度达到最大并开始下降时,然后每隔30秒记录一次温度值。当降温速率达到恒定时,继续记录5分钟后停止。

7. 关闭所有电源开关。取下温度计并打开量热计,拔掉引火导线,取出氧弹。通过放气阀将氧弹内气体缓缓放出。拧开氧弹盖并放于弹头支架上,检查氧弹内部(如氧弹内有积碳或有尚未燃烧的样品,须重新测定)。称量并记录剩余燃烧丝的质量,洗净并擦干量热计所有部件。

8. 称取0.7 g萘并重复以上操作。

实验记录和数据处理

1. 记录实验数据。

表1 **苯甲酸燃烧的数据记录结果**

苯甲酸质量 ________ g; 燃烧丝(I)质量 ________ g;实验温度 ________ ℃;大气压 ________ Pa

时间/time	
温度/℃	

表2 **萘燃烧的数据记录结果**

萘的质量 ________ g;燃烧丝(II)质量 ________ g

时间/time	
温度/℃	

2. 作温度-时间图,确定每次燃烧引起量热计温度的变化值。

3. 利用苯甲酸的实验数据,计算量热计的热容(C)。燃烧丝的燃烧热是$-6680\ J\cdot g^{-1}$。

4. 利用萘的实验数据及量热计的热容计算萘的摩尔燃烧焓(Q_p)。

5. 比较实验值和文献值,计算相对误差。

注意事项

1. 要保证燃烧丝与样品充分接触,不要使燃烧丝接触任何金属表面。

2. 移动量热计盖子时要小心不要损坏温度探头。

3. 当氧弹浸于内桶水浴中时,如果气泡成串冒出,需移去氧弹,重新充氧。

思考题

1. ΔH 和 ΔU 有什么区别,本装置测定的是哪个量?

2. 为什么每次实验都要在内桶中加入相同体积的水?

3. 虽然燃烧过程中由于大量热的放出导致温度达到了较高值,但我们计算热量时却只需要知道始态和终态温度,为什么?

4. 影响本次实验精度的主要因素是什么?

Experimental project: Measurement of heat of combustion

Objective

1. To learn structure of bomb calorimeter and operation.

2. To measure the enthalpy of combustion of an organic substance.

3. To determine the temperature change by Renault graphical method.

Principles

In a combustion reaction, a hydrocarbon sample combines with oxygen to form CO_2 and H_2O_2. In this experiment, the reaction will be carried out using a bomb calorimeter. The bomb is pressurized with oxygen to ensure complete combustion, and is sealed to prevent escape of the combustion products. The bomb is immersed in water in a bucket and the system is separated by an air gap from surroundings. The sample is ignited by passing a current through a fuse wire within the bomb.

The calorimeter can be considered insulated from its surroundings, so there is no heat exchange between the system and the surroundings, i. e., $Q=0$. It is clear that the energy released in the reaction must be transferred entirely as heat to the water bath and the other parts of the calorimeter causing the temperature to rise. Hence, if we know the heat capacity of the calorimeter and the temperature change, we can calculate the heat of the combustion reaction.

$$\frac{m}{M}\cdot Q_v + m_{丝}\cdot q_{丝} = -C\cdot \Delta T \tag{1}$$

Where m: mass of the sample, g; M: molar mass of the sample, $g\cdot mol^{-1}$; Q_v: heat combustion of the sample at constant volume, $J\cdot mol^{-1}$; m_{wire}: mass of the fused wire, g; q_{wire}: heat combustion of the fused wire, $J\cdot g^{-1}$; C: heat capacity of the calorimeter, $J\cdot K^{-1}$; ΔT: the change in temperature, K.

The heat capacity of the calorimeter, C, can be determined using another reaction with known heat of combustion conducted under the same conditions. High purity benzoic acid is often used for such purposes since its enthalpy of combustion at a constant volume has been accurately measured (Q_v=-3238.6kJ $\cdot mol^{-1}$). The reaction is

$$C_6H_5COOH\ (s) + (15/2)\ O_2(g) == 7\ CO_2(g) + 3\ H_2O(l)$$

Because the reaction in the bomb occurs at a constant volume, the heat released, Q_v, is equal to the internal energy change, ΔU. The enthalpy change for the process is:

$$\Delta H = \Delta U + \Delta(pV) \tag{2}$$

Using the perfect gas law and neglecting the $\Delta(pV)$ terms of solids and liquids gives

$$Q_p = Q_v + \Delta n_g RT \tag{3}$$

Where Δn_g: the change in the number of moles of gas in the reaction, mol.

In this experiment, the bomb is immersed in a water bath which contains a high resolution thermometer and a stirrer. In an ideal calorimeter, the temperature of the bath would be constant before and after combustion. In reality, heat leaks and stirrer work will produce some drift in the temperature. If the rate of energy transfer can be assumed to be constant, the rate of change of the temperature will be constant. The temperature of the calorimeter is measured as a function of time before, during and after ignition of the sample. A graph of temperature vs. time called a thermogram is shown in Fig. 1.

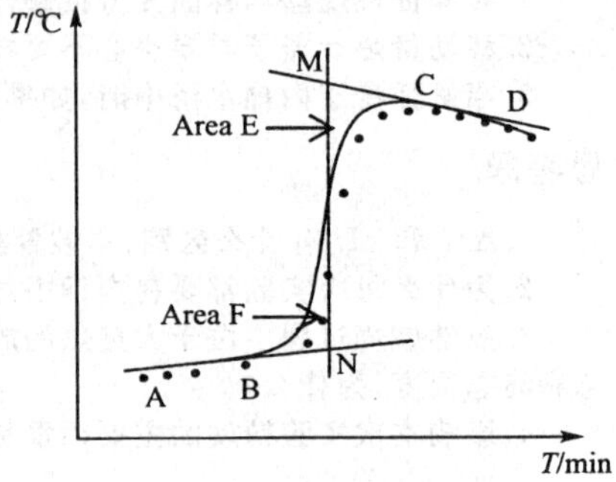

Fig. 1 Thermogram

The calorimeter is assembled with the temperature of the water bath slight below water jacket temperature, and a slight positive slope is observed as line *AB* in Fig. 1. After the reaction has occurred, the heat transferred from the bomb to the water will cause its temperature to rise sharply (line *BC*). After this period of time, this heat exchange will be completed and temperature of the water is usually above water jacket temperature. Then a slight negative slope is observed in the thermogram (line *CD*). The two lines, *AB* and *CD* are extrapolated. A vertical line *MN* is selected which will make the two areas *E* and *F* approximately equal. The temperature change (ΔT) will be determined from the final (*M*) and initial (*N*) temperatures on this line.

Apparatus and Chemicals

Apparatus: Bomb calorimeter; Pellet press; Oxygen cylinder; Thermometer; Iron wire; Balance; Scissors; Volumetric flask, 1000 ml.

Chemicals: Benzoic acid (AR); Naphthalene (AR).

Procedures

1. Weigh about 0.7 g of benzoic acid and form it into a pellet with the press. Weigh the pellet and record the exact mass. Place the pellet in a clean sample cup.

2. Cut and weigh a 10 cm of ignition wire. Place the bomb cover in its holder. Thread the wire through the holes on the electrodes and wrap the ends around the posts. Place the cup with the pellet in the O-ring of the electrode. Form a bend in the middle of the wire and use the point of the bend to make a firm contact of the wire to the pellet.

3. Place the cover with pellet into the body of the bomb and tighten it by hand only. Measure the electrical resistance from the electrode terminals on the lid. It should be about the same as the resistance of 10 cm of fuse wire.

4. Close the vent valve. Connect the gas line to the inlet valve on the bomb. Open the main valve on the cylinder and make sure the quick-release valve is closed. Slowly open the quick-release valve. When the pressure reads between 1.5 MPa, close the main valve. Release the residual pressure and close the quick-release valve. Open the vent valve and repeat the pressurizing step three times to remove all nitrogen from the bomb.

5. Check the electrical resistance again and the read should be about the same as before. Put the bomb on the raised portion of the bucket. Attach the two ignition wires to the bomb. Fill the bucket

with 2000 ml tap water (about 1 ℃ lower than water jacket temperature) using volumetric flask. Check for leaks. Place the calorimeter lid in place and carefully insert the temperature probe through the small hole. Put the drive belt in place and start the stirring motor. Connect the calorimeter to the firing box.

6. To obtain the initial temperature, record temperature in the precision of 0.001 ℃ every 30 seconds. Continue taking temperature readings until a constant rate of temperature change has been observed for 5 minutes. Then fire the bomb and obtain temperature readings every 15 seconds until the temperature seems to be stable. Then continue the readings every 30 seconds until a constant rate of temperature change is observed for 5 minutes.

7. Switch off all electrical circuits and take the calorimeter apart. Lift the bomb out of the bucket while disconnecting the ignition wires. Open the vent valve gently to release the pressure and open the bomb. Inspect the inside of the bomb (if there is a carbon deposit or pieces of unburned sample inside, the run will have to be repeated). Clean off unburned wire and weigh it. Clean and dry all parts of bomb calorimeter.

8. Repeat the above procedures with about 0.7 g of naphthalene.

Records and Analysis of Data

1. Present the results in the following table.

Tab. 1　　Experimental data for the combustion of benzoic acid

Mass of benzoic acid ________　Mass of ignition wire (I) ________

Experimental temperature ________　Atmosphere pressure ________

t/min	
T/℃	

Tab. 2　　Experimental data for the combustion of naphthalene

Mass of naphthalene ________　Mass of ignition wire (I) ________

t/min	
T/℃	

2. Plot temperature vs. time to determine ΔT of each combustion.

3. From the data of benzoic acid, calculate the heat capacity of the calorimeter (C). The heat of combustion for the fuse wire is $-6680\ J \cdot g^{-1}$.

4. Use the data of naphthalene and the heat capacity of the calorimeter to calculate the molar enthalpy of combustion for naphthalene.

5. Compare the measured value with a literature value and calculate the percent error.

Notes

1. Make sure that the fuse wire is in contact with the sample without touching any metal surface.

2. Be careful when moving the calorimeter lid, since the thermometer is very vulnerable.

3. If a stream of bubbles rises from the bomb when the bomb is immersed in water, remove and depressurize it.

Questions

1. What is the difference between ΔH and ΔU? What are we measuring in this apparatus?

2. Why is it important to have the same amount of water in the bucket?

3. Although the temperature reached during the combustion is very high, reflecting the large amount of energy released during the reaction, why do we need to know only the initial and final temperatures in order to calculate that energy change?

4. What are the main factors that limit the accuracy of the results?

第四章　化工原理实验

Chapter 4　Principles of Chemical Engineering Experiment

生词表

文丘里流量计 venturi meter [ven'tuəri // 'mi:tə]	[kəui'fiʃənt]
孔板流量计　orifice meter ['ɔrifis // 'mi:tə]	压力传感器 pressure transducer
转子流量计　rotameter ['rəutə, mi:tə]	水库，蓄水池 reservoir ['rezəvwɑ:]
流量系数　discharge coefficient [dis'tʃɑ: dʒ]	校准 calibrate ['kælibreit]

实验项目：流体流量的测定

实验目的

掌握流量计的标定方法，并对不同的测定方法进行比较。

实验原理

伯努利方程

伯努利方程描述的是流速与静压力的关系，而且是对于一个稳定、不可压缩的一维流动而言的：

$$\frac{u^2}{2g}+\frac{P}{\rho g}+Z=\text{常数} \tag{4.1}$$

式中：u 为流体流速，P 为静压力，Z 为相对于基准线的高度，ρ 为流体的密度，g 为重力加速度。

该方程可以应用到所测静压力下、一维流动的流体两点之间流速变化的测定。该方程表明，如果静压力沿流动方向增加，且相对于基准线的高度保持不变(即：$z=$ 常数)，则速度必然减少。使用伯努利方程重要的是要注意四个限制，流体必须：

1. 稳定
2. 不可压缩流体(密度为常数)
3. 无摩擦力(在流体和管壁间没有摩擦损失)
4. 流体做一维流动

质量守恒定律

$$m=\rho Q=\rho Au=\text{常数} \tag{4.2}$$

式中：m 为质量流量，ρ 为流体密度，Q 为体积流量，A 为流体流动的横截面积，u 为流体流速。

伯努利方程(方程 4.1)与质量守恒方程（方程 4.2)结合，可以测定流速。

实验装置

本装置包括一个孔板流量计，一个文丘里流量计和一个涡轮流量计，实验流程见图 4-1。管路上任意一个标有刻度的转子流量计都可以测量流速。该装置有一个 110L 的蓄水池和一个用来收集流量的 40L 的水槽。共有九个阀门用来控制流量和调节流速。孔板流量计连接一个压力传感器，此压力传感器能直接测量出通过孔板的压差。孔板流量计还连接两个压差管，也能测出通过孔板的压差。剩下的八个压差计连接文丘里流量计，能读出文丘里流量计流量轴上不同位置上的压力读数。关于流速的测量，压力表 1 和 4 显示的压力读数分别代表文丘里流量计入口和喉部的静压力。涡轮流量计则用来直接测量流量。

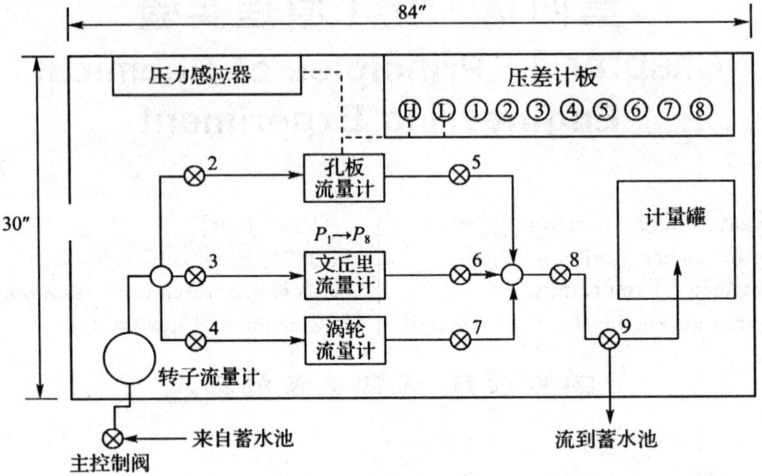

图 4-1 实验装置平面图

A 部分:三种流量测定方法的比较

实验步骤

1. 确定阀门 1 到 8 都是关闭的。
2. 调节阀门 9,使蓄水池(110L 的水槽)内的流体全部流空。
3. 打开阀门 4 和 7。
4. 轻轻地打开阀门 1 和 8,启动主离心泵。
5. 调节阀门 1 和 8,使转子流量计的读数达到 6L/min 左右。
6. 记录压力计和涡轮流量计的读数,填入表格 4-1 中。
7. 确定 40L 的小水槽是空的,按“Tare”键调零。调节阀门 9,使流体重新注入带重量刻度的水槽中。用提供的秒表记录注入该水槽一半体积的时间。在关闭秒表的同时调节阀门 9,使流体重新流入到蓄水池中。记录水槽中流体的重量。开启排水泵抽空小水槽中的水。
8. 重复步骤 4~7,依据上一次转子流量计的读数,再测量另外四组流速的读数,并且使每组值均增加 6L/min。
9. 关闭离心泵。

表 4-1 部分数据记录表(确保记录测量单位)

流量设置	转子流量计	涡轮流量计	重量刻度		
	Q(体积流量) 单位:	Q(体积流量) 单位:	质量 单位:	时间 单位:	Q(体积流量) 单位:
1					
2					
3					
4					

结果与讨论

假定涡轮流量计对流量的测量是正确的,上述“转子流量计”和“流体重量与时间”这两种测流量方法，最大的误差是什么？

B 部分：孔板流量计的校准

本实验所用的孔板流量计如原理部分所示。进口直径 DH 为 1.75 英寸，且孔板直径 D_0 为 0.5 英寸。

根据前面的讨论，可得到方程：

$$Q_{孔板} = KA_O\sqrt{\frac{2(P_H-P_L)}{\rho}} = KA_O\sqrt{\frac{2\Delta P}{\rho}} \tag{4.3}$$

$$Q_{孔板} = KA0\sqrt{2g(hH-hL)} = KA0\sqrt{2g\Delta h} \tag{4.4}$$

实验步骤

1. 确保所有阀门(1～8)都关闭。
2. 调节阀门 9，使水流入储水槽(110L)。
3. 打开阀门 2 和 5。
4. 缓慢打开阀门 1 和 8，开启主泵。
5. 调节阀门 1 和 8，尽可能使压力计(H)和(L)之间产生较大的△h。测量时注意使数据保持在可读范围内。为赶走压力计中的气泡，阀门 8 可能需要部分关闭几秒钟，但要确保没有完全关闭。
6. 记录压力计(H)和(L)的高度、微分压力传感器$\triangle h_{传感器}$值，以及转子流量计的读数(注：数据记录表已在下文提供)。
7. 重复操作 4～6 得到另外四组流量，流量调节间隔要均匀合理。
8. 关闭泵。

表 4-2　　B 部分数据记录表（确保记录测量单位）

流量设定	压力计读数			压力传感器	转子流量计
	上游压力计(H) 单位：	下游压力计(L) 单位：	Δh 单位：	$\Delta h_{传感器}$ 单位：(cm 水柱)	体积流量(Q) 单位：
1					
2					
3					
4					
5					

结果与讨论

1. 绘制 Q—2gΔh 图，确定流量计系数，得出的 Q—2gΔh 图应为线性关系。
2. 评价用压力计获得的 Δh 计算流量计系数的合理性，而不要用微分压力传感器所获得的 $\Delta h_{传感器}$来评价。你认为哪种测量方法合适，是试验中获得的经验还是课堂上获得的信息？(证实你的答案)。

Experimental project: Fluid Flow Measurements

Objective

To investigate the procedure of calibrating fluid flow meters as well, as compare different methods of flow measurement.

Theory

The Bernoulli's Equation

Bernoulli's equation relates the flow velocity to the static pressure, for a steady, incompressible flow along a streamline.

$$\frac{u^2}{2g} + \frac{P}{\rho g} + Z = Const \tag{4.1}$$

where: u — Velocity of the flow

P — Static pressure

Z — Elevation of streamline

ρ — Density of fluid

g — Acceleration due to gravity

This equation can be applied between two points on a streamline to determine the change in velocity given the measured static pressures. Conceptually, this equation indicates that if the static pressure along a streamline increases, while the elevation of the streamline remains constant (i. e. , z = constant), the velocity must then decrease. It is important to note that there are four restrictions for using the Bernoulli's equation, i. e. the flow must be:

1. Steady flow
2. Incompressible flow ($\rho = Const$)
3. Frictionless flow (no friction losses between the fluid and pipe walls)
4. Flow along a streamline.

Conservation of Mass

$$m = \rho Q = \rho Au = Constant \tag{4.2}$$

where, m=mass flow rate, r=fluid density, $Q=Au$=volume flow rate, A=cross-section area of flow passage, u=velocity of the flow.

Bernoulli's equation (Eq. 4. 1) combined with the equation of conservation of mass (eq. 4. 2), allows for the determination of the flow rate.

Experimental Apparatus

The apparatus contains an orifice meter, a Venturi meter, and a turbine meter in a parallel configuration as shown in Figure 4-1. The flow rate can be measured in any configuration using the calibrated Rotameter attached to the supply line. There is a 110L fluid reservoir as well as a 40L tank which is used for collecting fluid. There are a total of nine valves which are used to direct the flow and adjust the flow rate. The Orifice meter is connected to a differential pressure transducer, from which the pressure difference across the orifice plate can be directly determined. The orifice meter is also connected to two manometer tubes which can also be used to determine the pressure difference across the orifice plate. The Venturi meter is connected to the eight remaining manometers, which take pressure readings from different positions along the Venturi meter's flow axis. For flow rate measurements the pressure readings from manometers 1 and 4 represent the static pressure at the Venturi inlet and the Venturi throat, respectively. The turbine meter can be used to take direct flow measurements.

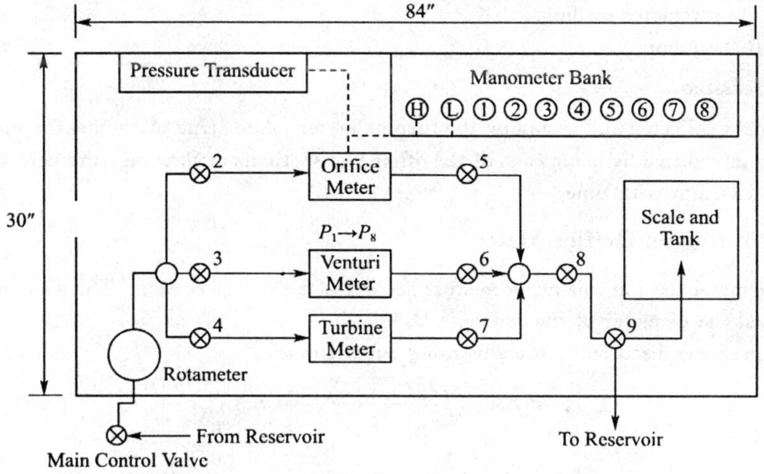

Figure 4-1　Plan view of the experimental apparatus

Part A: Comparison of Three Flow Measurement Methodologies

Procedure

1. Make sure that valves 1 to 8 are closed.
2. Adjust valve 9 so that the run-off is directed to the reservoir (110L tank).
3. Open valves 4 and 7.
4. Open valves 1 and 8 slightly and turn on the main pump.
5. Adjust valves 1 and 8 in order to achieve a rotameter reading of approximately 6 liters per minute.
6. Record the rotameter reading and the reading from the turbine meter in Table 4-1.

Table 4-1　Data collection table for Part A. Make sure to record the your measurement units

Flow Setting	Rotameter	Turbine	Weight Scale		
	Q (Volume flow rate) units:	Q (Volume flow rate) units:	Mass units:	Time units:	Q (Volume flow rate) units:
1					
2					
3					
4					

7. Making certain that the small tank (40L) is empty, zero the scale using the "Tare" button. Adjust valve 9 such that the flow is redirected into the tank on the weight scale. Record the filling time using the stop watch provided (fill the tank to approximately half its total capacity). Adjust valve 9 (while concurrently stopping the stop watch) to redirect the fluid back into the reservoir. Record the weight of the fluid in the tank. Empty the small tank by turning the drain pump on.
8. Repeat steps 5 through 7 for four additional flow rates increasing at increments of approximately

6L/min as per the rotameter reading.

9. Turn off the pump.

Results and Discussion

From the data collected and assuming the turbine meter takes "true" flow measurement, what is the maximum error introduced by using each of the other two methods of flow measurement (using the Rotameter, and fluid weight with time).

Part B: Calibrating an Orifice Meter

The orifice meter used in this experiment is shown in the theory section. The diameter of the inlet is $DH=1.75in$ and the diameter of the orifice is $D_0=0.5in$.

From the previous discussion, the governing equations are:

$$Q_{orifice} = KA_O\sqrt{\frac{2(P_H - P_L)}{\rho}} = KA_O\sqrt{\frac{2\Delta P}{\rho}} \tag{4.3}$$

and

$$Q_{orifice} = KA_O\sqrt{2g(h_H - h_L)} = KA_O\sqrt{2g\Delta h} \tag{4.4}$$

Procedure

1. Make sure that all valves (1 to 8) are closed.
2. Adjust valve 9, so that the run-off is directed to the reservoir.
3. Open valves 2 and 5.
4. Open valves 1 and 8 slightly and turn on the main pump.
5. Adjust valves 1 and 8 in order to achieve a large Δh between manometers (H) and (L) as possible. Be careful to stay within a range that is readable when trying to take measurements. Valve 8 may need to be partially closed for a few seconds in order to force out any air bubbles in the manometers. However, make sure not to fully close valve 8.
6. Record the heights of manometers (H) and (L). Also record Dh_{TRANS} from the differential pressure transducer. Record the rotameter reading (Note: a data collection table has been provided, below, for this purpose).
7. Repeat steps 4 to 6 for four additional flow rates. Adjust the additional flow rates such that they are reasonably evenly spaced.
8. Turn off the pump.

Table 4-2 Data collection table for Part B. Make sure to record your measurement units

Flow Setting	Manometer Readings			Transducer	Rotameter
	Upstream manometer (H) units:	Downstream manometer (L) units:	Δh units:	Δh_{TRANS} units: (cm of H_2O)	Volumetric Flow Rate (Q) units:
1					
2					
3					
4					
5					

Results and Discussion:

1. Plot Q versus $2g\Delta h$. Determine the meter coefficient, given that a plot of $Q-2g\Delta h$ should yield a linear relationship.

2. Comment on the appropriateness of using Δh from the manometers to compute the meter coefficient rather than using Δh_{TRANS} from the differential pressure transducer. Given your experience in the lab and the information taught in class which measurement would you suggest using (justify your answer).

第五章　生物化学实验
Chapter 5　Biochemistry Experiment

实验项目 1：劳瑞氏法测定蛋白质含量

实验目的

1. 学习和掌握常用测定蛋白质的原理和方法。
2. 熟练掌握可见分光光度计的操作。

实验原理

劳瑞氏法又称为 Folin-酚法，测定蛋白质含量的过程包括两步反应：第一步是在碱性条件下蛋白质与 Cu^{2+} 作用生成络合物，第二步是此络合物中的酪氨酸和色氨酸残基还原 Folin 试剂中的磷钼酸和磷钨酸，生成深蓝色的化合物。在一定范围，吸光度与蛋白质的含量成正比。所以，通过分光光度的方法我们能测定蛋白浓度。Folin-酚试剂法操作简便，灵敏度高，样品中蛋白质含量高于 5μg 即可测得，是测定蛋白质含量应用最广泛的方法之一。

实验器材

可见分光光度计、离心机、水浴、试管、小烧杯、漏斗及架、分析天平、吸管、容量瓶、滤纸、研钵、离心管

试剂

(1) 0.5mol/L NaOH

(2) Folin-酚 A 试剂：称取 10 g Na_2CO_3，2 g NaOH 和 0.25 g 酒石酸钾钠，溶解后用蒸馏水定容至 500ml。

(3) Folin-酚 B 试剂：在 1.5L 容积的圆底具塞烧瓶中加入 100g 钨酸钠（$Na_2WO_4 \cdot 2H_2O$）、25g 钼酸钠（$Na_2MO_4 \cdot 2H_2O$）及 700ml 蒸馏水，再加 50ml 的 85％磷酸、100ml 浓盐酸，充分混合，回流 10h。回流结束后，加入 150g 硫酸锂、50ml 蒸馏水和数滴液体溴，开口继续沸腾 15min，驱除过量的溴，冷却后稀释容量瓶定容至 1000ml，过滤，若滤液呈绿色，加入数滴液体溴直至被氧化成淡黄色。滤液置于棕色试剂瓶中暗室保存。使用时适当稀释（约 1 倍），使最终浓度相当于 1mol/L 盐酸。

过程

1. 绘制标准曲线

(1) 标准蛋白溶液：在分析天平上精确称取 25mg 牛血清白蛋白，加入少量蒸馏水溶解后转入 100ml 容量瓶中，最后准确稀释定容至 100ml，使蛋白的浓度为 250μg/ml。

(2)系列标准蛋白溶液：编号 6 支试管，按表 5-1 加入试剂，迅速摇匀混合。

表 5-1

试剂	管号					
	1	2	3	4	5	6
250μg/ml 牛血清白蛋白(ml)	0	0.2	0.4	0.6	0.8	1.0
H_2O(ml)	1.0	0.8	0.6	0.4	0.2	0
蛋白质含量(μg)	0	50	100	150	200	250
Folin-酚 A 试剂	5	5	5	5	5	5

再加 0.5ml Folin-酚 B 试剂，立即混合均匀。30min 后测定吸光值。以编号 0 号试管为对照，记录各试管内溶液 500nm 波长下的吸光度。标准曲线的绘制：以吸光值作 y 轴、标准蛋白浓度作 x 轴绘制标准曲线。

2. 测定样品溶液中的蛋白浓度

(1)称取绿豆芽下胚轴 1g 于研钵中，加蒸馏水 2ml，匀浆。转入离心管。4000r/min 离心 20min。弃去沉淀，上清液转入容量瓶，精确稀释定容至 100ml。

(2)取试管 2 支，吸入样品溶液 1ml，加入 Folin-酚 A 试剂，室温下放置 10min，再加入 Folin-酚 B 试剂，迅速混匀，静置 30min，于 500nm 波长下测定吸光值。

结果

根据未知样品的吸光值，从标准曲线上查出样品溶液中蛋白质的含量 X(μg)，然后计算样品中蛋白质的百分含量。

Experimental project 1: Quantitation of Protein by Lowry Method

Objective

1. Study and master the methods and principles to quantitate protein.
2. Skillfully master the operational method of visible spectrophotometer.

Theory

Lowry methods also called Folin-phenol method. The reaction process to quantitate protein can be divided into two steps. The first step is that the peptide bonds of protein can bind with Cu^{2+} to form protein Cu^{2+} coordination compound in the basic solution. While the second step is that the Tyr or the Trp residue in the coordination compound can deoxidize the phosphotungstic acid and phosphomolybdic acid of the Folin reagent, and then form the dark blue product. In a definite range, the relationship between absorbance and protein concentration is a straight line. So we can determine protein concentration by spectroscopic method. Folin-phenol method is easy to perform and has high sensitivity when the protein in the sample weighs up more than 5μg. It is one of the most widely used methods in the quantitation of protein.

Apparatus

Visible spectrophotometer, Centrifuge, Water bath, Test tubes, Small beaker, Funnel and Test tube shelf, Analytical balance, pipes, 100ml Volumetric flasks, Filter paper, Mortar, Centrifuge tube

Reagents

1. 0.5mol/L NaOH

2. Folin-phenol reagent A: Take 10 g Na_2CO_3, 2 g NaOH and 0.25 g sodium potassium tartrate, add distilled water to 500ml.

3. Folin-phenol reagent B: Add l00g of $Na_2WO_4 \cdot 2H_2O$, 25g of $Na_2MoO_4 \cdot 2H_2O$, 700ml of distilled water, 50ml of 85% H_3PO_4 and 100ml of thick HC1 into a 1.5L round bottom stopper flask, mix well and reflux for 10h. After reflux, add 150g of Li_2SO_4, 50ml of distilled water, several drops of liquid Br_2, boil 15min without cap to get rid of excessive Br_2. After cooling, dilute to a constant volume of 1000ml by volumetric flask, filtrate; add several drops of bromic water to be oxidated to pale yellow while the filtrate is green. Keep the filtrate in brown flask in darkroom. Dilute properly (about one time) before using to give 1 mol/L acid.

Procedures

1. Draw calibration curve

(1) Standard protein solution: Accurately weigh bovine serum albumin 25mg with analytical balance, dissolved in minimum volume of distilled water, transfer to a 100 ml flask, accurately dilute to 100ml, which makes the concentration of protein 250μg/ml.

(2) Series standard protein solution: Number six test tubes, add reagents as the following table, and shake them up right now.

Table 5-1

Reagent	Tube number					
	1	2	3	4	5	6
250μg/ml bovine serum albumin (ml)	0	0.2	0.4	0.6	0.8	1.0
H_2O(ml)	1.0	0.8	0.6	0.4	0.2	0
protein content(μg)	0	50	100	150	200	250
Folin-phenol reagent A (ml)	5	5	5	5	5	5

Add Folin-phenol reagent B, shaking immediately, keep 30 minutes to determine absorbance. Write down the absorbance at 500nm of each test tube. The test tube numbered 0 is contrast. Draw the calibration curve while the absorbance is the y axis, the standard protein concentration is the x axis.

2. Determine protein concentration of sample solution:

(1) Take 1 g mung bean hypocotyl in mortar, add 2 ml distilled water, grinding into homogenate. And then transfer to a centrifuge tube, centrifuged at 4000r/min for 20min. The precipitate is abandoned, and the clear supernatant extract was transferred into a volumetric flask, and then accurately dilute to 100ml.

(2) Take two test tubes, imbibe 1ml sample solution truly, add Folin-phenol reagent A, shake, and keep 10 minutes at room temperature, add 1ml Folin-phenol reagent B, shaking immediately, keep 30 minutes then determine absorbance at 500nm.

Results

According to the absorbance of unknown sample solution, check out the protein content X (μg) in the sample solution from the calibration curve. And then calculate of the percentage of protein in the samples.

实验项目 2:酶的基本性质

实验目的

酶是活细胞内具有催化化学反应的蛋白质。生物体内的酶具有各种功能。为了保证生命活动的高效进行,生物体内的化学反应基本上都是在酶催化下进行的。本实验即为了解酶催化的特异性,掌握测定酶性质的方法与原理以及 pH、温度、抑制剂和激活剂对酶活力的影响。

实验原理

酶的特异性是指酶对它所催化的反应以及对称之为底物的反应物选择的特异性。一种酶只催化一种化学反应或一类相似的反应。催化发生在被称为“活性部位”的特殊部位。酶的特异性是由于底物与酶活性位点精确的相互作用。这种精确性是酶蛋白三维空间构象的结果，我们称之为酶的特异性。

例如，淀粉酶和蔗糖酶虽然都催化糖苷键的水解，但是淀粉酶只对淀粉起作用，蔗糖酶只水解蔗糖。可用本乃狄试剂鉴定其结果。

与大多数化学反应一样，温度对酶活性有显著影响。在一定的温度范围，酶催化速率通常随着温度的升高而升高。反之，下降。只要酶稳定，活性充足，通常温度每升高 10℃，反应速度就加快一倍左右。由于酶是一种蛋白质，具有高温变性，因此，每种酶都有其最适温度，即酶促反应速度达最大值时的温度。如果反应温度超过最适温度，则很多酶活性直线下降。

酶具有保持其最大活性的最适 pH 值范围。高于或低于此 pH 值，酶活力降低，氢离子浓度在几方面影响酶活性。首先，氢离子浓度的变化影响酶分子活性部位的离子化；另外，也影响底物，如果底物含有解离的官能团，pH 值的变化能够改变其与活性中心结合的能力。其次，解离状态的官能团能改变影响酶的空间构象，pH 值的剧烈变化甚至会使酶变性。

许多物质都能影响酶的催化活性，能加速或抑制酶的催化作用。前者称为激活剂，后者称为抑制剂。本实验分别考察 NaCl 和 $CuSO_4$ 对碱性蛋白酶的激活和抑制作用。

本实验测定不同温度对唾液淀粉酶对淀粉作用的影响。通过与碘溶液反应的颜色变化确定唾液淀粉酶催化淀粉水解生成各种不同大小分子糊精及麦芽糖的能力。

实验材料

1. 仪器

试管及试管架；移液管；烧杯；恒温水浴锅；温度计；白瓷板

2. 试剂

(1) 1% 淀粉溶液：将 1g 淀粉溶解于 100ml 蒸馏水中。

(2) 唾液淀粉酶溶液：漱口后收集唾液，用小漏斗加脱脂棉过滤，用蒸馏水稀释 5～20 倍，试管内混匀。

(3) 2%蔗糖溶液：加 2g 蔗糖于 100ml 蒸馏水中。

(4)蔗糖酶溶液：取 1g 干酵母放入研钵中，加入少量石英砂和水研磨至匀浆。加 50ml 蒸馏水，静置片刻，过滤，弃去沉淀，上清液备用。

(5)本乃狄试剂

试剂 A：173g 柠檬酸钠和 100g $Na_2CO_3 \cdot 2H_2O$，溶于 600ml 热水中。

试剂 B：17.3g $CuSO_4 \cdot 5H_2O$ 溶于 100ml 热水中，冷却备用。

将试剂 A 慢慢加到试剂 B 中，边加边搅匀，最后定容至 1000ml。

(6)碘-碘化钾溶液：称取碘 10 克，碘化钾 20 克，同溶于 100ml 蒸馏水中，贮于棕色瓶内。使用前稀释 10 倍。

(7)磷酸缓冲液：

A 液：0.2mol/L Na_2HPO_4：

称取 28.40g Na_2HPO_4（或 71.64g $Na_2HPO_4 \cdot 12H_2O$）溶于蒸馏水中，定容至 1000ml。

B 液：0.1mol/L 柠檬酸

称取 21.01g 柠檬酸（$C_6H_8O_7 \cdot H_2O$）溶于蒸馏水中，定容至 1000ml。

pH5.0 磷酸缓冲液：10.30ml A 液＋9.70ml B 液

pH7.0 磷酸缓冲液：16.47ml A 液＋3.53ml B 液

pH8.0 磷酸缓冲液：19.45ml A 液＋0.55ml B 液

(8)1%$CuSO_4 \cdot 5H_2O$ 溶液。

(9)1%NaCl 溶液。

实验过程

1. 本实验通过唾液淀粉酶和蔗糖酶催化不同底物即淀粉和蔗糖观察酶的特异性。取 6 支试管，编号，按表 5-2 加入各试剂。

表 5-2

操作项目	管号					
	1	2	3	4	5	6
1%淀粉(ml)	1	1	0	0	1	0
2%蔗糖(ml)	0	0	1	1	0	1
唾液淀粉酶原液(ml)	1	0	1	0	0	0
蔗糖酶溶液(ml)	0	1	0	1	0	0
蒸馏水(ml)	0	0	0	0	1	1
酶促水解	摇匀,37℃水浴中保温 10min					
本乃狄试剂(ml)	2	2	2	2	2	2
反应	摇匀,沸水浴中加热 5～10min					
现象						
解释实验现象						

2. 本实验通过不同温度下唾液淀粉酶和蔗糖酶催化不同底物即淀粉和蔗糖观察温度对酶活性的影响。取 3 支试管,编号 1～3,按表 5-3 加入各试剂。

表 5-3

操作项目	管号		
	1	2	3
唾液淀粉酶原液(ml)	1	1	1
pH=7 磷酸缓冲液(ml)	2	2	2
不同温度预处理 5min	0℃	37℃	70℃
1%淀粉溶液(ml)	2	2	2

保持各自温度继续反应数分钟后,每隔 1 到 2 分钟从第 2 号管吸取 1 滴反应液于白瓷板上,加 1 滴碘液至反应液中(检查反应进行情况),直至反应液不再变色,只有碘液的颜色,立即从恒温水浴锅中取出所有试管,流水冷却 2min,各加 1 滴碘液至反应管中,观察反应现象。

3. 本实验通过不同 pH 值下唾液淀粉酶和蔗糖酶催化不同底物即淀粉和蔗糖观察 pH 值对酶活性的影响。取 3 支试管,按表 5-4 加入各试剂。

表 5-4

操作项目	管号		
	1	2	3
pH5.0 磷酸缓冲液(ml)	3	0	0
pH7.0 磷酸缓冲液(ml)	0	3	0
pH8.0 磷酸缓冲液(ml)	0	0	3
1%淀粉溶液(ml)	1	1	1
处理	混匀,37℃处理 2min		
唾液淀粉酶溶液(ml)	1	1	1
过程	保持各自温度继续反应数分钟后,每隔 1 到 2 分钟从第 2 号管吸取 1 滴反应液于白瓷板上,加 1 滴碘液至反应液中(检查反应进行情况),直至反应液不再变色,只有碘液的颜色,立即从恒温水浴锅中取出所有试管,流水冷却 2min。		

（续表）

操作项目	管号		
	1	2	3
碘液(滴)	1	1	1
现象			
现象解释			

4. 酶的抑制与激活

本实验考察 NaCl 和 $CuSO_4$ 对碱性蛋白酶的激活和抑制作用。取 3 支试管，按表 5-5 加入各试剂。

表 5-5

操作项目	管号		
	1	2	3
1% NaCl (ml)	1	0	0
1% $CuSO_4$(ml)	0	1	0
蒸馏水 (ml)	0	0	1
唾液淀粉酶溶液 (ml)	1	1	1
1%淀粉溶液(ml)	3	3	3
处理	混匀，37℃处理 1 分钟		
过程	保持各自温度继续反应数分钟后，每隔 1 到 2 分钟从第 2 号管吸取 1 滴反应液于白瓷板上，加 1 滴碘液至反应液中（检查反应进行情况），直至反应液不再变色，只有碘液的颜色，立即从恒温水浴锅中取出所有试管，流水冷却 2min。		
碘液(滴)	1	1	1
现象			
解释实验现象			

注意事项

1. 不同人的唾液淀粉酶活性并不相同，有时差别很大，因此稀释倍数可因个人情况而定。

2. 少量的激活剂或抑制剂就能影响酶的活性。但激活剂和抑制剂不是绝对的，有些物质在低浓度时为某种酶的激活剂，而在高浓度时则为该酶的抑制剂。对一种酶是激活剂，对另一种酶是抑制剂。

3. 反应时间要准确，器皿必须保持洁净和干燥。

思考题

1. 什么是酶的最适温度、最适 pH 值？有何实际意义？

2. 影响酶的催化活性的因素有哪些？简要说明作用原理。

3. 酶学实验必须注意控制哪些条件？为什么？

Experimental project: Basic Properties of Enzyme

Objective

Enzymes are proteins that catalyze chemical reactions in living cell. Enzymes serve a wide variety of functions inside living organisms. Almost all chemical reactions in a biological cell need enzymes in order to occur at rates sufficient for life. Intensify specificity of properties of enzyme by this test. Master the methods and principles of testing enzyme properties and effect of pH, temperature, inhibitors and activators on enzymatic activity.

Theory

Enzymes are highly specific both in the reactions that they catalyze and in their choice of reactants, which are called substrates. An enzyme usually catalyzes a single chemical reaction or a set of closely related reactions. Catalysis takes place at a particular site on the enzyme called the active site. The specificity of an enzyme is due to the precise interaction of the substrate with the active site of the enzyme. This precision is a result of the intricate three-dimensional structure of the enzyme protein, and we can observe specificities of enzyme.

For example, glycosidic bond can be hydrolyzed under the action of amylase enzymes and sucrose, but the digestion of starch only begins with the action of amylase enzymes, and the digestion of sucrose only begins with the action of sucrase, we can identify the results can be testified by Benedict reagent.

Like most chemical reactions, temperature affects enzyme activity remarkably. In the stated temperature range, the rates of enzyme catalyzed reactions generally increase with increasing temperature and decrease contrarily. Most enzymatic reactions double in rate for every 10℃ rise in temperature as long as the enzyme is stable and fully active. However, enzymes are proteins that become denatured at high temperature. Each enzyme has an optimum temperature at which it operates with maximal efficiency. If the temperature is raised beyond the optimal temperature, the activity of many enzymes declines abruptly.

Enzymes have an optimum pH or pH range in which their activity is maximal, and at higher or lower pH their activity decreases. Hydrogen ion concentration affects enzymes in several ways. First, catalytic activity is related to the ionic state of active site. Changes in hydrogen ion concentration can affect the ionization of active site groups. In addition, substrates may also be affected. If a substrate contains an ionizable group, a change in pH may alter its capacity to bind with the active site. Secondly, changes in ionizable groups may change the tertiary structure of the enzyme. Drastic changes in pH often lead to denaturation.

Many substances can affect enzymatic activity. They can agitate or inhibit catalysis of enzyme. The former is called agitator, and the latter is called inhibitor. NaCl and $CuSO_4$ provide examples of the experiment of alkaline protease agitation and inhibition.

We can observe the effect of temperature on enzyme activity by the reaction of ptyalin catalyzing starch in different temperatures in the experiment. We can identify degree of starch hydrolysis by the color reaction of iodine solution and the hydrolysis produces of starch such as dextrin and maltose.

Materials

1. Apparatus

Test tubes and test tube rack; Pipette; Flask; Constant temperature water boiler; Thermometer, White porcelain

2. Reagents

(1) 1% Starch solution: Add 1g starch in 100ml distilled water.

(2) Saliva amylase: Gather saliva after gargling, and filtrate by a small funnel and absorbent cotton. Dilute the filtrate to 5～20 folds volume. Swirl the tube to mix up the content.

(3) 2% Sucrose solution: Add 2g sucrose in 100ml distilled water.

(4) Sucrase solution: Add 1 g dry active yeast, a small amount of quartz sand and water in a mor-

tar, and grind to homogenate. Then add 50 ml distilled water, static for a moment, and filter, abandon the precipitate, the supernatant was for the following test.

(5) Benedict reagent

Reagent A: 173g sodium citrate and 100 g $Na_2CO_3 \cdot 2H_2O$ in 600 ml of hot H_2O.

Reagent B:. Dissolve 17.3g $CuSO_4 \cdot 5H_2O$ in 100 ml hot water, cool and prepare for the next step.

Add reagent A to reagent B slowly with constant stirring. Make the final volume to 1000ml.

(6) Iodine and potassium iodide solution: Dissolve 10g of iodine and 20g of potassium iodide in 100ml of distilled water, and keep it in brown flask. Add 10 times diluted water before using.

(7) Phosphate buffer:

Solution A: 0.2mol/L Na_2HPO_4

Weigh28.40g Na_2HPO_4 (or 71.64g $Na_2HPO_4 \cdot 12H_2O$). Dissolve it into distilled water and set the volume at 1000ml.

Solution B: 0.1mol/L citric acid solution

Weigh 21.01g $C_6H_8O_7 \cdot H_2O$. Dissolve it into distilled water and set the volume at 1000ml.

pH5.0 Phosphate buffer: Sip 10.3ml of solution A, and then add in 9.70 ml Solution B.

pH7.0 Phosphate buffer: Sip 16.47ml of solution A, and then add in 3.53ml Solution B.

pH8.0 Phosphate buffer: Sip 19.45ml of solution A, and then add in 0.55ml Solution B.

(8) 1% $CuSO_4 \cdot 5H_2O$ solution.

(9) 1% NaCl solution.

Procedures

1. We can observe enzyme specificities by using effect of a saliva amylase and sucrase catalyzing different substrates including the starch and sucrose in the experiment. Take six test tubes, number them, and add reagents as the following table:

Table 5-2

Item	Numbers of tubes					
	1	2	3	4	5	6
1% starch (ml)	1	1	0	0	1	0
2% sucrose (ml)	0	0	1	1	0	1
saliva amylase (ml)	1	0	1	0	0	0
sucrase (ml)	0	1	0	1	0	0
saliva amylase (ml)	0	0	0	0	1	1
procedure	mixed, keep for 10 min at 37℃					
Benedict reagent (ml)	2	2	2	2	2	2
procedure	mixed, keep for 5～10min at 100℃					
phenomena						
explanation for the phenomena						

2. We can observe the effect of temperature on enzyme activity by the reaction of saliva amylase catalyzing starch in different temperatures in the experiment. Take 3 test tubes, number them 1～3, and add reagents as the following table:

Table 5-3

Item	Numbers of tubes		
	1	2	3
saliva amylase solution (ml)	1	1	1
pH7 Phosphate buffer (ml)	2	2	2
treated for 5min at different temperature	0℃	37℃	70℃
1% starch solution (ml)	2	2	2

Keepreaction at the different temperature for several minutes, sip a drop of reaction liquid from tube 2 on the white porcelain every 1~2 minutes, and then add a drop of iodide solution into the reaction liquid on the white porcelain, the process was continuously carried until the reaction liquid only showed color of iodide solution. Then all of tubes removed from the constant temperature water boiler, cooled immediately with running water for 2 minutes, and add a drop of iodide solution into the reaction tubes, observe the phenomena.

3. We can observe the effect of pH on enzyme activity by the reaction of saliva amylase catalyzing starch in different temperatures in the experiment. Take three test tubes, number 1~3 and add reagents as the following table:

Table 5-4

Item	Numbers of tubes		
	1	2	3
pH5.0 Phosphate buffer (ml)	3	0	0
pH7.0 Phosphate buffer (ml)	0	3	0
pH8.0 Phosphate buffer (ml)	0	0	3
1% starch solution(ml)	1	1	1
treated	mixed, treated for 2min at 37℃		
saliva amylase solution (ml)	1	1	1
procedure	Keep reaction at the different temperature for several minutes, sip a drop of reaction liquid from tube 2 on the white porcelain every 1~2 minutes, and then add a drop of iodide solution into the reaction liquid on the white porcelain, the process was continuously carried until the reaction liquid only showed color of iodide solution. Then all of tubes removed from the constant temperature water boiler, cooled immediately with running water for 2 minutes.		
iodide solution (drop)	1	1	1
phenomena			
explanation for the phenomena			

4. NaCl and $CuSO_4$ provide examples of the experiment of alkaline protease agitation and inhibition. Take three test tubes, number 1~3 and add reagents as the following table:

Table 5-5

Item	Numbers of tubes		
	1	2	3
1% NaCl (ml)	1	0	0
1% $CuSO_4$ (ml)	0	1	0
distilled water (ml)	0	0	1
saliva amylase solution (ml)	1	1	1
1% starch solution(ml)	3	3	3
treated	mixed, treated for 1 min at 37℃		
procedure	Keep reaction at the different temperature for several minutes, sip a drop of reaction liquid from tube 1 on the white porcelain every 1～2 minutes, and then add a drop of iodide solution into the reaction liquid on the white porcelain, the process was continuously carried until the reaction liquid only showed color of iodide solution. Then all of tubes removed from the constant temperature water boiler, cooled immediately with running water for 2 minutes.		
iodide solution (drop)	1	1	1
phenomena			
explanation for the phenomena			

Attentions

1. The ptyalin activity in saliva of different people is not the same, sometimes is very different, so the

diluted times can be defined by every one.

2. Very few inhibitors and agitators can affect enzyme activity. Inhibitors and agitators are not absolute.

Some of them are enzyme agitators in low concentration and are inhibitors of the enzyme in high concentration.

They are agitators for some enzymes and are also inhibitors for other enzymes.

3. The reaction time must be exact. Utensils used must be clean and dry.

Questions

1. What are the optimum temperature and optimum pH of enzyme? What actual meaning do they have?
2. Several factors contribute to enzyme catalyses. What are they? Briefly explain the effect of each.
3. What conditions must be paid attention to while doing enzyme experiment? Why?

第六章　微生物学实验
Chapter 6　Microbiology Experiment

实验一　革兰氏染色

实验目的

1. 理解革兰氏染色的生物化学原理
2. 理解特异染色过程的基本原理
3. 完成符合要求的革兰氏染色
4. 区分混合细菌的革兰氏阳性和阴性

实验原理

革兰氏染色法是细菌学中广泛使用的一种非常有用的鉴别染色法，革兰氏染色法可将所有的细菌区分为革兰氏阳性菌(G^+)和革兰氏阴性菌(G^-)两大类。细菌先经碱性染料结晶紫染色，再经碘溶液媒染后，用酒精脱色，在一定条件下有的细菌紫色不被脱去，有的可被脱去，因此可把细菌分为两大类，前者叫作革兰氏阳性菌(G^+)，后者为革兰氏阴性菌(G^-)。为观察方便，脱色后再用一种红色染料如碱性番红等进行复染。阳性菌仍带紫色，阴性菌则被染上红色。

材料

结晶紫溶液、革兰氏碘液(2 g 碘化钾溶于 300 ml 含 1 g 结晶碘的蒸馏水)、95%乙醇和/或异丙醇-丙酮混合液(3∶1 V/V)、番红染液

洁净的载玻片

接种环

胶头滴管

本生灯

吸水纸

显微镜

拭镜纸和抹镜水

浸镜油

菌株

生长在 NB 液体培养基中的金黄色葡萄球菌、大肠杆菌和枯草芽孢杆菌

实验步骤

1. 用马克笔分别在三个载玻片左边标记上三个菌株的名字。
2. 用接种环取 1～2 环无菌水到载玻片的中间。
3. 加入 1～2 环菌液，并充分混匀。
4. 使用接种环将混合物涂布形成一个一端厚而另一端薄的涂层，这被称为薄涂片。
5. 让涂片自然晾干或者在本生灯火焰上方烘干固定。
6. 将涂片放在染色架上。
7. 加结晶紫染色液浸过涂片，染色 30 秒。
8. 用水冲洗 5 秒。
9. 滴加革兰氏碘液覆盖，媒染 1 分钟。
10. 用水冲洗 5 秒。

11. 连续滴加95%乙醇脱色15～30秒。不要脱色时间过长，滴加脱色液直到无结晶紫流下。另一种选择，异丙醇-丙酮混合液(3∶1 V/V)脱色30～60秒。

12. 用水冲洗5秒。

13. 番红复染60～80秒。番红染色的强度变化很大，对于每一批菌株要求不同的染色时间。(如果你是色盲，使用俾斯麦棕色染料代替番红。)

14. 用水冲洗5秒。

15. 将染好的涂片用吸水纸吸干，用油镜观察。革兰氏阳性菌染为蓝到紫色；而革兰氏阴性菌染为粉到红色。不需要在涂片上放盖玻片。

16. 将浸过油的镜头按下述方法擦拭干净，a. 先用擦镜纸将油镜头上的油擦去。b. 用擦镜纸蘸少许二甲苯将镜头擦2～3次。c. 再用干净的擦镜纸将镜头擦2～3次。注意擦镜头时向一个方向擦拭。

Experiment 1 Gram Stain

Objectives

Each student should be able to

1. Understand the biochemistry underlying the Gram stain
2. Understand the theoretical basis for differential staining procedures
3. Perform a satisfactory Gram stain
4. Differentiate a mixture of bacteria into gram positive and gram negative cells

Principles

The **Gram stain** is the most useful and widely employed differential stain in bacteriology. It divides bacteria into two groups—**gram negative** and **gram positive.**

The first step in the procedure involves staining with the basic dye crystal violet. It is followed by treatment with an iodine solution, which increases the interaction between the bacterial cell and the dye so that the dye is more tightly bound or the cell is more strongly stained. The smear is then decolorized by washing with an agent such as 95% ethanol. Gram-positive bacteria retain

the crystal violet-iodine complex when washed with the decolorizer, whereas gram negative bacteria lose their crystal violet-iodine complex and become colorless.

Finally, the smear is **counterstained** with the safranin, different in color than crystal violet. The safranin will stain the colorless, gram negative bacteria pink but does not alter the dark purple color of the gram positive bacteria. The end result is that gram positive bacteria are deep purple in color and gram negative bacteria are pinkish to red in color .

Materials

solutions of crystal violet, Gram's iodine (2 g potassium iodide in 300 ml distilled water plus 1 g iodine crystals), 95% ethanol and/or isopropanol-acetone mixture (3:1 v/v), and safranin

clean glass slides

inoculating loop

Glue-head dropper

Bunsen burner

bibulous paper

microscope

lens paper and lens cleaner

immersion oil

Strains

Staphylococcus aureus, *Escherichia coli*, and *Bacillus subtilis* grown in NB broth

Procedure for Gram-Stain Technique

1. With a mark pen, label the left-hand corner of three glass slides with the names of the respective bacteria.

2. Use an inoculating loop to apply 1 to 2 loops of sterile water to the middle centre of a clean microscope slide.

3. Add 1 to 2 loops of the bacteria and mix thoroughly.

4. Spread the mixture over the slide using a inoculating loop, and form a smear that is thick at one end and thin at the other. This is known as a **thin smear**.

5. Allow the smear to air dry, or heat-fix with Bunsen burner.

6. Place the slides on the staining rack.

7. Flood the smears with crystal violet and let stand for 30 seconds.

8. Rinse with water for 5 seconds.

9. Cover with Gram's iodine mordant and let stand for 1 minute.

10. Rinse with water for 5 seconds.

11. Decolorize with 95% ethanol for 15 to 30 seconds. Do not decolorize too long. Add the decolorizer drop by drop until the crystal violet fails to wash from the slide. Alternatively, the smears may be decolorized for 30 to 60 seconds with a mixture of isopropanol-acetone (3 : 1 V/V).

12. Rinse with water for 5 seconds.

13. Counterstain with safranin for about 60 to 80 seconds. Safranin preparations vary considerably in strength, and different staining times may be required for each batch of stain. (If you are color-blind, use Bismark brown stain rather than safranin.)

14. Rinse with water for 5 seconds.

15. Blot dry with bibulous paper and examine under oil immersion. Gram positive organisms stain blue to purple; gram negative organisms stain pink to red. There is no need to place a coverslip on the stained smear.

16. After you are finished with the microscope, place the low-power objective in line with the ocular, lower the tube to its lowest position, clean the oil from the oil immersion lens with lens paper and lens cleaner, cover, and return the microscope to its proper storage place.

实验二 细菌的生化活性

材料

奇异变形杆菌(ATCC 14273)和表皮葡萄球菌(ATCC 14990)胰蛋白胨大豆肉汤培养物

含 3%细菌脂肪酶试剂的酒清蓝琼脂平板。

接种环

35℃保温箱

蜡笔

本生灯

学习目的

1. 理解脂肪水解的生化过程。

2. 测定细菌通过特定的脂肪酶水解脂肪的能力。
3. 解释通过颜色变化反应检测脂肪的水解是可能的。
4. 完成一次脂肪水解测试。

原理

脂肪是含有大量储能物质的高分子量化合物。细菌通常能分解两类脂肪，分别是甘油三酯和磷脂。甘油三酯是脂肪酶的一种酶水解成甘油和游离的脂肪酸，甘油和游离的脂肪酸能被细菌细胞吸收，通过糖酵解、β-氧化和三羧酸循环进一步代谢。这些脂肪也可以进入其他用于细胞膜磷脂合成的代谢途径。由于磷脂是所有细胞的功能组分，细菌水解宿主细胞磷脂的能力是病原微生物传播的重要因素之一。另外当产脂肪酶的细菌污染食物产品时，分解脂肪的细菌水解脂肪产生腐败被称为酸败。

当水解脂肪的细菌在添加这些相同脂肪的琼脂固体培养基上培养时，由于脂肪酸的释放而使周围变为酸性。添加 pH 指示剂到培养基，通过颜色变化来检测脂肪的水解。例如含细菌脂肪酶试剂的酒清蓝琼脂平板呈淡紫色，分解脂肪的细菌菌落周围由于酸性 pH 而变为品蓝。

步骤

第一阶段

1. 用蜡笔在酒清蓝琼脂平板底部划成两部分，分别标上菌名 *P. mirabilis* 和 *S. epidermidis*。在平板上写上自己的名字和日期。
2. 将不同菌种涂布接种到酒清蓝琼脂平板上。
3. 平板 35℃倒置培养 24～48 小时。

第二阶段

4. 检查平板上脂肪水解的结果。在细菌生长的周围若有一个明显的蓝色区域则证明有水解作用。如果没有水解作用，则菌落周围保持淡紫色。
5. 测量水解作用区域，在报告中记录结果。

Experiment 2　Biochemical Activities of Bacteria

Lipids: Lipid Hydrolysis

Materials

tryptic soy broth cultures of *Proteus mirabilis* (ATCC 14273) and *Staphylococcus epidermidis* (ATCC 14990)

petri plate containing spirit blue agar with 3% Bacto lipase reagent (Difco)

inoculating loop

incubator set at 35℃

wax pencil

Bunsen burner

Learning Objectives

Each student should be able to

1. Understand the biochemical process of lipid hydrolysis
2. Determine the ability of bacteria to hydrolyze lipids by producing specific lipases
3. Explain how it is possible to detect the hydrolysis of lipids by a color change reaction
4. Perform a lipid hydrolysis test

Principles

Lipids are high molecular weight compounds possessing large amounts of stored energy. The two common lipids catabolized by bacteria are the **triglycerides** (**triacylglycerols**) and phospholipids. Triglycerides are hydrolyzed by the enzymes called **lipases** into glycerol and free fatty acid molecules. Glycerol and free fatty acid molecules can then be taken up by the bacterial cell and further metabolized through reactions of glycolysis, β-oxidation pathway, and the citric acid cycle. These lipids can also enter other metabolic pathways where they are used for the synthesis of cell membrane phospholipids. Since phospholipids are functional components of all cells, the ability of bacteria to hydrolyze host-cell phospholipids is an important factor in the spread of pathogenic bacteria. In addition, when lipase-producing bacteria contaminate food products, the lipolytic bacteria hydrolyze the lipids, causing spoilage termed **rancidity**.

When these same lipids are added to an agar solidified culture medium and are cultured with lipolytic bacteria, the surrounding medium becomes acidic due to the release of fatty acids. By adding a pH indicator to the culture medium, it is possible to detect the hydrolysis of lipids by a color change. For example, spirit blue agar with Bacto lipase reagent has a lavender color. It turns royal blue around lipolytic bacterial colonies due to the acid pH.

Procedure

First Period

1. With a wax pencil, divide the bottom of a spirit blue agar plate in half and label half the plate *P. mirabilis* and the other half *S. epidermidis*. Place your name and date on the plate.
2. Spot-inoculate the spirit blue agar plates with the respective bacteria.
3. Incubate the plate in an inverted position for 24 to 48 hours at 35℃.

Second Period

4. Examine the plate for evidence of lipid hydrolysis. Hydrolysis is evidenced by a blue zone around the bacterial growth. If no lipid hydrolysis has taken place, the zone around the colony will remain lavender.
5. Measure the zone of hydrolysis and record your results in the report.

实验三　淀粉水解试验

材料

枯草芽孢杆菌(ATCC 6051)、大肠杆菌(ATCC 11229)和普通变形杆菌(ATCC 13315)的24～48小时的胰蛋白胨大豆琼脂斜面

淀粉琼脂平板

革兰氏碘液(1克碘、2克碘化钾、300毫升去离子水)

蜡笔

接种环

本生灯

35℃保温箱

学习目的

每个学生都应该掌握

1. 理解淀粉水解作用的生化过程。
2. 完成一次淀粉水解试验。

原理

许多细菌产生的酶叫作水解酶。水解酶在水溶液中催化有机大分子迅速分解为小分子。这个试验是碳水化合物淀粉的水解作用。

淀粉分子由两种结构组成：直链淀粉，没有分支的葡萄糖聚合物（200～300 单元）和支链淀粉，大的支链聚合物。直链淀粉和支链淀粉都能被某些细菌使用它们的 α-淀粉酶迅速水解，产生糊精、麦芽糖和葡萄糖。

革兰氏碘液能用于检测淀粉的存在，当它接触淀粉时，就会形成一个蓝色化合物。水解了的淀粉不能产生颜色变化。如果滴加革兰氏碘液到菌株生长的淀粉培养基时，出现透明的区域，说明该菌产生 α-淀粉酶。如果没有透明区，说明淀粉没有被水解。

第一阶段：淀粉水解作用试验

1. 用蜡笔在淀粉琼脂平板底部划成三部分，分别标上将接种的菌名。在平板上写上自己的名字和日期。

2. 无菌操作分别划线接种到平板的直线区域内。

3. 35℃培养 24～48 小时。

第二阶段：

1. 在淀粉琼脂平板的线形条状菌落上滴几滴革兰氏碘液。如果生长物的周围是透明的，说明淀粉已经被水解了，试验呈阳性；而如果不是透明的或者整个培养基都变成了蓝色，说明试验呈阴性。

2. 在报告中记录结果。

Experiment 3　Starch Hydrolysis

Materials

24- to 48-hour tryptic soy agar slant cultures of *Bacillus subtilis* (ATCC 6051), *Escherichia coli* (ATCC 11229), and *Proteus vulgaris* (ATCC 13315)

starch agar plate

Gram's iodine (1 g I_2, 2 g KI, 300 ml distilled water)

wax pencil

inoculating loop

Bunsen burner

incubator set at 35℃

Learning Objectives

Each student should be able to

1. Understand the biochemistry of starch hydrolysis
2. Perform a starch hydrolysis test

Principles

Many bacteria produce enzymes called **hydrolases.** Hydrolases catalyze the splitting of organic molecules into smaller molecules in the presence of water. This exercise will present the hydrolysis of the carbohydrate starch.

The starch molecule consists of two constituents: **amylose**, an unbranched glucose polymer (200 to 300 units) and **amylopectin**, a large branched polymer. Both amylopectin and amylose are rapidly hydrolyzed by certain bacteria, using their **α-amylases**, to yield dextrins, maltose, and glucose.

Gram's iodine can be used to indicate the presence of starch. When it contacts starch, it forms a blue to brown complex. Hydrolyzed starch does not produce a color change. If a clear area appears after adding Gram's iodine to a medium containing starch and bacterial growth, α-amylase has been produced by the bacteria. If there is no clearing, starch has not been hydrolyzed.

Procedure

First Period: Starch Hydrolysis Test

1. With a wax pencil, divide a starch agar plate into three straight sections. Label each with the bacterium to be inoculated. Add your name and date to the plate.

2. Using aseptic technique, streak the respective bacteria onto the plate in **a straight line** within the section.

3. Incubate the plate for 24 to 48 hours at 35℃.

Second Period:

1. Place several drops of Gram's iodine on each of the line streaks on the starch agar plate. If the area around the line of growth is clear, starch has been hydrolyzed, and the test is positive; if it is not clear or the entire medium turns blue, starch has not been hydrolyzed, and the test is negative.

2. Record your results in the report.

实验四 三糖铁琼脂试验

材料

粪产碱杆菌(ATCC 8750)、大肠杆菌(ATCC 11229)、普通变形杆菌(ATCC 13315)、铜绿假单胞菌(ATCC 10145)和福氏志贺菌(ATCC 12661) 的 24～48 小时的胰蛋白胨大豆肉汤培养物

5 支三糖铁琼脂斜面

本生灯

接种针

35℃保温箱

试管架

学习目的

每个学生都应该掌握

1. 理解三糖铁琼脂试验的生化反应原理。
2. 能分辨肠杆菌科不同的细菌。
3. 能区分肠杆菌科和其他肠道细菌。
4. 完成一次 TSI 试验。

原理

三糖铁琼脂试验最早在 1911 年被拉塞尔使用,通常用于肠道细菌的鉴定。还能通过细菌催化葡萄糖、乳糖或者蔗糖,利用硫酸亚铁铵或者硫代硫酸钠释放硫化物的能力来从其他革兰氏阴性肠道细菌中鉴定肠杆菌科。TSI 琼脂斜面含有 1%乳糖和蔗糖、0.1%的葡萄糖。还有 pH 指示剂酚红也被加入培养基中来检测碳水化合物产酸。

通常使用的克氏双糖铁琼脂(1917 年由克勒命名)是一种类似于 TSI 的鉴定培养基,适用于获得大约相同的信息。TSI 斜面通过划线曲折来回接种在斜面表面,然后用接种针穿刺琼脂的深处。培养 18～24 小时检测糖的发酵、气体产生和 H_2S 的产生。TSI 试管中可能出现以下反应:

1. 由于葡萄糖的发酵使柱状部分为黄色和斜面部分为红色(柱状部分由于酸的形成而使酚红指示剂变成黄色)。因为培养基中的葡萄糖极为有限,因此产生的酸也很有限,不能持续作用,所以斜面仍为

红色(碱性)。

2. 柱状和斜面部分均为黄色是由于发酵了乳糖和/或蔗糖(黄色的斜面和柱状部分归于这些高浓度的糖)导致整个培养基中产生过量的酸。

3. 琼脂柱断开表示有气体产生。

4. 琼脂变黑表示有气体(H_2S)产生。

5. 红色柱状和斜面表明糖没有被利用,也没有 H_2S 和其他气体产生。

步骤

第一阶段

1. 每个 TSI 琼脂斜面上分别标上将接种的菌名。以一个试管作为对照。在平板上写上自己的名字和日期。
2. 无菌操作划线接恰当的菌种到斜面上,穿刺接种到柱状部分。拧上试管盖,但不要太紧。
3. 35℃倒置培养 18～24 小时。试管在培养时为了观察是否变黑,应该每天观察,直到第七天。

第二阶段

1. 观察所有斜面位置和柱状位置培养物的颜色变化,培养基中是否有变黑现象。
2. 在报告中记录结果。

Experiment 4 Triple Sugar Iron Agar Test

Materials

24 to 48-hour tryptic soy broth cultures of *Alcaligenes faecalis* (ATCC 8750), *Escherichia coli* (ATCC 11229), *Proteus vulgaris* (ATCC13315), *Pseudomonas aeruginosa* (ATCC 10145), and *Shigella flexneri* (ATCC 12661)

5 triple sugar iron agar slants
Bunsen burner
inoculating needle
incubator set at35℃
test-tube rack

Learning Objectives

Each student should be able to

1. Understand the biochemical reactions involved in the triple sugar iron agar test
2. Differentiate among members of the family *Enterobacteriaceae*
3. Distinguish between the*Enterobacteriaceae* and other intestinal bacteria
4. Perform a TSI test

Principles

As originally described in 1911 by F. F. Russell, the triple sugar iron (TSI) agar test is generally used for the identification of enteric bacteria (*Enterobacteriaceae*). It is also used to distinguish the *Enterobacteriaceae* from other gram-negative intestinal bacilli by their ability to catabolize glucose, lactose, or sucrose, and to liberate sulfides from ferrous ammonium sulfate or sodium thiosulfate. TSI agar slants contain a 1% concentration of lactose and sucrose, and a 0.1% glucose concentration. The pH indicator, phenol red, is also incorporated into the medium to detect acid production from carbohydrate fermentation.

Often Kligler Iron Agar (named after I. J. Kligler in 1917), a differential medium similar to TSI, is used to obtain approximately the same information. TSI slants are inoculated by streaking the slant surface using a zig-zag streak pattern and then stabbing the agar deep with a straight inoculating needle. Incubation is for 18 to 24 hours in order to detect the presence of sugar fermentation, gas production, and H_2S production. The following reactions may occur in the TSI tube:

1. **Yellow butt (A) and red slant (A)** due to the fermentation of glucose (phenol red indicator turns yellow due to the persisting acid formation in the butt). The slant remains red (alkaline) (K) because of the limited glucose in the medium and, therefore, limited acid formation, which does not persist.

2. **A yellow butt (A) and slant (A)** due to the fermentation of lactose and/or sucrose (yellow slant and butt due to the high concentration of these sugars) leading to excessive acid formation in the entire medium.

3. **Gas formation** noted by splitting of the agar.

4. **Gas formation** (H_2S) seen by blackening of the agar.

5. **Red butt (K) and slant (K)** indicates that none of the sugars were fermented and neither gas nor H_2S were produced.

Procedure

First Period

1. Label each of the TSI agar slants with the name of the bacterium to be inoculated. Use one of the tubes as a control. Place your name and date on each tube.

2. Using aseptic technique, streak the slant with the appropriate bacterium and then stab the butt. Screw the caps on the tubes but do not tighten!

3. Incubate for only 18 to 24 hours at 35℃ for changes in the butt and on the slant. Tubes should be incubated and checked daily for up to seven days in order to observe blackening.

Second Period

1. Examine all slant cultures for the color of the slant and butt, and for the presence or absence of blackening within the medium.

2. Record your results in the report.

第七章　发酵工程实验
Chapter 7　Fermentation Engineering Experiment

实验项目：葡萄酒的制作过程

· 请在开始之前阅读这份说明

用无味的葡萄酒洗涤剂(经过零售商推荐的)洗净所有仪器，之后用热水彻底地漂洗残渣。

通过用亚硫酸盐溶液漂洗以洗净仪器。需要将50克(3汤匙)的粉末状亚硫酸盐溶解在4升(大概1加仑)的冷水里，以配置偏亚硫酸氢盐溶液。必须用亚硫酸盐溶液浸泡或者喷射清洗仪器的每一块地方，然后干燥或者用冷水漂洗干净。剩余的溶液可以贮存在密闭的容器里保存两个月。

· 将质量好的饮用水和本套原料用品混合，如果不能确定水的质量，可以考虑用瓶装水。

本套原料用品包括了多种原料包，请按指导添加各种原料。

开始的温度是起决定性作用的。如果酵母菌加入到过冷的原料中，它将不会发酵，仍是清澈透明。在添加酵母菌之前，仔细地检查培养汁的温度，使其保持在22～24℃(72～75℉)。

请将箱子顶部的地区代码标签撕掉，并贴上本说明书，或者制作葡萄酒的记录单。零售商将会从附有你意见或说明的便贴中获得有关信息。在便贴上同时写下葡萄酒的种类和生产日期。

· 仪器要求

主发酵罐(最小容积30升或者7.9美制加仑)，带长搅拌勺的盖(塑料或是不锈钢)、量杯、比重计测试瓶、取酒样桶、虹吸棒和软管、23升(6美制加仑)大玻璃瓶(玻璃或是塑料)、瓶塞和气阀、

用来清洗的无味葡萄酒洗涤剂，用来消毒的焦粉、30个酒瓶子、30个软木塞和一个装罐机。

· 在每批发酵之前，要准备以下物品：

(1)用于发酵的原料。一般来讲，也意味着就是葡萄汁，但是几乎所有的水果都可以制酒，尤其是海棠果和接骨木果。酿酒用品店会卖8～13升的葡萄汁，类似于雷司令、美乐、雪当利这样的葡萄汁都可以买到。如果你对葡萄品种没有要求，可以直接用未消毒的普通葡萄汁，但是这样做不但会增加成本，而且结果难以预测。

(2)高品质的水。一般指蒸馏水，含氯太多的水会破坏酒的品质。

(3)葡萄酒酵母。所用的酵母种类直接影响到最终酒品质的好坏。面包酵母在迫不得已的时候可以用。如果你的条件还可以，最好选择适合的葡萄酒酵母或者是其他合适的果酒酵母。

(4)处理原酒的材料。你可以选择皂土、明胶，如果实在条件不好，用蛋清也可以。头两个材料可以在酿酒用品店买到。

(5) 用于停止发酵的材料。纯抗坏血酸(维生素C)是很好的停发材料，但是一定要注意不要用含糖或香精的VC片，当然在酿酒用品商店买一瓶山梨酸钾也是不错的选择。

(6) 瓶子和软木塞。除了做香槟以外，你可以任意选择二手瓶子，只要保持瓶子是清洁的即可。清洁瓶子最好的方式：首先，清理掉瓶子里面的残留物质；其次，用充足强效的清洁液浸泡瓶子，以除去瓶子的标签和胶体；再次用清水洗净，洗掉清洁剂，清洗干净并消毒。价格合适，也可以直接用新瓶子，新瓶子可被重复使用。大概30个瓶子可以装23升的酒。木塞无论是人造的或是天然原料的都可以(天然的更便宜)。如果你想存放葡萄酒超过6个月，一定要选择长点的木塞。

(7) 用于灭菌消毒的焦亚硫酸钠。注意如果你周围或者你的家人有哮喘，不要用这个东西。可用碘消灵或者漂白剂替代。如果你不想你的酒最后变成醋，那么焦亚硫酸钠是最便利的消毒方式。如果你必须使用碘消灵，一定要按照商品的使用方法和剂量。如果用漂白剂，用5%浓度的漂白剂浸泡10分钟后，一定要清洗器具6次以上。对于一般人，焦亚硫酸钠足够安全，但是对于有哮喘的人来说，一定要慎之又慎。

(8) 清洁剂。餐具清洁剂效力不够强，酿酒用品商店卖的清洁剂效力较强。无异味的清洁剂可较

好地去除标签和异味等。

• **葡萄酒制作**

实际上，做葡萄酒的过程简单，甚至有点乏味。一般可以按照如下步骤进行：

(1)确保酿酒的环境温度恒定，气温可以保持在23℃或75℉以下。温度越高，酒转醋的风险就越大。

(2)在接触酒之前，所有的器具必须洗净消毒。包括漏斗、杯子、碗和其他的东西。如果连续使用，在每次使用之前，都要洗净消毒。优质的酿造酵母可以压制部分外来的细菌，但是一些厌氧菌却不好控制，容易把酒转变为醋。

(3)把果汁倒入一个大塑料容器。如果你用的是浓缩果汁，将之稀释到23升。将醪液控制在22～24℃或72～75℉。当发酵液接近20升的时候，要用灭过菌的温度计对温度进行测量，之后通过添加热或者冷果汁或者无菌水来调节温度。

(4)疯狂搅拌。

(5)添加酵母。

(6)将消毒的液体比重计放入发酵液中监控液体的比重。开始时显示数值大概为1.010或稍大。当比重小于这个数，最好加入些糖浆并搅拌，然后再次检测。然后移去比重计。

(7)盖上盖子。用大塑料布盖住容器，或者用绳子将塑料布固定在容器上。

(8)在1～2天内，发酵液就会开始冒泡。很快，酵母会沉到容器底部，这个时候你可能看不到酵母在工作，但这个时候酵母在发酵。

(9)一周后，洗净并消毒大玻璃瓶、塞子、气阀和塑料管。将发酵液虹吸入大玻璃瓶。尽量多地在容器底部留下沉淀物。然后将瓶子加上塞子和气阀，并确保气阀装一半水。

(10)如果在有阳光射入的地方，玻璃瓶应该用干净的深色薄布罩着。

(11)10天后测定酒的比重。最好的方法是用取酒器取少量酒放入量筒，然后用比重计进行测量。如果初始比重在1.010，那么10天后理想比重在0.998或者更低。之后每天监测，直到比重不在减少。

(12)移去塞子和气阀，加入使其停止发酵的材料，促进沉降。直接搅拌5分钟，确保瓶底沉淀物被搅拌均匀。重新装塞子和气阀。

(13)14天后，洗净并消毒瓶子和管子。将酒虹吸入瓶子并且用软木塞塞好。可以买一个瓶塞机或者从友善的葡萄酒店租或借瓶塞机，很多葡萄酒店可以免费为顾客提供。销售的大多数瓶塞预先灭过菌。如果你不能买到准备好的瓶塞，可以用亚硫酸溶液浸泡灭菌。

(14)保持瓶子竖立3天，然后整齐平放至少1个月。如果是红葡萄酒，最好放置一年以上。如果想要把酒放置6个月以上的，最好在酒中稍微加一点亚硫酸，这样可以很好防止变色。家里如果有人有哮喘病一定要注意使用。

Experimental projects: Wine making process

• **Please read all instructions before starting.**

Clean all equipment with an unscented winemaking detergent (recommended by your retailer) and rinse thoroughly with hot water to remove all residues.

Sanitise your equipment by rinsing it with a metabisulphite solution. To make this solution, dissolve 50 grams (3 tablespoons) of metabisulphite powder in 4 litres (approximately one gallon) of cool water. You must dip or spray every piece of equipment with this sulfurdioxide solution. Drip dry or rinse with cool water. Leftover solution can be stored in a tightly sealed container for two months.

• **Use good quality drinking water with this kit. If you're not sure of your water quality, consider using bottled water.**

This kit may contain multiple packages of each ingredient, please add all packages when directed.

The starting temperature of the wine is critical. If yeast is added to a kit that is too cold, it will not ferment or clear properly. Double check that the juice temperature is between 22°-24℃ (72°-75℉) before adding the yeast.

Please remove the code number sticker from the box top and attach it to these instructions, or to your winemaking record book. Your retailer will require the information from this sticker should you have any questions or comments. Write down the type of wine and the date started as well.

• **Equipment required**

Primary fermenter (minimum cabage 30 litres/7.9 US gallon), with lid Long stirring spoon (plastic or stainless steel), Measuring cup, Hydrometer and test jar, Wine thief, Siphon rod and hose, Carboy (glass or plastic) 23 litres (6 US gallon), Bung and airlock, Unscented winemaking detergent for cleaning, Metabisulphite powder for sanitizing, Thirty wine bottles, thirty corks and a corking machine.

• **For every batch, you will also need the following:**

(1) Something to ferment. Generally, that means grape juice, but wine has been made from almost all fruits, especially crabapples and elderberries. Winemaking stores sell kits containing 8～13 liters of concentrated grape juice varieties such as Riesling, Merlot and Chardonnay. If you are not into a specific variety, you can use unpasteurized grape juice, but it tends to be more expensive and your results might not be as predictable.

(2) Good quality water. Generally this means distilled water. Heavily chlorinated can skunk your wine.

(3) Wine yeast. The type of yeast you use will affect the flavor of the finished wine. Bread yeast can be used in a pinch. If you're not in dire circumstances you want to use a wine yeast tailored for the type of grape or other fruit you're using.

(4) Something to settle the wine. You can use sterilized bentonite, isinglass, or (in a pinch) egg whites. The first two are readily available at any winemaking store.

(5) Something to stop fermentation. You can use a little pure ascorbic acid (Vitamin C, but do not use vitamin pills containing sugar or flavoring), or you can buy a packet of potassium sorbate at the wine store.

(6) Bottles and corks. You can reuse bottles, and unless you are making champagne, it does not really matter what type of bottle you use as long as you get it clean. The best way to do this is to remove any residue in the bottles, soak them in water with strong detergent to remove the labels and the glue, wash them again in new water and detergent, rinse well, and sanitize. New bottles are also available at a reasonable price and can be reused indefinitely. You will need 30 bottles for a standard 23-liter recipe. Both natural and synthetic corks work well (although the natural is cheaper); you will need longer corks if you plan to keep the wine for longer than 6 months.

(7) Sodium metabisulfite for sanitizing. Warning — Do not use this under any circumstance if you or anybody in your family has asthma. Use iodophor or bleach instead. Sodium metabisulfite is the most convenient form of sanitizing, and sanitize you must if you want your wine to turn out as wine and not as vinegar. If you must use iodophor, follow the instructions on the label and rinse well. If you use bleach, soak your items for 10 minutes in a 5% bleach dilution, and then rinse at least 6 times thoroughly. Sodium metabisulfite is perfectly safe for the average person, but it can cause a serious reaction in asthmatics, so heed the warning.

(8) Detergent. Dish detergent is probably not strong enough. Wine stores sell strong, unscented detergents that do a better job at removing labels, old smells, etc.

• **Making the Wine**

The process of actually making the wine is simple but somewhat tedious. the following are good rules:

(1)Make sure the area in which you are making your wine is at a stable temperature below 23℃ or 75℉. At higher temperatures, the chances of your wine turning into vinegar are increased.

(2) Clean and sanitize everything just before it touches the wine. That includes any funnels, cups, bowls and other items. If you have to use them on two consecutive days, clean and sanitize before every use. A good yeast will outgrow the odd bacterium that gets in, but the sloppier you are the more likely something like acetobacter is going to get in and turn your wine to vinegar.

(3) Pour your juice into the large plastic container. If you are using a concentrate, top up with filtered water to 23 liters. You want the finished product to be between 65 and 75℉ or 18 to 23℃. Test with your sanitized dairy thermometer when you're up to 20 liters, and add hot or cold juice or water at the end to adjust the temperature.

(4) Stir like mad.

(5) Add yeast.

(6)Check your specific gravity by putting the sanitized hydrometer right in the container. It should show a specific gravity of 1.010 or greater. If it is less, you probably want to add a bit of sugar syrup and stir, then check again. Remove the hydrometer.

(7) Put the lid on. Cover the container with a big plastic sheet and tie it closed with some string or twine around the middle of the container.

(8) In a day or two, your must should be bubbling away. Soon after that the yeast will drop to the bottom of the container and you will not see it working, but it will still be fermenting.

(9) After a week, clean and sanitize the carboy, the bung, the airlock and the plastic tubing. Siphon the must into the carboy. Leave out as much of the sedimentation on the very bottom as you can. Attach the bung and the airlock, making sure to fill the airlock halfway with water.

(10) Cover the carboy with a dark, clean, thin sheet if you are doing this in an area where light can get to it.

(11) Wait 10 days and check the specific gravity. The best way to do this is to use the wine thief to transfer a little wine from the carboy to the measuring tube, then use the hydrometer to check the specific gravity. If you started out at 1.010, you are looking for a specific gravity of 0.998 or lower. Check this once a day until the specific gravity is the same on two separate days.

(12) Remove the bung and airlock and add whatever you are using to stop fermentation and to encourage settling. Stir for 5 minutes straight. Make sure to stir up all the sedimentation on the bottom. Replace the bung and the airlock.

(13) 14 days later, clean and sanitize your bottles and your tubing. Siphon the wine into the bottles and cork them. You can buy a corking machine, or you can rent or borrow one from your friendly wine store - many stores lend them free of charge to their customers. Most corks are sold pre-sanitized, but you can soak them in a sulfite solution if you cannot buy them already prepared.

(14) Let the bottles sit upright for 3 days, then let them sit on their side for at least a month. If you have made red wine, you should leave it for a year. If you do hold wine for over 6 months, you may wish to add a bit of sulfite to prevent discoloration, unless, of course, someone in the house has asthma.

第八章　细胞生物学实验
Chapter 8　Cell Biology Experiment

实验一　差速离心法鼠肝脏的大量线粒体制备

原理

差速离心经常用于初步制备细胞膜或者亚细胞器片断的步骤，仅作通过密度梯度进一步纯化之前的一步。这一技术根据其沉降速度(v)来分离颗粒，尽管有些情况颗粒密度(ρ_p)(例如细胞核)也是很重要的，但沉降速度主要依赖于颗粒的直径(d)。其他参数 ρ_l(溶液密度) 和 μ(溶液黏度)是不变的，g 是离心力。

$$v = d^2(\rho_p - \rho_l)/18\mu * g$$

由于 v 与 d^2 成正比，从公式可以看出多数细胞核的沉降速率比其他颗粒大很多，而只有最大的线粒体被从如溶酶体和过氧化物体等细胞器中分离开。线粒体比溶酶体和过氧化物体密度大，这将会加强这一趋势。因此，匀浆在 500～1000g 离心 5～10 分钟能分离出所有的细胞核，相对高一点的离心力(RCF)(3000×g 10 分钟)能将重一点的线粒体片断分离出来，所有最小线粒体需要的非常高的离心力(如 10000 至 15000g)也能将溶酶体和过氧化物体分离到。

不同大小细胞器的异质性必然会影响基于沉降速率的颗粒分离，但差速离心强加了一个更深层次问题，问题起因于转子的沉降路径长度和转子直径增加了旋转转子的离心力。如果匀浆添满了离心试管，样品最顶端的离心力就相对小于样品底部。另一方面，样品顶端的颗粒有更大距离形成稳定颗粒。

上面等式显示直径 5 微米的颗粒沉淀速率将是直径为 2.5 微米密度相同颗粒的四倍。在这种情况下，在试管四分之一位置低的所有较小颗粒的沉降时间，与最初在试管顶部的较大颗粒的沉降时间相同。因此差速离心如果选择的转速和离心时间能让所有的细胞核沉降，那么一些线粒体也会向试管底部沉降。这揭示出高离心力和短距离都沉降。正是由于这个原因，“洗”的差速离心颗粒是常见的做法；培养液重新悬浮颗粒并再次离心减少小颗粒的污染。当然这也会减少目的颗粒的产量。

样品在固定角转子的顶部和底部之间离心路径长度、离心力的差异，都小于同样试管体积的摆动斗转子。差速离心应始终保持一个固定角转子的最大分辨率，管角度越低越好。但是随着角度的降低，颗粒趋向在管壁分散，变得不紧凑，有时还不稳定，移出上清液也更困难了。所以在实践中 30°的试管角度可能是一个好的折中。

对于氧化磷酸化的研究，用简单差速离心制备的线粒体常常是更好的，而不是用梯度离心进行的纯化。这不是为了得到高回收，而是为了快速、尽可能轻柔地制备来提高溶液中耦合态的高度保留。因此最好避免使用耗时的密度梯度，还可能使线粒体接触潜在的高力量和不适合媒介。牛的心脏和鼠的肝脏广泛地应用于这一目的。

通常试验动物是生长龄较小的成年雄性动物，能取出湿重 10 克的肝脏。这个方案虽然是这样设计的，但可以按比例放大或者缩小应用到不同量的肝脏。动物通常都会断食过夜，从而减少肝内的糖原含量，这有利于分离过程。

材料

150～200 克雄性大白鼠
冰浴保存的肝匀质培养基(LHM)
解剖工具
波-埃氏均质器
带有浮桶式转头和合适试管的低速离心机
带有固定角转子和 40～50 毫升聚碳酸酯管真空泵的高速离心机

带有宽大杵的杜恩氏匀浆器(30～40 毫升)
肝匀质培养基(LHM)
250 毫升水中加入:

18.2 克	甘露醇(终浓度 0.2 M)
8.55 克	蔗糖(终浓度 5 0 mM)
0.37 克	氯化钾(终浓度 10 mM)
5.0 毫升	100mM Na2EDTA (终浓度 1 mM)
50 毫升	100mM HEPES (终浓度 10 mM)

KOH 调 pH 到 7.4
加水至 500 毫升
4℃保存 1～2 天

肝的分离

1. 150～200 克大白鼠断食过夜。
2. 用颈椎脱臼法或者断头法处死动物。

这必须由经验丰富的动物技师指导或者直接操作完成。

3. 立即剪开腹部取出肝脏,浸入预冷的 20 毫升 LHM 溶液中。
4. 倒掉溶液,细致地剪碎肝。肝的小块应该小于 25 mm^3。
5. 摇动肝的碎块加入 30 毫升冰浴的 LHM 溶液,让小块沉淀下来。
6. 倒掉溶液,重新加入 40 毫升新鲜溶液。

肝的研磨匀浆

7. 转移一半的悬浮物到波-埃氏均质器的钢化玻璃容器中。
8. 将预冷的碾锥连接到上端高扭矩电子马达上,碾锥以每分钟 500 转、上下 5～6 次搅拌切碎肝使其均匀。匀浆倒到一个放置在冰上的烧杯里。
9. 用溶液冲洗掉均质器和碾锥上粘的结缔组织。另一半悬浮物重复以上过程。

线粒体的分离

10. 用吊桶式转头的低速离心机在 4℃ 1000×g 离心匀浆 10 分钟。
11. 吸取上清转移到 40～50 毫升的聚碳酸酯管中。

方便的是使用一个 20～30 毫升连接在金属填充套管(内径 0.8～1.0 毫米)上的注射器去吸出上清液,以便再次离心。

12. 上清 4 ℃ 3000×g 带有固定角转子的高速离心机离心 10 分钟。
13. 使用连于真空泵的玻璃巴斯德移液管从每个试管中吸取上清液,保持移液管头在弧形液面来移除尽可能多地上层漂浮的脂肪层。此外还要尽可能多地去除覆盖于棕色线粒体之上的松散粉红色层。
14. 清除许多黏附在管壁上的黏糊组织。

清除脂肪是必要的,因为游离的脂肪酸强力的磷酸化作用解偶联剂。

15. 加入 8 毫升 LHM 溶液到每个小球上,用玻璃棒自然搅拌悬浮。然后在 Dounce 均质器中轻柔打 3～4 次使其重新充分悬浮。
16. 用 LHM 补足原有的溶液体积,转移到一个新的试管中,高速离心机 3000×g 再次离心 10 分钟。
17. 不少于两次地重复 13～16 步。

对于大量的制备,在第 2、3 次冲洗时用于悬浮小球的 LHM 总体积可以减少 50%

18. 重新在缓冲液中悬浮精制的线粒体,缓冲液的组分与接下来的分析或者处理过程相匹配;许多时候 LHM 是最好的。

注意事项：

均质培养基包含甘露糖、蔗糖、EGTA 或者 EDTA 的螯合物和缓冲液（通常为 HEPES 或者 MOPS），适用于呼吸研究。

所有的溶液、玻璃器皿、离心管和器材都应该在 0～4 ℃预冷，并一直存放于冰浴中。使用波-埃氏均质器的玻璃器皿时，应该使用热绝缘手套或者硅橡胶手保护器，避免热烫伤，还能在极少情况下破碎器皿时，保护不对手形成伤害。

Experiment 1　Preparation of the Heavy Mitochondrial Fraction from Rat Liver by Differential Centrifugation

Differential centrifugation is often used merely as a preliminary procedure to prepare a crude fraction of membranes or subcellular organelles prior to further purification in a density gradient. This technique separates particles according to their velocity of sedimentation (v), which depends principally on the diameter of the particle (d), although particle density (ρ_p) can also be important in some cases (e. g., nuclei). For other parameters, ρ_l (the density of the liquid) and μ (its viscosity), are constant, and g is the applied centrifugal force.

$$v = d^2(\rho_p - \rho_l)/18\mu \times g$$

Since v is proportional to d^2, it is clear from these values that the sedimentation rate of the majority of nuclei is much greater than that of other particles, while only the largest mitochondria are likely to be separable from organelles such as lysosomes and peroxisomes. This is enhanced by the tendency of the larger mitochondria to be slightly more dense than lysosomes and peroxisomes. Thus centrifugation of a homogenate at 500 to 1000g for 5 to 10 min will pellet essentially all of the nuclei, a slightly higher relative centrifugal force (RCF) of 3000$\times g$ for 10 min will pellet the "heavy" fraction of the mitochondria, and the considerably higher RCFs (e. g., 10000 to 15000$\times g$) required to pellet all of the smallest mitochondria will also pellet all lysosomes and peroxisomes.

The heterogeneity in the size of different organelles is bound to compromise separations of particles on the basis of sedimentation rate, but differential centrifugation (as opposed to rate-zonal centrifugation) imposes a further problem that arises from the sedimentation path length of the rotor and the radial increase g in force in the spinning rotor. If the homogenate fills the centrifuge tube, then the RCF at the top of the sample is considerably less than that at the bottom of the sample. On the other hand, particles at the top of the sample have the furthest distance to travel to form stable pellets.

The above equation shows that a particle with a diameter of 5 μm will sediment four times faster than one with a diameter of 2. 5 μm and equal density. In such cases, all the smaller particles in the lower quarter of the centrifuge tube will sediment in the time taken for the larger particles, initially at the top of the tube, to pellet. Thus, in differential centrifugation, if a rotational speed and centrifugation time are chosen to allow all of the nuclei to pellet, then some of the mitochondria towards the bottom of the tube, which are exposed to the highest RCF and have the shortest distance to travel, will also pellet. It is for this reason that "washing" of a differential centrifugation pellet is a common practice; resuspending any pellet in medium and repeating the centrifugation will reduce contamination by the smaller particles. However, it will also reduce the yield of the particle of interest.

The sedimentation path length and the difference in RCF between the top and bottom of the sample are both less in a fixed-angle rotor than in a swinging-bucket rotor of the same tube volume. Thus, differential centrifugation should always be carried out in a fixed angle rotor for maximum resolution, and the lower the tube angle the better. However, as the tube angle decreases, the pellet tends to spread up the wall of the tube and become less compact and sometimes less stable, making removal of the supernatant more difficult. In practice, therefore, a tube angle of 30° is probably a good compromise.

For studies on oxidative phosphorylation, mitochondria prepared by simple differential centrifugation are often preferred to those purified by gradient centrifugation. Here the aim is not to achieve high recoveries, but to prepare the organelles as rapidly and as gently as possible in a medium known to promote the retention of a highly coupled state. Hence, it is best to avoid the use of time-consuming density gradients, which may also expose the mitochondria to potentially damaging high forces and unsuitable media. Bovine heart and the heavy fraction from rat liver are widely used for this purpose. See Rice and Lindsay for a review of methods for the measurement of functions associated with electron transport and oxidative phosphorylation and the preparation of submitochondrial particles.

Young adult male animals (150 to 200 g) are routinely used, providing livers of 10g wet weight. This protocol is designed for one such liver, but can be scaled up or down proportionally for different amounts of liver. The animals are normally deprived of food overnight to reduce the glycogen content of the liver; this facilitates the separation process.

Materials

150- to 200-g male Sprague-Dawley rat
Liver homogenization medium (LHM; see recipe), ice cold
Dissecting tools
Potter-Elvehjem homogenizer (0.09-mm clearance; 25-ml working volume)
Overhead high-torque electric motor (thyristor-controlled)
Low-speed centrifuge with swinging bucket rotor and appropriate tubes
High-speed centrifuge with fixed-angle rotor and 40- to 50-ml polycarbonate tubes Vacuum pump
Dounce homogenizer (30- to 40-ml volume) with loose-fitting pestle (Wheaton type B)

Liver homogenization medium (LHM)

To 250 ml H_2O add:
18.2 g mannitol (0.2 M final)
8.55 g sucrose (50 mM final)
0.37 g KCl (10 mM final)
5.0 ml 100 mM Na2EDTA (1 mM final)
50 ml 100 mM HEPES (10 mM final)
Adjust pH to 7.4 with KOH
Add H_2O to 500 ml
Store up to 1 to 2 days at 4℃

Isolate liver

1. Deprive a 150- to 200-g male Sprague-Dawley rat of food overnight.
2. Sacrifice the animal by cervical dislocation or decapitation.

This must be supervised or carried out by an experienced animal technician.

3. Open the abdominal cavity and transfer the liver to a chilled beaker containing 20ml LHM.
4. Decant the medium and finely mince the liver using scissors.

The pieces of liver should be no more than 25 mm^3.

5. Agitate the minced tissue in 30ml ice-cold LHM and allow the pieces to settle out.
6. Decant the medium and replace with 40ml fresh medium.

Homogenize liver

7. Transfer half the suspension to the chilled glass vessel of a Potter-Elvehjem homogenizer.

8. Attach the cold pestle to an overhead high-torque electric motor and homogenize the minced liver using five to six up-and-down strokes of the pestle, rotating at 500 rpm. Decant the homogenate into a beaker on ice.

9. Rinse the homogenizer with medium and wipe the pestle to remove any adhering connective tissue. Repeat the procedure with the other half of the suspension.

Isolate mitochondria

10. Centrifuge the homogenate 10 min at 1000×*g*, 4℃, in a swinging-bucket rotor using a low-speed centrifuge.

11. Aspirate the supernatant and transfer to 40- to 50-ml polycarbonate tubes.

It is convenient to use a 20- to 30-ml syringe attached to a metal filling cannula (i. d. 0. 8 to 1. 0 mm) to aspirate supernatants that are to be recentrifuged.

12. Centrifuge the supernatant 10 min at 3000×*g*, 4℃, in a fixed-angle rotor using a high-speed centrifuge.

13. Using a glass Pasteur pipet attached to some form of vacuum pump, aspirate the supernatant from each tube, keeping the tip of the pipet at the meniscus to remove as much of the floating lipid layer as possible. Also remove as much as possible of the loose-packed pinkish layer that overlies the brown mitochondria.

14. Wipe away any remaining lipid adhering to the wall of the tube with a paper tissue.

Removal of this lipid is essential, as free fatty acids are potent uncouplers of phosphorylation from electron transport.

15. Add a small amount of LHM (8 ml) to each pellet and crudely resuspend the pellet with a glass rod. Then resuspend fully using 3 to 4 very gentle strokes in a Dounce homogenizer.

16. Make up to the original volume with LHM, transfer to new tubes, and recentrifuge 10 min at 3000×*g* in the high-speed centrifuge.

17. Repeat steps 13 to 16 twice more.

For large-scale preparations, the total volume of LHM used to resuspend the pellet can be reduced by 50% for the second and third washes.

18. Resuspend the purified mitochondria in a buffer whose composition is compatible with any subsequent analysis or processing; in many instances, LHM will be satisfactory.

Note:

Subcellular Fractionation and Isolation of Organelles

A homogenization medium containing mannitol and sucrose, a chelating agent (either EGTA or EDTA), and a buffer (normally HEPES or MOPS) is best suited to respiratory studies.

All solutions, glassware, centrifuge tubes, and equipment should be precooled to 0 to 4℃ and kept on ice throughout. When handling the glass vessel of the Potter-Elvehjem homogenizer, a thermally insulated glove or silicone rubber hand protector should be used, not only to avoid heat transfer from the skin, but also to protect the hand in the unlikely event that the vessel breaks.

第九章　基因工程实验
Chapter 9　Gene Engineering Experiment

实验项目：大肠杆菌感受态细胞的制备

实验目的

1. 通过本实验学习氯化钙法制备大肠杆菌感受态细胞的技术。
2. 了解感受态细胞的概念及其在分子生物学研究中的意义。

实验原理

受体细胞经过一些特殊方法（如：$CaCl_2$ 等化学试剂法）的处理后，细胞膜的通透性发生变化，成为能容许携带异源 DNA 的载体分子通过的感受态细胞（competent cell）。将异源 DNA 分子引入某一细胞株系，使受体细胞获得新的遗传性状的手段叫做转化（Transformation），是基因工程等研究领域的基本实验技术。

仪器与试剂

仪器：移液器、离心机、酒精灯、可见分光光度计

试剂：$CaCl_2$、$MgCl_2$、甘油、*E. coli*、LB 液体培养基

实验步骤

1. 取 1 份 100μL 于－80℃冻存的 *E. coli* 细胞接种到 500～1000ml 灭菌的 LB 液体培养基中。在 LB 中不加任何抗生素，因为这些菌不含抗菌基因的质粒。操作在无菌条件下进行。
2. 在摇床中 37℃培养直到 OD_{600} 值介于 0.3～0.4（1cm 比色皿）。
3. 用 Sorvall GSA（250 毫升离心管）转子以 5000 rpm 4℃下离心 10 分钟。此时将 100 mM $CaCl_2$ 和 100 mM $MgCl_2$ 溶液置于冰中预冷。
4. 用 1/4 体积冰冷的 $MgCl_2$ 轻轻重悬细菌沉淀，该过程需要在 3 至 5 分钟内完成。菌悬液 4000 rpm 离心 10 分钟。
5. 在冰上用 1/20 体积冰冷的 $CaCl_2$ 重悬沉淀，再加入 9/20 体积的 $CaCl_2$。在冰上冷却 20 分钟。
6. 悬浮液 4000 rpm 离心 10 分钟，沉淀以 1/50 体积保存于 15% 甘油中的预冷无菌的 85 mM $CaCl_2$ 溶液中重悬，并分成 100μL 小份于－80℃冻存.

注意事项

1. 需要在冰上完成的操作必须充分低温，否则容易影响感受态细胞质量。
2. 相关试剂必须保证灭菌彻底。

Experimental projects: generation of competent cells of *E. coli*

Objective

1. To learn how to make competent cells of *E. coli* by $CaCl_2$ method.
2. To understand the concept of competent cell and the significance to molecular biological research.

Principles

Permeability of receptor cells membrane will change when some special chemical reagents treated (such as $CaCl_2$) and become competent cells that permit DNA molecules from different source through. The method which made receptor cells get new genetic traits when different source DNA was transfered into cells was called Transformation. This is the basic experimental technology in the field of gene engineering.

Apparatus and reagents

Apparatus: micropipettor, centrifuge, alcohol burner, visible spectrophotometer

Reagents: $CaCl_2$, $MgCl_2$, glycerol, *E. coli*, LB broth

Procedures

1. Take an 100μL aliquot *E. coli* cells saved at −80℃ and inoculate into 500 ml to 1000 ml sterile LB broth. DO NOT ADD antibiotic, since these cells do not have plasmid with antibacterial gene. Work as sterile as possible.
2. Incubate the cells on a shaker at 37℃ until they reach an OD_{600} 0.3 to 0.4 (1 cm path length).
3. Centrifuge in the Sorvall GSA rotor (250 ml centrifuge bottle) at 5000 rpm for 10 min at 4℃. 100 mM $CaCl_2$ and 100 mM $MgCl_2$ solutions was precooled in ice.
4. Gently resuspend the bacteria pellet on ice in 1/4 volume of ice cold $MgCl_2$, taking 3 to 5 minutes for this procedure. Centrifuge the cell suspension at 4000 rpm in the Sorvall GSA rotor for 10 minutes.
5. Resuspend the bacteria pellet on ice in 1/20 volume of ice cold $CaCl_2$ and then add an additional 9/20 volume of $CaCl_2$. Keep this suspension on ice for at least 20 minutes.
6. Centrifuge the cell suspension at 4000 rpm in the GSA rotor for 10 minutes and resuspend the cell pellet in 1/50 volume precooled sterile 85 mM $CaCl_2$ in 15% glycerol w/v. Dispense in 100 μL aliquots and freeze at −80℃.

Notes

1. The operation on ice must be strict to finish, or the efficiency will be bad.
2. All reagents must be sterile。

第十章　解剖学实验
Chapter 10　Anatomy Experiment

实验一　躯干骨实验

一、填图题

1. 椎体；2. 横突孔；3. 上关节突；4. 椎孔；5. 棘突

二、绘图题：请绘出胸椎侧面观，并标出以下结构：

1. 椎体；2. 棘突；3. 横突肋凹；4. 上肋凹；5. 下肋凹；6. 椎下切迹

三、名词解释：肋角和肋沟

肋骨的结构变化很大，但前10对都有肋头和与脊椎连接的肋结节。最后两对只有肋头没有肋结节。另外，12对肋骨都有肋颈、肋角和肋体，后面通过关节与两个胸椎的肋凹连接。结节是一个球形作用，只是侧头部，它与胸椎的横突肋凹连接。肋颈是肋头和肋体之间的溢缩部分。肋角位于肺后缘的对面，肋体是肋骨弯曲的主体。沿着肋体的内表面是一个内陷的沟叫做肋沟，能保护肋血管和神经。

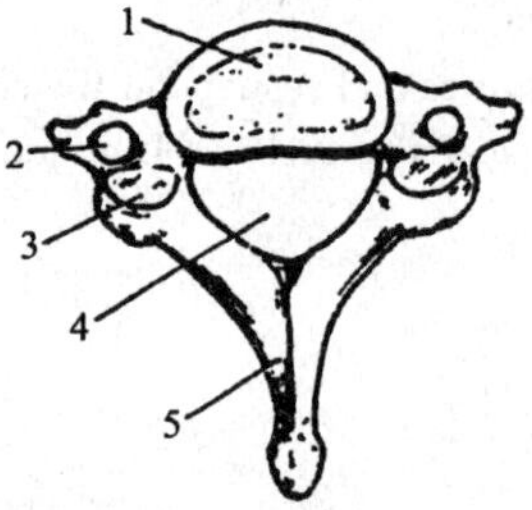

Sixth vertebrae cervicales
(superior view)
第7颈椎上面观

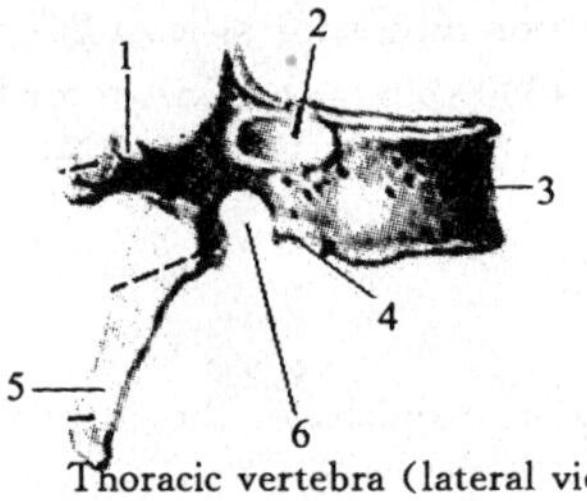

Thoracic vertebra (lateral view)
胸椎(侧面)

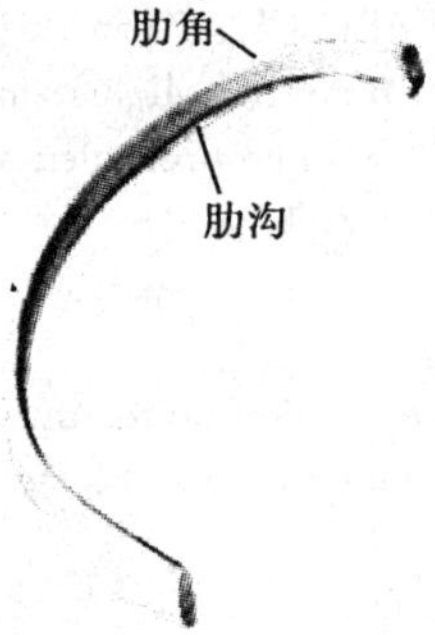

Experiment 1　Truncal Bones Experiment

Look at the picture and fill in the blanks

1. Vertebral body; 2. Transverse foramen; 3. Superior articular process; 4. Vertebral foramen; 5. Spinous process

Mapping problem

Drawing out the thoracic vertebra (lateral view) and mark the following structure

1. Vertebral body; 2. Spinous process; 3. Transverse costal fovea; 4. Superior costal fovea; 5. Inferior costal fovea; 6. Inferior vertebral notch

Explain the glossary: costal angle and costal groove

Although the ribs vary structurally, each of the first 10 pairs has a head and a tubercle for articulation with vertebra. The last two have a head but no tubercle. In addition, each of the 12 pairs has a neck, angle, and body. The head projects posteriorly and articulates with the adjacent two costal foveae of thoracic vertebrae. The tubercle is a knoblike process, just lateral to the head. It articulates with the transverse costal fovea of the thoracic vertebra. The neck is the constricted area between the head and the tubercle. **The angles** are located opposite to the posterior border of lungs. The body is the curved main part of the rib. Along the inner surface of the body is a depresses sulcus called the **costal groove** that protects the costal vessels and nerves.

实验二　感觉器官实验

一、填图题

1. 前骨半规管；2. 骨壶腹；3. 外骨半规管；4. 后骨半规管；5. 蜗窗；6. 前庭窗；7. 前庭；8. 耳蜗

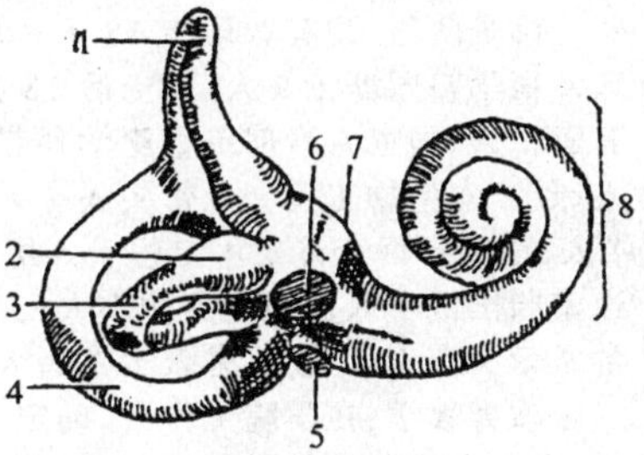

二、绘图题：请绘出眼球水平切面并标示下列结构

1. 晶状体；2. 睫状小带；3. 视网膜；4. 巩膜；5. 角膜

三、名词解释

黄斑：

视神经盘是视神经连接眼球和其纤维的分布于视网膜神经层的地方。视神经盘的正横向是一个小的椭圆形，淡黄色的命名为黄斑。黄斑中间凹陷的部分叫做中央凹，是最灵敏的视力区域。

Experiment 2　Sense Organ Experiment

Look at the picture and fill in the blanks

1. anterior semicircular canal ; 2. born ampulla; 3. lateral semicircular canal; 4. posterior semicircular canal; 5. fenestra cochlea ; 6. fenestra vestibuli; 7. vestibule; 8. cochlea

Mapping problem: Drawing out the eye horizontal section and mark the following structure

1. crystalline lens; 2. ciliary zonule; 3. retina; 4. sclera; 5. cornea

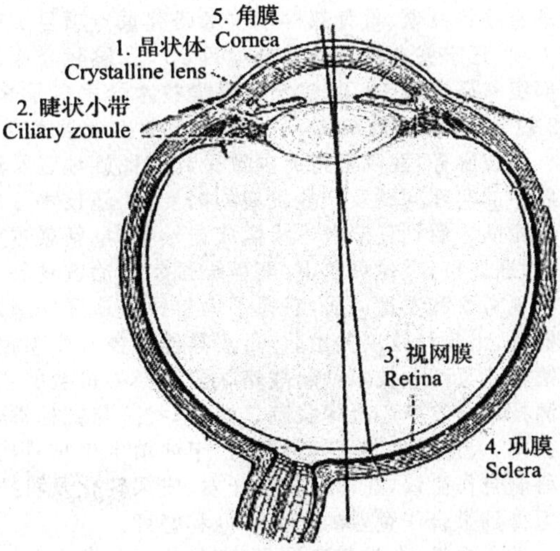

Explain the glossary: macula lutea

The optic disc is where the optic nerves connects the eyeball and its fibers spread out in the neural layer of the retina. Just lateral to the optic disc is a small oval, yellowish point termed as the macula lutea. The central depressed part of the macula, called the fovea centralis, is the area with the most acute vision.

第十一章　生命科学学院简介

Chapter 11　Introduction to College of Life Science

生命科学学院下设生物工程、化学工程、食品工程3个系，拥有生物工程、化学工程与工艺、应用化学、食品科学与工程、食品质量与安全、制药工程6个本科专业，拥有生物工程研究中心、食品工程与质量安全研究中心两个校级学科建设平台。设有化学基础实验教学中心、生物基础实验教学中心、综合实验教学中心、实验室建设与管理中心四个实验教学管理机构。

师资队伍：现有教职工72人，其中特聘教授2人、教授11人、副教授26人，具有博士学位教师40人。留学归国博士9人、博士后13人。有博士生导师3人、硕士生导师25人。国务院政府特殊津贴享有者1人、国家人事部跨世纪学科带头人1人、教育部高等学校食品与营养科学教学指导委员会委员1人、辽宁省生物工程专业带头人1人、辽宁省教学名师1人、辽宁省优秀教师1人、辽宁省青年骨干教师4人、辽宁省百千万人才工程千人层次2人、辽宁高层次科技专家库专家1人、大连市优秀专家2人、大连市归国留学人员创业英才1人、大连市化学化工产业特聘专家1人、校优秀学科带头人2人、校优秀学术带头人4人、校优秀教学带头人1人。

教育教学：在学院工作中，确定了教学工作的中心地位、教学质量的首要地位和人才培养的优先地位，全面实施教学质量工程，明确了现有专业的定位与内涵，突出了专业建设特色，取得多项建设成绩。生物工程专业为教育部高等学校特色专业建设点、辽宁省首批示范性专业、辽宁省紧缺本科生物工程人才培养基地，生物工程专业教学团队为省级优秀教学团队，化学基础实验教学中心为辽宁省省级示范中心，有机化学和普通化学为辽宁省省级精品课。近五年，主持和承担各类省部级教改项目12项，获得国家级教学成果二等奖2项，省部级教学成果奖6项。生命科学学院拥有实验与研究空间8440平方米，拥有教学实验室35个，研究室42个，学生创新工作室24个，仪器分析室11个，信息与网络教学工作室1个。

科学研究：学院确定了以科研立项与科技服务为科研工作的核心，发挥人才资源和科研平台优势，依托重点实验室和研究中心，组建了多个学科专业交叉的科研团队，努力提高全院教师的科技创新力和社会贡献力。近五年，主持和承担科研课题220余项，其中国家"十二五"科技支撑计划项目1项、国家自然科学基金项目28项、国家"十一五"科技支撑计划子课题1项、教育部新世纪优秀人才支持计划项目1项、国家人力资源和社会保障部留学人员科技活动项目择优资助1项、教育部留学回国人员科研启动基金项目9项、教育部科学技术研究重点项目2项、国家民委重点科研课题8项、辽宁省科技攻关项目1项、辽宁省自然科学基金项目25项、国际合作项目2项、横向课题29项，总经费2400余万元。获省部级科研成果10项，辽宁省科学技术学术成果奖26项。发表论文680多篇，其中SCI、EI、ISTP收录182篇，申请和授权专利20余项。主编出版学术专著及教材共20部，译著1部。

科技服务：我院坚持为少数民族和民族地区及地方经济发展服务的理念，针对民族地区和地方经济发展中的实际问题及产业发展的需要，实施技术集成创新与开发，开展科技攻关和科技服务。多次组织博士团队先后到辽宁省本溪满族自治县，阜新蒙古族自治县，内蒙古自治区呼和浩特市，鄂尔多斯市、赤峰市、通辽市、呼伦贝尔市，新疆维吾尔自治区乌鲁木齐市、塔城地区、吐鲁番市、石河子市等民族地区进行实地考察和实地调研，获得了大量民族地区经济发展中的特色资源开发与需求信息，学院通过整合科技资源，组建科技攻关团队，落实具体科技开发项目的对接与实施。现在已获科技支疆项目1项，作为内蒙古蓝莓产业技术创新战略联盟专家委员会的主任委员单位，获自治区科技计划项目1项，辽宁省葡萄酒技术开发中心授牌我院。同时，我院与吉林省金塔实业(集团)股份有限公司、吉林鹿业生物制品有限公司、大连富生制药有限公司、宁波甬牛生物科技有限公司、河北华容制药有限公司等多家企业签订了科研合作协议，在产品联合开发、成果转化及科技服务等方面发挥了重要作用，科技创新力与社会贡献力得到了合作企业的充分认可和好评。

学科建设：在学科建设中，充分依托重点实验室、工程研究中心等高端学科建设平台，发挥优秀人才团队作用，以生物学、化学以及食品科学为基础，形成大思维，构建大学科，搭建大平台，组建大团队，在生命科学学院的6个本科专业建设与发展中进行有效交叉与融合，实现学科专业共建，平台共用，项目共研，建设生物工程、化学工程与技术、食品科学与工程三个一级学科，实现学科专业的建设与发展目标。2011年获批服务国家特殊需求人才培养项目—生物工程领域专业硕士研究生培养立项。

拥有生物化工国家民委重点学科、辽宁省重点培育学科。拥有生物技术与资源利用国家民委—教育部重点实验室、国家民委生物化学工程重点实验室、辽宁省生物催化与仿生化学重点实验室、辽宁省省级高校生物工程下游技术重点实验室、辽宁省食源性病原微生物快速检测与控制工程技术研究中心、大连市生物化学工程技术中心、辽宁葡萄酒技术开发中心、大连开发区食品安全评价与检测中心等高端研究平台。拥有万元以上仪器设备446台套、10万元以上仪器设备80台套，仪器设备总价达5000多万元。

学术交流：我院已与清华大学、北京大学、大连理工大学、山东大学、南京工业大学及中国科学院大连化学物理研究所、微生物研究所、过程工程研究所、植物研究所等100多个知名院所建立了密切学术交流与合作关系。教师参加国际国内学术会议达200多人次，接待各级领导、国内外专家来访达500多人次，举办学术报告会80多场。先后与中科院大连化学物理研究所、大连理工大学、辽宁师范大学、大连工业大学、沈阳农业大学等联合招生并独立培养研究生78人，其中博士研究生9人。

我院与美国康奈尔大学、日本九州大学、奈良先端科学技术大学院大学、筑波大学、韩国江原道大学校大学、瑞士霍夫曼-罗氏有限公司等国外院校和公司进行广泛的国际学术交流与合作。这些学校的相关学科领域的教授来院讲学与访问100人次。我院教师出访美国、瑞士、英国、加拿大、日本、韩国、新加坡、比利时等国家进行学术交流或做博士后达30多人次。2007年主办了中韩生物科学与生物技术国际研讨会，2006和2008年与清华大学共同承办了两届生物工程与科学国际学术会议，2009年主办了辽宁省生物技术学术报告会。

人才培养：以科学发展观为指导，全面践行教学、科研与学生工作一体化的人才培养新模式，有效提高了人才培养质量，全面推进大学生的素质、创新、考研、就业四项工程。生命科学学院现有48个教学班，在校本科生1351名。近几年，毕业生一次性就业率保持在90%以上，毕业生平均考研率32%，尤其是2009年考研率达44%。考取的主要院校有复旦大学、中国海洋大学、天津大学、吉林大学、华东理工大学、大连理工大学、大连海事大学、北京理工大学、北京化工大学、北京科技大学、东北大学以及中国科学院相关研究所等知名院所。在学生创新能力培养方面取得了显著成绩，近四年，大学生科研立项151项，发表论文59篇，其中SCI收录1篇。获国际奖励5人次，国家级奖励174人次，省级奖励209人次，市级奖励58人次。近几年，我院培养的本科生先后到美国、日本、英国、韩国、澳大利亚等国家留学达25人。

生命科学学院在学校党委的正确领导下，全面贯彻党的教育方针，坚持学校党委确定的为少数民族和民族地区服务的办学宗旨，领导班子成员认真履行了分工职责，坚持改革创新、民主决策，秉承【励志博学 明德拓新 笃行卓越】院训，提倡敬业和奉献精神，全面提高教师的健康指数、幸福指数和工作指数，有效促进了学院协调与快速发展。

College of Life Science includes three departments, i. e. biological engineering, chemical engineering and food engineering, six bachelor's degree programs, i. e. biological engineering, chemical engineering and technology, applied chemistry, food science and engineering, food quality and safety and pharmaceutical engineering, two school-level discipline construction platforms, i. e. biological engineering research center, food engineering and quality safety research center, as well as four experimental teaching management organizations, i. e. fundamental chemistry experimental teaching center, fundamental biology experimental teaching center, comprehensive experimental teaching center, and laboratory construction and management center.

Faculties: There are 72 teaching and administrative staffs in the college, including 2 specially-appointed professors, 11 professors, 26 associate professors, and 40 teachers with doctorate. The college has 9 doctors returning from abroad, 13 post doctorates, 3 doctorial tutors, and 25 master's tutors, 1 person enjoying the special government allowances of the State Council, 1 person served as the leader on the cross-century subject of the Ministry of China, 1 person served as the member of the steering committee for food and nutrition science education of the colleges and universities of Ministry of Education of

China, 1 person served as the leader on the biological engineering of Liaoning Province, 1 person served as the renowned teacher of Liaoning Province, 1 person served as the prominent teacher of Liaoning Province, 4 persons served as the young backbone teacher of Liaoning Province, 2 persons chosen for the thousand level in the millions of talents project of Liaoning Province, 1 person served as the expert of the high-level technology expert database of Liaoning Province, 2 persons served as the outstanding expert of Dalian City, 1 person served as the entrepreneurial talent among the personnel returned from studying abroad of Dalian City, 1 person served as the specially-appointed expert of Dalian chemistry and chemical industry, 2 persons served as the leader on the excellent disciplines of the university, 4 persons served as the leader on the excellent academics of the university, and 1 person served as the leader on the outstanding teaching of the university.

Education and Teaching: In the work of the college, we determine the central status of teaching, the primary status of teaching quality and the priority status of personnel training, fully implement the teaching quality engineering, clarify the orientation and connotation of the existing specialties, highlight the characteristics of specialty construction, and achieve a number of construction achievements. Specifically, the biological engineering specialty is known as the featured specialty construction point for the colleges and universities of Ministry of Education, the first batch of exemplary specialty of Liaoning Province, and the personnel training base of shortage of biological engineering (bachelor) of Liaoning Province; the professional teaching team of biological engineering is a provincial excellent teaching team, the fundamental chemistry experimental teaching center is known as the provincial demonstration center of Liaoning Province, and the organic chemistry and general chemistry are known as the provincial excellent courses of Liaoning Province. In the past five years, we presided and assumed 12 provincial and ministerial education reform projects, and won 2 second prizes at the national teaching achievement award and 6 teaching achievement prizes at the provincial and ministerial levels. College of Life Science has an experiment and research space of 8440m^2, 35 teaching laboratories, 42 research laboratories, 24 students' innovation studios, 11 instrumental analysis rooms and 1 information and network teaching studio.

Scientific Research: The college determines scientific research project initiation and technology services as the core of scientific research, and establishes a research team with intersectional multiple disciplines to work hard to improve the scientific and technological innovation and social contribution for the teachers of the whole college by playing the advantages of human resources and research platform and relying on the key laboratories and research centers. In the past five years, we presided and assumed more than 220 scientific research projects, including a national science and technology support project in the "Twelfth Five-Year Plan" period, 28 projects of Natural Science Foundation of China, a sub-subject of national science and technology support program in the "Eleventh Five-Year Plan" period, a project in the program for new century excellent talents in university of the Ministry of Education, a preferential support project in the science and technology activity project for students studying abroad of the Ministry of Human Resources and Social Security of China, 9 projects in the scientific research starting foundation for returned overseas Chinese scholars of the Ministry of Education, 2 scientific research key projects of the Ministry of Education, 8 key research projects of the State Ethnic Affairs Commission, a scientific and technological project of Liaoning Province, 25 projects for the Natural Science Foundation of Liaoning Province, 2 international cooperation projects, and 29 transverse projects, wherein the total funding is more than 24 million Yuan. The college won 10 provincial and ministerial scientific achievements and 26 science and technology academic achievements of Liaoning Province, published more than 680 papers (182 papers were included in SCI, EI and ISTP), and applied and authorized more than 20 patents. Moreover, the college totally edited and published 20 academic monographs and textbooks, and translated 1 book.

Technology Service: In allusion to the practical problems in the ethnic minority areas and local economic development and the requirements of industrial development, our college implements technology integration innovation and development, tackles hard-nut problems in science and technology and provides technology services while adhering to the concept of servicing for ethnic minorities, ethnic minority areas and local economic development. We organized Doctor teams for several times to successively implement field survey and field research in Benxi Manchu Autonomous County of Liaoning Province, Fuxin Mongolia Autonomous Country, Hohhot City of Inner Mongolia Autonomous Region, Ordos City, Chifeng City, Tongliao City, Hulun Buir City, Urumqi City of Xinjiang Uygur Autonomous Region, Tacheng Prefecture, Turpan City, Shihezi City and other ethnic minority areas, obtained a great deal of information for the characteristic resources development and demand in the economic development of ethnic minority areas, and carried out the butt joint and implementation of concrete science and technology development projects by integrating science and technology resources and building a team for tackling hard-nut problems in science and technology. At present, we have acquired a science and technology supporting project in Xinjiang as the chairman unit in the expert committee for the technology innovation strategy alliance of blueberry industry in Inner Mongolia, acquired a science and technology plan project in the autonomous region, and awarded by the grape wine technology development center of Liaoning Province. At the same time, our college signed scientific research cooperation agreements with many enterprises, such as Jilin Gold Tower Industrial Group Co., Ltd., Jilin Deer Biological Products Co., Ltd., Dalian Fusheng Pharmaceutical Co., Ltd., Ningbo Yongniu Biotechnology Co., Ltd., Hebei Huarong Pharmaceutical Co., Ltd. and so on, played an important role in joint product development, achievement transformation, technology services and other aspects, and gained the full recognition and praise of the cooperative enterprises in the science and technology innovation and social contribution.

Discipline Construction: In the discipline construction, we fully rely on the key laboratories, engineering research centers and other high-end discipline construction platforms, play the role of talents team, form large thoughts, construct large disciplines, build large platforms and build large teams on the basis of biology, chemistry and food science, effectively intersect and integrate the six bachelor's degree programs of the College of Life Science in the construction and development process, realize the co-construction of the disciplines and specialties, the share of platforms and joint research of projects, build three first-level disciplines, i. e. biological engineering, chemical engineering and technology and food science and engineering, and achieve the construction and development targets of disciplines and specialties. In 2011, our college was approved for serving in the personnel training project of national special needs, i. e. the training project for the candidates of professional master degree in the field of biological engineering.

We have the key discipline of biochemical engineering of the State Ethnic Affairs Commission, the key cultivation discipline of Liaoning Province, the key laboratory of Ministry of Education for the biotechnology and resource utilization of the State Ethnic Affairs Commission, the key laboratory for the biochemical engineering of the State Ethnic Affairs Commission, the key laboratory for the biocatalysis and biomimetic chemistry of Liaoning Province, the key laboratory for the downstream processing of provincial college biological engineering of Liaoning Province, the technology research center for the rapid detection and control engineering of foodborne pathogenic microorganism in Liaoning Province, the technology center for biochemical engineering of Dalian City, the grape wine technology development center of Liaoning, the food safety evaluation and testing center in development district of Dalian and other high-end research platforms. We have 446 units of equipment of more than 10,000 Yuan and 80 sets of equipment of more than 100,000 Yuan, wherein the total price of the equipment is more than 50 million Yuan.

Academic Exchange: We establish close academic exchange sand cooperation relationships with more than 100 well-known institutes, such as Tsinghua University, Peking University, Dalian University of Technology, Shandong University, Nanjing University of Technology, and Dalian Institute of Chemical Physics, Institute of Microbiology, Institute of Process Engineering and Institute of Botany of the Chinese Academy of Sciences (CAS), and so on. Furthermore, more than 200 teachers of our college participated in domestic and international academic conferences, we received more than 500 visits of the leaders at all levels and the experts at home and abroad, and held more than 80 academic reports. We successively implemented joint recruitment with CAS Dalian Institute of Chemical Physics, Dalian University of Technology, Liaoning Normal University, Dalian Polytechnic University, Shenyang Agricultural University and the like, and independently cultivated 78 post graduate students, including 9 doctoral candidates.

Our college carried out wide international academic exchange and cooperation with Cornell University (U. S.), Kyushu University (Japan), Nara Institute of Science and Technology, Tsukuba University, Gangwon-do University (Korea), Switzerland F. Hoffmann-LA Roche Ltd. and other foreign universities and companies, wherein the professors for the relevant disciplines of these universities accessed to our college for 100 times. There are more than 30 teachers of our college visited to the United States, Switzerland, Britain, Canada, Japan, Korea, Singapore, Belgium and other countries for the academic exchanges or post doctorate. We hosted the international seminar for the bioscience and biotechnology on China and Korea in 2007, undertaken two international academic conferences for the biological engineering and science with Tsinghua University in 2006 and 2008, and hosted the symposium for the biotechnology of Liaoning Province in 2009.

Personnel Training: Based on the guide of scientific development, we comprehensively practice the new personnel training mode with integrated teaching, research and student work, effectively improve the quality of personnel training, and comprehensively promote the four projects for the quality, innovation, postgraduate-entrance examination and employment of college students. At present, College of Life Science has 48 teaching classes and 1351 undergraduate students. In recent years, the one-time employment rate of the graduates is kept in the level of more than 90%, the average success ratio in the postgraduate-entrance examination is 32%, especially the success ratio in the postgraduate-entrance examination is 44% in 2009, and the main colleges and universities involved therein comprises Fudan University, Ocean University of China, Tianjin University, Jilin University, East China University of Science and Technology, Dalian University of Technology, Dalian Maritime University, Beijing Institute of Technology, Beijing University of Chemical Technology, Beijing University of Science and Technology, Northeastern University, the relevant institutions of Chinese Academy of Sciences, and other well-known institutions. We have made remarkable achievements in the cultivation for the innovation capability of the students. In the past four years, there are 151 scientific research projects of college students, 59 papers published (one of the papers was included in SCI), 5 persons won international awards, 174 persons won national awards, 209 persons won provincial awards, and 58 persons won municipal awards. In recent years, 25 undergraduates cultivated by our college went to the United States, Japan, Britain, Korea, Australia and other countries to study successively.

College of Life Science, under the correct leadership of the college Party committee, fully implements the Party's education policy, and adheres to the school mission of serving for ethnic minorities and ethnic minority areas determined by the college Party committee, seriously carries out the division of responsibilities by the members of the leading group, persists in reform and innovation and democratic decision-making, adheres to the college motto of [encouragement and erudition, virtue and development, persistence and excellence], promotes professionalism and dedication, comprehensively improves the health index, happiness index and performance index of the teachers, and effectively promotes the coordination and rapid development of the college.

第十二章　模拟国际学术会议
Chapter 12　Simulative International Conference

Dear Teachers and Students,

The First Simulative International Conference on Life Science (SICLS) will be held at the Room ZC109, Multi-functional Building, on May 29th, 2011. The Chairmen of the conference would like to cordially invite you to this exceptional event.

亲爱的老师和同学们：

2011 年 5 月 29 日首届生命科学模拟国际会议将于综合楼 ZC109 室召开,欢迎你们光临会场。

Presentations 会议通知

Poster presentations will be displayed on 1 meter high by 2 meter wide (100 cm×200 cm) boards. The suitable size for poster is: 80cm high×60 cm wide.

海报展板大小 100 cm(高)×200 cm(宽),海报最佳尺寸 80cm(高)×60 cm(宽)

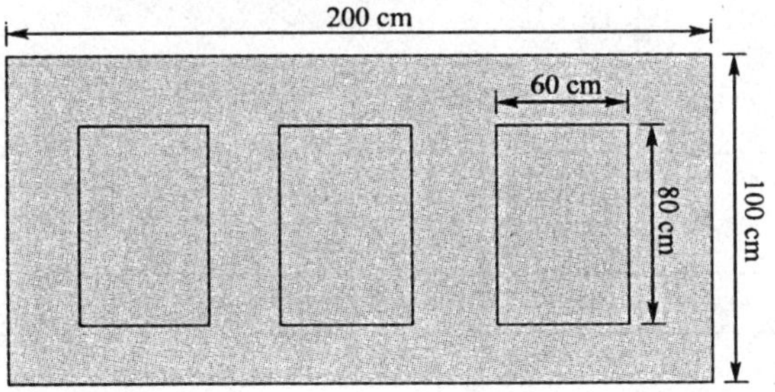

PPTfiles for oral presentations can be uploaded at a designated desk at the conference venue at Room ZC109 on the 29th May. If you would prefer to email them in advance, please email it to ×××@yahoo.cn before 12:00 of May 29th. The conference language is English.

进行大会口头报告的 PPT 请在 5 月 29 日至会议指定地点综合楼 ZC109 室提前拷贝,Email 请上传至×××@yahoo.cn

Organizers 组织者:×××

College of Life Science 生命科学学院

Sponsors 发起者:×××

Bilingual Education Research Center 双语教学中心

Chairmen 主席:×××

(President 校长,××× University ×××大学)

Vice-chairmen 副主席:×××

(College of Life Science 生命科学学院, ×××University ×××大学)

Secretary-in-General 秘书长: ×××

Conference Group members 会议组成员:×××

Program 会议日程

<table>
<tr><td colspan="2">FIRST SIMULATIVE INTERNATIONAL CONFERENCE ON LIFE SCIENCE
首届生命科学模拟国际会议</td></tr>
<tr><td colspan="2">Room ZC109, Multi-functional Building, ××× University on May 29th, 2011
2011 年 5 月 29 日于×××大学综合楼 ZC109 室</td></tr>
<tr><td colspan="2">Congress Program & Schedule at a Glance 大会程序和时间表一览</td></tr>
<tr><td colspan="2">Opening ceremony 开幕式</td></tr>
<tr><td>Chairman 主席</td><td>×××</td></tr>
<tr><td>Distinguished Guests 贵宾</td><td>×××,×××,×××</td></tr>
<tr><td>Review Committee Member
评委会成员</td><td>×××,×××,×××</td></tr>
<tr><td rowspan="2">Leader Oration
报告主持人
14:10～14:30</td><td>Title 1:……</td></tr>
<tr><td>Title 2:……</td></tr>
<tr><td>14:30～16:10</td><td>Title 1:……
Title 2:……</td></tr>
<tr><td>16:10～16:20</td><td>Coffee/Tea break and posters 茶歇 浏览海报</td></tr>
<tr><td colspan="2">Closing Ceremony 闭幕式</td></tr>
<tr><td>Chairman 主席</td><td>×××</td></tr>
<tr><td>Summary Speech
总结性发言
16:20～16:40</td><td>×××</td></tr>
<tr><td colspan="2">Prize Giving 颁奖</td></tr>
<tr><td>17:00～17:20</td><td>1. Academic Report First Prize 学术报告一等奖
Awards presenters 颁奖嘉宾:×××
2. Academic Report Second Prize 学术报告二等奖
Awards presenters 颁奖嘉宾:×××
3. Academic Report Third Prize 学术报告三等奖
Awards presenters 颁奖嘉宾:×××
4. Prize for the excellent paper 最佳论文奖
Awards presenters 颁奖嘉宾:×××
5. Prize for the excellent POSTER 最佳 POSTER 奖
Awards presenters 颁奖嘉宾:×××</td></tr>
<tr><td>17:20～17:30</td><td>Group Photo 合影</td></tr>
</table>

第十三章 面试
Chapter 13 Interview

第一节 公务员面试
Section 1 Civil Servant Interview

尊敬的各位考官，早上好。今天很荣幸能在这里参加面试，向各位考官请教和学习。同时通过这次面试也可以把我自己展现给大家，希望你们能记住我，下面介绍一下我的基本情况。

本人是×××学院的本科毕业生，平时我喜欢看新闻和上网浏览信息，性格活泼开朗，能关心身边的人和事，和亲人朋友融洽相处，能做到理解和原谅，我对生活充满信心。我现在是一名在×××村任职的大学生村官，已经任职近两年，在工作中我先后在不同的岗位工作过。开始我在村里工作，随后因为工作出色，在做好村里本职工作的同时，我被镇党委借调到镇党委办公室接受锻炼，通过两年的努力工作，在工作上取得一些的成绩，同时也得到县委组织部和人事局的认可，在每年度的考核中我都取得了非常优异的成绩，得到主管部门的嘉奖。

通过近2年的工作我学到了很多知识，同时还培养了我坚忍不拔的意志和顽强拼搏的精神，使我能够在工作中不断地克服困难、积极进取。加入公务员的行列是我多年以来的一个强烈愿望，同时我认识到人和工作的关系是建立在自我认知的基础上，我认为我有能力也有信心做好这份工作。

公务员是一个神圣而高尚的职业，它追求的是公共利益的最大化，它存在的根本目的是为人民服务，为国家服务。这也是我对公务员认知的最好诠释。所以，这份工作能够实现我的社会理想和人生价值，希望大家能够认可我，给我这个机会！

Good morning, every respected interviewer. It is my great honor to be here today to attend the interview. It is a good opportunity to consult and learn from every interviewer. Meanwhile, I hope I can leave you good impression all through the interview. Next, I will introduce myself a little bit.

I graduated from ××× College. In my spare time, I like watching news and browsing information on the Internet. Being out-going and active, I care about people and issues all around me, capable for getting well along with my relatives and friends, and understand and forgive others; also, I am full of confidence to life. Now I have been working as an undergraduate village officer in ××× village for almost two years. I was in different positions to offer service in my work: at the beginning, I worked in the village; then due to outstanding performance, I was partially transferred to Party Committee Office of the town by Party Committee of the town for accepting training at the same time to implement my duties in the village; by two-year hard working, there is some achievement on my work and I am recognized by Organization Department and Personnel Bureau of County Party Committee; I always get very excellent score in every annual assessment and am rewarded by competent department.

Based on the two-year work, I have learned a lot and cultivated my gritty will and hard-working spirit as well, which enables me to overcome difficulty and actively strive for improvement in my work. To be a civil servant is my strong wish for many years, and at the same time, I know relationship between human beings and work is based on self-cognition, so I believe I am capable and confident to do this job well.

Civil servant is a holy and noble job; what it seeks for is maximum public profit and the primary purpose of its existence is to serve people and nation. That is my best interpretation to cognition of civil servant. Therefore, this job can realize my social ideal and life value and I hope all of you can recognize me and offer me this chance.

第二节 普通面试

Section 2 Normal Interview

我叫×××，专业生物工程。我是一个性格开朗、充满自信的阳光女孩。开朗的性格使我充满活力，善于与人交往；从容自信使我勇于挑战自我去尝试学习新的事物。在大学期间我不断地完善自己，提高自身素质。在学好专业课和公共课的基础上，我博览群书，为自己打下了坚实的专业基础。我阅读了大量课外书籍，不断增加新知识，陶冶情操，开阔视野。不仅如此，我经常参加学校举办的英语角活动，锻炼英语口语能力，也结交了许多外国朋友。

在校期间我积极参加学校组织的社团活动，锻炼了一定的沟通组织能力。我曾任×××文学社组织部长。由于特别擅长文艺和体育，我曾任系文艺部长、院文艺干事和校舞蹈协会拉丁舞教练，和系足球队啦啦队队长，组织并参加了校内外举办的多场文艺演出活动和体育比赛活动。

在校期间我还积极参加社会实践活动，提高我的社会适应能力，扩展我的知识面。由于受过礼仪培训，作为礼仪小姐，我参加了多次国际、国内会议，并接待过国内外宾客。×××年××月还参加了第二届亚洲生物技术经济论坛年会的接待工作。

我性格开朗、大方，有较强的沟通组织能力和实际动手能力，善于人际交往。掌握良好的礼仪知识和接待工作经验，头脑灵活、反应敏捷，能够灵活处理工作中的突发事件。

My name is ××× with major of biological engineering. Being active and out-going, I am a sunshine girl with confidence. Out-going characteristic makes me full of energy and good at communicating with others; being calm and confident enables me to bravely challenge myself to try and study something new. During the period of university, I improved myself continuously and prompted my own quality. On the basis to study professional and public classes well, I read various kinds of books, which laid solid professional foundation. I read large quantity of extracurricular books so as to constantly accumulate new knowledge, cultivate sentiment and broaden horizon. Moreover, I often took part in activity of "English Corner" that was held by university, not only to practice oral English but also to make many foreign friends.

During my stay in university, I actively took part in club activities that were held by the university, which practiced some certain ability of communicating and organizing. I was as Minister of Organization for ××× Literary Association. Due to excellence at literature, art and sports, I was in the position of Minister of Literature and Art for the Department, officer of Literature and Art for the College, Latin Coach for Dance Association of the University, and Cheering Squad Leader for Football Team of the Department; also I organized and attended many in and out-of university literature and art shows and sport matches.

In my university, I also participated in social practice activities actively to improve my ability to adapt to society and extend my knowledge scale. Since I accepted etiquette training, I attended many times of international and home conferences and received abroad and domestic guests as Miss Etiquette. Also, I attended reception working for the 2nd Asian Biological Technology and Economy Forum Annual Conference on ×××, ×××.

Being out-going and generous, I am with relatively strong ability of communicating, organizing and practical operation and good at interpersonal communication. I master good etiquette knowledge and reception working experience, and have flexible mind and quick reflection, so that I am capable to deal with urgent issues accordingly.

第三节 专业面试

Section 3 Speciality Interview

老师们好。今天的天气不错，很高兴在这儿认识大家。首先，我先做一下自我介绍。我叫×××，23岁，来自古老而美丽的城市 — 北京。

我毕业于大连民族学院生命科学学院的生物工程专业，我很喜欢这个专业。

其次，介绍一下我在大学所学的专业。在大连民族学院我所学的专业是生物工程专业。生物工程专业介绍现代生物学及其有关规律和交叉产品集成的发展状况。其核心是DNA重组技术、基因工程中心，包括微生物工程、生化工程、细胞工程以及其他生物产品。通过四年的专业学习，我掌握了现代生物技术的基本理论、基础知识和基本技能。并通过基本研究能力和基本技能的训练，培养了我良好的科学品质、浓厚的创新精神和较高水平的生物技术实验技能。

四年的大学教育使我学到很多东西、尝试了很多东西，也使自己得到很大的锻炼和提高。使我不仅知道学什么、怎么思考，还知道实践能力的重要性(诸如尽可能多做实验)。大学生活还使自己交了很多朋友。他们帮助我提高学习和研究能力，像一个男人一样做事，也经常为我树立好的榜样。

除此之外，我空闲时间也有很多爱好，我喜欢踢足球，我认为踢足球提高了我的身体素质，也教会我如何进行团队合作和与他人共处。绘画和写作是我另外的爱好。

综上所述，我选择学习生物工程专业是为了扩大我在生物科学领域的视野，提高我的研究能力。我将尽最大努力融入新的团队，并在研究生学习过程中表现良好。

Hello, My teachers. It's a fine day today, and I'm very pleased to meet you here. First of all, I'd like to introduce myself to you. My name is ×××, I am 23 years old. My hometown is Beijing, which is a really ancient beautiful city.

I graduated from Life Sciences College of Dalian Nationalities University. My speciality is Bioengineering. I like it very much.

Secondly, I will introduce my major in the university. My major is bioengineering in Dalian Nationalities University. Bioengineering is the development of modern biology and its related disciplines and cross-product integration, its core is based on recombinant DNA technology, genetic engineering centers, including microbial engineering, biochemical engineering, cell engineering and bio-products and other fields. I have mastered the basic theory, basic knowledge and basic skills of biotechnology by four-year professional learning, accessing to the application of basic research and technological development of the preliminary training, I have scientific quality of a good, strong sense of innovation and practical ability of high-level expertise in biotechnology.

Four years' university education gives me a lot of things to learn, a lot of chances to try, and a lot of practices to improve myself. It teaches me not only what to study and how to think, but also to see the importance of practical ability (such as doing experiment as much as possible). In the university life, I have made many good friends. They help me improve my study and research ability, do everything just like a man, and often give me good example to follow.

Besides what I have introduced myself above, I also have many interests in my spare time. I like playing football, which is an effective way I think to improve my body health, and it can teach me how to join in a group and get along with other people. Drawing and writing is another favor to me.

Above all, I choose the major in order to broaden my view in biology science, and enhance my research ability. I will do my best to join the new group and be good at postgraduate study.

第十四章　推荐信
Chapter 14　Recommendation

第一节　来自专业教授的推荐信
Section 1　Recommendation from Speciality Professor

×××，教授
大连民族学院 生命科学学院
中国　辽宁 大连
2011 年 7 月 1 日

尊敬的录取委员会：

我很高兴推荐我特别喜欢的学生——×××，希望她能够接受你们著名的研究生教育。

2008 年，×××以优异的高考成绩进入大连民族学院生物工程系学习，回顾以往，我们学校因拥有这样优秀的学生而骄傲。

×××一进入大学就使我印象深刻。那时，我负责讲授生物工程原理课程，×××是班级里表现出色的学生。她总是坐在班级第一排，认真记笔记，她也是班级中经常和我交流的学生之一，有时会提出奇怪的问题。由于兴趣，她也选修了我的植物病原微生物课程。课后她总是到我的办公室问一些学术问题，有时也问一些他未来职业发展的问题。我总是向她强调调查研究对于生物专业意味着什么，并鼓励她要不遗余力多做调查研究。她也是一位听话的女孩，我定期地带她到我的实验室，教她一些基本实验技能。她在期末考试中名列前茅已不令人吃惊。

在我和×××相处的过程中，她给我深刻印象的不仅是她的聪明，还有她的雄心与执着。我确信，×××在硕士研究生学习中也将是一名出色的学生，所以，考虑到她的申请愿望，我很愿意强力推荐，非常希望你能实现她的愿望。

此致

×××　教授
大连民族学院 生命科学学院

From: ×××, Professor
College of Life Science
Dalian Nationalities University
Dalian, Liaoning Province, P. R. China
July 1, 2011

Dear Admission Committee:

I take great pleasure in recommending ×××, one of my favorite students, for admission into your distinguished graduate program.

××× was admitted in 2008 into Dalian Nationalities University, Department of Bioengineering by her good performance in the Chinese College Entrance Examination. In retrospect, our school was proud of enrolling such a talented student.

She impressed me almost as she entered into my university. At that time, I was in charge of lecturing them the course Experimental Biology and Engineering Principles and she was performing excellently in the class. She always sits in the first row of the class and carefully takes notes and also, she is one of the students who usually interact with me in the class and sometimes she can pose really tricky questions. I also appreciate her hands-on ability and the ability to solve problems. Due to great interest, she also elected my course Plant Pathogenic Microorganisms. After class, she always came to my office to ask academic questions and sometimes, questions concerning her future career development. I always emphasize to her what research means to biology majors and encourage her to spare no effort in that. Also, she is quite a tractable girl so that I regularly took her to my lab to teach her some basic experimental skills. Unsurprisingly, she scored highly in the final exam.

In my experience with ×××, I was impressed with not only her extraordinary intelligence but also her ambitions and persistence. I am sure that ××× will be an outstanding student in any master program that she may care to enroll in. So I would like to support her firmly in her quest recommendation when considering her application. I would greatly appreciate it you decide to accept her as she wishes.

Yours sincerely
××× Professor
College of Life Science
Dalian Nationalities University

第二节　来自专业教师的推荐信
Section 2　Recommendation from Speciality Teacher

敬启者：

这封信的目的是推荐×××同学，一位值得接受为硕士研究生的候选者。

• ×××同学充满热情。×××同学学过我的生物分离工程课程。她给我很深印象，不仅因为她取得好的成绩"A"，还因为她主动和诚恳的学习态度。有时，她到我办公室问一些课堂中没有提到的有趣的问题，和我讨论一些工程理论在科学与工业中的发展与应用。她想象着这个学科是一个令人惊异的学科。如果人们能够很好地理解与应用它，这个世界也将变得更好。"很好"，我想，"这个女孩真的很热爱她的专业"。经过我的系统培训后，她已经学会了如何使用旋转蒸发仪、紫外光谱仪和索氏萃取仪等。

• ×××同学富有创造性。由于她在班级中的出色表现，我同意她成为我的科研助手。她参加了我关于组织培养方面的研究，帮助我处理一些实验事宜。从她日常工作的一些细节中，可以看出她对工作的积极和热心。当有了新想法或发现一些问题时，她总是告诉我她将如何解决问题，改进实验结果。在经过多次尝试—失败—改进后，我们最终能得到满意的结果。

×××同学在其专业基础课程中也取得优异成绩，我真的相信她的学习能力很强，学习的积极性很高。所以，我毫不怀疑她在贵校也将成为一名优秀学生。如果还需要关于×××同学的进一步信息，请联系我。

此致！

×××（签名）

To whom it may concern:

The purpose of this letter is to recommend×××, a worthy candidate for admission to your graduate program.

• *She is really passionate*. Ms. ×× has ever taken my course "Biological Separation Engineering Experiment". I was deeply impressed with her not only the good score "A" she got in the class but also the highly active and devout attitude when she learn Bioengineering. Sometimes, she came to my office to ask me some interesting questions not mentioned in my class and to discuss the development and application of bioengineering theory in Science and Industry with me. She imagined this subject is truly an amazing subject. If people understand and use it very well, this world will get better. "Well!" I thought, "This young lady really likes her major." After systematic training in my class, she has now learned how to use Rotary Evaporator, UV spectrophotometer and Soxhlet extractor, etc..

• *She is truly creative*. Due to her good performance in the class, I agreed that she was to be my research assistant. She took part in my research about tissue culturing to help me on some lab routines. From some details of her daily work, she's shown me the enthusiasm and hardily spirit in her work. When getting a new idea or finding some questions, she always shared her opinion about how to overcome the error or improve the experiment results to me. After try-and-error and ameliorating the experiment many times, we finally got a satisfactory result.

Ms. ×× also got admirable grades in a couple of core courses of her major. I truly believe she has strong learning capability and enthusiasm in this field. So I have no doubt that she will become a good student in your esteemed department. Should you need any further information about Ms. ××, please do not hesitate to contact me.

Sincerely yours,

×××(signature)

第十五章　自荐信
Chapter 15　Self-recommendation Letter

第一节　生物工程专业学生的自荐信
Section 1　Self-recommendation Letter for Bioengineering Students

申请人:×××
申请系及学位:生物工程 硕士学位
申请学年:2012 年秋季学期

简介

我想申请×××大学生物工程硕士学位课程。下面是目前我的教育背景、研究经历和项目的陈述。

教育背景

我将于 2012 年 7 月在中国大连民族学院完成生物工程专业学士学位学习。在大学学习期间,我完成了生物工程专业的基础知识和实践技能训练。在四年的大学生活期间,我也同样关注人文科学教育和自身的全面发展。一方面,我非常努力学习并出色完成了很多核心课程如工程制图(87 分)、化学实验(90 分)、计算机科学(86 分)和现代生物技术(85 分)等的学习;另一方面,我也不想成为书虫,积极参加一系列社会活动以提高我的实践能力,并曾经获得大学暑期社会实践活动突出贡献奖。学术兴趣和所做的努力极大地激励我进一步探索生物工程领域,所以,我决定在研究生学校继续我的学习。

相关的研究和项目

在本科阶段我的学术兴趣是植物组织培养。因此,我自觉地锻炼了这方面的能力。从 2010 年 5 月开始,我自愿参与协助×××教授进行的吸收 Pb^{2+} 的放线菌菌株筛选项目,该项目的目的是为了筛选获得 2～3 株具有高效吸收重金属离子的放线菌菌株。作为一名研究助手,我主要完成了吸收实验的一部分。从这个实验中,我不仅学会了基本的植物组织培养技术,提高了实验技能,最重要的,我还掌握了一套系统科学的研究方法,这将有助于我未来的学术生活。

申请的研究领域

我特别感兴趣动物和植物的学习和研究。当我还是一个小女孩的时候,我总是好奇于每一个花朵、每一只飞鸟和我周围活跃的世界。我对大自然的美丽和所有生物都充满热情。我热心于去探索这个领域,并任由我的兴趣无限放大。由于对生物工程及其他相关学科的广泛爱好,因此,我喜欢学习生物工程的每一门专业课程。

结论

尽管在大连民族学院的学习为我奠定了很好的生物职涯,但我仍能深切地感受到在美国学习的紧迫与责任。美国是一个除了教育之外均高度商业化的国家。首先,我坚信,在美国我能够受到纯粹学术精神的启迪;其次,美国的生物工程领域比世界上其他国家更加先进,大多数的原始理念及技术均来自于美国。我希望追随更先进的研究,促进提高我国生物工程的研究现状。当我完成学业后,无论前面的道路如何曲折,我都将努力使自己成为中国最先进实验室的一名研究人员。我梦想着自己成为一名通过科学方法改变中国生物工程现状的人,这将有助于我们的下一代。最后,感谢你对我的关注。

此致

×××(签名)

Applicant:×××
Proposed Department & Degree: M. S. in Bioengineering
Semester and year applying for: Fall Semester, 2012

Introduction

I am applying for admission to the M. S. program in Bioengineering, at ××× University. Here I present my education background, research experiences and study objectives in this statement of purpose.

Education Background

I will complete my B. S. degree in Bioengineering from Dalian Nationalities University, P. R. China, in July 2012. During my undergraduate study period, I' ve acquired fundamental knowledge and practical application of Bioengineering. During the four years' university life, I paid equal attention to academic study and all-round development of myself. On the one hand, I studied hard and performed well in many of my core courses like *Engineering Drawing* (87), *Chemistry Experiment* (90), *Basic Computer Science* (86) and *Modern Biotechnology* (85), etc.. On the other hand, not wanting to become a bookworm, I actively participated into a series of social activities to improve my practice ability and have ever been honored as outstanding individuals of summer social practice of my university. All of these academic interests and efforts were strong motivations for me to further explore into the bioengineering field, so I decide to continue my studies at graduate school.

Related Research and Projects

My academic interest in undergraduate study was on the plant tissue culture. Therefore, I consciously exercised my ability in this aspect. I voluntarily assisted in the research project of Screening of Pb2+ Ion Adsorption Type Actinobacillus Strains, under the supervision of Professor ××× from May 2010. The purpose of our project is to screen out 2 to 3 Actinobacillus Strains with high adsorption capacity to common heavy metal ions. As a research assistant, I have been primarily working on the part of the adsorption test. From this experience I not only learned the basic operations of culturing plant tissues and improved my experimental skills, but more importantly, I grasped a whole set of systematic and scientific research methodologies, which I think will benefit me to the greatest extent in my future academic life.

Proposed Research Areas

I am specifically interested in the study and research on animals and plants. When I was just a little girl, I was so curious about every blooming flower, every flying bird, and the whole lively world around me. I feel I have infinite passion towards the natural beauty and all living things. Thus I eager to further probe into this area, and to give full play to my interest. However, due to my broad interest in biological engineering and other branches of learning, I like learn each major courses about bioengineer.

Conclusion

Although studying at Dalian Nationalities University has solidly grounded me for a biological career, I have for long felt the urgency and responsibility to study in America. The United States is a highly commercialized country, luckily enough with the exception of its education. I firmly believe I can receive the edification of pure academic spirit in US. Secondly, the field of biological engineering in America is much more advanced than the other countries in the world. Most of the original ideas and methodologies come from America. I wish to pursue more advanced studies and then improve my country's biological engineering situation. After completing my program, I will try my best to be a researcher at the most cutting-edge laboratory in China, no matter how tortuous the road ahead is. I dream to be the one who can take scientific way to radically change the situation of bioengineering in China, to further benefit our next generations. Moreover, I thank you for your attention to this statement.

Sincerely yours,

×××(signature)

第二节 普通专业学生的自荐信
Section 2 Self-recommendation Letter for Normal Speciality Students

自荐信Ⅰ

首先感谢您在百忙之中抽出时间来翻阅我的自荐材料，事业有成的您的审视，将是我的镜鉴。在您翻开第一页时，就已经为我开启了一扇希望与成功的大门，同时希望我也能对您有所助益。

我是大连民族学院生命科学学院生物工程专业××届毕业生。在几年的大学生活中，我本着勤奋刻苦、踏实认真的态度，完成了我的本科学业，且成绩较为优秀。在认真学习专业课知识的同时，我还通览了中外的诸多名著，学习了网页制作和网站策划、广告摄影等辅助知识。不断地充实自己，提高自己的能力是我一贯的宗旨，我深信这些在以后的日子里将会逐步地体现出来。

在这几年的大学生活中，我除了认真学习之外，在平时的日常生活中还积极地参与学校组织的各项活动，以及社会实践，这更加培养了我的实践能力和动手能力。对于学校的各项政策，规章制度，我都能严格遵守，政治觉悟较高，为人诚实可信，自信严谨，在同学中有较好的人际关系。

我相信自己便是您最好的选择，并承诺一定会发挥自己最大的潜能为公司的发展贡献力量，当然如果我的申请没有成功说明我尚有不足之处，还需不断地改进，毕竟这是一个终身学习的年代。

Self-recommendation Letter Ⅰ

Firstly, thank you for your time of reading my self-recommendation material. I will take your view comments as precise advises to me. Your opening of the first page has started a gate with hope and success for me. And I hope I can do you a favor as well.

I graduated from College of Life Science in Dalian Nationalities University with major of biological engineering. During my college days, with diligent, assiduous, steadfast and earnest attitude, I finished my bachelor study with relatively outstanding scores. I also took overall view of many Chinese and foreign masterpieces and learned auxiliary knowledge, such as webpage making, website planning, advertising photography, etc. at the same time to work hard to study knowledge of professional class. My consistent aim is to enrich myself and improve my ability continuously and I firmly believe that will be reflected gradually in the future.

In these years of university life, besides working hard to study, I actively took part in every kind of activity held by university and social practice in the daily life, which furthered to cultivate my practice and operation ability. I can strictly follow every policy, regulation and system of university; I have relatively high awareness of politics; I am honest, trustful, confident and rigorous; also, I have good interpersonal relationship among students.

I believe I am your best choice and promise that I must release my maximum potential to make efforts to the development of company; if my application is not successful, it means I still need to improve myself. After all, it is a time to learn all the time.

自荐信Ⅱ

尊敬的领导：

我叫××，是大连民族学院生物工程专业毕业生，将于××年××月结束大学生活，步入社会。四年的大学经历，令我深刻地体会到，加入一支优秀的团队，不仅可以锻炼充实自己，提高自身各方面的能力，更有机会充分实现自身的价值，为单位集体做出更大的贡献。我相信贵单位正是我理想中的精英团队，因此真诚希望您吸收我成为这个集体的一员。

通过大学四年中的努力学习，我扎实地掌握了生物工程专业的本领，成绩优异；在课余时间，我也广泛摄取了其他领域的知识。××年×月，我曾在××有限公司实习，在工作中我讲究方法，尽职尽责，赢得了领导和同事们的一致好评和信任。通过组织和参与各项活动，我提高了自身的组织协调能力，提升了集体的凝聚力和团队精神，并为我今后参加其他工作做好了充分的准备。

诚诚恳恳做人，踏踏实实做事是我的人生准则，复合型知识结构使我能胜任社会上的多种工作。在莘莘学子中，我不敢自诩明珠宝玉，更不能自诩才学超群，但我拥有不懈奋斗的意志、愈战愈强的精神和踏实肯干的作风，这才是最重要的。

与其他求职者相比，也许身为应届毕业生的我在工作经历方面会处于一定的弱势，但相反也正是一个优势，此时的我在工作上就像一张白纸，有很强的可塑性，同时我的优势是学习能力强，不管是专业知识还是人际沟通。我相信如果贵公司给我一个工作机会，结合我以上的两点优势，这将会是一个高回报率的投资。此外，对于适合自己的工作我会长期工作，不轻易跳槽。

我由衷地希望能成为贵单位的一员，强健的体魄已经为我认真务实的工作提供了保障，真心希望能凭借自己的努力和热情为单位的发展做出贡献。最后，再次感谢您阅读此信。

此致

敬礼

自荐人：××

Self-recommendation Letter Ⅱ

Respected Leader,

I am ×× and a graduate with major of biological engineering from Dalian Nationalities University. I will graduate from university in ××, ×× and step into society. Four years' university experience let me realize deeply that joining a good team not only can practice and enrich myself and improve my ability in all aspects but also can be offered chance to realize my own value and make more contribution to the unit. I believe your unit is my ideal elite team, so I sincerely hope you will absorb me to be one of your members.

By four years, hard working in university, I solidly mastered skill of biological engineering with good score; in my spare time of class, I also widely accumulated knowledge in other fields. As a trainee in ×× Limited Company in ××, ××, I paid attention to method and was responsible and accountable, which won good assessment and trust of all leaders and colleagues. By organizing and attending different kind of activities, I improved my own ability of organizing and coordinating and prompted my collective cohesion and team spirit, which got abundant preparation for me to take other jobs in the future.

Being an honest and sincere individual and being steadfast to do things is my life principle; compound knowledge structure makes me take many jobs in society successfully. Among so many students, I cannot dare to praise myself to be bright pearl and precious jade or even intelligent over others, but I have insistent and striving will, spirit that growing in strength in fighting, and steadfast style, which would be the most important.

Compared with other candidates, for me, such kind of fresh graduate, I have some certain disadvantage in aspect of working experience, but on the contrary, it can be regarded as advantage because I am just like a piece of vacant paper in my job and easy to be trained; meanwhile my advantage is the strong ability to learn something new in both professional knowledge and interpersonal communication. I believe if your company offers me a chance and combines above advantages of me, it will be an investment with high rate of return. Moreover, for the job that is suitable for me to do a long term, I will not resign easily.

I sincerely hope I can join your unit. Strong physique provides good guarantee for me to work hard and practically. I sincerely hope I can contribute to the unit with my own effort and enthusiasm. Last but not least, thanks again for reading my letter.

Best regards,

××

第十六章 投稿说明(英文)

Chapter 16 Instructions to Authors(In English)

Introduction of the journal 杂志简介

For example: Published since 1954, the Canadian Journal of Microbiology publishes contributions by recognized scientists world-wide and has an international readership in more than 58 countries. Journal topics include applied microbiology and biotechnology; microbial structure and function; fungi and other eukaryotic protista; infection and immunity; microbial ecology; physiology, metabolism, and enzymology; virology, genetics, and molecular biology; and microbial genomics and bioinformatics.

《加拿大微生物学报》于 1954 年开始发行,并已得到世界科学家的广泛认可,读者遍布世界 58 个国家。学报内容涵盖应用微生物和生物技术,微生物结构与功能,真菌及其他原生生物,侵染与免疫,微生物的生态、生理、代谢及酶学,滤过性微生物学,遗传学,分子生物学,微生物基因组学及生物信息学等内容。

Types of papers 文章类型

For example: The Canadian Journal of Microbiology (Can. J. Microbiol.) is a widely subscribed-to international journal of general microbiology and publishes articles, notes, minireviews, reviews, and letters in English or French.

《加拿大微生物学报》是有关普通微生物学的国际性杂志,文章类型包括原始研究论文(article),研究结果的简要总结(notes),小综述,综述及快报,可用英语或法语书写。

Page charges 版面费

There are no page charges for publication in the Canadian Journal of Microbiology.

《加拿大微生物学报》不收取版面费。

Other information

Cover letter 附信

The corresponding author should send a cover letter with the submission, signed by all authors.

通讯作者应该发送一封含所有作者签名的附信。

Copyright 版权

The submission package must include copyright release forms signed by all authors.

上传包裹中应该含有一封具有所有作者签名的版权确认书。

Reviewers 评阅人

Lists the names, addresses, telephone and fax numbers, and e-mail addresses of four to six persons who are qualified to act as referees.

推荐 4～6 名评阅人,列出他们的姓名、地址、电话、传真及 e-mail 地址。

Writing your paper 论文书写格式

Title 题目

Both titles and abstracts provide information for contemporary alerting and information retrieval services, and should therefore be informative but brief.

题目和摘要要提供前沿的信息和供检索服务的信息，因此，信息面应宽而简单。

Title page 首页

The title page should contain the following. (i) The full title of the paper. (ii) Authors listed in the order in which they are to appear at the head of the printed article. (iii) Affiliation and address (including e-mail address) for each author. (iv) Name, address, telephone number, fax number, and e-mail address of the author responsible for correspondence.

首页应该包含下列内容：(i) 完整的论文题目，(ii) 按论文排序所列的作者姓名，(iii) 每位作者的联系方式及地址(包括 e-mail 地址)，(iv) 负责联络的作者姓名、地址、电话、传真和 e-mail 地址。

Abstract 摘要

An abstract is required for every contribution and should contain accurate descriptive words that will draw the reader to the content. It should not be more than 200 words and should appear on a separate page. References should not be cited in the abstract unless they are absolutely essential.

每份稿件必须有摘要部分，且文字描述准确并能引起读者关注。摘要字数在 200 字以内，单独设页。除非常必要，摘要中不列参考文献。

Key words 关键词

Key words should not exceed 3—5 and should be placed directly below the abstract.

关键词一般 3～5 个，写在摘要的下面。

Text 正文

Manuscripts should be compiled in the following order: Introduction; materials and methods, results, discussion or conclusion, acknowledgments; appendixes (as appropriate); references; table(s) with caption(s) (on individual sheets); figure caption(s) (as a list).

论文正文书写顺序为：前言、材料与方法、结果、讨论或结论、致谢、附录(可无)、参考文献、带标题的表格(单独设页)、带标题的图(按图的目录列出)。

第十七章 生物类仪器设备中英文名录
Chapter 17 Bio-instrument Names

20 升不锈钢自动发酵罐　20-litre stainless automatic fermentor
DNA 测序仪　DNA sequencer
DV800 核酸/蛋白分析仪　DV800 nucleic acid protein analyzer
PCR 操作台　PCR chamber
PX2 多用途 PCR 仪　PX2 multi-purpose PCR equipment
RCCS-批量式旋转细胞培养系统　RCCS-batch and rotary cell culture system
氨基酸自动分析仪　auto amino acid analyzer
比浊仪　turbidity comparator
不锈钢板式滤器　stainless steel filter
超滤器　ultra filter
垂直电泳系统　vertical electrophoresis system
超声波细胞破碎仪　ultrasonic cell disruption system
成像检测系统　chemiluminescent detection system
蛋白纯化系统　protein purification system
蛋白核酸分析仪　protein nucleic acid spectrophotometer
蛋白纯化工作站　protein purification apparatus
等电聚焦双向电泳系统　isoelectric focal bidirection electrophoresis
电化学工作站　electrochemical workstation
电击转化仪　pulse cell transfection system
电细胞融合仪　cell electro fusion device
电泳槽　electrophoresis cells
动物细胞培养反应器　animal cell bioreactor
多肽合成仪　peptide synthesis apparatus
光学生物成像系统　spectral bio image system
核酸提取仪　nucleic acid isolation machine
基因枪　particle delivery system (PDS)
近红外成分分析仪　near-infrared content analyzer
玻璃微电极拉制器　glass microelectrode puller
流式细胞分析仪　flow cytometer
脉冲电泳仪　pulse field electrophoresis spectrometer
毛细管电泳系统　capillary electrophoresis system
酶标仪　micro plate spectrophotometer
凝胶成像系统　gel imaging system
凝胶染色系统　gel stainer system
全波长扫描酶标仪　pan-wavelength micro plate reader
全自动 dna 测序仪　automatic dna sequencer
全自动固相萃取系统　automatic solid phase extract system
全自动菌落计数仪　automatic colony counter
全自动凝胶成像分析系统　automatic gel image analyzing system
实时荧光定量 pcr 系统　real-time fluorescence quantitative pcr
索式抽提仪　soxhlet extractor
梯度 pcr 仪　gradient pcr instrument
微生物快速自动鉴定系统　auto identification system for microorganism
洗板机　plate cleaning instrument
细胞融合仪　cell fusion system
细胞遗传学工作站　cytogenetics work station
小动物行为活动记录分析系统　video camera and analysis system for animal observation
小型双垂直电泳槽　mini double-sided vertical cell
液氮罐　liquid nitrogen container
真空浓缩机　speed vacuum concentration apparatus
转移电泳槽　trans-blot transfer electrophoresis
紫外交联仪　UV cross linking machine (UV)
高速离心机　high speed centrifuge
XRD 粉末衍射仪　X-Ray diffraction powder meter
一氧化碳分析仪　portable ir co analyzer
台式离心机 benchtop centrifuge
双波长飞点薄层扫描仪　dual-wavelength flying spot tlc scanning instrument
气相色谱仪　gas chromatography
水质监测仪　water quality monitor
水浴摇床　water baths shaker
加热磁力搅拌器　heating magnetic stirring appa-

ratus
半自动生化分析仪 semiautomatic biochemistry analyzer
半微量电子分析天平 electro analytical semi-micro-balance
可见风光光度计 vis spectrophotometer
生化培养箱 biochemistry incubators
真空干燥器 vacuum drying oven
电子天平 electronic balance
电热恒温鼓风干燥箱 electro-thermostatic blast oven
石墨炉原子化器 graphite furnace atomizer
全自动界面张力仪 automatic tension apparatus
冰箱(无霜) refrigerator (frost free)
扫描探针显微镜 scanning probe microscope
红外分光光度计 infrared spectrophotometer
自动双重纯水蒸馏器 automatical double pure water distillatory
自动电位滴定仪 automatic titrimeter
自动血球计数仪 automatic blood cell counter
低速大容量离心机 low speed large capacity centrifuge
实验室用制冰机 ice machine for laboratory
表面吸附仪 surface and aperture meter
表面活性测定仪 exterior live determiner
原子吸收分光光度计 atomic absorption spectrophotometer
火焰分光光度计 flame emission
核酸蛋白纯化系统 nucleic acid protem auto-purification system
高速冷冻离心机 high-speed refrigerated centrifuge
循环水式多用真空泵 vacuum pump with circulated water system
超声波细胞粉碎机 ultrasonic cell crusher
超纯水系统 ultra purified water system
超滤装置 ultrafilter instrument
微颗粒测定仪 micro particulate monitor
熔点仪 melting point apparatus
精密电子天平 precision electronic balance
酸度计 pH meter
薄层色谱扫描分析仪 thin-layer chromatogram scanning

第十八章　生物工程学科领域中的最新研究成果
Chapter 18　Recent Research Results in Bioengineering

第一节　酶工程
Section 1　Enzyme Engineering

仿生信号转导体系的构建:囊泡表面磷脂信号分子诱发的乳酸脱氢酶的激活

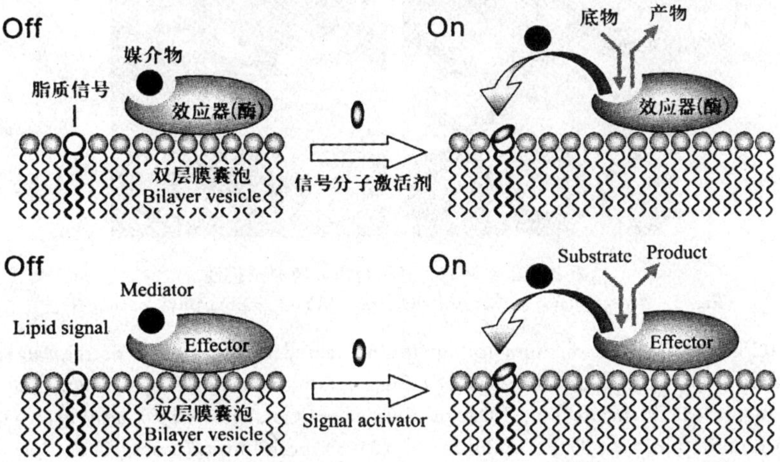

图 1　仿生信号转导体系示意图

Off,酶活性被抑制的状态;On，酶活性恢复的状态

Fig. 1 Schematic representation of a bio－inspired signal transduction system

An artificial signal transduction system, in which an enzymatic activity was controlled by a lipid signal embedded in a biomembrane model, was constructed. The system was inspired by the biological phosphoinositide (PI) － triggered signal transduction. For this propose, hybrid bilayer vesicles were formed by combination of dimyristylphosphatidylethanolamine as a lipid signal and a cationic synthetic lipid as a matrix. The aggregate morphology was confirmed by means of transmission electron microscopy and dynamic light scatting measurements, and the vesicles showed a phase transition from gel to liquid － crystalline state as evaluated with differential scanning calorimetry. On the cationic vesicular surface, the lipid signal specifically recognized pyridoxal 5'－phosphate as a signal activator to form a signal－signal activator complex. The hydrophobic and the electrostatic interactions were much enhanced in the microenvironment on the vesicular surface to form the binary complex. The resulting binary complex exhibited extremely high binding affinity toward Cu^{2+} ions. Thus, the catalytic activity of lactate dehydrogenase, which was inhibited by binding of Cu^{2+} ion to the enzyme active site on bilayer vesicles, was effectively recovered upon formation of the signal - signal activator complex, reflecting the binding affinity of the signal－signal activator complex with Cu^{2+} ion overcame that of the enzyme.

以细胞肌醇磷脂信号转导途径为原型,在合成肽脂囊泡(人工细胞膜模型)上,利用天然磷脂为信号分子,成功地激活了处于囊泡表面的乳酸脱氢酶。为此,将 1,2-二-十四烷基磷脂酰乙醇胺等天然磷脂嵌入合成肽脂 *N*,*N*-二-十六烷基-N^{α}-6-三甲氨基己酰基-L-甘氨酰胺囊泡中,制备了稳定的混合双层膜囊泡,用透射电子显微照相、动态光散射及差示扫描量热等手段确认了混合囊泡的形态及粒径分布. 以磷酸吡哆醛等维生素 B_6 类化合物为信号分子激活剂，利用它们与天然磷脂形成复合体，进而与 Cu^{2+} 形成强的金属配合物的性质，实现了对处于囊泡表面、被 Cu^{2+} 抑制的乳酸脱氢酶的激活，构建了一个新的仿生信号转导体系。

第二节　微生物工程
Section 2　Microbiology Engineering

龟裂链霉菌菌株 MY02 的抗真菌活性及其活性代谢产物的纯化

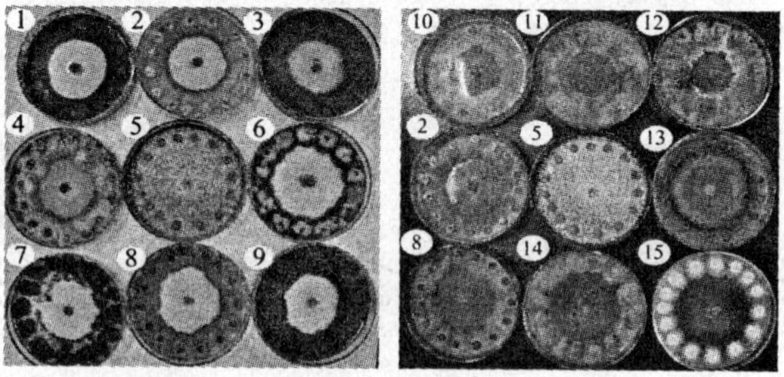

图 2　菌株 MY02 对植物病原菌的拮抗性

Fig. 2　Antagonistic activity of the strain MY02 against phytopathogen

Strain MY02 grown in PDA medium for up to 96h tested for antifungal activity against 14 phytopathogens respectively. (1) Bipolaris maydis, (2) Fusarium oxysporium f. sp. cucumarinum, (3) Fulvia fulva, (4) F. solani, (5) Plain disks on PDA medium as controls showed no activity, (6) Alternaria alternata, (7) Curvularia lunata, (8) F. moniliform, (9) Exserohilum tucicum, (10) Pythium aphanidermatum, (11) Botrytis cinerea, (12) Sclerotinia sclerotiorum, (13) Rhizoctonia solani, (14) Coniothyrium diplodiella, (15) Verticillium dahliae, respectively.

在 PDA 培养基上培养 96h 的菌株 MY02 对 14 种植物病原真菌的抗菌活性：(1)玉米小斑病菌，(2)黄瓜枯萎病病菌，(3)番茄叶霉病病菌，(4)辣椒根腐病病菌，(5)PDA 培养基上的空白培养基片对照，(6)烟草赤星病病菌，(7)玉米弯孢病病菌，(8)水稻恶苗病病菌，(9)玉米大斑病病菌，(10)瓜果腐霉病病菌，(11)番茄灰霉病病菌，(12)黄瓜菌核病病菌，(13)水稻纹枯病病菌，(14)葡萄白腐病菌，(15)茄子黄萎病病菌。

The genus of *Streptomyces*, a saprophytic Gram—positive bacterium, has been shown to have characteristics which make them useful as biocontrol agents against soil borne plant pathogen fungi. These characteristics include production of different kinds of secondary metabolites and biologically active substances, such as enzymes degrading the fungal cell wall directly and antibiotics which have high commercial value. The soil streptomyces are the major contributors to the biological buffering of soils and have roles in decomposition of organic matter conductive to crop production.

革兰氏阳性腐生细菌-链霉菌属具有作为抗植物土传病害生防因子的特征。这些特征包括产生各种次级代谢产物和生物活性物质，如降解真菌细胞壁的酶和抗生素，具有很高的商业价值。同时，土壤放线菌还是重要的土壤生物缓冲剂和土壤中有机物降解者。

A Gram-positive, moderately halophilic, facultatively alkalophilic, catalase- and oxidase positive, obligately aerobic, filamentous actinomycete strain, designated YIM 90022^T, was isolated from saline soil collected from the Qaidam Basin, north-west part of China. Phylogenetic analysis based on 16S rRNA gene sequences indicated that the new isolate was a member of the genus *Nocardiopsis* and the sequence similarities between the isolate and the type strains of members of the genus *Nocardiopsis* were in the range of 95.1%-98.7%. Phenotypic and chemotaxonomic properties of this organism also indicated that strain YIM 90022^T was a member of the genus *Nocardiopsis*. The strain grew well on most of the media tested, producing yellow-white to deep brown substrate mycelium and white aerial mycelium. Light gray to deep brown diffusible pigments were produced. The substrate mycelium was well developed and fragmented with age; the aerial mycelium produced long, straight to flexuous spore chains with non-motile, smooth-surfaced, rod shaped spores on them. The strain grew in the presence of 1%-15% (w/v) total salts (optimum, 3%-5%) and at pH 6. 0-10. 5 (optimum, pH 8. 5) and 10-45℃ (optimum, 30℃). Whole-cell hydrolysates of strain YIM 90022^T contained meso-diaminopimelic acid and no diagnostic sugars. The predominant menaquinones were MK-10(H_4), MK-9(H_8), MK-10(H_6) and MK-10 (H_8). Polar lipids comprised diphosphatidylglycerol, phosphatidylcholine, phosphatidylglycerol and phosphatidylmethylethanolamine. The major cellular fatty acids were iso-C16:0, anteiso-C17:0, 10-methyl-C18:0 and 10-methyl-C17:0. The DNA G+C content of strain YIM 90022^T was 71. 5 mol%. The combination of phylogenetic analysis, DNA-DNA relatedness data, phenotypic characteristics and chemotaxonomic data supported the suggestion that strain YIM 90022^T represents a new species of the genus *Nocardiopsis*, for which the name *Nocardiopsis terrae* sp. nov. is proposed. The type strain is YIM 90022^T (=CCTCCAA 208011^T=KCTC 19431^T).

从中国西北部的柴达木盆地的盐碱土中分离得到一株中度嗜盐、兼性嗜碱的过氧化氢酶和氧化酶阳性、专性需氧的丝状放线菌 $YIM90022^T$。基于 16S rRNA 基因序列的系统发育分析表明这个新菌株是拟诺卡氏属的成员并且它和拟诺卡氏属模式菌株的相似度为 95.1%～98.7%。该菌株的表型和化学特性也表明菌株 $YIM90022^T$ 是拟诺卡氏菌属的一个种。其在大部分的培养基上都生长良好并且产生黄白色到深褐色的基内菌丝和白色的气生菌丝。同时会产生从浅灰色到深褐色的扩散性色素。基内菌丝生长很好且随着年龄的增大而断裂,气生菌丝能产生直或弯曲的长孢子链,链上生有表面光滑不能动的棒状孢子。该菌株适合生长在含盐量为 1%～15%(最适为 3%～5%),pH 为 6.0～10.5(最适为 8.5),温度为 10～45℃(最适为 30℃)的培养基中。菌株 $YIM90022^T$ 的全细胞水解产物含有内消旋-二氨基庚二酸,无特征糖。最主要的甲基萘醌为 MK-10(H_4)、MK-9(H_8)、MK-10(H_4)、MK-10(H_8)。极性油脂是由二磷脂酰甘油、磷脂酰胆碱、磷脂酰甘油、磷脂酰甲基乙醇胺组成的。主要的细胞脂肪酸为 iso-$C_{16:0}$, anteiso-$C_{17:0}$, 10-methyl-$C_{18:0}$ 以及 10-methyl-$C_{17:0}$。菌株 $YIM90022^T$ 的 DNA(G+C)含量为 71.5%。系统发育树的分析、DNA-DNA 杂交、表型特征、化学分类学数据都表明菌株 $YIM90022^T$ 是拟诺卡氏菌属的一个新种,并命名为 *Nocardiopsis terrae* sp. Nov. 模式菌株为 $YIM90022^T$(=CCTCCAA 208011^T=KCTC 19431^T)。

第三节　分离工程
Section 3　Separation Engineering

Exoinulinases and endo-inulinases from *Aspergillus ficuum* JNSP5-06

Three exoinulinases (Exo-I, Exo-II, and Exo-III) and two endo-inulinases (Endo-I and Endo-II) were purified from the culture broth of *Aspergillus ficuum* JNSP5-06 by ammonium sulphate precipitation, DEAEcellulose column chromatography, Sepharose CL-6B column chromatography and preparative electrophoresis. The molecular weights of Exo-I, Exo-II, Exo-III, Endo-I, and Endo-II were determined to be 70 kDa, 40 kDa, 46 kDa, 34 kDa, and 31 kDa, respectively. Using inulin as the substrate, their Km values were 43. 1 mg/ml, 31. 5 mg/ml, 25. 3 mg/ml, 14. 8 mg/ml, and 25. 6 mg/ml, respectively. These five inulinases were stable below 50℃ with optimum activity at 45℃, and were stable at a pH range of 4 - 8 with an optimum pH at 4. 5 for exoinulinase and at 5. 0 for endo-inulinase. The inulinase activity was completely inhibited by Ag+ and strongly inhibited by Fe^{2+} and Al^{3+}, whereas K^+, Ca^{2+}, Li^{2+}, EDTA and urea had no significant influence on the inulinase activity.

来自 *Aspergillus ficuum* JNSP5-06 内切和外切菊粉酶

Aspergillus ficuum JNSP5-06 的培养液经硫酸铵沉淀、DEAE-cellulose 层析、CL-6B 凝胶过滤色谱和制备电泳技术分离纯化,得到三个外切菊粉酶(EXO-I、EXO-II、EXO-III)和两个内切菊粉酶(Endo-I

和 Endo-I)。经分子量测定 EXO-I、EXO-II、EXO-III、Endo-I、Endo-II 的分子量分别为 70 kDa、40 kDa、46 kDa、34 kDa 和 31 kDa。使用菊粉作为底物，其 Km 值分别为 43.1mg/ml、31.5mg/ml、25.3mg/ml、14.8mg/ml 和 25.6mg/ml。这五个菊粉酶在 50℃以下稳定，45℃为最适温度，pH 范围为 4-8，外切菊粉酶与内切菊粉酶的最适 pH 分别为 4.5 和 5.0。Ag^+ 能够完全抑制菊粉酶的活性，Fe^{2+} 和 Al^{3+} 对其有强烈的抑制作用，而 K^+、Ca^{2+}、Li^{2+}、EDTA 和尿素对菊粉酶的活性没有显著影响。

第四节 基因工程
Section 4 Gene Engineering

Aim: To establish a rapid, improved soil environmental DNA extraction and purification protocol.

Methods and Results: Three different soil DNA isolation and four purification strategies were compared on different soil samples with variable rates of success. Bead beating extraction gave significantly higher DNA yields than microwave-based and liquid nitrogen grinding DNA extraction methods. The inclusion of soil washing prior to cell lysis decreased the amount of purification steps required. Although these soil types differed, polyvinylpolypyrrolidone (PVPP)-sepharose 2B column elution was sufficient for all three samples, yielding DNA pure enough for successful application in molecular studies. One soil sample retained 80% of the initial DNA after successful purification. Conclusions: Optimization of a purification protocol confirmed that only a combination of previously described methods proved sufficient in yielding pure environmental DNA from humic-rich soils. Total processing time for DNA extraction and subsequent purification from multiple samples was considerably more rapid than the previously described methods.

Significance and Impact of the Study: This study developed a new optimized soil DNA extraction and purification protocol that is suitable for soil samples from different environmental sources that are rich in humic acid content.

目的：建立一个快速、改进的土壤中环境 DNA 的提取和纯化方法。

方法和结果：针对不同的土壤样品，对具有不同提取率的三种 DNA 提取方法和四种 DNA 纯化方法进行了比较。其中珠磨法的 DNA 提取率明显高于微波法和液氮研磨法。细胞裂解前先清洗土壤样品可以减少一些必要的纯化步骤。尽管土壤样品类型不同，但经聚乙烯吡咯烷酮-琼脂糖凝胶 2B 柱洗脱方法处理后，三种土壤样品的 DNA 纯度足以进行后续的分子研究。经纯化，一个土壤样品可以保留其最初 DNA 产量的 80%。

结果：通过对纯化方案进行优化，最终证实上述的组合纯化方法可从富含腐殖酸的土壤中提取纯度足够高的环境 DNA。而这种方法在提取和纯化 DNA 过程中的耗时也远少于此前提到的其他方法。

意义和影响：本研究建立了一个新的、优化的土壤 DNA 提取和纯化方案，该方法适用于不同来源的各种富含腐殖酸的土壤样品。

参考文献

[1] Boehncke W. H. Radeke H. H. Biologics in General Medicine. Springer-Verlag Berlin Heidelberg, New York, 2007

[2] 何维. 医学免疫学. 2 版. 北京:人民卫生出版社,2010

[3] 林慰慈,魏雪涛,薛彬,等. 免疫学(中译版). 2 版. 北京:科学出版社,2010

[4] 刘俏,范圣第. 基于 MATLAB 的化工实验技术(汉-英). 北京:中国轻工业出版社,2007

[5] Macaulay D. B. General, Organic and Biological Chemistry: Laboratory Experiments: An Integrated Approach. John Wiley & Sons Ltd, 2009

[6] Patrick G. L. Instant Notes Organic Chemistry . 2 nd Ed. Beijing:Science Press, 2002

[7] Paulson D. S. Biostatistics and Microbiology: A Survival Manual . New York: Springer Science, Business Media, LLC, 2008

[8] Peter M Lydyard, Alex Whelan, Michael W Fanger. 免疫学(影印版). 北京:科学出版社,2000

[9] 施巧琴. 酶工程. 北京:科学出版社,2005

[10] Solomons T. W. G. Organic Chemistry . Beijing:Chemical Industry Press, 2004

[11] Wang M. Organic Chemistry Experiments . Beijing: Higher Education Press, 2011

[12] 王沛. 制药工程. 北京:人民卫生出版社,2012

[13] 吴继洲. 天然药物化学. 北京:高等教育出版社, 2010

[14] 吴立军. 天然药物化学. 北京:人民卫生出版社, 2003

[15] 邬行彦,储炬,宫衡. 生物工程/生物技术专业英语. 北京:化学工业出版社,1999

[16] 吴昊,乔德阳. 生物化工与制药专业英语. 北京:化学工业出版社,2009

[17] 许实波. 海洋生物制药. 北京:化学工业出版社,2002

[18] 张永琴,刘福胜. 生物与制药工程专业英语. 北京:化学工业出版社,2007